Functional Human Anatomy

Functional
HUMAN
ANATOMY

JAMES E. CROUCH, PH.D.

Professor of Zoology
San Diego State College
San Diego, California

Second Edition

With 393 Illustrations, 113 in Color

LEA & FEBIGER · Philadelphia

First Edition, 1965
  Reprinted September, 1965
  Reprinted September, 1966
  Reprinted June, 1967
  Reprinted June, 1968
  Reprinted January, 1970
  Reprinted November, 1970

Second Edition, 1972
  Reprinted November, 1972
  Reprinted September, 1973

ISBN 0-8121-0341-6

Library of Congress Catalog Card Number 73-146029
Published in Great Britain by Henry Kimpton Publishers, London

PRINTED IN THE UNITED STATES OF AMERICA

Dedicated to my wife, Mary,
for her
patience and understanding

Preface

The second edition of *Functional Human Anatomy* retains the same philosophy and organization as the first edition which has enjoyed a good reception by the profession. The second edition also incorporates many minor and a few major changes and additions which were suggested to me by those who have used the book. I am very grateful to them for it is only through the constructive criticisms and suggestions of those who know the book well that its quality and usefulness can be improved. My students have been very helpful in their comments and in the general attitude which they express toward the book.

About one hundred of the illustrations have been redone or modified and many more are in color in the second edition. These are mostly in the chapter on the skeletal muscles; a few are on the skeletal and reproductive systems. Two illustrations are added in the chapter on general and special senses.

Modifications in the written text are relatively minor. The introductory Chapters 1 to 6 remain essentially the same with only minor changes in the chapter on tissues. In Chapter 10 on general myology a description of the minute structure of skeletal muscle tissue is added along with an electron micrograph. The innervations of skeletal muscles are added to the muscle table in Chapter 11. In Chapter 12 on the circulatory system a few of the illustrations have been corrected and some color has been added.

There have been some changes in the positioning of illustrations on the excretory and reproductive systems and two drawings on the female system and one on the male have been redone. In Chapter 18, The Nervous System, parts of the material on organization of the system have been modified especially in reference to the autonomic nervous system. In the last chapter on the organs of general and special sense the material on neuromuscular spindles and neurotendinous receptors has been rewritten in line with current concepts.

Miss Ursula Wolf-Rottkay has been the illustrator for the second edition of *Functional Human Anatomy* and to her I express my deep appreciation. She has been most helpful too in her suggestions relative to the general makeup and appearance of the book.

I am indebted to the editors and staff of Lea & Febiger for their cooperation and confidence and especially to John F. Spahr, George H. Mundorff, Thomas J. Colaiezzi and Miss Nancy Lewis with whom I have worked closely since the publication of the book in 1965.

As author I assume full responsibility for any errors or omissions in the book. They should reflect in no way upon those who are mentioned, in this or in the preface to the first edition, as having had a part in the project.

JAMES E. CROUCH

San Diego, California

Preface to First Edition

"Anatomy is destiny" is a phrase which came to mind frequently during the writing of this book. The truth in this statement of Sigmund Freud's is apparent to one who sees anatomy as encompassing the whole range of structure from subatomic particles to gross anatomical organization. While an elementary textbook of Human Anatomy certainly cannot presume to range over such a broad area, it should try to help students to see the potential and feel the challenge that is, in reality, there. In living anatomy, in the broad sense, is found the basis of all function. In living and dead bodies or in their fossil remains is revealed the life of both the present and the past, and, indeed, the means or mechanism for seeing into the future.

While any beginning course in Human Anatomy must be largely descriptive, engaging the student in dissection and memorization, it need not be only this. Certainly, the interrelationships of the structure of the body and the general functions of its organs and systems should receive repeated mention and emphasis. In addition, excursions into histology, embryology, and comparative anatomy at appropriate points in a course can add interest because they give meaning to the anatomy of man. Embryology and comparative anatomy in particular enable us to see man as continuous with all of nature and at the same time to emphasize his unique characteristics. If human anatomy can be taught with these overtones, it not only can give the student a knowledge and understanding of his own structure, but can help him to formulate a philosophy of life. It is the aim of this book to bring this added dimension to the study of human anatomy.

The systematic rather than the regional approach to anatomy is used in this book. It involves the students in fewer difficulties and is more appropriate for the majors in physical and health education, nursing, physical and occupational therapy, and art, as well as for others who take this course. Each system is discussed in sufficient depth to give a good basis for physiology, kinesiology, and other sciences into which the students may go for further training. Also, it should serve well medical and dental students as a means either for quick preview or review of the body. No effort has been made to emphasize one system more than another.

The first chapter, called "Point of View," sets the philosophical tone of the book, defines and classifies man, and states some of the problems of our day. It suggests how man, using his unique characteristics, has created these problems and how he might resolve them or be destroyed. Chapter Two is a consideration of basic terminology, while Chapter Three reviews in a very minimal way protoplasm and the cell as the basis of structure and function. Although students should have biology as a prerequisite to human anatomy, experience indicates that a brief review is helpful.

A short presentation of descriptive embryology constitutes the fourth chapter. It carries the individual through the period of the embryo and lays the foundation for further discussion of development in the chapters dealing with the body systems. If an instructor or student does not wish to include this material, it is so arranged that it can be easily excluded from assignments. It is my belief however, that some knowledge in this field enhances one's understanding of adult structure.

A chapter called "Organization of the Body" completes the introductory part of the book. It deals primarily with histology, although it describes the organization of cells into tissues, tissues into organs, and organs into systems. It names and describes briefly each system. Histology, like embryology, is also made a part of the discussion of the anatomy of each system, thus keeping it constantly before the student. It leads naturally into a consideration of the integumentary system.

Comparative anatomy receives no special chapter consideration, although it is woven into the whole fabric of the book. It makes the student constantly aware of his evolutionary past and suggests again his own position in the stream of life.

Each chapter is concluded by questions which are arranged in the same sequence as the material presented in the chapter. They thus serve as a review and a means of emphasizing important points, enabling the student to evaluate his progress in anatomy.

The terminology throughout the book, with few exceptions, is based upon the International Nomenclature adopted by the International Congress of Anatomists meeting at Paris in July 1955. The most important terms for the student are printed in boldface type for emphasis and in many cases alternate terms or the old terminology are given in parentheses and in italics after the more acceptable names. This seems a necessary service to the student since the old terminology so often is found in reference books which are still widely used.

The great majority of the illustrations were drawn specifically for this book and therefore relate closely to the text material. Some of them are quite complex, attempting to show important relationships; others, as many of those on the muscular system, are simplified to emphasize action. Where the labeling on a given illustration goes beyond the text material, those labels which are most important to the beginning student are printed in boldface type.

A glossary is placed at the end of the book. It provides help in pronunciation of terms and gives their derivation and meanings as well. Since anatomy is in part a study in language, the student should get into the habit of using regularly this section of the book.

It is with pleasure that I acknowledge my indebtedness to the many who have contributed directly or indirectly to the general body of knowledge of anatomy from which one draws so heavily in the writing of a book. Among them are the works of Gray, Cunningham, Morris, Sobotta, Spalteholz, and Grant. I also acknowledge the help, encouragement, and inspiration which I have received from my colleagues in zoology, biology, nursing, and physical education and from the many students who have studied anatomy with me.

Special recognition and thanks go to those who have read parts of the manuscript and have given the benefit of their constructive criticisms: the late Dr. Gordon Tucker, Dr. Charles L. Brandt, Dr. Mabel A. Myers, Dr. Fred W. Kasch, Dr. Gerald Collier, and Dr. Harry H. Plymale. Leon L. Gardner, M.D., read the entire manuscript and made many constructive suggestions. John R. Blake, M.D., read the chapter on the circulatory system. Their help is deeply appreciated.

Four of my students—Jo Ann Smith, Mary Bevington, Linda Wood, and Roger Marchand—read parts of the manuscript. Jo Ann Smith, Mary Bevington, and Marilynn Boland did most of the typing. To each of them I give my thanks.

For the illustrations I am indebted to Mr. Joseph M. Yuhasz, Mrs. Martha B. Lackey, Mrs. Loretta Douglass, Dr. James Koevenig, Mr. Kenneth Raymond, and Mr. Al Rowen for their excellent work. I am particularly appreciative of the efforts of Mrs. Lackey who entered the project late, but who enabled me to complete the book within a reasonable time. Mr. Bruce Lightheart and Mr. Roger Marchand provided the photomicrographs and I appreciate their contributions to the book.

The illustration on page 20 is from Vesalius, Andreas, *Icones anatomical*, 1934, New York, courtesy of the New York Academy of Medicine.

It was Dr. Charles Moritz who encouraged me to write this book and his company, Lea & Febiger, who furnished the necessary support, cooperation and confidence. I acknowledge with thanks the opportunity which they have made available to me.

If there are errors or omissions in the book, I alone assume full responsibility for them. They should reflect in no way upon those who are mentioned as having had a part in the project.

JAMES E. CROUCH

San Diego, California

Contents

Functional Human Anatomy

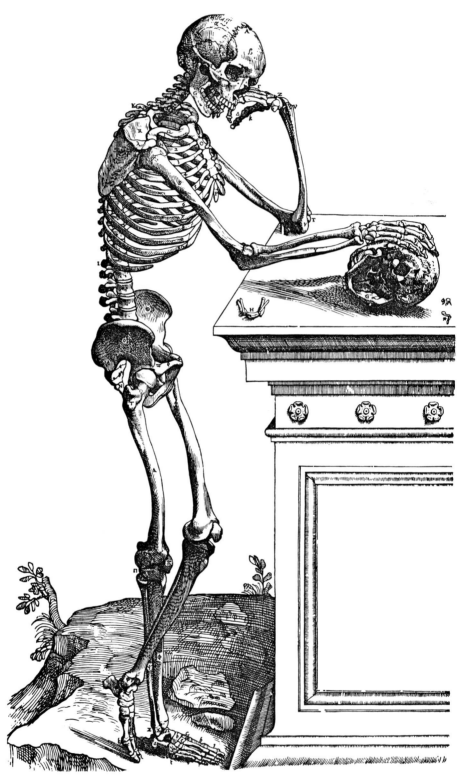

This illustration from the work of Andreas Vesalius (1514–1564), the "father of anatomy," is one of the most admired of his osteological series. It is the skeletal Hamlet soliloquizing beside the tomb upon some poor Yorick. On the tomb, to the right of the skull, are two of the ear ossicles, the malleus and incus; to the left is the hyoid bone. (Vesalius, Andreas, *Icones anatomical*, 1934, New York, courtesy of New York Academy of Medicine.)

> *"Thoughtful men will find in the lowly stock whence man has sprung, the best evidence of the splendor of his capacities; and will discern in his long progress through the past, a reasonable ground of faith in his attainment of a noble future."*
>
> —Huxley

Chapter 1

Point of View

MAN'S PLACE IN THE WORLD OF LIFE

Have you ever pondered the question of why you are engaged in this pursuit of knowledge—and in this study of human anatomy? You may say that you wish to gain a college degree to improve your chances in the economic world, that you wish to broaden your knowledge and understanding of your environment so as to better adapt to it or that you wish to improve your knowledge and appreciation of the physical body in which you live. Probably all of these things and others account for your being in college and for reading this book. But—basically, you are engaged in these activities because you are of the species man, an animal organism unique in its capacity to study, to learn, to understand, to gain wisdom in reference to himself and to record and pass on this knowledge to others of his kind.

Alfred Korzybski (1921) in his book **The Manhood of Humanity** expresses in an interesting way man's place in the world of life. He says that the plants are the "chemistry-binding" class of life, since they capture the energy of the sun to convert water and carbon dioxide into organic food. In doing this they become essential go-betweens in making the sun's energy available to all life. Animals, Korzybski called the "space-binders" because they need to move about in search of food. And then he said, "How shall we define man?" Because man moves about as other animals do he is a space-binder. However, man also keeps a record of his activities and has built a discipline that we call history. He digs into the earth's crust and reads the record of the past by studying the minerals, the rocks, the soil and the evidences of past life, the fossils. He builds a present on the basis of the experiences of this past, on his own accomplishments, and on his speculations as to the future. Korzybski, therefore, defines man in terms of mathematics and mechanics as the "time-binding" class of life. Man's is a space-time world.

Man himself is of a three-fold nature. He is an animal organism. He is an individual with an heredity which is his alone. He is a member of a society of men. The implications of these facts are multitude. Volumes have been written about them. And they are facts which should always be in the minds of those who study man—especially those who study in the science of the anatomy of man. The study of his anatomy, of all anatomy, can be dynamic. For in living anatomy we find the basis of all function, in fact, the basis of life itself. In the living and dead bodies of man and other animals and in their fossil remains, we find revealed the life of the past and the present, and indeed the means or mechanism for predicting about life in the future.

Classification of Man. Man, being an animal organism, is classified as such by the biologist. The biologist generally divides the living world into two kingdoms—the plant and animal. Within each kingdom he divides the animals or plants into major categories called **phyla.** A study of the phyla and their relationships one to the other is called **Phylogeny.** It is essentially a study in **Organic Evolution** or the "history of the race." In the animal kingdom the phyla range from the *relatively simple* single-celled members of the **Phylum Protozoa** to the highly complex back-boned animals of the **Phylum Chordata.** We should remember in using the terms simple and complex in reference to living things that no life is really simple—only that some organisms have a relatively simpler organization than others.

Man belongs to the **Phylum Chor-data.** To belong in this group animals must have the following characteristics which zoologists consider "diagnostic." They must have a **notochord,** a **dorsal hollow nerve** (*spinal*) **cord,** and **pharyngeal pouches** (Fig. 1–1).

A notochord is a flexible rod of tissue that runs down the mid-dorsal side of the body. It lies in the position which, in vertebrate animals, will be occupied later by the vertebral column. It is composed of vacuolated cells and covered by a fibrous sheath. It is easily seen in embryonic stages of man. In adult man it is represented only by a small mass of tissue, the **nucleus pulposus,** located inside of the discs between the bodies of the vertebrae (Fig. 1–2).

The hollow nerve cord of chordates lies dorsal to the notochord. It results from an invagination of the mid-dorsal part of the embryo. It gives rise to the spinal cord and the brain (Fig. 1–1).

The pharyngeal pouches are evaginations of the lateral pharyngeal walls. They are paired and segmentally arranged. At each point on the body wall which overlies one of the evaginations, an invagination develops. The evaginations and invaginations come into contact and may break through to form **gill slits** (Fig. 1–3). In fishes and some amphibians, gill slits are formed and gills

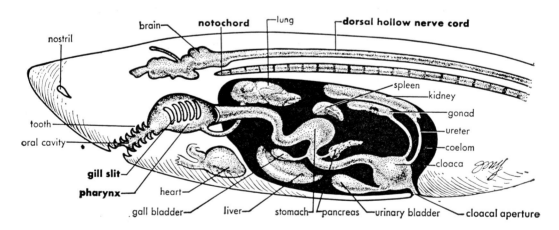

FIG. 1–1. Schematic diagram of a chordate.

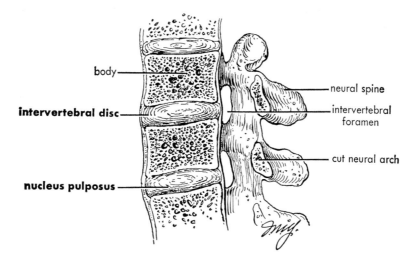

FIG. 1–2. Median section through vertebrae and intervertebral discs to show nucleus pulposus.

for external respiration appear in this area. In lung-breathing land-dwelling chordates, gills are not formed, nor are the slits except as transitory stages in development. In those chordates which do not use gills for breathing, these pharyngeal pouches may give rise to other structures, for instance, the **auditory** (*Eustachian*) **tube** and **middle ear cavity,** while corresponding invaginations of the body wall give rise to the external auditory canals. When the evaginations and invaginations meet and do not break through, they form the **tympanic membrane** or **ear drum.** When you break an ear drum you do, in a sense, have a gill slit. Occasionally, a human child is born with fistulae (gill slits) in the neck region. These may be closed surgically and thus this evidence of our proud evolutionary history is erased. Like the notochord the pharyngeal pouches are evident only during development.

It should be apparent from the discussion of these diagnostic chordate characteristics that it is not enough just to know the anatomy of the adult animal. To understand the anatomy of a species we need to study it at all stages of its

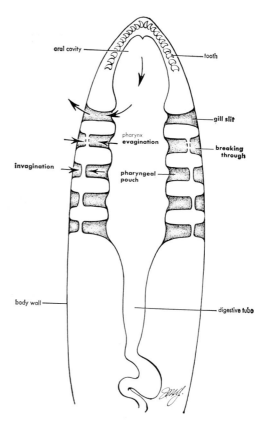

FIG. 1–3. Schematic drawing to show origin of gill slits.

life history. We need to know something about individual development (*Ontogeny*) and especially prenatal development or **Embryology**. Also we have found it helpful to refer to other animals than man. **Comparative Anatomy,** therefore, must be considered another approach to the study of **Human Anatomy.**

Other important characteristics of the Phylum Chordata, though not limited to this phylum, are **bilateral symmetry, segmentation,** and **body cavities.**

Bilateral symmetry refers to the condition whereby an animal can be sectioned into right and left halves, the one the mirror image of the other.

Segmentation means that an animal is made up of a longitudinal series of regularly arranged parts. This characteristic is quite obvious in an earthworm but quite obscure in the superficial anatomy of most chordates. Yet it is still suggested in the arrangement of vertebrae in the backbone, the sequence of paired spinal nerves, and perhaps best as a primary condition in the embryos of vertebrates in the early disposition of muscle segments or myotomes (Fig. 4–13).

Two main **body cavities,** dorsal and ventral, are present each with its subdivisions. The **dorsal cavity** is divided into **cranial** and **vertebral** portions which house the brain and spinal cord

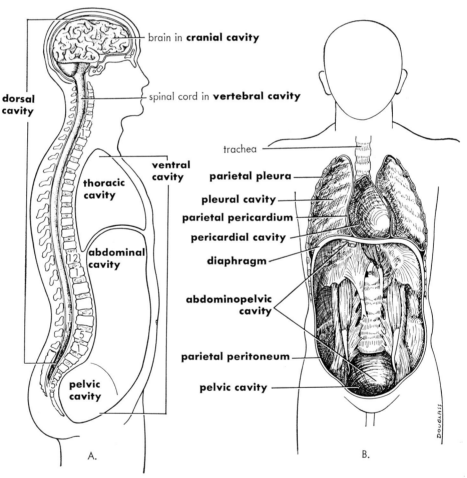

FIG. 1–4. Body cavities. *A.* Median section showing component parts of dorsal and ventral cavities. *B.* Frontal section showing subdivisions of ventral cavity.

respectively. The **ventral cavity** is divided by the diaphragm into **thoracic** and **abdominopelvic** cavities (Fig. 1–4).

The **thoracic cavity** is divided into a **pericardial cavity** around the heart and a separate **pleural cavity** around each lung. The part of the thoracic cavity between the pleural cavities is the **mediastinum** and besides the heart and pericardial cavity it houses the trachea, bronchi, esophagus, thymus, and major blood and lymphatic vessels.

The **abdominopelvic cavity** is arbitrarily divided into the upper **abdominal** and the lower **pelvic** cavities. The organs they house are known collectively as the **viscera.** The abdominal cavity contains the stomach, spleen, pancreas, liver, gallbladder, and the small and large intestines. The pelvic cavity contains the sigmoid colon, rectum, and urinary bladder. In the female the ovaries, uterine tubes, and uterus are also enclosed by the pelvic cavity; in the male the prostate gland, seminal vesicles, and a portion of the ductus deferens.

The **Phylum Chordata** is divided into a number of subphyla. The one to which man belongs is the Vertebrata— the back-boned animals. This characteristic is found in the several **classes** of Fishes, and in the **classes** of Amphibians, Reptiles, Birds and Mammals. Man belongs to the **class** of Mammals, char-

acterized by having **hair** and **mammary** glands. Man is in the **primate order** of mammals as are the monkeys and apes. This does not mean that he is descended from the present-day monkeys and apes, which is a common fallacy, but only that they all have a common ancestry. And, while he naturally shares many characteristics with these other **Anthropoids,** he differs too in many ways, not the least of which is his large brain. Man's brain weighs about 1350 grams, that of the highest apes less than 400 grams. Also man has a steeper facial angle, less prominent supraorbital ridges, a more prominent nose and chin, and less hair than the apes. He engages in articulate speech. He assumes a more erect posture. Finally, man belongs to the **family Hominidae,** the **genus Homo** and the **species sapiens.**

Individual Man. Oliver Wendell Holmes once said, "Heredity is an omnibus in which all our ancestors ride, and every now and then one of them puts his head out and embarrasses us." Indeed, it is the mechanism of heredity, the chromosomes and genes, which account for us as a species and as individuals today. It is this which binds us to our past, to our deep-rooted alliance with all of the living world, and this too which gives us an identity of our own.

Did not this biological fact contribute

To summarize, man's relationships to other animals may be shown in the following classification:

Kingdom: Animalia
   Phylum: Chordata—dorsal, hollow nerve cord; notochord; pharyngeal pouches.
      Subphylum: Vertebrata—vertebral column.
         Class: Mammalia—hair; mammary glands.
            Order: Primates—primitive, but highly developed cerebrum in some; five digits; nails on digits; thumb usually opposable.
               Family: Hominidae—large cerebrum; highly developed eyesight; terrestrial biped.
                  Genus: *Homo*—steep facial angle; nose and chin more prominent; less prominent supraorbital ridges.
                     Species: *sapiens*—largest cerebrum.
                     Scientific Name: *Homo sapiens.*

also in the development of our philosophy of democracy which ideally gives certain "inalienable rights" to the individual, invests him with dignity, and makes the state his servant? And is not the success of our democratic system of government dependent upon the level of innate ability, of education, and of responsibility of its individual citizens?

René Dubos in his book *So Human an Animal* says, "Each human being is unique, unprecedented, unrepeatable. The species Homo sapiens can be described in the lifeless words of physics and chemistry, but not the man of flesh and bone. We recognize him as a unique person by his voice, his facial expressions, and the way he walks—and even more by his creative responses to surroundings and events."

Social Man. But—man is not alone. From the day of his conception his biological individuality is subject to the influences of the environment—first of the limited environment of the womb and later, in ever-expanding progression the family, the society of men, nature as manifest on our planet and now beyond. Thus, man's biological nature becomes a biosocial nature.

Man's unique intellectual capacity has enabled him to alter his external environment as no other species and to take from it for his own use its abundant resources. As he has come into conflict with his fellow men or used up the available natural resources he has moved to other areas. Wars, famine and pestilence have been his adversaries and have kept his population under control. But in more recent times we have witnessed the control of contagious and other diseases through advances in medical science, and man's life expectancy has been lengthened. He has also gained control over almost unlimited power in the atom, and war is now suicidal. At the same time the old conflicts involving economics, politics, morals and ethics persist. To these is added a high rate of population growth which, with the lowered death rate, challenges our capacity to survive. The question is serious. Can we control the tremendous forces of nature of which we have gained knowledge, and can we control population growth? Or to put it another way, will we destroy ourselves by an atomic explosion or a population explosion? There is, of course, a third way. Man can use his intellectual resources to solve these problems. He could build a civilization unequalled in all of his long history. Will he do so?

These matters may seem remote from the title and avowed purpose of this book. Rather, I would consider them of primary importance, for they place man in a proper relationship to the physical and biological world. They suggest his uniqueness and his power. They should arouse one's interest in the physical basis of his remarkable being—in his anatomy in the broadest sense.

"Man is not the centre of the universe as was naively believed in the past, but something much more beautiful—Man the ascending arrow of great biological synthesis. Man is the last-born, the keenest, the most complex, the most subtle of the successive layers of life. This is nothing less than a fundamental vision. And I shall leave it at that."

—Teilhard de Chardin, P.

QUESTIONS

1. What are the basic characteristics an animal must have to be classified as a chordate?

2. Why is it important to know the complete life history of an animal?
3. Describe a notochord.   Does man have one?
4. What is the relationship of pharyngeal pouches to the ear?
5. What is man's scientific name?   What constitutes a scientific name?
6. What characteristics of man make him a member of the Class Mammalia?
7. How do you think the study of human anatomy can give you a better understanding of man as an individual and as a social animal?

"What a piece of work is man!" ... Shakespeare, Hamlet

Chapter 2

Basic Terminology

ANATOMY—A DEFINITION

ANATOMY is that science which deals with the structure of the body. Literally, the term anatomy, which comes from the Greek, means "to cut up," or as we would prefer to say, to dissect (dis-sect′). Dissection was the only method available to the early anatomists. The microscope and other refinements of technique which we use today were not known. Their studies therefore were classed as **Gross Anatomy.**

Gross Anatomy. Gross Anatomy, the primary concern of this book, can be approached in either of two ways. One is **Regional** in which the body is divided into a number of divisions such as head and neck, thorax, abdomen, pelvis, upper extremity and lower extremity. These are studied individually, learning all of their gross structures and their interrelationships. This is the method used in most professional school courses.

Surface Anatomy, another aspect of gross regional anatomy, is too often neglected. It gives an opportunity to study one's own living body. Bones can be palpated and their positions in relationship to soft structures noted. The outlines of muscles can be seen and felt, and their actions observed. The action of joints can be watched. The pulsations of arteries can be felt, and the valves of veins studied. Many of the superficial nerves can be located. The skin and its connection to underlying parts can be judged.

Systematic Anatomy is the other approach to the study of gross human anatomy. Using this method, the systems, such as the **skeletal system,** are studied in their entirety though they extend through more than one region of the body. The systematic approach is the one used in most undergraduate courses and serves our needs best since we wish to present anatomy as a basis for **Physiology.** Physiology is defined as the science of function. Anatomy and physiology have more meaning when studied together. Vesalius, the Father of Anatomy, who lived in the 16th century called his great treatise on anatomy "De humani corporis fabrica" which is often translated as the *"workings of the human body."*

Microscopic Anatomy. This deals with the minute structures of the body. It may be divided into three main areas: **Cytology,** the study of **cells; Histology,** the study of **tissues;** and **Organology,** the study of **organs** as they are developed from tissues. These again give us a better basis for understanding the functions of the body.

Developmental Anatomy. The study of development of the body from the fertilized egg to maturity may be divided into two aspects, **prenatal development** or **Embryology,** and **postnatal development.** The total development of an individual may be called **Ontogeny.** Through this approach to the study of human anatomy, not only is our knowledge of normal structure extended, but our understanding of anomalies is enhanced.

Comparative Anatomy. This science deals with the structure of other animals and sheds light on man and his relationships. It provides strong evidence in support of the principle of **Organic Evolution.** It deals with the history of the race which we have termed **Phylogeny.**

Homology and Analogy. The study of embryology and comparative anatomy often reveal to us the essential similarity of certain structures or organs among different animals. The wing of a bird and the forelimb of man appear superficially quite different; actually, their basic structures of bones, muscles, nerves, and blood vessels are similar. Their differences are easily understood in terms of adaptations for the functions they perform. Such basic correspondence of structures is due to an inheritance of these features from a common ancestry. They are called **homologous** structures. All homologies are not as obvious as the illustration above. The little bones of the middle ear, the malleus, incus, and stapes, are homologous to certain jaw structures of fishes, for example.

Analogy refers to similarity of the functions of structures. The wing of a butterfly and that of a bird both serve the function of flight and are analogous. Their structures, however, are quite unlike in origin and development. They are not homologous. Some structures may be both homologous and analogous, as the forelimbs of bats and birds.

A SENSE OF DIRECTION

Anatomical Position. In order to communicate effectively about the human body, anatomists have agreed upon the rule of referring all descriptions to the, so-called, **anatomical position.** In this position, the body is standing erect, the face is directed forward, the arms are hanging at the sides with the thumbs pointing away from the body (Fig. 2–1). It may be difficult at first to remember this rule because in dissection the body is usually prone. However, it represents an effort to make the study of human anatomy that of a living rather than a dead body.

Terms of Direction. Because man has assumed the upright or bipedal habit

TABLE 2–1

Term	*Quadruped*	*Man*
posterior	Refers to tail or caudal end	Refers to back
dorsal	Refers to back	Refers to back
anterior	Refers to head or cranial end	Refers to front or belly side
ventral	Refers to belly side	Refers to front or belly side
superior	Refers to dorsal side	Means upper or higher part of body
inferior	Refers to ventral side	Means the lower part of body

and most other land vertebrates are quadrupeds we experience some confusion in the use of certain terms of direction. Table 2–1 may prove helpful. Refer also to Figures 2–2 and 2–3.

When we wish to refer to the sole of the foot, we use the term **plantar**. The upper surface of the foot is called **dorsal**. The anterior surfaces of the hands or forearms are called **volar** or **palmar**. Their opposite surfaces are called **posterior**.

The **medial** structures of the body are those nearer the midline, those farther to the side are called **lateral**. The terms **external** and **internal** are used mostly in reference to the body wall, body cavities, and hollow organs. **Superficial** and

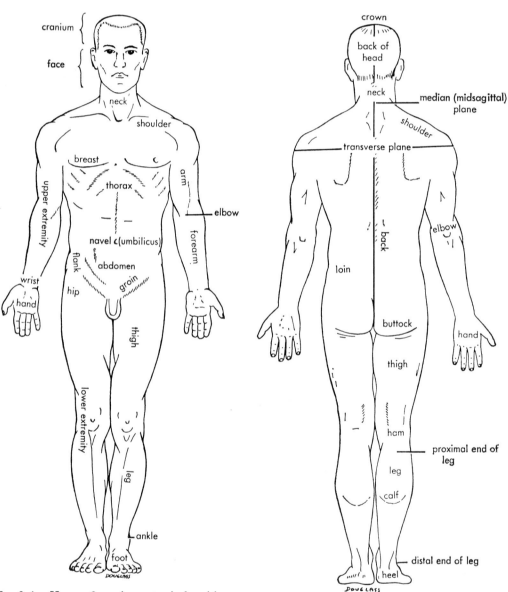

FIG. 2–1. Human figure in anatomical position—
anterior view.

FIG. 2–2. Human figure—posterior view.

deep refer to relative distances from the body surface.

The terms **proximal** and **distal** are used to denote distance from source or attachment of a structure or from the center of the body. The **distal** end of a limb is at the hand or foot. The end nearer the body is **proximal.** The **proximal** end of a finger is at its point of attachment to the hand; its free end is **distal.**

Planes of Reference. In studying human anatomy we frequently make use of sections cut through the various **planes** of the body. These studies are an aid to our understanding of the relationships of organs and parts.

The **median** plane, sometimes called **midsagittal,** divides the body into equal right and left halves. Any plane parallel to the median plane is called **sagittal** or by some, **parasagittal.** Vertical to the median plane is the **frontal** or **coronal** plane. It divides the body or part into anterior and posterior portions. The **transverse** or **horizontal** plane is at right angles to both the sagittal and frontal planes and divides the body into superior and inferior portions (Figs. 2–2 and 2–3).

UNITS OF MEASUREMENT

Though the **metric system** is used universally in the fields of science, it has not become a part of the language of most lay people in the United States. We still think and act in terms of feet, yards, pounds and quarts. Some of the more commonly used units of measurement are therefore given in the following list along with their approximate equivalents in the English system. The metric system is used in this book.

TABLE 2–2

LENGTH

1 micron (μ) = 0.001 millimeters = 1000 millimicrons (mμ) = 1/25000 inch.
1 millimeter (mm) = 0.001 meter = 1/25 inch.
1 centimeter (cm) = 10 mm. = 0.01 meter.
1 meter (m) = 100 cm. = 1000 mm. = 39.37 inches.
1 kilometer (km) = 1000 m.
1 inch = 25 mm. = 2.5 cm.
1 foot = 30.58 cm.
In electron microscopy we use the angstrom unit (Å) which is 0.1 millimicrons (mμ).

AREA

1 square inch = 6.45 square centimeters (cm^2).
1 square millimeter (mm^2) = 0.00155 square inch.
1 square centimeter (cm^2) = 0.155 square inch.
1 square meter (m^2) = 10.76 square feet.

VOLUME

1 cubic millimeter (mm^3) = .0001 cubic centimeters = 0.00061 cubic inches.
1 cubic inch = 16.38 cubic centimeters.
1 pint (pt.) liquid = 473.17 cubic centimeters.
1 quart (qt.) liquid = 946.35 cubic centimeters.
1 liter (l) = 1000 cubic centimeters = 1.056 qts.

WEIGHTS

1 ounce (oz.) = 28.34 grams.
1 pound (lb.) = 453.59 grams.
1 gram (g.) = 15.43 grains = 0.035 ounces.
1 kilogram (kg.) = 1000 grams = 2.2 pounds.

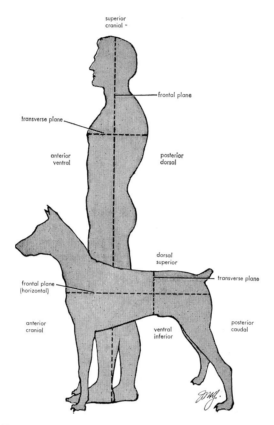

FIG. 2–3.   Planes of reference and terms of direction.

QUESTIONS

1. Distinguish between anatomy and physiology.
2. How can comparative anatomy and embryology help one in developing an understanding of human anatomy?
3. Define organic evolution.
4. What is phylogeny?   Ontogeny?
5. Why is it important to define an "anatomical position" for man prior to a study of his anatomy?
6. Are the following statements examples of the proper usage of the terms in *italics*?
    a. The diaphragm is *superior* to the stomach.
    b. The hand is at the *proximal* end of the radius and ulna.
    c. The backbone is on the *anterior* side of the body.
    d. The arm is *lateral* to the shoulder.
    e. The hip is *distal* to the thigh.
    f. A *frontal section* of the body divides it into anterior and posterior parts.
7. What is the plantar surface of the foot?   The volar surface of the hand?

"Over the structure of the chemical molecule rises the structure of the living substance as a broader and higher kind of organization."

— Hertwig

Chapter 3

The Basis of Structure and Function

PROTOPLASM

Definition. The living substance in all plant and animal bodies is **protoplasm.** It was called by T. H. Huxley, "the physical basis of life." It is a complex mixture of mineral salts and organic compounds in water. It contains no rare elements or compounds, but it does have an organization, as yet not fully understood, which enables it to manifest certain properties which we have come to identify with **life.** The most important of these properties are metabolism, irritability, contractility, conductivity, and reproduction and growth.

Metabolism. Metabolism encompasses all of the chemical changes by which protoplasm uses and transforms materials for growth, maintenance, repair, and the yielding of energy to do work. **Anabolism** is the constructive aspect of metabolism and involves the synthesis of materials which are built into protoplasm, stored there, or used to form secretions. **Catabolism** is destructive metabolism, but nevertheless essential, for it results in the release of energy for heat and work. Waste products result from this action. Both of these types of metabolism go on at all times. During em-

bryonic development and early life, anabolic processes predominate, insuring the growth and differentiation of the new individual.

Irritability refers to the capacity of protoplasm to respond to stimuli from the environment. The success of protoplasm is measured by its making favorable adjustments or adaptations to stimuli.

Contractility is the capacity to change form. In the human body the muscle tissues and the leukocytes are specialized in this property. The leukocytes of the blood stream, like an ameba, form pseudopods by which they change their form and move. Muscle fibers change form by contracting. They become shorter and thicker.

Conductivity is the power of protoplasm to carry the effect of a stimulation from one part to another. While all protoplasm can conduct, nervous tissue is particularly well specialized in this property.

Reproduction and growth refer to the ability of protoplasm to make more of its kind. The units of protoplasm may thus increase in size or in number. In either case, growth would be involved. Reproduction applies more strictly to the formation of new individuals.

3

CELLS

Cell Theory. In most living organisms protoplasm is divided into small units called **cells.** Robert Hooke, in 1665, first reported cells from his examinations of cork and other plant materials. Rene Dutrochet, in 1824, published on the cellular makeup of plants and animals; yet M. J. Schleiden, in 1838, a botanist, and Theodor Schwann, in 1839, a zoologist, are usually given credit for the formulation of the **cell theory.** Their contribution was more of a synthesis of ideas and observations than a discovery.

The doctrine states that cells are the units of structure and function in living organisms. Indeed, it means that we can understand life only to the extent that we can understand the cells. The cell theory, like Darwin's theory of evolution published in 1859, gave tremendous impetus and direction to biological thought and research.

A later important development came in the work of Rudolf Virchow (1858), a German physician, who discovered that cells come only from pre-existing cells. This fact, coupled with the later discovery that sperms and eggs are also

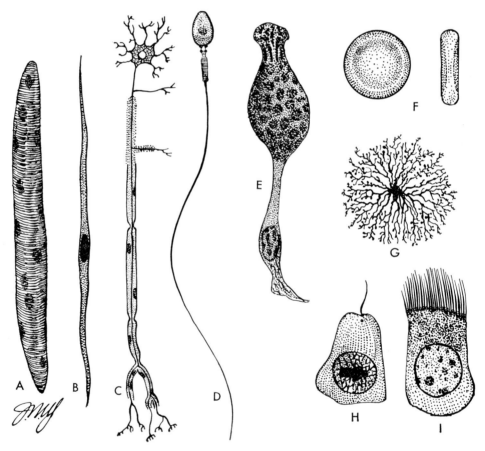

FIG. 3-1.   Animal cells of various shapes.

A, Skeletal muscle.
B, Smooth muscle.
C, Motor neuron.
D, Spermatozoon.
E, Secreting cell.

F, Red corpuscles, human—front and side.
G, Astrocyte—cortex of brain.
H, Flagellated epithelial cell.
I, Ciliated epithelial cell.

cells which unite in the formation of a new individual, and the discovery of chromosomes which pass on from one generation to the next, give a clear picture of the continuity of all life. The importance of the cell in all biological study is thus well established.

Shape, Size, Numbers of Cells (Fig. 3–1). Cells vary greatly in shape and size as we see them in the tissues of the human body. **Shape** variations depend in large measure upon mechanical forces exerted upon the cells and upon the function which they perform. Muscle cells contract, nerve cells conduct impulses and their length is an adaptation to this kind of activity. Red corpuscles are round and flattened discs for efficient exchange of oxygen and for passing through narrow channels, the capillaries. Certain white blood cells or **leukocytes** move about the body in the manner of an ameba and help to fight off the parasitic invaders of man's body. Their shape, therefore, is subject to momentary change.

Sizes of cells are equally variable. If we assume that certain cells have a common function, three factors would influence their size: (1) the relationship of cell volume to surface area; (2) the ratio of nucleus size to the amount of cytoplasm; (3) the rate of metabolism. The importance of these factors will be clarified later in our discussion of cells and tissues.

The **numbers** of cells in organisms is of some interest and importance. Of course, in single-celled organisms the cell and the organism are one. In the multicellular organisms, the number of cells is in large part related to the size of the organism. In most plants and animals there is therefore no definite number of cells for a given species although there are some notable exceptions such as the green alga, **Pandorina morum,** which has regularly 8 or 16 cells.

Man has a variable number of cells depending on the size of the individual, and the number of cells varies almost from moment to moment. Their numbers run into the trillions. Yet they all had their start in the single fertilized egg cell. Important, too, is the fact that the rate of cell production during development is not the same in all parts of the body. In some parts such as the nervous tissues the number of cells is complete at birth. They lose their capacity to multiply. Cells of the epidermis of the skin die, are sloughed off, and are replaced throughout life. Blood cells also fall into this category.

Cell Structure and Function. Our knowledge of cell structure is still quite meager. Yet some important discoveries have been made in recent years. We know now that a cell is a highly organized chemical system, not just a number of loosely related parts contained within a membrane. Figure 3–2 shows some of the more obvious features of a cell. Most "general purpose" cells will show these features. In others which are more specialized, certain characteristics are emphasized, lacking, or obscured. The basic components of a cell are the **cell membrane,** the **cytoplasm** and the **nucleus.**

The **cell membrane** is the living outer edge of the cell—a limiting structure. It is usually pliable or elastic in quality, but sometimes quite rigid. It is capable of growth and of limited repair. While it is easily demonstrated as a double-layered structure under the electron microscope, it is often invisible under the light microscope. The **cell membrane** is **selectively permeable** which means that it determines what materials enter or leave the cell. Also, this permeability changes from time to time. It determines to a high degree the cell's relationships to its external environment.

The **cytoplasm** lies inside the cell membrane and between it and the nucleus. (The cell membrane may be considered a part of the cytoplasm.)

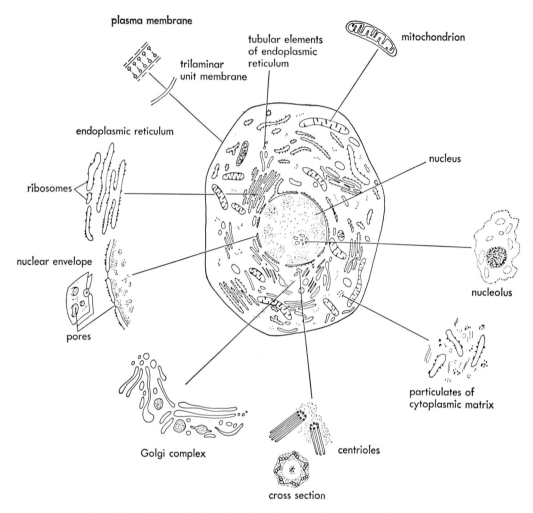

FIG. 3-2.   Schematic drawing of a cell as seen in electron micrographs.

The ground substance of the living cytoplasm appears homogeneous and has a viscous or "white of egg" quality. Careful killing, fixing, and staining help to bring out some of the details of structure. With the optical microscope certain formed constituents can be seen in the ground substance of cytoplasm of properly prepared cells. These may be classified as living particles, the **organelles,** or as "non-living" **inclusions.** Among the organelles are the **mitochondria, centrioles, Golgi material** and **fibrils.** The inclusions are such materials as **pigments, crystals, secretory granules, vacuoles** and **oil droplets.**

Mitochondria, referred to by Sievitz (1957) as the "powerhouse of the cell" are tiny rods or spheres which may change from one of these shapes to another. They are found clustered where cellular activity is very high, such as at the center of contraction of muscles, at nerve synapses, or at the absorptive surfaces of the cells of the intestine. Like the cell itself they have a selective membrane and exchange materials with other parts of the cytoplasm. They are the centers of catabolic enzyme action. In them, molecules of carbohydrates, proteins, and fats are broken down, yielding heat and energy for cellular activities.

The **centrioles** are usually located near the nucleus and in the center of a condensed portion of the cytoplasm called the **cell center** (*attraction sphere*). They play a prominent role in mitosis and will be considered later on pages 20 and 21.

The **Golgi material** is a network of fibrils or, according to some authors, a series of membranous structures (Swanson, 1960). It may be widely dispersed in the cell, but more commonly is found close to the cell center. Since it is especially prominent in secreting cells, some have felt that it may be involved in synthesis. Its exact function is still in doubt.

Fibrils are fine threads seen in many cells where they may give stability—as in certain epithelial cells. The fibrils of muscle cells (*fibers*) are more evident and are involved in contraction.

In plant cells the **chloroplasts** are in the cytoplasm. They are the centers for food manufacture (*photosynthesis*).

The **nucleus,** the controlling center of the cell, is composed of a **nuclear membrane,** one or more **nucleoli,** and **chromatin** all bathed in nuclear fluid. The nuclear membrane is probably similar to the cell membrane regulating relations with the rest of the cell. The nucleoli are prominent in resting cells, but disappear during mitosis. It has been suggested that they may store and pass genetic information to the cytoplasm (Swanson, 1960). They do make proteins, but our knowledge is very limited as to their total contributions.

Chromatin appears in resting cells as a fine network of beaded threads. During mitosis it becomes organized into the **chromosomes** which are in turn made up of **genes.** The definition of genes is a difficult matter. While they are the agents of heredity, they have a far wider significance in that they govern the synthesis of specific proteins including enzymes and hormones which in turn means that they control the whole life of the cell and of the total organism. Because they also have the capacity to mutate (*change*) they are the key to evolution.

Due to work in recent years it is now possible to give a fairly specific biochemical definition of a gene. Chromosomes consist of nucleoproteins which are complexes of **deoxyribonucleic acid (DNA), ribonucleic acid (RNA),** and various proteins. *The deoxyribonucleic acid or DNA fraction of the nucleoprotein is the material of the gene.* The DNA is a kind of pattern or template that is capable of self-duplication as in the duplication of chromosomes in mitosis (below). It also serves as a pattern for the formation of RNA which determines the specificity of the proteins synthesized in a cell or organism. Figure 3–3 shows schematically the above sequence of events.

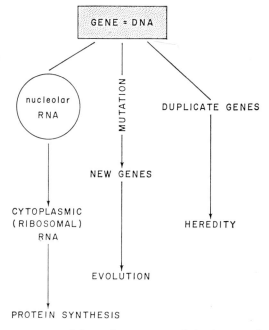

Fig. 3–3. Schematic summary of the pattern of gene action.

The consequences of protein specificity are quite well known. The introduction of a protein of one organism into the cells of another may cause marked disturbance, disease, or even shock and death. Thus, foreign proteins as pollens may cause allergy in man; certain bacteria produce disease; one part of an organism grafted onto another organism normally does not heal or take; and the mixing of the blood of one person with that of another, if their bloods are incompatible types, may cause shock and even death.

MITOSIS

As suggested earlier the size to which a cell can grow and remain viable depends upon its surface-volume relationship, the relationship of nucleus size to the volume of cytoplasm, and its rate of metabolism. Since a cell has to get all of its food and oxygen and excrete its waste products through the cell membrane, this membrane has to present adequate surface area. Further, since in a growing cell the surface area increases by the square of the radius and the volume by the cube of the radius, it is evident that a cell would soon grow beyond the capacity of its surface to service it. Also the nucleus bears somewhat the same relationship to the cytoplasm. A cell can function only for a short time without the controlling nucleus and the nucleus can control only a limited amount of cytoplasm. Finally, both of these situations involve the rate of metabolism in the cell—more activity means more food and oxygen and more surface area is required to supply these needs.

The situation can be met by a change in the shape of the cell to provide more surface such as a folding of the membrane or a flattening or lengthening out of the cell. Or—the cell can divide into two smaller daughter cells to establish a favorable surface-volume relationship. Most cells follow this latter course. This duplication of themselves is usually accompanied by the process of **mitosis.** The cells multiply by dividing—and the organism of which they are a part grows or renews its parts.

The process of mitosis has been known since about 1880. Only in recent years, however, has the mechanism of mitosis been isolated from other parts of the cell for critical study. Our knowledge has increased as has our understanding, but, as always in science, new and perhaps more difficult problems appear. We still do not know what causes the chromosomes to divide or to move apart, and these are central factors in the process of mitosis. They should challenge the imagination, ingenuity and skill of our young scientists. These are the kinds of problems which emphasize the need for a training, not just in biology, but in the sciences of chemistry, physics, and mathematics.

Only the more obvious features of mitosis and cell division will be described and illustrated here (Fig. 3–4). As a nucleus prepares to divide, it enlarges and the chromosomes, invisible before this time, appear as thin **paired** threads. Each half of a paired thread (chromosome) is called a **chromatid.** The chromosomes then shorten and swell and the nucleoli and nuclear membrane disappear (*prophase*). The centrioles, which lie to one side of the nucleus, move apart and come to lie at opposite poles of the nucleus. They are responsible for organizing the spindle to which the chromosomes attach. The chromosomes are now lined up on the spindle midway between the centrioles and their asters (*metaphase*). Each chromatid now begins to separate from its partner, each moving to opposite poles of the cell (*anaphase*). As this movement progresses, the cell membrane constricts midway between the two groups of chromosomes and the cell divides. Nuclei reorganize within the daughter cells thus formed and mitosis is complete (*telophase*). The whole process takes about thirty minutes, or at most a few hours.

Mitosis has important implications in biology. Each chromosome which appeared as a double thread in the early stages of mitosis actually represents daughter chromosomes—formed by the

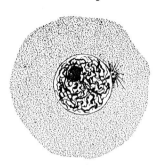

"Resting cell"—centriole single; individual chromosomes not visible.

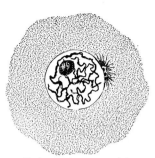

Early prophase—centrioles divided; chromatin thread appears double.

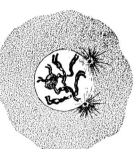

Late prophase—centrioles moving to opposite poles; chromosomes with chromatids distinct; nuclear membrane disappearing.

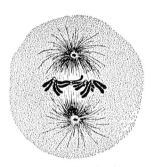

Metaphase—centriole with asters are at opposite poles; spindle connecting them is complete, chromosomes are lined up on equatorial plane.

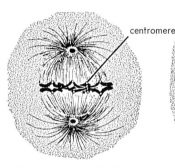

centromere

Early Anaphase—daughter chromosomes separate centromeres leading.

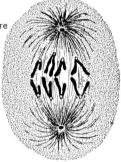

Anaphase—chromosome sets move toward opposite poles with centromeres leading; arms of chromosomes trail behind.

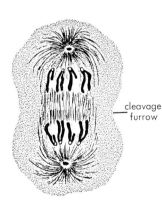

Early telophase—cleavage furrow developing; full species number of chromosomes to each daughter cell.

cleavage furrow

Telophase—migration of chromosomes complete; asters subsiding; nuclei reforming.

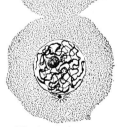

Mitosis complete— nuclei reconstituted; cytoplasm also divided.

FIG. 3–4.   Stages in mitosis.

self-duplication of a parent chromosome. The more obvious aspects of mitosis which have been described insure that these daughter chromosomes will be properly distributed to daughter cells. The daughter cells in turn will be equal in terms of number and quality of chromosomes (*hereditary potential*) and other capacities to their mother cell. The integrity of the cell has been maintained.

QUESTIONS

1. What are the unique characteristics of protoplasm which most concern the biologists?
2. What relationship do anabolism and catabolism have to metabolism?
3. Define irritability. Why is it sometimes called a diagnostic characteristic of living things?
4. What are the differences between growth and reproduction? How are they alike?
5. What is the cell theory?
6. Why are the names of Schleiden and Schwann so often associated with the cell theory? Did they discover cells?
7. Name the basic components of a cell and state briefly their functions.
8. Diagram and label a generalized cell.
9. Why is the volume of a cell so important in relationship to the extent of the cell surface? How is this related to cell division—mitosis?
10. Diagram the main steps in the process of mitosis. Label each one fully.
11. What is accomplished by mitosis other than the mere formation of two nuclei from one?
12. What does mitosis have to do with human anatomy?

"The student of Nature wonders the more and is astonished the less, the more conversant he becomes with her operations, but of all the perennial miracles she offers to his inspection, perhaps the most worthy of admiration is the development of a plant or animal from its embryo."

—Huxley

Chapter 4

Origin and Development of the Individual—Embryology

MEIOSIS AND FERTILIZATION

Orientation and Definition. In Chapter 3 the process of mitosis was briefly considered. This process, you should recall, involves the self-duplication of the chromosomes of the parent cell and their equal distribution to daughter cells. The result is that the daughter cells have the same chromosome number and hereditary potential as the parent cell. The cytoplasm of the cell is also about equally divided. By this process, in generation after generation of cells, the hereditary material is kept constant.

The reproduction of individuals by the sexual method, however, raises another question. In this process two cells, a **spermatozoon** (*sperm*) and an **ovum** (*egg*), come together to form a **fertilized egg** (*zygote*). If both of these cells, spermatozoon and ovum, carried a full complement of chromosomes, fertilization would result in a cell with two times the proper number for the species. Since the fertilized egg divides by mitosis, as do all subsequent generations of cells, the new individual would possess in all of its cells twice as many chromosomes as it should have. Such doubling of numbers of chromosomes would also continue in successive generations of individuals. This situation is avoided by a special process called **meiosis** which the sex cells undergo during their development (*maturation*) in the **gonads** (*sex organs*). It results in the reduction of the chromosome number of the spermatozoa and ova by one-half (*haploid number*) so that when they join in **fertilization** the chromosome number (*diploid*) of the species is re-established. Other outcomes of meiosis and of fertilization become apparent as the processes are briefly described.

Meiosis (Fig. 4–1). Since **meiosis** is much the same in male and female it will be described only once. In the early stages of meiosis there is a coming together of like (*homologous*) chromosomes, a process called **synapsis.** This is in

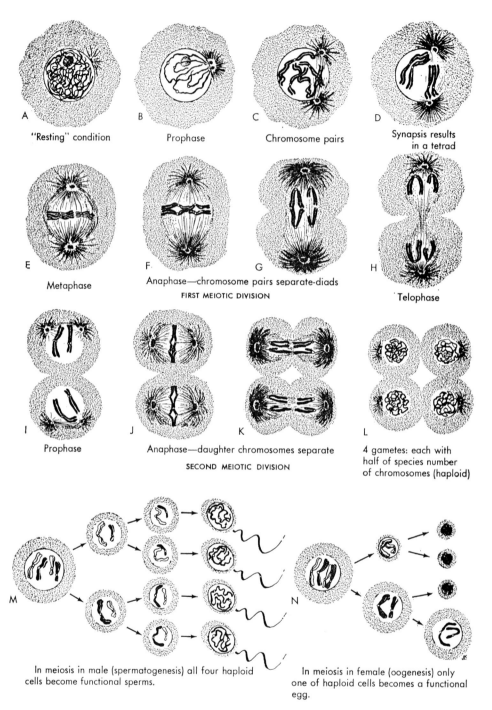

A   "Resting" condition

B   Prophase

C   Chromosome pairs

D   Synapsis results in a tetrad

E   Metaphase

F· G   Anaphase—chromosome pairs separate-diads

FIRST MEIOTIC DIVISION

H   Telophase

I   Prophase

J   K   Anaphase—daughter chromosomes separate

SECOND MEIOTIC DIVISION

L   4 gametes: each with half of species number of chromosomes (haploid)

M   In meiosis in male (spermatogenesis) all four haploid cells become functional sperms.

N   In meiosis in female (oogenesis) only one of haploid cells becomes a functional egg.

FIG. 4–1.   Schematic representation of meiosis. For simplification the species number of chromosomes was taken as four rather than the forty-six characteristic of man.

contrast to mitosis when each individual chromosome remains separate from its homologue. The paired chromosomes, each of which has duplicated itself longitudinally, shorten and thicken. Thus, each pair gives rise to a four-parted body or **tetrad,** each unit of which is called a **chromatid.** The nuclear membrane and nucleolus disappear and a spindle forms. This ends the first stage or **prophase.**

The tetrads now line up on the equatorial plane of the spindle (*metaphase*) and soon the duplicated homologous chromosomes separate from each other and move to opposite poles (*anaphase*). The nucleus now reforms, the cytoplasm divides and two daughter cells, each with the haploid number of chromosomes, are formed (*telophase*). Since each chromosome at this stage is made up of two chromatids they are sometimes called **diads.**

Following this first meiotic division the two new cells may go into an **interphase** or they may go immediately into a second meiotic division. In this division, the already divided chromosomes line up on the equatorial plane and at anaphase the **chromatids** separate and move to opposite poles of the spindle. The cell completes its division and four haploid cells result. These are the **gametes** or sex cells, the **sperms** and **eggs.**

The **sperm** or **spermatozoon** is one of the smallest cells of the human body and one of the most specialized (Fig. 4–2). It consists of a head and tail. The **head** contains the condensed nuclear material and is covered on its anterior two-thirds by a thin **head cap.** At the anterior margin of the head cap is a tiny **acrosome** which, when the sperm contacts an ovum, produces lysins and enzymatic materials which dissolve the surface coverings of the ovum. The sperm is thus aided in its penetration of the ovum in fertilization.

The **tail** of the spermatozoon consists of a middle piece, principal piece, and end piece. It contains a core of longitu-

dinal filaments. In the **middle piece** the filaments are surrounded by a spiral **mitochondrial sheath** and in the **principal piece** by a **fibrous sheath.** The **end piece** has only its plasma membrane which covers the entire spermatozoon.

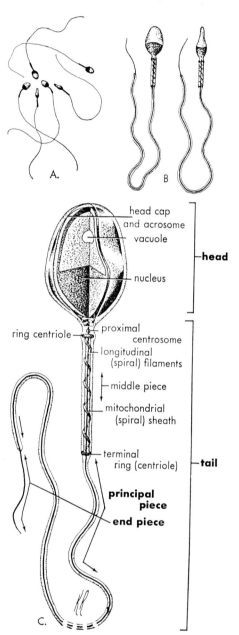

Fig. 4–2. *A,* A swarm of spermatozoa. *B,* Face and profile views. *C,* Schematic representation of spermatozoon.

One of the centrosomes, the **proximal centrosome,** is located between the head and middle piece; centrosome material

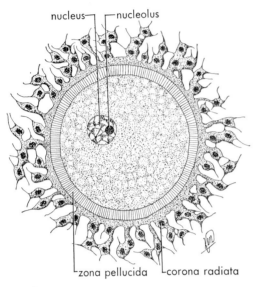

nucleus — ┌ nucleolus

zona pellucida   └ corona radiata

FIG. 4–3.   Ovum with zona pellucida.

is also represented in the **ring centrioles** at proximal and distal ends of the middle piece.

In contrast to the spermatozoon the **ovum** or egg is a generalized-appearing cell (Fig. 4–3). It is also large, for in the process of meiosis the cytoplasm was divided unequally, most of it being retained by the one cell which was to become the **functional** egg. The smaller cells, so produced, have a full complement of chromosomes but a minimal amount of cytoplasm, and are called **polar bodies.** They play no further role in reproduction (Fig. 4–1). The egg is covered externally by a thick, tough capsule, the **zona pellucida.**

Fertilization.   As the sperm penetrates the egg and their nuclei are joined in **fertilization,** a new individual is formed (Fig. 4–4). It has received half of its chromosomes from the male, half

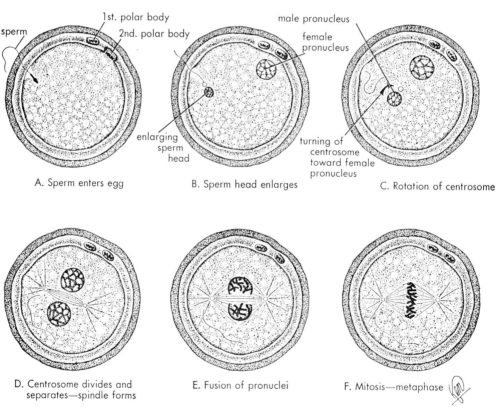

1st. polar body

sperm

2nd. polar body

male pronucleus

female pronucleus

enlarging sperm head

turning of centrosome toward female pronucleus

A. Sperm enters egg

B. Sperm head enlarges

C. Rotation of centrosome

D. Centrosome divides and separates—spindle forms

E. Fusion of pronuclei

F. Mitosis—metaphase

FIG. 4–4.   Schematic diagrams showing fertilization.

from the female. The diploid number is restored and the new individual is a genetic reflection of both of its parents. The sperm in penetrating the egg leaves its tail outside, and only the nucleus (*pronucleus*) and one centriole enter. Once a sperm has entered an egg other sperms are barred by changes that take place in the outer surface of the egg. The sperm also serves to initiate embryological development.

Genetic Importance. Besides the now obvious fact that meiosis produces germ cells with the haploid number of chromosomes there are other important implications. When the homologous chromosomes, one maternal and one paternal, line up on the equatorial plane at the first metaphase of meiosis their orientation is entirely random. Therefore, when they segregate at anaphase there will result in the haploid cells a random distribution of maternal and paternal chromosomes. In man, with 23 pairs of chromosomes, this could result in 8,388,608 possible chromosome combinations in the gametes (Swanson, 1960). While it would be theoretically possible for a gamete to contain only paternal or maternal chromosomes, it is very unlikely.

Also, when the homologous chromosomes are lined up together and form chromatids there may be an exchange of segments between the chromatids of the two homologues. This is often called **crossing over** (Fig. 4–5). Add to this an occasional **mutated** (*changed*) gene, and it is clear that no two gametes are apt to be genetically alike.

Sexual reproduction, therefore, involving as it does meiosis and fertilization, insures the production not only of new individuals but of individuals which vary widely among themselves. Natural selection operates upon these variables and the evolution of organisms results. As William Patten (1925) has said, "In this way there is produced sufficient stability

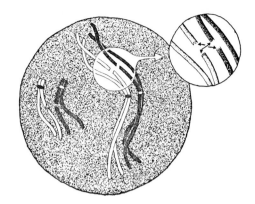

A, Chromatids split—arrows show direction of crossing.

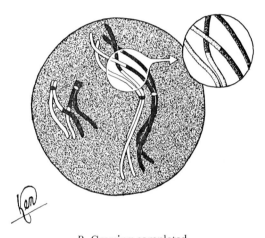

B, Crossing completed.

Fig. 4–5. Schematic representation of crossing over.

to insure continuity and at the same time sufficient variety to insure progress."

EMBRYOLOGY

Definition. Embryology, you should recall, deals with the period of development prior to birth. The treatment of it will be mostly descriptive although much of the interest at the present time is in experimental embryology which deals with the forces and mechanisms that govern developmental processes.

The descriptions of the early stages of development are based mostly upon

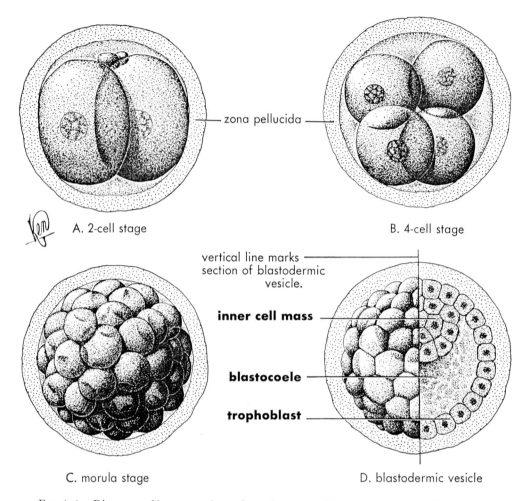

zona pellucida

A. 2-cell stage

B. 4-cell stage

vertical line marks
section of blastodermic
vesicle.

inner cell mass

blastocoele

trophoblast

C. morula stage

D. blastodermic vesicle

Fig. 4–6.   Diagrams of human embryos from cleavage to blastodermic vesicle (blastocyst).

embryos of mammals such as the rabbit and pig because comparable material is not readily available for man. However, as increasing numbers of human embryos become available they give us no reason to believe that early development of man is very different from that of other mammals. The Carnegie Institution of Washington now has a fairly complete collection of human embryos from the 2-cell (5-day) stage to the end of the embryonic period (Science, 1956).

Cleavage (Fig. 4–6). The spermatozoon which fertilizes the egg also initiates its division (*mitosis*). The early cell divisions in development are called

cleavage and result in the formation of a small ball of cells, the **morula** (*little mulberry*) inside of the zona pellucida of the original egg. These cleavages take place in such rapid succession that no time for the growth of the individual cells is allowed, and hence the morula is little larger than the original fertilized ovum and its individual cells are proportionately smaller. The embryo during this early period is moved down through the uterine tube (*Fallopian*), by active contraction of its muscular walls and by ciliary action, into the body of the uterus. The time involved is probably about three days.

Differentiation — blastocyst formation (*blastodermic vesicle*). The term cleavage is seldom applied to the cell divisions following morula formation. Cell division now becomes involved with **differentiation** as cells shift their positions, segregate, and display special functions in keeping with their different potencies. The morula soon becomes modified into two parts, an outer layer of cells, the **trophoblast** and an **inner cell mass** (Fig. 4–7). The trophoblast secretes fluid into the interior and grows rapidly, pressing against the zona pellucida, causing it to stretch, thin out, and finally to disappear. The inner cell mass is now attached to one side of the trophoblast and a large cavity, the blastocoele, filled with fluid is present inside of the trophoblast. This is the **blastocyst** or **blastodermic vesicle**. It probably remains free in the uterine cavity for a few days before implanting itself into the wall of the uterus.

Implantation. Implantation takes place when the blastocyst contacts and adheres to the uterine mucosa (*endometrium*) which has become highly vascular, glandular, and thick (Fig. 4–8). The trophoblast "digests" its way into the endometrium and the blastocyst gradually becomes completely buried. The

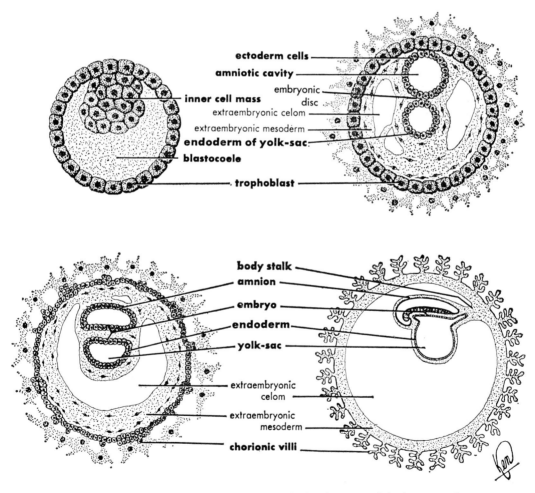

ectoderm cells
amniotic cavity
embryonic disc
inner cell mass
extraembryonic celom
extraembryonic mesoderm
endoderm of yolk-sac
blastocoele
trophoblast

body stalk
amnion
embryo
endoderm
yolk-sac
extraembryonic celom
extraembryonic mesoderm
chorionic villi

FIG. 4–7. Schematic representation of the early development of the human embryo. (Modified from Arey and Prentiss.)

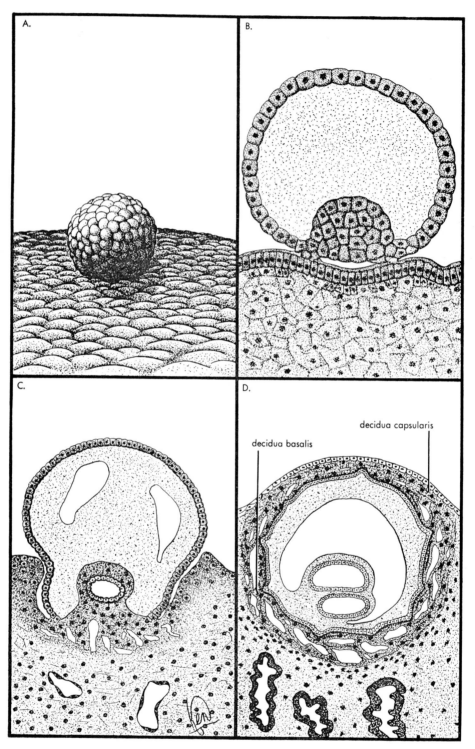

FIG. 4–8.   Schematic representation of the implantation of the embryo.   *A*, Blastocyst settles upon the endometrium.   *B*, Blastocyst enlarged and sectioned to show inner cell mass.   *C*, Embryo "digesting" its way into the endometrium.   *D*, Endometrium enclosing the embryo.

embryo, therefore, develops within the wall of the uterus and not in its cavity. Implantation is probably complete by the eleventh day after fertilization and the embryo gradually establishes a functional relationship with the maternal blood supply so that it may be nourished, receive oxygen, and have its waste products carried away. These relationships will be described later.

Germ Layer Formation (Fig. 4–7, 9). Germ layer formation takes place while the blastocyst is implanting in the endometrium. There are three primary germ layers, the ectoderm, endoderm, and mesoderm. The cells of the inner cell mass differentiate into two groups. The ones nearest to the trophoblast become the **ectoderm** and those which separate from its inner surface (*delamination*), form the **endoderm** (*gastrulation*). Within the mass of ectoderm cells a cavity appears, the **amniotic cavity,** and its thin lining is the **amnion.** The endoderm grows and fashions itself into the **yolk sac** which

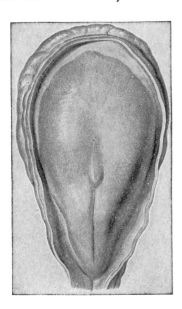

FIG. 4–9. Embryonic disc of sixteen-day-old human embryo surrounded by cut edge of amnion. Body stalk cut off at posterior end. Primitive groove overlies primitive streak. Primitive node at anterior end of primitive groove. × 50. (Heuser.)

4

pushes deep into the blastocoele. At the point of contact of the ectoderm and endoderm is the **embryonic** (*germinal*) **disc,** where the **embryo proper** will continue to develop. While the differentiation of ectoderm and endoderm was taking place scattered cells appeared arising from the trophoblast and the area around the embryonic disc. These spread through the remaining blastocoele and form the **extraembryonic mesoderm.** The extraembryonic mesoderm then reorganizes to form layers which come to lie close to the yolk sac and the trophoblast. The cavity which appears between these layers is the **extraembryonic coelom.** The trophoblast and its mesoderm are now called the **chorion.** **Villi** grow out from it into the surrounding substance of the uterus. As the amnion expands, the yolk sac and embryo are moved away from the trophoblast wall until they are held to it only by a stalk of concentrated extraembryonic mesoderm, the **body stalk** (Figs. 4–7, 19).

Embryonic Disc and Mesoderm Formation (Fig. 4–9). The area of contact of the ectoderm and endoderm, the embryonic disc, is flat and leaf-like in shape. If the trophoblast and amnion dorsal to the embryo proper are cut away, the dorsal side of the embryonic disc can be seen. The smaller end of the disc is continuous with the body stalk and is the **caudal end** of the embryo. The midline of the embryonic disc at the caudal end is marked by a **primitive streak,** an elongated mass of proliferating cells lying at the bottom of a **primitive groove** (Fig. 4–9). Cells proliferate rapidly from the primitive streak and spread anteriorly and laterally between the ectoderm and endoderm as the third germ layer, the **mesoderm** (Fig. 4–10*A*). This is not to be confused with the extraembryonic mesoderm mentioned earlier. This new mesoderm is a part of the embryo proper, and plays a vital role in its development.

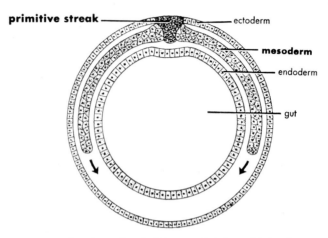

A, Section through primitive streak (about 16 days).

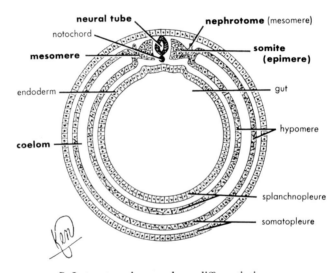

B, Later stage in mesoderm differentiation.

Fɪɢ. 4–10.   Diagrammatic representation to show spread and differentiation of the mesoderm.
(Modified from Arey.)

At the cephalic end of the primitive streak is a thickened mass of cells—the **primitive knot** or **node** (Fig. 4–9). It produces mesoderm also which pushes forward between the ectoderm and endoderm as the **head process.** The roof of the head process becomes the **notochord** and as the process continues to grow the primitive streak shortens.

At about the sixteenth day a constriction appears between the amnion and the yolk sac at the margins of the embryonic disc. This constriction continues and the flat, embryonic disc is gradually changed into a tubular embryo (Figs. 4–7, 19). Since the region of the head process grows more rapidly than the rest of the embryo, this part tends to be thrown up into a fold. The constricting also gradually decreases the diameter of the opening into the yolk sac and the **digestive tube** or **gut** of the embryo

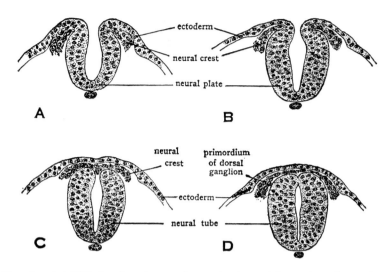

FIG. 4–11.   Drawing ($\times$ 135) showing closure of neural tube and formation of neural crest.   From pig embryos of: *A*, 8 somites; *B*, 10 somites; *C*, 11 somites; *D*, 13 somites.   (From *Human Embryology* by Bradley M. Patten, 2nd ed., copyright, 1953.   Used by permission.)

takes form.   From the posterior end of the gut an endodermal diverticulum pushes out into the mesoderm of the body stalk.   This is the **allantois, a fetal membrane** (Figs. 4–7, 19).

Neural tube (Figs. 4–11, 12).   Neural tube formation is also apparent at this time.   In the area between the primitive streak and the forward edge of the embryonic disc, thickenings appear in the ectoderm to either side of the midline.   As these thickenings grow, a groove is formed between them.   The thickened areas are called the **neural folds,** the groove the **neural groove.**   This process continues, and as the folds rise up they approach each other medially and finally join to form a **neural tube** (Fig. 4–12, 13).   The thin body ectoderm separates from the tube and becomes continuous over it.   The neural tube develops subsequently into the **brain** and **spinal cord.**   As the neural tube forms, masses of cells called the neural crests appear on either side.   They later become the posterior root ganglia of spinal nerves and sensory ganglia of certain cranial nerves (Fig. 4–11).

Somites.   While the neural tube is forming the embryonic mesoderm thick-ens to either side of the tube and the noto-chord.   Clefts soon appear in these thickened areas, forming a series of blocks which are called **somites** (Figs. 4–13, 14).   These somites represent the primary **segmentation** of the body and in turn influence segmentation in the muscular, skeletal, and even in the nervous system.   The somites are at first hollow, but gradually their cavities (*myocoels*) fill with cells.   Usually 36 to 38 pairs of somites develop.

Coelom Formation.   Lateral to the somites the mesoderm, as it spreads between the ectoderm and endoderm, splits into two layers, and the cavity so formed is the **embryonic coelom** (Fig. 4–10*B*).   The further development of the coelom is described on pages 361 and 410.   The outer wall of the mesoderm, together with the ectoderm against which it lies, is called the **somatopleure,** while the inner wall of mesoderm with the adjacent endoderm forms the **splanchnopleure.**   The first blood vessels form in the meso-derm of the splanchnopleure and others follow in other mesodermal areas.   The heart appears on the ventral side of the embryo.   Also a constricted mass of

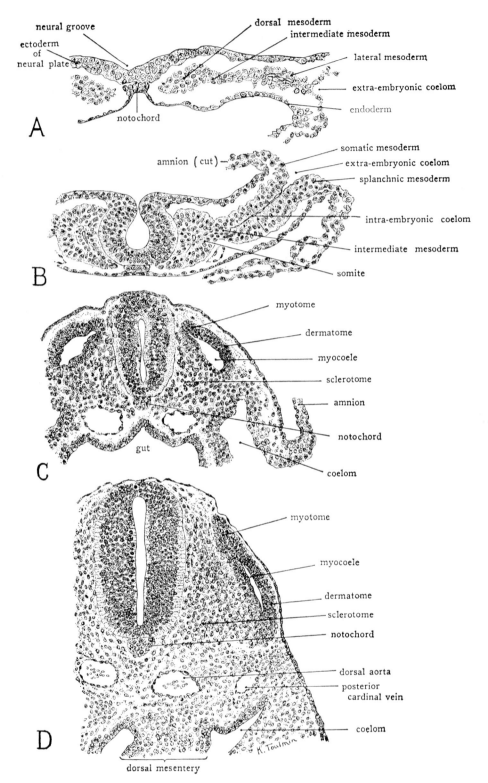

neural groove

dorsal mesoderm

intermediate mesoderm

ectoderm
of
neural plate

lateral mesoderm

extra-embryonic coelom

endoderm

notochord

A

amnion (cut)

somatic mesoderm

extra-embryonic coelom

splanchnic mesoderm

intra-embryonic coelom

intermediate mesoderm

somite

B

myotome

dermatome

myocoele

sclerotome

amnion

notochord

gut

coelom

C

myotome

myocoele

dermatome

sclerotome

notochord

dorsal aorta

posterior
cardinal vein

coelom

D

K. Toulmin

dorsal mesentery

Fig. 4–12.   Drawings (× 150) from transverse sections of pig embryos of various ages to show formation and early differentiation of somites.   (From series in the Carnegie Collection, courtesy of Carnegie
Embryological Institute.)   *A*, Beginning of somite formation.   *B*, Seven-somite embryo.   *C*, Sixteen-
somite embryo.   *D*, Thirty-somite embryo.

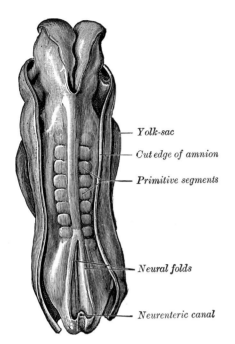

FIG. 4–13.   Dorsum of human embryo, 2.11 mm.
in length.   (After Eternod.)

- *Yolk-sac*
- *Cut edge of amnion*
- *Primitive segments*
- *Neural folds*
- *Neurenteric canal*

mesoderm appears between the somites and the lateral mesoderm, the **intermediate cell mass,** from which reproductive and urinary organs will later develop.

External Form (Figs. 4–14, 15). Important changes in the **external form** of the embryo also take place during this period of development. During the fourth and fifth weeks the face begins to take on form with the development of a series of cartilaginous arches in the head and neck region. These are the **branchial arches** and correspond to the gill arches of fishes. Between the branchial arches **pharyngeal pouches** appear, and, in the ectoderm over these, corresponding indentations form, the **outer branchial grooves** (Fig. 1–3). Anterior to these is a **frontonasal process** and between it and the first branchial arch the **eyes** develop. **External nasal orifices** are present but far apart, and the nose is not yet lifted

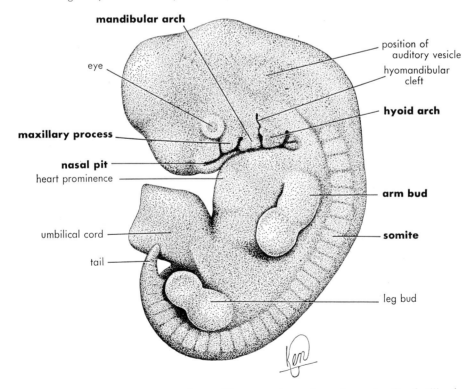

FIG. 4–14.   Development of body form.   Human embryo about six weeks after fertilization.
(Modified from Patten, *Human Embryology.*)

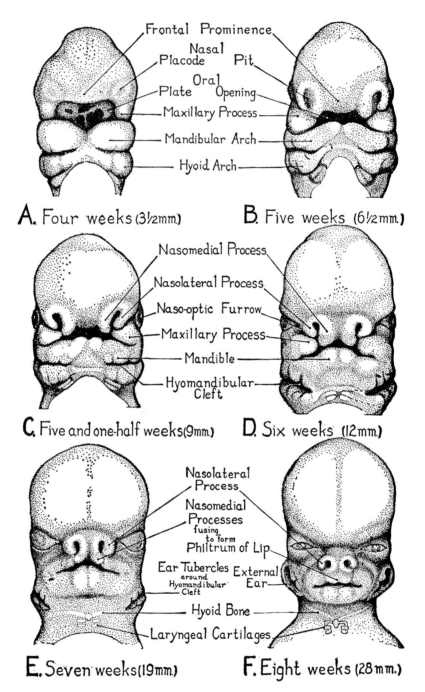

FIG. 4–15.   Drawings showing, in frontal aspect, some of the important steps in the formation of the face. (After William Patten, from Morris, *Human Anatomy*, courtesy of McGraw-Hill Book Co.)

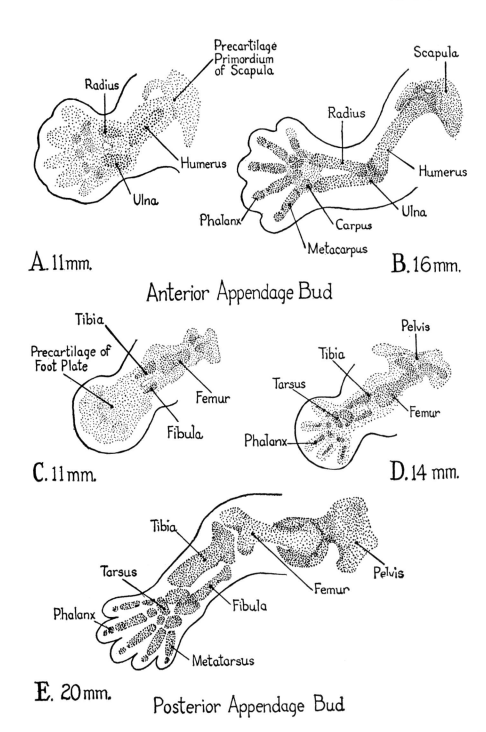

A. 11mm.

B. 16mm.

Anterior Appendage Bud

C. 11mm.

D. 14 mm.

E. 20mm.

Posterior Appendage Bud

FIG. 4–16. Early stages in development of the skeleton of the appendages. The fine stippling in the drawings of the younger stages indicates precartilage concentrations of mesenchyme; the more sharply circumscribed and coarser stippling of the older stages represents cartilage. (Modified from Bardeen and Lewis, Am. J. Anat., Vol. 1, 1901, in Patten, *Human Embryology*.)

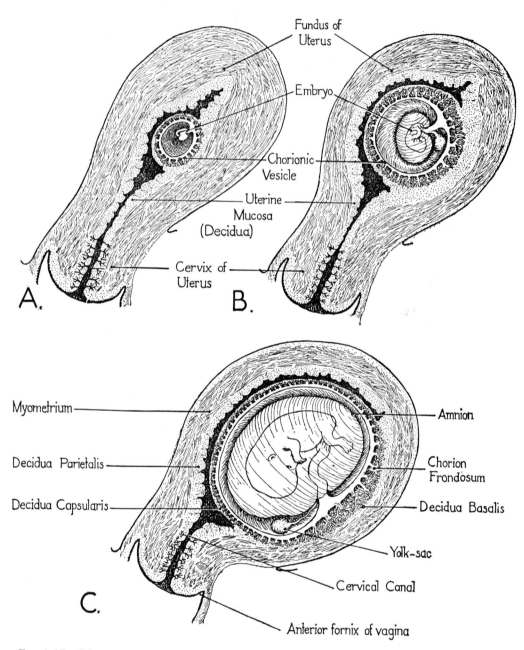

FIG. 4–17.   Diagrams showing uterus in early weeks of pregnancy.   Embryos and their membranes are drawn to actual size.   Uterus is within actual size range—about correct for a small primipara. *A*, At three weeks.   *B*, At five weeks.   *C*, At eight weeks fertilization age.   (From *Human Embryology*, by Bradley M. Patten, 2nd ed., copyright 1953, McGraw-Hill Book Co.)

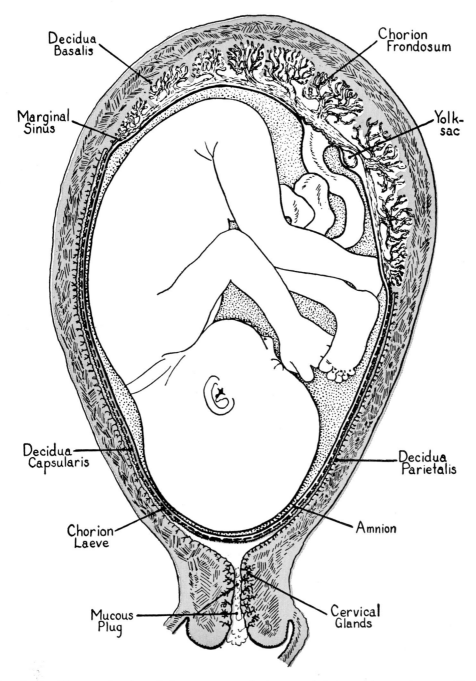

FIG. 4–18.   Diagram showing relations to uterus of a five-month fetus and its membranes.   Uterine structures have been colored red to contrast with fetal structures, shown in black and white.   Amnion is drawn as a solid black line, amniotic cavity is stippled, and chorion laeve is represented by a broken line. (From *Human Embryology* by Bradley M. Patten, 2nd ed., copyright 1953, McGraw-Hill Book Co.)

up from the face. Soon, however, due to the narrowing and elongation of the frontonasal process the eyes are brought forward, the nostrils closer together, and the nose elevates above the face surface. The first branchial arch develops a **maxillary process** from its lateral side. This process grows forward to become the maxilla while the medial portion of the arch forms the **mandible**.   The second branchial arch is the hyoid arch, and the other arches contribute to the formation of the cartilages of the larynx.

Limb buds (Figs. 4–14, 16) also appear during the fourth and fifth week as small swellings at the sides of the trunk. Somatic mesoderm pushes into these buds and by division of the cells forms masses of tissue, the **mesenchyme**. By the sixth week the three divisions of the limbs can be seen, the arm, forearm, and hand; or the thigh, leg, and foot.   Within the developing limbs the central part of the cell mass condenses to form the cartilage of the limb skeleton, the peripheral part forms the intrinsic muscles of the limbs. The forelimbs differentiate a little more rapidly than the hindlimbs.   Longitudinal grooves develop in the hand and foot portions of the limbs to define the fingers and toes. If they fail to appear, the individual has a "webbed" hand or foot.

Fetal Membranes and Placenta (Figs. 4–17 to 20).   The fetal membranes, **amnion, chorion, allantois** and **yolk sac,** and the **placenta** house, protect, nourish, provide oxygen, and get rid of waste materials for the developing individual. Recall now that at about the eleventh day of development the embryo had become implanted in the wall of the uterus, "digesting" its way into the mucosa to take a position in its deeper layers.   The portion of the endometrium (*mucosa*) which closes over the implanted embryo is called the **decidua capsularis,** and the deeper part, between the embryo and the muscular wall of the uterus, is the **decidua basalis** (*decidua placentalis*). As the embryo (*first two months*) and later the **fetus** (*third–tenth month*) grows, the decidua capsularis is stretched and pushed out into the cavity of the enlarging uterus to completely fill it.   Soon the decidua capsularis and the mucosa which it contacts atrophy.   The decidua basalis contributes to the **placenta** (Figs. 4–8, 17, 19).

The trophoblast of the early blastocyst

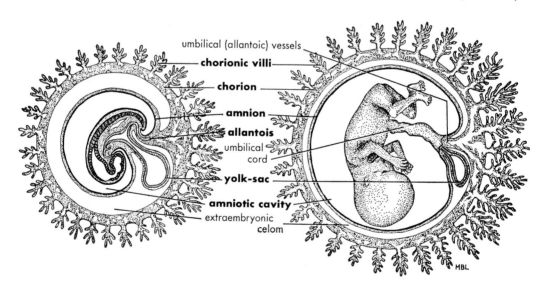

Fig. 4–19.   The development of fetal membranes in man.   (Modified from Patten.)

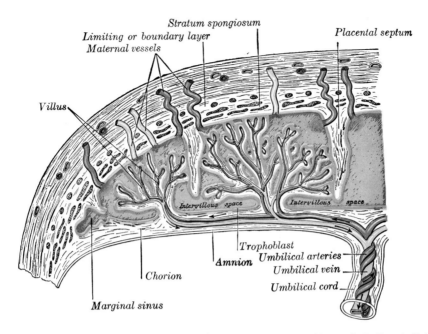

Stratum spongiosum

Limiting or boundary layer
Maternal vessels

Placental septum

Villus

Intervillous space *Intervillous space*

Trophoblast
Amnion *Umbilical arteries*
Umbilical vein

Chorion
Umbilical cord

Marginal sinus

FIG. 4–20. Scheme of placental circulation. (Gray, *Anatomy of the Human Body*, Lea & Febiger.)

consisted of a single layer of cells. As it implants in the mucosa of the uterus, it develops two layers, an outer one showing no cell boundaries and an inner one in which cells are well defined. Villi grow out from the cellular trophoblast and penetrate deeply into the decidua and maternal blood seeps into the intervillous spaces. This structure, the trophoblast and its villi, plus the lining extraembryonic mesoderm, is the outer fetal membrane, the **chorion** (Fig. 4–19). It contributes a part of the placenta. The placenta, therefore, is a double structure derived in part from maternal tissue, the decidua basalis (*placentalis*), and in part from the embryo, the chorion. **Umbilical arteries,** carrying waste materials from the embryo, lead into capillaries in the chorionic villi where the blood is brought into close relationship with maternal blood. **Umbilical veins** return this blood, now laden with oxygen and nutrient materials to the embryo (Fig. 4–20).

The amnion, another fetal membrane, arose early in development as the ectodermal part of the inner cell mass

hollowed out. At the same time the endodermal cells of the inner cell mass gave rise to the yolk sac. As development progressed, the blastocoele became filled with extraembryonic mesoderm which later formed an extraembryonic coelom pushing the mesoderm back against and around the trophoblast, the yolk sac and the amnion (Fig. 4–7). As the embryo grew, the amnion became very much expanded, and in so doing pushed the now-shrunken yolk sac close to the body stalk which it joins, and together they form the **umbilical cord.** The amnion comes to rest against the inside of the chorion. Also, as these changes were taking place the head and tail folds of the embryo were developing and the body stalk (*umbilical cord*) was shifted from its former attachment in the tail region to the midventral side of the embryo (Fig. 4–19).

The last fetal membrane has also been mentioned before—the **allantois** (Figs. 4–7, 19). It is derived as an outgrowth from the endoderm into the mesoderm of the body stalk at the caudal end of the embryo. While it becomes a large expanded structure in reptiles, birds, and many mammals, it remains rudimentary in primates. It contributes only its blood vessels which are the important umbilical vessels which bring the embryo and the maternal structures into functional association in the placenta.

After the child is born, the placenta and fetal membranes are discharged from the uterus as the "afterbirth." The separation of the placenta from the uterine wall ruptures blood vessels and some blood is lost. However, this hemorrhage is controlled by the contraction of the muscles of the uterine wall.

Our story of **descriptive embryology** will terminate at the end of the eighth week of development which is the end of the period of the embryo. The basic plan of the vertebrate body—in this case the human body—has been established. Further development—that of the fetus— involves the continuing differentiation of the three germ layers, the ectoderm, mesoderm, and endoderm. A wandering embryonic tissue, the **mesenchyme,** derived primarily from the mesoderm, is also of great importance in development.

The story of how the organs and body systems are constructed from these building materials will be taken up as we study each system of the body. Table 4–1 shows in outline the tissues which derive from the various germ layers.

Dynamics of Development. The human "mind" is not, or at least should not, be satisfied with a mere description of events. Causes or stimuli for such events must be sought to give understanding.

Perhaps the most obvious part of our story so far is that development is an "unfolding" of a pattern of structure. One event leads to another in a regular sequence. Indeed, one event appears to **cause** the next one. This is not a new concept. It is the theory of **epigenesis** of **Aristotle** and is accepted by the embryologists of our time. Modern biologists, of course, through the development of the sciences of genetics and evolution understand it better than did Aristotle. It is evident that genes and to some extent cytoplasmic factors control this sequence of events that we call development, as they control the whole life cycle which is the individual. Biochemistry, as applied to embryology and genetics, can be ex-

TABLE 4–1.   TISSUES DERIVED FROM GERM LAYERS IN MAN

Ectoderm	Mesoderm (including mesenchyme)	Endoderm

1. Epidermis of skin, including:
   Skin glands
   Hair and nails
   Lens of eye
2. Epithelium of:
   Nasal cavities
   Sinuses
   Mouth, including:
      Oral glands
      Enamel
   Sense organs
   Anal canal
3. Nervous tissues
   Hypophysis
	1. Muscle:
	   Skeletal, cardiac, smooth
	2. Connective tissue including:
	   Cartilage
	   Bone
	   Notochord
	   Blood
	   Bone marrow
	   Lymphoid tissue
	3. Epithelium of:
	   Blood vessels
	   Lymphatics
	   Coelomic cavities
	   Kidney and ureters
	   Gonads and ducts
	   Adrenal cortex
	   Joint cavities
		1. Epithelium of:
		   Pharynx
		   Auditory tube
		   Tonsils
		   Thyroid
		   Parathyroid
		   Thymus
		   Larynx
		   Trachea
		   Lungs
		   Digestive tube and
		      its glands
		   Bladder
		   Vagina and
		      vestibule
		   Urethra and
		      glands

pected to supply some of the answers as to how this stimulus-response mechanism actually works. Progress in these fields is very slow, however, because of the complex nature of the problem.

Recent research has demonstrated that this "unfolding" in development depends also on communication systems outside of the cells (*extracellular*) between one part of the embryo and another. Spemann in 1938, studying the **dorsal lip** of the **blastopore** of embryos, showed that this area had marked influence in organizing the embryo. The blastopore lip is called an **organizer** and the process is **induction**. To cite another example, **induction** is shown in the development of the lens of the eye of the frog (Fig. 4–21). As

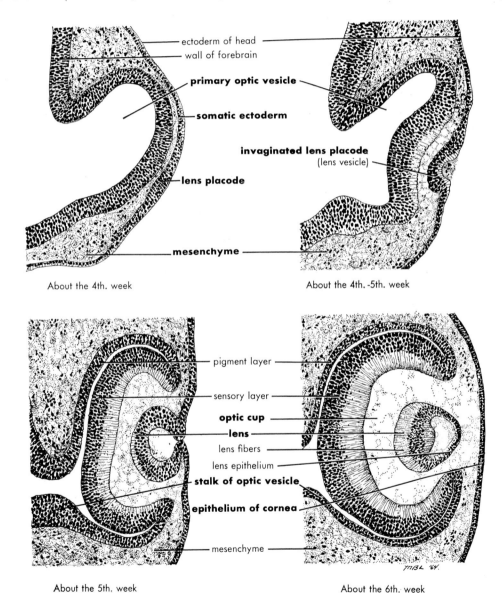

ectoderm of head
wall of forebrain
primary optic vesicle
somatic ectoderm
invaginated lens placode
(lens vesicle)
lens placode
mesenchyme

About the 4th. week

About the 4th. -5th. week

pigment layer
sensory layer
optic cup
lens
lens fibers
lens epithelium
stalk of optic vesicle
epithelium of cornea
mesenchyme

About the 5th. week

About the 6th. week

FIG. 4–21.   Early stages in the development of the optic cup and lens.
(Modified from Patten.)

the frog's brain develops, it sends outward a pair of structures called optic vesicles which indent at their distal ends to form the optic cups which come close to the skin. At this point in the skin a thickening occurs which breaks away from the skin and moves into the optic cup and becomes the lens of the eye. If the developing optic cup is removed and placed under the skin of the belly of the frog, it will induce there the formation of a lens. This lens, therefore, comes from tissue which does not ordinarily produce lens material but has been **induced** to do so by the presence of the optic cup.

Other agents than genes and organizers have an influence on development. Among these are the hormones. If thyroxine is lacking during development, the individual becomes a **cretin. Sex hormones** and those of the **anterior pituitary** play important roles.

As the embryo differentiates into many kinds of cells, the cells tend to lose their broad potencies—their capacity to differentiate in many directions becomes much more limited. The fertilized egg could be described as fully potent because from it come all of the cells of the body; the germ layers are less potent; and a completed organ like the stomach has lost its potency for differentiation.

Anomalies (Fig. 4–22). Since the

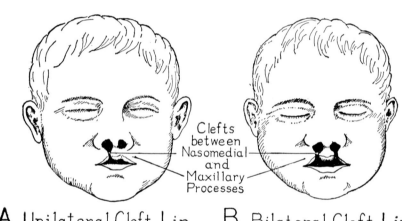

A. Unilateral Cleft Lip   B. Bilateral Cleft Lip

Clefts between Nasomedial and Maxillary Processes

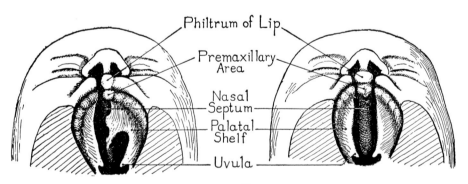

Philtrum of Lip
Premaxillary Area
Nasal Septum
Palatal Shelf
Uvula

C. Palate in a case of Unilateral Cleft Lip

D. Palate in a case of Bilateral Cleft Lip

FIG. 4–22.   Cases of cleft lip and cleft palate. (Modified from Corning.)

growth of the embryo is such a complex, precise, carefully balanced, yet flexible process, there is ample opportunity for imperfections to appear. These may be simple, like a mole, or more serious as a cleft palate, or very marked as in so-called **monsters.** Children born without a brain, spinal cords that have failed to close, and a wide range of other abnor- malities are known. Many of these can be better understood if one is familiar with the normal sequence of develop- ment. The causation of anomalies is not to be found in any simple, general expla- nation. Each type needs to be studied intensively. Heredity, disease, chemical agents, physical factors, and irradiation are among the possible causes.

"The most significant period of an individual's life is spent in his mother's womb."

—Chinese Philosopher

QUESTIONS

1. Where does meiosis take place in man?
2. List some of the ways in which meiosis and mitosis differ as processes.
3. What does meiosis accomplish for a species?
4. Meiosis in the male and female are much the same. How do they differ?
5. What is the significance of fertilization in terms of genetics and evolution?
6. Define the following: polar body, chromatid, homologous chromosomes, tetrad, gamete, gonad, pronucleus, zona pellucida, haploid, crossing over of chromosomes.
7. What is the process called by which the fertilized egg divides?
8. What is the morula and where might it be found in the female reproductive system?
9. What do we mean by differentiation as used in embryology?
10. Describe by words and diagram the structure of the blastocyst.
11. Describe briefly the implantation of the embryo.
12. Define the embryonic disk and tell how embryonic mesoderm is formed.
13. How is the notochord formed? The neural tube?
14. What are the somites and how are they related to the segmentation of the body?
15. Describe the formation of the embryonic coelom.
16. Name the three germ layers and recall how each came about in the embryo. Why are they important?
17. What is the source of mesenchyme and what part does it play in embryology?
18. Describe the early formation of limbs.
19. Name the fetal membranes and recall their origin in the embryo. What are their functions?
20. What is the placenta and why is it often described as having a double origin?
21. What are the deciduae?
22. What constitutes the umbilical cord?
23. What germ layer gives rise to the heart and blood vessels?
24. Where do the first blood vessels appear in the embryo?
25. What is the afterbirth?
26. Distinguish between embryo and fetus.
27. What is epigenesis? Is it a new idea?
28. How do modern biologists explain epigenesis?
29. What are organizers or inductors? Give examples.
30. What are anomalies? How can embryology and comparative anatomy help us to understand them?

> "*Over the structure of the cell rises the structure of plants and animals, which exhibit the yet more complicated, elaborate combinations of millions and billions of cells coordinated and differentiated in the most extremely different ways.*"
>
> —Hertwig

Chapter 5

The Organization of the Body

TISSUES, ORGANS AND SYSTEMS

DEFINITION AND SCOPE

CELLS have been briefly described and their importance as structural and functional units of the body has been emphasized. We have seen how they divide by mitosis and how special cells, the ova and spermatozoa, are prepared by meiosis for fertilization and the development of a new individual. The early development of man has been outlined and the capacity of cells to group or organize themselves into germ layers was observed. It is apparent that in multicellular animals isolated cells are not common.

Instead, similar cells are joined together, loosely or closely, and with the intervening non-cellular materials form the **tissues** of the body. The study of tissues and how they form the organs constitutes the science of **histology.**

The tissues of the body are derived from the germ layers and from an intermediate embryonic tissue, the mesenchyme, by complex processes of **differentiation** and **specialization.** The resulting tissues perform different functions so that a division of labor is established in the organism. There are four groups of primary tissues found in the human body, and each group contains a number of varieties. (1) **Epithelial tissues** are characterized by having the cells closely joined one to another and by being found on the free surfaces of the body, externally and internally. Glands and other structures are derived from them. (2) **Connective tissues** are those in which the cells are separated by large amounts of intercellular materials. These intercellular substances are non-living products formed by the cells. The classification of connective tissues is based upon the kinds of cells and of intercellular materials, and they range from fluid blood to hard bone. (3) **Muscle tissues** are all characterized by the high degree of **contractility** of their cells or fibers. Their function is **movement,** ranging from movement of the skeleton to the contracting of the stomach or the beating of the heart. (4) **Nervous tissue** is composed of cells specialized in the properties of **irritability** and **conductivity.** Impulses move swiftly over the nerves resulting in the integration of the many body functions.

As the tissues are forming they are also being combined to form the **organs** of the body. Organs may therefore be defined as structures composed of two or more tissues integrated in such a manner as to perform certain functions. The heart, for example, has its chambers lined with a special kind of epithelium, its walls are predominantly muscle, and there are also connective tissues present. It is a pump which maintains the flow and pressure of blood in the body.

It is going beyond the usual limits of histology to consider the matter of **body systems,** but it is logical to define and list them at this point in our discussion. Body systems represent combinations of organs of similar or related functions which work together as a unit. The heart organ is useless without the other organs of the **circulatory system,** the arteries, capillaries, veins, and lymphatics. Together they form a system for the

transport and exchange of materials throughout the body through the medium of blood, lymph, and tissue fluid.

The systems of the human body are as follows:

1. **Integumentary System** (Fig. 6–1). This consists of the skin and its derivatives such as the hair, nails, and glands. Among its functions are protection, reception of stimuli, secretion, and temperature regulation.

2. **Skeletal System** (Fig. 7–1). This system includes the bones, cartilages, and connective tissues which bind these organs together to form the joints. Its functions are support, movement, protection, and blood formation.

3. **Muscular System.** The skeletal muscles are the organs of this system. The muscles are the active agents in movement working in conjunction with the skeletal system, which is passive.

4. **Circulatory System** (Fig. 12–1). This is the transportation system of the body and provides the cells with food, oxygen, and other materials and carries away their waste products. It is involved also in maintaining equilibrium in the internal environment, in temperature regulation and protection. Its organs are the heart, blood vessels, and lymphatics.

5. **Endocrine System** (Fig. 17–1). This is a system of glands which secrete their products, the **hormones,** into the blood for distribution to the tissues and organs which are influenced by them. These glands go by various names such as **endocrine, ductless glands,** or **glands of internal secretion.** The pituitary, thyroid, parathyroid and adrenal are all endocrine glands. Parts of such organs as the ovary, testis, and pancreas also belong to this system. Many body functions are controlled and coordinated by these organs.

6. **Digestive System** (Fig. 13–2). The main organs of this system form the alimentary canal beginning at the mouth and ending at the anus. The mouth,

5

pharynx, esophagus, stomach, and small and large intestines are the principal organs. The salivary glands, liver, and pancreas belong to this system and empty their secretions into it, and there are many microscopic glands in the walls of the various organs. The system reduces the food to particles of small size so that it can be absorbed into the blood or lymph. Indigestible materials are eliminated at the anus.

7. **Respiratory System** (Fig. 14–1). The essential organs of this system are the lungs. They are served by the nose, pharynx, larynx, trachea, and bronchi which constitute the rest of the system. By insuring a constant flow of air through the lungs, oxygen is made available and carbon dioxide can be expelled from the body. Voice production, loss of heat and moisture, and olfaction are other functions of the system.

8. **Excretory System** (Fig. 15–1). The kidneys are the essential organs of excretion. They also serve in the regulation of the internal environment in terms of acid-base balance and the concentration of body fluids (*osmoregulation*). Accessory organs of this system are the ureters, urinary bladder, and urethra. These conduct, store, and eliminate the urine.

9. **Reproductive System** (Figs. 16–3, 4). The preceding systems serve for the support, protection, movement, and maintenance of the individual. The reproductive system insures the perpetuation of the species. The essential organs are the gonads: testes in the male; ovaries in the female. They produce the germ cells (*sperm and eggs*) and certain hormones. The accessory organs of the male include the ductus deferens, seminal vesicles, ejaculatory ducts, prostate gland, urethra, bulbourethral glands, and penis. They serve to transmit the sperm and to introduce them into the reproductive tract of the female. In the female the uterine tubes, uterus, vagina, and vulva are the principal accessory organs. They transmit the egg, house the embryo and fetus, and at term expel the fetus.

10. **Nervous System** (Fig. 18–1). The brain, spinal cord, nerves, and ganglia form the nervous system. The first two organs constitute the central nervous system, the latter two the peripheral system. The autonomics are a part of the peripheral system. This system orients the individual, controls and coordinates his organs and systems, and accounts for his intelligence. The **effectors** which respond to the impulses carried to them by the nerves are the skeletal, smooth, and cardiac muscles and the various glands of the body.

11. **Sense Organs.** The sense organs may be considered a part of the nervous system. They are its **receptors.** They pick up the stimuli from the environment which result in impulses being set up in the sensory nerves. The eye, ear, olfactory epithelium, and taste buds are specialized receptors. Others occur in the skin such as those for temperature, pain, and touch. Nerve endings in muscles and tendons are subject to stretch and the **proprioceptors** around the joints enable us to know the position of the body and its parts.

The Whole Body

While it is necessary in descriptive anatomy to discuss separate organs, separate systems, individual cells and tissues, we never should lose sight of the unity of the body and of the interdependence of its parts. Indeed, the most remarkable fact about the human body is that many billions of cells, put together in so many complicated ways, can result in such a well-integrated individual. The nervous and endocrine systems are primarily responsible for this integration, but they, like other systems, are dependent. They, too, depend upon the circulatory system, for example, to bring food and oxygen to their cells from the free surfaces of the body and to remove their waste products. All cells, in fact, live in an environment of fluid, the **tissue fluid,** which is constantly renewed from the circulating **blood** and returned to it by way of the blood capillaries or the **lymph** in the lymphatic vessels. These fluids are the life-lines of the organism—of each and every one of its cells. Though we live in an environment of air as organisms, our cells are still aquatic, and when they do come into contact with the drying air they die to form protective coats for underlying cells as on the outer layers of the skin, or they are provided with glands whose secretions keep them moist, as in the nose and mouth.

If you can discipline yourselves to think in terms of interrelationships as you study anatomy, your knowledge will result in understanding. Understanding will increase your interest and make learning much easier.

Let us turn now to a closer look at the tissues. As the tissues develop from the relatively undifferentiated cells of the germ layers the cells multiply rapidly and undergo change in form and in their relationships to one another. Differentiation is more apparent in the cytoplasm than in the nuclei of the cells. Beyond the realm of the visible are the chemical changes which take place within the cells. These changes become apparent to some extent in the secretion products which are formed and which constitute an important part of some tissues. All of these changes, obvious or subtle, equip the tissues to carry out their special functions. The broad capacities of the germ layers are lost in the demands of specialization.

EPITHELIAL TISSUES (Figs. 5–1, 2)

Structure and Functions

These are the tissues of the free internal and external surfaces of the body and of the structures derived embryologically from such surfaces. They have their cells closely joined, hence there is a minimum of intercellular material. The cells may be held together by a cement substance, by the adhesion of cell membranes, or by interdigitating structures. In some cases no membranes separating the cells can be demonstrated and the condition is called a **syncytium.** With few exceptions the epithelium lies upon a basement membrane which may be a product of the epithelium though it has, for a long time, been considered as representing a condensation of the upper surface of the underlying connective tissue.

Epithelial cells show a polarity due to the different environmental relationships of the two sides or ends of the cells. The end of the cell, in contact with a free surface, meets more varied conditions than the end associated only with neighboring cells. While the polarity is apparent in the internal arrangements of the cell, we will concern ourselves only with the free or distal surface. On some surfaces of the body, one of the most apparent features is the presence of **cilia,** microscopic hair-like projections, which by their movements propel materials over epithelial surfaces. These are common in the respiratory passageways and in the efferent ducts of the testis. In other areas, the electron microscope has revealed even smaller surface-projecting structures—the microvilli. On the surface cells of the intestinal epithelium these microvilli are called **striated borders** and the proximal convoluted tubules of the kidney have similar structures designated **brush borders.** Striated and brush borders probably aid in absorption.

Epithelial tissues are not penetrated ordinarily by **blood vessels** but are serviced by the vessels of the underlying connective tissues. Nerve endings do penetrate between the cells of epithelium and may actually enter some cells.

The epithelia serve numerous functions. They form **protective** layers over the body. They form expansive surfaces

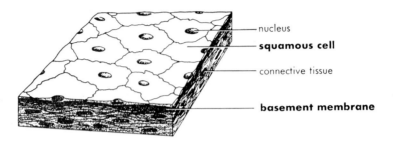

A. Simple Squamous

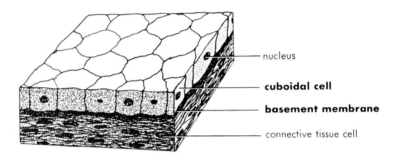

B. Simple Cuboidal

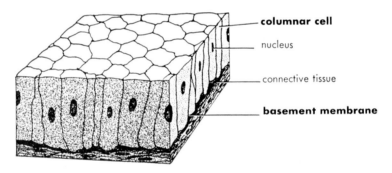

C. Simple Columnar

Fig. 5–1.   Types of simple epithelia.

for the **absorption** of materials into the internal environment, and conversely for the **excretion** of waste products. They produce **secretions** and form important parts of **sense organs.**

CLASSIFICATION (Figs. 5–1, 2)

Most epithelial tissues are named on the basis of (1) the number of layers of cells in the tissue and (2) the shape of the surface cells. If an epithelium consists of only one layer of cells it is a **simple epithelium,** if it has two or more layers it is a **stratified epithelium.** Epithelial cells are of three shapes (Fig. 5–1). 1. **Squamous** cells are thin and plate-like; so flat, in fact, that the nucleus causes a slight bulge in the cell surface. 2. **Cuboidal** cells are about equal in height and width. In a section perpendicular to the surface they appear square. 3. **Columnar** cells are taller than they are wide and in perpendicular section are rectangular.

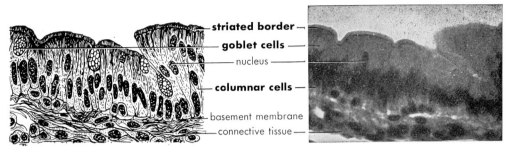

striated border
goblet cells
nucleus
columnar cells
basement membrane
connective tissue

A, Simple columnar epithelium with a striated border.

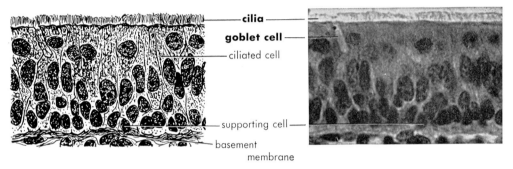

cilia
goblet cell
ciliated cell
supporting cell
basement
membrane

B, Pseudostratified ciliated columnar epithelium.

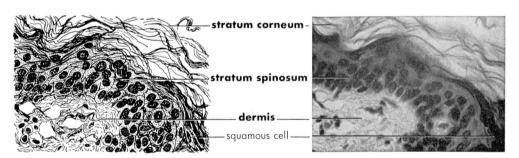

stratum corneum
stratum spinosum
dermis
squamous cell

C, Stratified squamous epithelium.

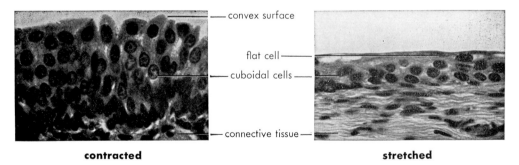

convex surface
flat cell
cuboidal cells
connective tissue

contracted   **stretched**

D, Transitional epithelium.

FIG. 5–2.   Sketches and photomicrographs of different types of epithelia.

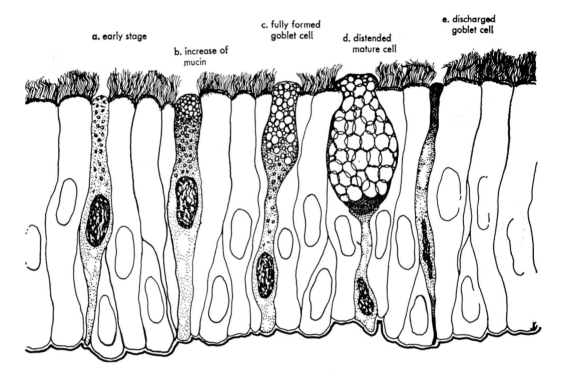

Fig. 5–3.   Sequence of events in goblet cell secretion in pseudostratified ciliated columnar epithelium

Transitional forms do occur among these basic forms and, of course, their free surfaces may be provided with cilia or other structures.   Some of these features, such as cilia, are brought into the naming of the tissues.

If we now apply our first and second basic rules together we can identify the following tissues.   Study the diagrams carefully (Figs. 5–1, 2).

1. **Simple Squamous Epithelium** (Figs. 5–1*A*; 15–5*e*, 6).   This tissue is found in Bowman's capsule and Henle's loop of the renal tubule, in the smallest excretory ducts of many glands, in the rete testis, the inner wall of the membranous labyrinth of the ear, and on the inner surface of the tympanic membrane.

Three tissues having the same appearance as simple squamous epithelium but differing from it in origin and developmental potencies might be listed here. They differ from the usual epithelia when under certain circumstances they transform to fibroblasts, a type of connective tissue cell.   They are as follows:

Endothelium is the simple layer of squamous cells which forms the inner lining of the heart, blood vessels, and lymph vessels.   It arises from mesenchymal cells.

Mesothelium is the simple squamous epithelium which lines the coelomic cavities forming the free surfaces of all the serous membranes.   It is derived from the mesoderm.

Mesenchymal epithelium is simple squamous epithelium lining the subdural and subarachnoid spaces, the chambers of the eye, and the perilymphatic spaces of the inner ear. It is formed by the flattening of fibroblast cells.

2. **Simple Cuboidal Epithelium** (Fig. 5–1*B*).   This tissue is widely distributed in the body.   It covers the free surface of

the ovary. It is found in many glands such as the thyroid, on the choroid plexus, on the inner surface of the capsule of the lens, in a part of the renal tubules, and as the pigmented epithelium of the retina.

3. **Simple Columnar Epithelium** (Figs. 5–1C, 2A). This epithelium is most readily observed in the lining of the alimentary canal from the cardia of the stomach to the anus. It is also in the excretory ducts of many glands.

4. **Simple Columnar Ciliated Epithelium.** This is like number three except that cilia are present on the free surfaces of the cells. This tissue is found in the central canal of the spinal cord, in the uterine tube and uterus, the small bronchi, and the nasal sinuses.

5. **Pseudostratified Columnar Epithelium.** Pseudo, meaning false, is applied to this tissue because, while it appears to consist of many layers of cells when seen in perpendicular section, it actually has only one layer. The appearance of stratification is due to some of the cells being short while others that are taller overlap them. Also, the nuclei are at different levels in the cells and the cells are quite irregular in shape. All of the cells do touch the underlying connective tissue. This tissue is found in the male urethra and in the ducts of several glands, like the parotid.

6. **Pseudostratified Columnar Ciliated Epithelium** (Fig. 5–2B). This is like number five above except for the presence of cilia. It is found in the auditory (*Eustachian*) tube, tympanic cavity, lacrimal sac, epididymis, ductus deferens, and respiratory passageways.

7. **Stratified Squamous Epithelium** (Fig. 5–2C). The cells of the many layers of this tissue vary in shape. The ones next to the underlying tissue are cuboidal or sometimes columnar. Outside of these the cells are polyhedral and of varied shape, while near the surface they become very flat or squamous. As the cells of the deep layer undergo

mitotic division the outer cells flatten, die, and finally slough off from the surface. In this way the tissue renews and maintains itself. This kind of epithelium makes up the epidermis of the skin. It lines the mouth, esophagus, vagina, and a part of the female urethra. It covers the outer surface of the cornea of the eye.

8. **Stratified Columnar Epithelium.** This is similar to number seven in its lower layers but its surface cells are, of course, columnar. It is relatively scarce in the body. Small surfaces of it appear on the pharynx, on the epiglottis, and in the anal region. It is also in the cavernous urethra and in a few of the large ducts of some glands.

9. **Stratified Columnar Ciliated Epithelium.** Similar to eight, but with cilia, it, too, is of very limited distribution. It is found in the larynx and on the nasal surface of the soft palate.

10. **Transitional Epithelium** (Fig. 5–2D). With transitional epithelium we depart from our system of naming. This tissue was thought at one time to represent a transition between stratified squamous and columnar epithelia, hence the name transitional. It is a stratified epithelium and is found on the inner walls of hollow organs like the urinary bladder and ureter which are subject to great mechanical change. As the bladder contracts, for example, the epithelium piles up into many layers, but when the bladder is stretched, only about two or three layers of cells can be seen. In the stretched condition the surface cells are squamous, but when contracted they are thicker and with convex surfaces (Fig. 5–2D)

GLANDS (Figs. 5–3, 5)

The metabolic functions of epithelium involve, among other things, **secretion.** Glands develop from the epithelium as specialists in this function. They produce products mostly in aqueous solution which differ from the blood plasma, the lymph or tissue fluid, either in concentration of materials or in the kinds of materials present. In all cases, the gland cells perform work in secreting; work beyond that required for carrying on their own metabolism. Sweat, milk, hydrochloric acid, cerebrospinal fluid, hormones, enzymes, and mucins are some of the secretions of the glands.

As epithelial cells differentiate into gland cells they may remain within the epithelium or push into the underlying connective tissue. In the latter case, the connective tissue becomes the supporting framework of the gland. If the gland retains its connection with the epithelial surface, this connection serves as its duct or ducts and it is classified as an **exocrine gland,** a gland of **external** secretion. If the gland tissue loses its connection with the epithelium at the surface it empties its secretions into the blood and is an **endocrine gland,** a gland of **internal** secretion, or ductless gland.

Beyond being classified as exocrine or endocrine, glands may be grouped on the basis of the manner in which they produce their secretions. Three types are generally recognized.

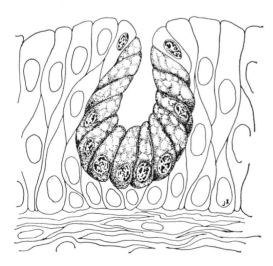

Fig. 5-4.   Intra-epithelial gland. (Modified from Bloom and Fawcett.)

1. **Merocrine glands** are those in which there is no cell destruction in the production of the secretion. The secretion is formed, discharged, and then formed again in a cyclic manner. The **pancreas** and **salivary glands** are in this group.

2. **Holocrine glands** are those where the secretion products accumulate in the cell bodies, then the cells die and are discharged as the secretion of the gland. New cells are constantly formed to replace those lost. **Sebaceous glands** are of this type.

3. **Apocrine glands** are intermediate between 1 and 2. Their secretions gather at the outer ends of the gland cells. These ends of the cells then pinch off to form the secretion. The remaining nuclei and cytoplasm, after a short period of recovery, repeat the process. The **mammary glands** belong to this group.

Classification of exocrine glands on the basis of the arrangement of their epithelial components is our main concern at the present time. As we study the organs and systems of the body many types of glands will be discussed and it is helpful to have a framework of classification in which to place them. The exocrine glands fall into two main groups: (1) **unicellular** and (2) **multicellular.**

Unicellular (*one-celled*) **glands** are represented by **mucous** or **goblet** cells which occur in the epithelial lining of digestive, respiratory, and urogenital systems. In lower animals such as fish and amphibians they are common in the skin. They produce a protein material **mucin** which with water produces mucus to lubricate the free surfaces of membranes.

The mucous cells are shaped like a goblet and hence the name, goblet cells. The outer end of the cell is swollen, the inner or basal end is slender and contains the nucleus. A goblet cell may empty its contents gradually and retain its shape, or empty suddenly and become collapsed. It again refills and repeats the

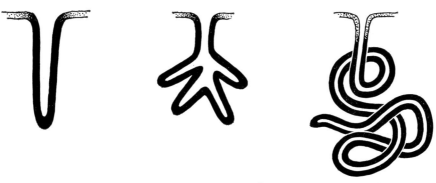

A. Simple tubular   B. Simple branched tubular   C. Simple coiled tubular

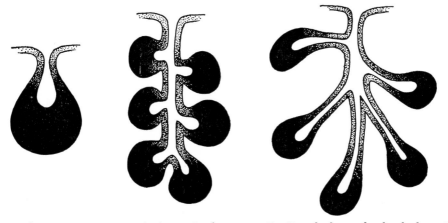

D. Simple acinar   E. Simple branched acinar   F. Simple branched tubulo-acinar

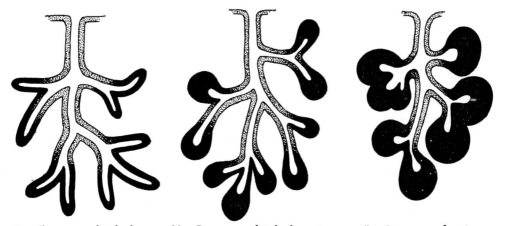

G. Compound tubular   H. Compound tubulo-acinar   I. Compound acinar

Fig. 5–5.   Schematic diagrams of exocrine glands.

cycle. Periodically, these cells die and are shed (Fig. 5–3).

Multicellular glands occur in a variety of forms. The simplest are in the form of flat plates of secreting cells or as groups of secreting cells which form a small pit within the epithelium and secrete through a common opening (Fig. 5–4). The latter are called **intra-epithelial glands** and are found on the laryngeal surface of the epiglottis.

All of the other multicellular glands during their development push into the underlying connective tissue from which they gain support. Also the blood, lymph, and nerve supply is supported by the connective tissue and is brought into close relationship with the secreting epithelium.

The following classification of **multicellular glands** with examples should be studied in conjunction with the illustrations (Fig. 5–5).

ponent. They consist only of connective tissue and will be studied later.

Two kinds of epithelial membranes are important in the human body. They are the **mucous** and **serous** membranes.

Mucous Membranes (*mucosa*). Mucous membranes line all of the hollow organs and cavities which open upon the skin surface of the body. Hence, they line most of the organs of the digestive, respiratory, urinary, and reproductive systems. Their surface epithelium may vary in type such as stratified squamous in the esophagus and simple columnar in the intestine, but it is always kept moist by mucus. The connective tissue beneath the epithelium is the **lamina propria.** Variations in the mucous membranes will become more apparent as we study the separate organs.

Serous Membranes (*serosa*). Serous membranes line the coelomic cavities, contribute to mesenteries and omenta,

Simple
- Tubular
  - Crypts of Lieberkühn of the intestine
- Coiled tubular
  - Sweat glands of the skin
- Branched tubular
  - Glands of stomach, uterus, and oral cavity
- Acinar (saccular)
  - Seminal vesicles
- Branched acinar (saccular, alveolar)
  - Sebaceous

Compound
- Tubular
  - The pure mucous glands of the oral cavity, some glands of Brunner, and bulbo-urethral glands
- Acinar (alveolar)
  - Mammary glands
- Tubuloacinar
  - Pancreas, parotid glands

EPITHELIAL MEMBRANES

These membranes consist of a surface layer of epithelium and an underlying layer of connective tissue. They should not be confused with **fibrous membranes** of the joints, bursae, and tendon sheaths which have no epithelial com-

and cover the surfaces of related organs. Their surface epithelium is of **mesothelium** placed over a thin layer of loose connective tissue. They liberate fluid which serves to moisten and lubricate. Some cells are also released into the body cavities. Pleura, pericardium, and peritoneum are examples.

CONNECTIVE TISSUES

GENERAL

The connective tissues represent wide-ranging types, both in their variety and distribution. Many of them form a continuum through the body and, as would be expected, the various types are not always distinct and many transition forms occur. They are all characterized, however, by large amounts of inter-cellular materials (*matrix*) through which the cells are found singly or in groups. They are mostly of mesodermal origin through mesenchyme.

The functions of connective tissues are as varied as the tissues themselves. Binding, supporting, and protecting are the most obvious and purely mechanical ones. But, they are also involved in the vital problems of circulation of body fluids, the storage of excess food materials, and in inflammation some of them play an essential role. Other functions will become apparent as connective tissues are studied in relationship to the organ systems.

CLASSIFICATION

Whereas in the epithelial tissues cell shapes and arrangements were the chief means of classification, in connective tissues the **intercellular substances** are of primary importance. Cells, of course, are also considered, though the various types are sometimes difficult to distinguish except by special techniques of staining or even tissue culture.

The connective tissues may be grouped as follows:

    Connective tissue proper (Fig. 5–6)
      Loose connective tissue (*areolar*)
      Dense connective tissue
      Regular connective tissue
        Tendon (*white fibrous connective tissue*)
        Fibrous membranes
        Lamellated connective tissue

    Special connective tissue
      Mucous
      Elastic
      Reticular
      Adipose
      Pigment
    Cartilage (Fig. 5–7)
    Bone (Fig. 5–8)
    Blood and lymph (Fig. 5–9)

Connective Tissues Proper. The **loose connective tissue** is a good one to start with and to describe in some detail as it contains many of the cell types and intercellular materials that make up the other connective tissues. It is also widely distributed in the body. It makes up the **subcutaneous tissue** or **superficial fascia** and penetrates between the organs to fill space and bind structures together. It is found in the walls of blood vessels that penetrate other tissues of the body. It is ideal for supporting muscles because its loose nature allows for their movements in relationship to one another and neighboring organs. Also, it contains many potential spaces that can fill with fluid and is involved in the metabolism of the cells. Because of these spaces and the sponge-like nature of loose connective tissue it was called by the early anatomists "areolar" tissue. Other functions associated with this tissue are its capacity for repair after injury and for combating infection. It is because of this tissue that it is relatively easy to separate organs in dissection—as the removal of the skin from the underlying muscles or the separation of the muscles themselves.

The more important intercellular substances of loose connective tissue are the (1) collagenous or white fibers, (2) the elastic or yellow fibers, (3) the reticular fibers, and (4) the amorphous ground substance (Fig. 5–6C).

Collagenous (*white*) **fibers** are the most characteristic features of most types of connective tissue. In loose tissue they

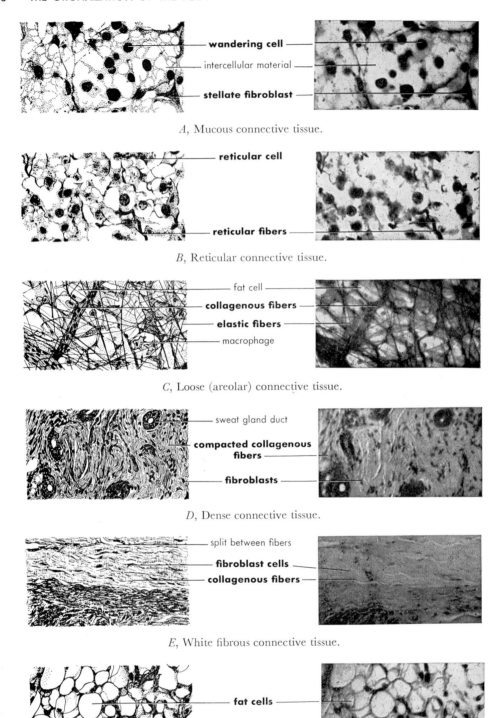

A, Mucous connective tissue.

B, Reticular connective tissue.

C, Loose (areolar) connective tissue.

D, Dense connective tissue.

E, White fibrous connective tissue.

F, Adipose tissue.

FIG. 5–6.   Sketches and photomicrographs of different types of connective tissues.

appear as wavy ribbons with longitudinal striations. The striations are due to the parallel arrangement within the fibers of many fine fibrils. While the fibrils do not branch individually, groups of them may separate to give a branched appearance to the fiber as a whole. White fibers are strong, inelastic, and yield a collagen when boiled in water. They are digested by pepsin in an acid environment.

Elastic (*yellow*) **fibers** are fairly long, cylindrical threads or flat ribbons. They have a brilliant appearance, branch freely, often forming networks, and show no fibrillar structure. In large masses they have a yellowish tinge. They contain elastin and hence stretch readily and return to their original form when the pressure is released.

Reticular fibers are very fine, highly branched, and form networks concentrated in areas where the connective tissue is adjacent to other structures, as along blood vessels, in basement membranes, lymph nodes, and around muscle, nerve fibers, and fat cells. They can be stained black with silver by special methods. They seem to be continuous with collagenous fibers and, like them, are inelastic. Some scientists consider them immature collagenous fibers. Unless they are specially stained they usually cannot be seen.

The **amorphous ground substance** in which the fibers are imbedded is difficult to see. If the tissue is deeply stained, it may show as a pale film among the fibers and cells. It, like the fibers, is believed to be derived from the connective tissue cells. Its viscosity varies from fluid to gel and the tissue fluids which originate from the blood plasma are closely associated with it.

The **cells** of loose connective tissue are of many and varied types (Fig. 5–6C). **Mesenchymal cells** can usually be seen along the blood vessels. **Fibroblasts,** thought by some to give rise to fibers, are common and usually close to the collagenous fibers. They appear as flat, long, or star-shaped cells. They may send out long, spear-shaped processes. In profile they appear as flat spindles. The nuclei are large ovals. **Histiocytes** (*macrophages*) may be as numerous as fibroblasts and have somewhat the same shape, or may be oval, rounded or elongate. Their nuclei are smaller and more irregular in shape. They have more coarse, dark-staining chromatic particles than do fibroblasts. They ingest and store certain kinds of microscopic particles. Normally, they do not move about. In cases of inflammation they move by ameboid action and engulf bacteria and other particles. In this situation, they are called **free macrophages. Lymphoid wandering** cells are apparently identical with the non-granular leukocytes of the blood, are irregular in form and are highly ameboid. **Mast cells** are small, oval cells. It is believed they secrete a substance similar to heparin which inhibits the coagulation of blood. Fat cells may occur singly or in large groups. They accumulate fat and as the cells gradually fill, the cytoplasm and nuclei are pressed close to the cell wall. This gives them their characteristic appearance. When fat cells become so abundant as to crowd out other tissue components, it is called **adipose tissue.** Other cells appear in the loose connective tissues, but the ones listed should serve our purpose.

Dense connective tissue contains essentially the same elements as loose connective tissue but the cells are fewer and the collagenous fibers more numerous. The collagenous fibers are so interwoven and compacted as to form a dense matting. The dermis of the skin and parts of the walls of urinary ducts, digestive organs, and blood vessels are composed of dense connective tissue. The dermis of the skin of some animals is our source of leather (Fig. 5–6D).

Regular Connective Tissues. Regular

connective tissues are those in which the fibers, especially the collagenous fibers, are abundant and arranged in a definite fashion. **Tendons,** for example, have the collagenous bundles running parallel to one another and in between the bundles are rows of **fibroblasts**—the only connective tissue cells present (Fig. 5–6E). **Ligaments** are similar to tendons, but the collagenous fibers are not so regularly arranged and there may be some elastic fibers present. In the **ligamentum nuchae** of the necks of grazing mammals, the elastic fibers predominate.

The **fibrous membranes** include such structures as the aponeuroses, fasciae, perichondrium, periosteum, sclera and cornea of eye, dura mater, capsules of certain organs, and the tunica albuginea of the testis.

In the **aponeuroses** and **fasciae,** the white fibers are arranged in layers or sheets. In each layer the fibers tend to follow parallel courses. The direction of the fibers may be the same or different from one layer to the next. Some fibers cross between layers. A few elastic fibers are found around the bundles of collagenous fibers and fibroblast cells are found between the bundles. Aponeuroses are essentially broad, flat tendons. The fasciae, as described here, are **deep fasciae** as distinguished from the superficial fasciae which is composed of loose connective tissue. The ramifications of the deep fasciae are extensive and complex. Reference will be made to them at pertinent points in the book.

In the remaining fibrous membranes listed above, the fibers and cells are somewhat less regularly arranged. There are areas of transition where fiber arrangement becomes very irregular and dense tissues are formed. At the other extreme they blend into loose connective tissue. These membranes will be considered individually as we develop the story of the body systems.

Lamellated connective tissue forms the thin, resistant sheaths of small, usually cylindrical organs. It is similar to loose connective tissue in composition, but in a condensed or compacted form. It is found as the perineurium around the nerve bundles in a nerve trunk, in the walls of the seminiferous tubules of the testes, and in the sheaths of Pacinian corpuscles (Fig. 6–1).

Special Connective Tissues. The special connective tissues will be only briefly outlined at this time. Some of them will receive further consideration where they contribute significantly to the structure of the body systems.

Mucous connective tissue is essentially an embryonic tissue found in many parts of the developing body (Fig. 5–6A). It is quite like the loose connective tissue. It is commonly studied as it occurs in the human umbilical cord where it is known as Wharton's jelly. Characteristically, it has large fibroblast cells with long processes which contact those of neighboring cells.

Elastic connective tissue has a preponderance of yellow or elastic fibers with a few white fibers intervening. It is found as ligaments where some stretch is important as in the **ligamenta flava** of the vertebrae. In this form it may be classed as a regular connective tissue. Elastic fibers are also found in true vocal cords, the walls of the larger arteries, in the trachea and bronchi, and many other locations.

Reticular connective tissue is associated with the **hemopoietic** or blood-forming tissues and forms a part of the framework of organs such as the lymph nodes, spleen, and liver (Fig. 5–6B). Accompanying the reticular fibers are **primitive** and **phagocytic reticular** cells. The primitive reticular cells can give rise to many types of blood and connective tissue cells among which are the phagocytic reticular cells or fixed macrophages.

Adipose tissue is one made up predominantly of fat cells which have been described (Fig. 5–6F). Other cell types such as fibroblasts are present and both white and yellow fibers run in all directions among the fat cells. Argyrophile (*reticular*) fibers are present and form net-like baskets around each of the fat cells. Fat is a storage tissue and, as such, is a reflection of the metabolic character of the individual. It forms pads around various organs such as the kidneys and is abundant in the subcutaneous tissues and in the greater omentum.

Pigment tissue is made up of special pigment cells (*chromatophores*) in association with loose or dense connective tissue. It may be found under the epidermis of the skin and between the sclerotic and choroid coats of the eye (Fig. 19–13).

The **lamina propria** is a difficult connective tissue to classify. In some situations it resembles loose connective tissue. In some it resembles dense connective tissue, while in still others it is like reticular tissue. It will be considered in connection with those organs such as the intestine, uterus, lungs, and others where it occurs (Fig. 13–15).

Cartilage (Fig. 5–7). This is a variety of fibrous connective tissue in which the matrix is abundant and firm. The cells, called **chondrocytes,** lie within spaces in the matrix, the **lacunae.** Each lacuna may contain one or several cells. Cartilage is a **non-vascular** tissue though some vessels may pass through it in going to other structures. When boiled, it yields its characteristic substance, **chondrin.** Except for its naked surfaces in joint cavities, cartilage is covered externally by a fibrous membrane, the **perichondrium,** from which it receives its nutritive supplies.

Cartilage occurs widely in the body. In the early fetus, the greater part of the skeleton is cartilaginous but it is mostly replaced later by bone. Cartilage remains, however, on the articulating surfaces of bones and in some of the joints as well as in costal cartilages, certain respiratory organs, and the external ear and related canals. Since so much of the cartilage is replaced during development, it is called **temporary** in contrast to that which remains, the **permanent** cartilage.

There are three kinds of cartilage, based upon features of its minute structure. They are (1) **hyaline,** (2) **fibrocartilage,** and (3) **elastic cartilage.**

Hyaline cartilage (Fig. 5–7A) is the most abundant in the adult body and is also the type which precedes bone formation. It forms a mass of firm consistency sometimes called gristle. It has considerable elasticity and the homogeneous matrix often has a bluish color. The cells well fill the lacunae and are flattened where they contact each other, rounded where they contact the walls of the lacunae. The lacunae walls, the **capsules,** often stand out from the surrounding matrix, due to their different staining reaction or their concentric striations.

Hyaline cartilage makes up the costal cartilages, tracheal, and most laryngeal cartilages, the cartilages of the nose, and the articular cartilages on the ends of bones. It differs somewhat in appearance in these various locations but can always be recognized as hyaline.

Fibrocartilage (Fig. 5–7C) has a limited distribution in the body. It is found as the **intervertebral discs** of the vertebral column, at the **symphysis pubis** and a few other joints. It differs in appearance from hyaline cartilage due to the presence of thick, compact, parallel collagenous bundles between which the encapsulated cells are found. Little can be seen of the ground substance. This cartilage may be a transitional form between hyaline cartilage and dense connective tissue.

Elastic cartilage is found in the external ear, the auditory tube, epiglottis and to a lesser extent in the corniculate and cuneiform cartilages of the larynx

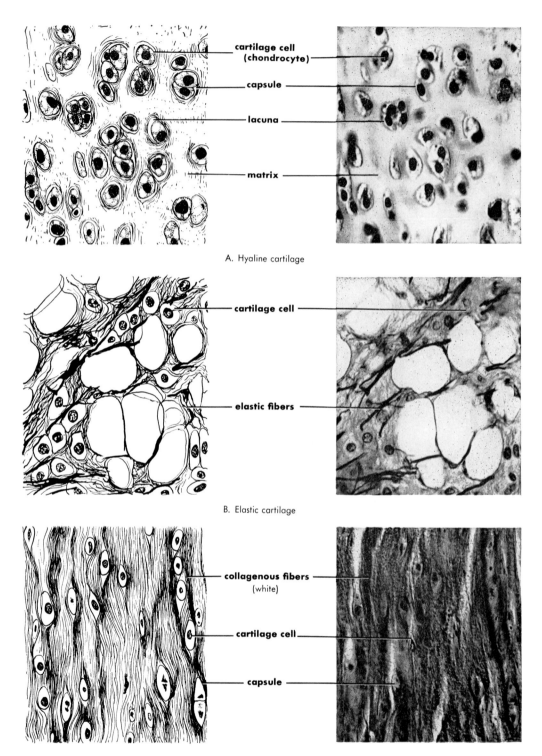

cartilage cell
(chondrocyte)

capsule

lacuna

matrix

A. Hyaline cartilage

cartilage cell

elastic fibers

B. Elastic cartilage

collagenous fibers
(white)

cartilage cell

capsule

C. Fibrocartilage

FIG. 5–7.   Sketches and photomicrographs of types of cartilage.

(Fig. 5–7*B*). It is more opaque, flexible and elastic than hyaline cartilage. It has a yellowish color. Microscopically, one can see the abundant, branching, elastic fibers penetrating in all directions and often concentrating in the walls of the lacunae around the cells. Some of the elastic fibers continue into the perichondrium.

Bone (Fig. 5–8). Bone tissue forms the greater part of the skeleton. It differs from cartilage in having (1) a blood supply, (2) **lacunae** connected by microscopic canals, the **canaliculi,** through which processes of the bone cells (*osteocytes*) connect, and (3) the **matrix** impregnated with inorganic salts, which make up about 30 per cent of the weight of bony tissue. Moreover, in much of bone tissue there is a characteristic arrangement of parts called **Haversian systems.** These consist of a central canal for the passage of blood vessels, the **Haversian canal,** around which are arranged concentric circles of bony matrix, the **concentric lamellae.** Between the lamellae are the lacunae containing the osteocytes which connect through the canaliculi. This relationship of parts allows for a circulation of tissue fluid from the blood vessels through the canaliculi to nourish the osteocytes. Between neighboring Haversian systems are other lamellae lacking Haversian canals and called **interstitial lamellae.** Beneath the outer covering of bone, the **periosteum,** and the bone's inner lining, the **endosteum,** are found **circumferential lamellae** which also lack Haversian systems. Finally, Haversian systems are not found in the **spongy** or cancellous parts of bones, but only in the compact areas. It is interesting to note that among mammals the Haversian systems, though of the same general pattern, show sufficient variation to be used in the identification of species.

Blood and Lymph (Fig. 5–9). These are peculiar among the connective tissues because the intercellular material is liquid and the cells are free in it. The blood flows through the blood vessels and heart in which it is normally confined. The lymph is found in the lymphatic vessels. Blood and lymph are often considered separately as the **vascular tissues.**

The fluid or intercellular part of the blood is called the **plasma** and makes up 50 to 60 per cent of the blood by volume. It is about 90 per cent water, 7 to 9 per cent proteins, 0.9 per cent inorganic salts, and contains small quantities of other substances as urea, uric acid, creatine, creatinine, ammonia salts, amino acids, sugar, and fat. It transports hormones.

The formed elements of the blood, the corpuscles, make up 40 to 50 per cent of the blood by volume. They consist of the **red corpuscles** or **erythrocytes,** the **white corpuscles,** or **leukocytes,** and the **platelets.**

The **erythrocytes** are tiny biconcave discs which lose their nuclei before they enter the blood stream. Their primary function is the transport of oxygen which they pick up in the lungs and release in the tissues of the body. The oxygen is carried in combination with hemoglobin, an important constituent of the erythrocytes. There are from 4,500,000 to 5,000,000 erythrocytes in each cubic millimeter of blood. Reduced numbers of erythrocytes, or erythrocytes with reduced amount of hemoglobin, result in **anemia.**

Leukocytes are far fewer in number than erythrocytes, only about 6,000 to 10,000 per cubic millimeter of blood. They occur, however, in greater variety and all retain their nuclei. There are two general groups of leukocytes, the **non-granular** and **granular.** The non-granular leukocytes are of two kinds, **lymphocytes** and **monocytes** which have a clear cytoplasm. The lymphocytes make up 20 to 25 per cent of the leuko-

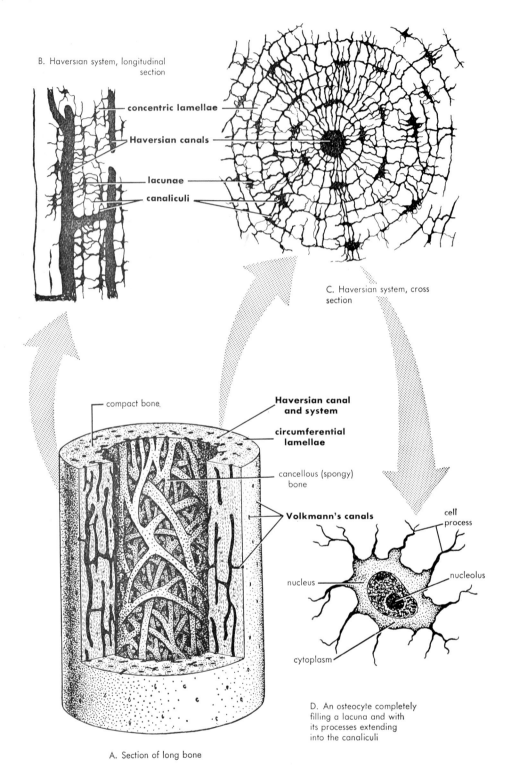

B. Haversian system, longitudinal section

concentric lamellae

Haversian canals

lacunae

canaliculi

C. Haversian system, cross section

compact bone

Haversian canal and system

circumferential lamellae

cancellous (spongy) bone

Volkmann's canals

cell process

nucleus

nucleolus

cytoplasm

D. An osteocyte completely filling a lacuna and with its processes extending into the canaliculi

A. Section of long bone

FIG. 5–8. Bone tissue.

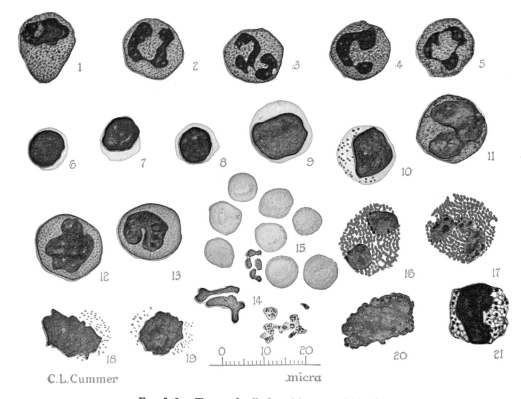

C.L.Cummer

micra

FIG. 5–9.   Types of cells found in normal blood.

All cells drawn with the same magnification and outlined with the camera lucida for the purpose of comparing sizes.   Each division of the scale at the bottom of the plate represents one micron.   Magnification, 1150.

Stained with Wright's stain.   Nos. 1 to 5, inclusive, polymorphonuclear neutrophilic leukocytes; 6, 7 and 8, lymphocytes; 9, 10, large mononuclear leukocytes; 11, 12, 13, transitional leukocytes; 14, a group of platelets; 15, a group of red blood cells; 16, 17, polymorphonuclear eosinophilic leukocytes; 18, 19, 20, "basket cells," degenerated leukocytes; 21, "mast-cell," basophilic leukocyte. (From Finerty and Cowdry, *A Textbook of Histology*, Lea & Febiger.)

cytes.   They are small cells but with a large nucleus which leaves only a narrow rim of cytoplasm.   They are believed to contribute to the repair of injured tissues. **Monocytes,** are large mononuclear leukocytes.   They are among the largest of the leukocytes, possibly developing from the lymphocytes.   They have phagocytic properties.   They make up 3 to 8 per cent of the leukocytes.

Granular leukocytes are all characterized by the lobulated nature of their nuclei and the heavily granulated cytoplasm.   They are sometimes called the **polymorphonuclear leukocytes.** Three types are recognized, based upon the characteristics of their cytoplasmic granules and nuclear form—the **eosinophils, basophils** and **neutrophils.   Eosinophils,** making up from 2 to 5 per cent of the leukocytes, have a nucleus of two lobes and cytoplasmic granules which stain bright red with Wright's stain. **Basophils,** the rarest of the leukocytes, 0.5 per cent, have a nucleus which is often bent into an S-shape and granules which are very large and less abundant and stain purplish blue with Wright's stain.   The **neutrophils** are the most abundant leukocytes, 65 to 75 per cent, and have a nucleus with 3 to 5 lobes and fine cytoplasmic granules which, with Wright's stain, have a light orchid hue.

The granular leukocytes help to protect the body against invasion by foreign bacteria or other particles. They move through the capillary walls between endothelial cells into the tissue fluid to carry out their **phagocytic** activity. A marked increase in the leukocyte count suggests a focus of infection somewhere in the body. In the malignancy known as leukemia, the leukocyte count is also greatly increased.

Blood platelets are small disc-shaped structures which lack nuclei. By some they are considered to be fragments of cytoplasm of large cells, the **megakaryocytes** which are present in the red bone marrow. There are about 250,000 in each cubic millimeter of blood. They play a vital role in the coagulation of blood.

Lymph has constituents similar to those in blood plasma but in different concentration. Plasma proteins, for example, are in much lower concentration.

Lymph contains numerous lymphocytes for they are formed in the lymph nodes which are situated in the pathways of the lymphatic vessels. An occasional granular leukocyte may be found, but no erythrocytes. Lymph is derived from the **tissue fluid** which, in turn, was derived from the blood through the capillary walls. Together they provide a circulation of materials necessary for the lives of the multitude of cells in a complex body.

Hemopoietic tissues give rise to the various types of blood corpuscles. The non-granular leukocytes are formed mainly in the **lymphoid tissues.** Granular leukocytes and erythrocytes are derived from the **red bone marrow** or **myeloid tissue.** During embryonic development erythrocytes are formed in the mesenchyme and liver.

MUSCLE AND NERVOUS TISSUES

Discussion of these tissues will be postponed until the systems in which they are involved are studied.

QUESTIONS

1. Define a tissue and name the four groups of primary tissues of the human body.
2. Compare epithelial and connective tissues in terms of their structure, location, and relationship to the blood and nerve supply.
3. Compare the functions of epithelial and connective tissues.
4. How are glands related to epithelium?
5. What is a secretion?
6. Are you able to name and give an example of each one of the epithelial tissues? Probably not, but you should be able to do so by the end of the course. They will be referred to frequently and you will see many of them in the laboratory.
7. What characteristics of epithelial tissues are used in giving them names?
8. What are endothelium, mesothelium, and mesenchymal epithelium? How do they differ from ordinary simple squamous epithelium?
9. What are mucous and serous membranes?
10. Distinguish between endocrine and exocrine glands.
11. Glands may be classified by the manner in which they produce their secretions. Name three such types of glands and explain their differences.
12. You should be able to give examples of some of the simple and compound glands of the body. You will learn them gradually as the body systems are studied.
13. Describe loose connective tissue. It gives you a good basis for understanding the more specialized varieties of connective tissue.
14. How do you classify tendons and ligaments among the connective tissues?

15. Describe mucous connective tissue.
16. What are some of the special characteristics of reticular cells?
17. What and where is the lamina propria?
18. Compare cartilage and bone in terms of relationships and nature of component parts.
19. What are the kinds of cartilage and where are they found?
20. Name the formed components of blood tissue and the functions of each.
21. How do blood and lymph differ and how are they related to the tissue fluid?

> *"Sunlight, not innate aging, is mainly responsible for the worst manifestations of senile skin."*
>
> —Albert H. Kligman, 1969.

Chapter **6**

The Integument (Skin)

GENERAL

Definition. The **integument** or **skin** forms a protective, pliable covering over the entire exterior of the body. At the various openings on the body surface, the mouth, nares, anus, urethra and vagina, the skin is continuous with mucous membrane linings. The integument is composed of a layer of closely packed cells, the **epidermis,** which rests upon an inner fibrous layer, the **dermis** or **corium.** The dermis is sometimes called the true skin. There are also accessory skin structures, the **hair, nails,** and **glands** (Fig. 6–1).

The skin of an adult has a surface area of about 1.8 square meters (*3000 square inches*) and a variable thickness. It is thickest on the palms, soles and back where it reaches about 6 mm. At the other extreme on the tympanic membrane and over the eyelids it is only 0.5 mm. The over-all average is 1 to 2 mm.

The attachment of the skin varies from loose to tight. It is separated from the underlying structures in most parts of the body by a subcutaneous tissue called the **superficial fascia** (*hypodermis*) (Fig. 6–1). This tissue allows the muscles which underlie most of the skin to move freely. At points where the superficial fascia is minimal and no muscles underlie the skin, as on the anterior surface of the tibia, the skin attaches to the periosteum of the bone. Also on the external ear the skin attaches quite directly to the underlying cartilage and over the joints of the digits to the capsules of the joints. If you examine your own fingers, you can see how tightly the skin attaches at the joints and the **flexion creases** which appear there, and how loose the skin is between the joints on the palmar surface.

Careful examination of the skin surface, on the back of the hand, for example, will reveal delicate lines which form a definite pattern. These are **flexion lines,** and around the joints where movement is greater they become deeper and are of a permanent nature—the **flexion creases.** On the finger and toe pads and on the palms and plantar surfaces generally are

68

seen alternating ridges and grooves which have a constant pattern—the "finger prints" used for identification of individuals. They are called **friction ridges** because they prevent slippage when we grasp objects. The sweat glands open along the summits of these ridges but they are devoid of hair and sebaceous glands.

Skin color is caused in part by the pigment **carotene** (*yellow*) which is in the surface layer and, particularly in thin skin, by the color of the circulating blood which shows through to give the "flesh" appearance. The intensity of the "flesh" color depends upon the state of contraction or dilation of the superficial vessels and upon the extent of oxygenation of the blood. The difference in color among individuals and ethnic groups is due to the concentration of the pigment **melanin** which will be discussed later.

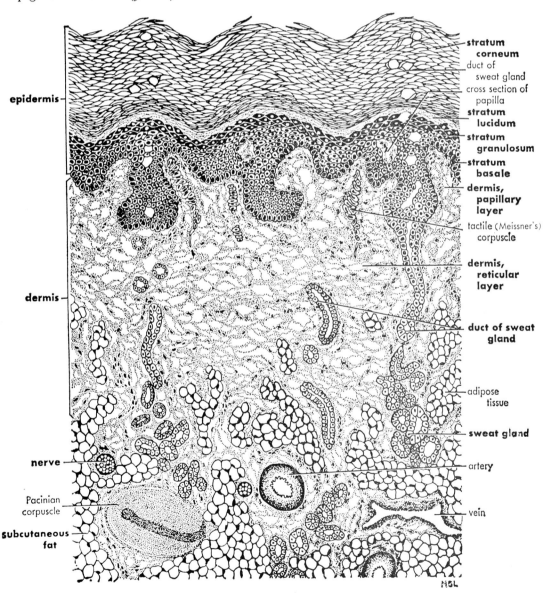

epidermis

stratum
corneum
duct of
sweat gland
cross section of
papilla
stratum
lucidum
stratum
granulosum
stratum
basale
dermis,
papillary
layer
tactile (Meissner's)
corpuscle

dermis

dermis,
reticular
layer

duct of sweat
gland

adipose
tissue

sweat gland

nerve

artery

Pacinian
corpuscle

vein

subcutaneous
fat

Fig. 6–1.   Diagram of a section of skin and subcutaneous tissue. Highly magnified.

Functions of the Integument. The functions of the skin are many and some of them have been suggested by the preceding discussion. Basically, it stands as a protective barrier against the ever-changing and often adverse conditions of the external environment and adapts itself to them. Its receptors provide an awareness of this external environment. It adapts to wear by the thickening of the stratum corneum, to form calluses, if necessary. It prevents the body from drying. Its oiled surface sheds water, and its pigment helps to protect the body from harmful ultraviolet radiations. It is an effective first line of defense against infectious organisms.

The integument is an important regulator of the body temperature. When the outside temperature is high or during body exercise much heat is lost through the skin. This is accomplished through the coordinated activity of nerve endings, blood vessels, and sweat glands. Integumentary vessels are dilated, sweat is produced, and its evaporation from the body surface has a cooling effect. Conversely, when the outside air is cool the blood vessels of the skin contract and heat is kept within the body.

Secretion, and to a limited extent excretion, are functions of the skin. During copious sweating as much as 1 gram of non-protein nitrogen per hour may be excreted. Sodium chloride (*salt*) is also an important constituent of sweat and its loss from the body has marked effect upon osmotic balance. The skin has a limited capacity for absorption, especially when materials like hormones, vitamins, and drugs are placed on the skin in a proper vehicle.

STRUCTURE OF THE INTEGUMENT

As you study the skin, recall what has been said in the previous chapters about germ layers and their differentiation into tissues which, in turn, give rise to organs and systems. These processes are well illustrated by the skin.

Epidermis. This is the outer and thinner layer of the skin. It ranges from 0.07 to 0.12 mm. over most of the body to about 0.8 mm. on the palmar surfaces of the hands and is thickest on the plantar surfaces of the foot where it may be 1.4 mm. External pressures and other factors may cause a further thickening of the outer horny layers. **Calluses** and **corns** are the result of localized pressures.

The epidermis is a **stratified squamous epithelium** which comes about through the differentiation of the outer germ layer, the **ectoderm.** Where the skin is **thick** four layers can be distinguished in the epidermis. These layers, from the outside inward, are the **stratum corneum, stratum lucidum, stratum granulosum,** and **stratum spinosum** (*mucosum, Malpighii, germinativum*). In thin skin, only the strata corneum and spinosum are obvious. There may be a thin stratum granulosum in some cases (Fig. 5–2C).

The **stratum spinosum** is the inner and actively growing layer of the epidermis. Its basal cells, the **stratum basale,** are cylindrical or columnar in shape and rest upon the papillary surface of the dermis. Here they are close to the blood supply from which they derive their nourishment and oxygen. The other cells of the stratum spinosum become polyhedral to flat as they approach the granular layer. Mitotic divisions take place continually in these cells, especially those of the basal layer. These cells are gradually crowded outward to be lost as dead scale-like plates from the corneal surface. The surfaces of the polyhedral cells are marked by cytoplasmic extensions which connect with similar extensions on adjacent cells, hence the name **"prickle cells"** and stratum spinosum.

The cells of the stratum spinosum contain varying amounts of the brown pigment, **melanin.** This pigment is pro-

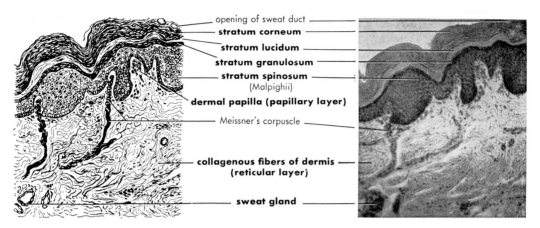

opening of sweat duct
stratum corneum
stratum lucidum
stratum granulosum
stratum spinosum
(Malpighii)
dermal papilla (papillary layer)
Meissner's corpuscle
collagenous fibers of dermis
(reticular layer)
sweat gland

FIG. 6–2.   Section of thick skin.

duced by stellate **melanocytes** which occur among the basal cells of the spinosum overlapping into the dermis. The melanin granules are then picked up by the cells of the spinosum, especially by those of the basal layer. As these cells are crowded toward the surface the melanin granules are fragmented to dust, and some are still apparent in the corneal cells. In Negroes and other dark-skinned people, the skin color is due to the greater amount of melanin in all layers of the epidermis. Only in albinos is the skin without melanin.

The **stratum granulosum** consists of three to five layers of flattened cells whose cytoplasm contains many keratohyalin granules. As the granules accumulate, especially around the nuclei, the nuclei gradually fade and disintegrate.

The **stratum lucidum** is a clear, homogeneous band outside of the stratum granulosum. It is composed of rows of flattened cells containing **eleidin**, a substance believed to be formed from the granules of the granulosum layer. Cell boundaries are seldom seen and only occasionally a small **pyknotic** nucleus.

The **stratum corneum**, the outermost layer, consists of flat, scale-like, dead, and cornified cells. They contain a fibrous protein called **keratin**. The surface cells are constantly being lost and are as con-

stantly replaced from the moving up of cells from the underlying layers. They form a protective covering and one which helps to prevent the excessive loss of moisture from the body surface.

Dermis or **corium** (Fig. 6–1). The **dermis** is a layer of **dense connective tissue** derived from the **mesoderm** germ layer through the medium of **mesenchyme**. It consists of an outer **papillary** layer which fits intimately into the under side of the epidermis, and an inner **reticular** layer. The line of demarcation between these two layers is not distinct and the reticular layer also blends gradually into the underlying **hypodermis** or **superficial fascia**. The dermis is thicker than the epidermis, ranging from 0.6 mm. on the eyelids to as much as 3 mm. on the soles and palms.

The **papillary** layer is softer and contains more elastic and reticular fibers and fewer white fibers than the reticular layer. The connective tissue cells are more common also in this layer. Fat cells are scarce throughout the corium though abundant in the hypodermis.

The **papillary layer** gets its name from the numerous projections, the papillae, which extend from its upper surface into the epidermis. There may be as many as 100 papillae per square mm. in some areas. These papillae form no special pattern over most of the body but in the thick skin of the soles and palms they are regularly arranged on a system of parallel ridges corresponding to and forming a base for the **friction ridges** of the epidermis. As a result of long-continued rubbing and friction the intimate connection of dermis and epidermis may be broken and a blister formed. Many papillae contain loops of capillaries which bring blood close to the epidermis, others have specialized nerve endings, such as the **tactile** (*Meissner's*) **corpuscles** (Fig. 6-1). Pigment produced in the melanocytes of the epidermis is thought by some to be picked up by **phagocytic macrophage** cells in the papillary corium and stored there. When these cells die, the pigment is released and again picked up by other macrophages. These pigment-storing cells may be called **chromatophores.** Tattooing is done by placing pigment through the epidermis into the dermis where it is taken up by the chromatophores. The effect is a fairly permanent one.

The **reticular layer** of the dermis has more bundles of collagenous fibers than the papillary layer, many running parallel to each other and more or less parallel to the skin surface. They form a dense felt-work which is strong and resistant. It is the reticular layer of the dermis of some animals which is used for leather and suede.

The **subcutaneous tissue** or **hypodermis** (Fig. 6-1) while not a part of the skin is so closely associated with it that it should be described at this point. It also goes by the name of **superficial fascia** and is an example of **loose connective tissue.** Where fat is abundant in it, it is called the **panniculus adiposus.** It is derived embryologically from the mesoderm germ layer through the medium of mesenchyme.

The subcutaneous layer varies in thickness from one part of the body to another. It is usually thicker than the dermis. It binds the skin loosely to underlying structures. In joint areas, as around the elbow and knee, discrete fluid-filled compartments are found in it, the **bursae.** In the regions of the face, neck, and scalp, subcutaneous **voluntary muscles** insert into the dermis. The facial muscles are the ones by which we so often reflect our emotions, our deeper feelings. Subcutaneous voluntary muscles are more widely distributed in other mammals and enable these animals to move the skin more extensively than is possible in man. **Smooth muscles** are also present in some areas of the subcutaneous tissue. The **dartos** muscles of the scrotum and the muscles of the areola and nipple are examples. Blood vessels and lymphatics travel through the subcutaneous tissue in going to and from the skin and many sweat glands and hair follicles extend down into it.

HAIR AND NAILS

Hair (Fig. 6-3). Man's hair is more widely distributed over the body than many realize. Much of it is fine and soft and goes unnoticed. Only the palms, soles, dorsal sides of the terminal portions of the digits, lips, nipples of the breasts, umbilicus and the skin portions of male and female genitalia are truly hairless. The hair on the head, and after puberty in the axillary and pubic regions, is conspicuous. In the male, a beard develops and other parts of the body may become conspicuously hairy.

Hair in man cannot be said to play the important roles that it does in most mammals in protection and temperature regulation. It is of some importance in tactile perception. Also in cases of extensive burning or laceration of the skin the epithelium of the hair follicles may play an important part in regeneration of epidermis. Economic man finds the cutting and grooming of the hair of his fellow citizens a good source of income,

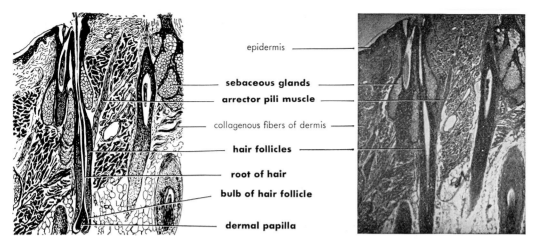

epidermis

sebaceous glands

arrector pili muscle

collagenous fibers of dermis

hair follicles

root of hair

bulb of hair follicle

dermal papilla

FIG. 6–3. Section of skin from scalp.

particularly that of the female. It is a secondary sexual characteristic of some importance in our "civilized" world.

Hair varies in thickness and length and also in the shape of the cross-sections, ranging from round to oval. Oval hair tends to be curly. Hair color is due to pigment in the hair cells and to some extent to air in the shafts. Melanin is the common pigment and it, of course, occurs in various concentrations to give the many shades from light to dark brown or black. A soluble red pigment is found in some individuals.

The first or primary hair to appear in the fetus is called the **lanugo.** It is succeeded by the secondary or **down** hair and this in turn is partially replaced by the **terminal** hair. The life of a hair is from two to five years in most parts of the body, three to five months in the case of the eyelashes. Failure of hair to replace itself results in baldness which is common in the male, rare in the female.

Hair is a product of the epidermis and therefore of the ectoderm. Buds of the epithelium grow down into the mesenchyme (*embryonic tissue*) to form the **hair follicles** from which the hair develops. The mesenchyme gives rise to dermis and subcutaneous tissue which cover the outside of the follicle. At the inner

expanded end of the follicle, the **bulb,** the dermis pushes up to form the **dermal papilla** through which the blood and nerve supply reach the hair. The epithelial cells of the bulb are the **germinative cells** which give rise to the hair. As the hair grows outward through the follicle, it finally comes to extend beyond the free surface of the skin. This free part is called the **shaft,** that within the follicle, the **root.**

A cross-section of a thick hair consists of an inner **medulla** of loosely arranged, relatively soft horny cells and many air spaces. Outside of this and constituting the main part of the hair shaft is the **cortex** of hard, closely placed, horny cells. The outermost layer, the **cuticle,** has the hardest horny cells tightly joined by serrate edges.

The hairs insert obliquely into the skin and hence form an obtuse angle on one side and an acute angle on the other with the skin surface. Each hair has one or more **sebaceous** (*oil*) **glands** whose ducts empty into the neck of the follicle. A smooth muscle, the **arrector pili,** attaches to the connective tissue of the hair follicle below the sebaceous glands and crosses the obtuse angle to the papillary layer of the skin. Its contraction straightens the hair, puts pressure upon the sebaceous

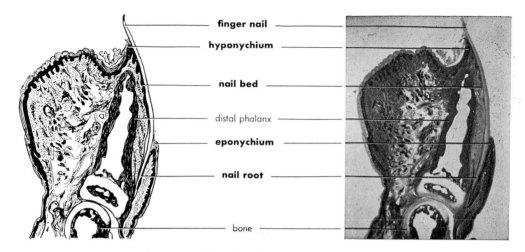

finger nail

hyponychium

nail bed

distal phalanx

eponychium

nail root

bone

FIG. 6–4.   Fingertip of monkey showing nail.

glands, squeezing out their secretions into the hair follicles, and depresses the skin near the hair to cause the appearance of "goose pimples" (Fig. 6–3). The elevation of the hair of furred animals creates more air spaces within the fur and hence gives better insulation against the cold.

Nails (Fig. 6–4). The **nails** are modifications of the **strata corneum** and **lucidum** on the dorsal surfaces of the terminal phalanges of the digits. They rest upon the **nail bed** consisting of the germinal layer of the epidermis and the underlying dermis. The nail is bordered proximally and laterally by a fold of skin, the **nail fold** or **wall.** Between the fold and the nail is the **nail groove.** The visible part of the nail is the **body of the nail;** the proximal edge, hidden under the nail wall, is the **root.** The distal margin of the nail is free and is the part that we cut. The pink color of the nail is due to the highly vascular layers that show through the semitransparent nail structure. This area is also well supplied with sensory nerve endings. At the proximal end of the nail there is a whitish area, the **lunula,** which may be hidden by the nail fold. At the nail fold the stratum germinativum and stratum corneum turn back into the nail groove,

and at the **root** of the nail the stratum corneum turns back over the free surface to form the **eponychium.** The stratum germinativum grows under the nail to become a part of the nail bed. At the distal end of the digit under the free edge of the nail the stratum corneum again appears as a thickened layer to form the **hyponychium.** This is where the dirt accumulates in "dirty nails."

Under the proximal portion of the nail and extending forward to the distal edge of the lunula the germinal epithelium is very thick and is called the **nail matrix.** It is here, in the deepest layer, that the cells divide and push into the upper layers which gradually flatten and cornify to form the nail. From this area of growth the nail gradually moves forward over the nail bed. Nails grow about 1 mm. per week.

GLANDS OF THE INTEGUMENT

The integumentary glands are derived by invaginations of cords of the epithelium (*ectodermal*) into the underlying mesenchyme. These cords hollow out to form **lumina** and the epithelial cells differentiate into duct cells, secretory cells, and in some cases into myoepithelial

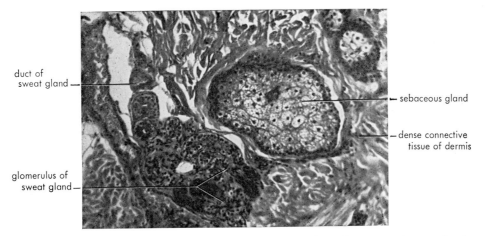

duct of
sweat gland —

sebaceous gland

dense connective
tissue of dermis

glomerulus of
sweat gland —

Fig. 6–5. Photomicrograph of section of dermis of skin showing sebaceous and sweat glands.

cells. The latter cells are contractile and help to discharge the secretions from the glands by their sphincter-like action.

The important glands of the skin are the sebaceous, sudoriferous (*sweat*) and mammary glands.

Sebaceous Glands (Figs. 6–3, 5). The sebaceous glands, with a few exceptions, develop from the epithelium in the necks of hair follicles. Sebum is their secretion and it is emptied into the hair follicles from which it flows out onto the skin surface and the hair. Sebaceous glands are **holocrine** and of the **simple** or **branched acinar** type. There are a few compound sebaceous glands in the skin of the nose and the external ear. The upper eyelids have special glands the **tarsal** or **meibomian** which consist of numerous alveoli branching from a single duct. A few sebaceous glands open on to the surface directly, as in the hairless skin of the lips, labia minora, glans penis, and clitoris. No sebaceous glands are found in the skin of the palms and soles or on the dorsal sides of the distal segments of the digits.

The ducts of the large sebaceous glands of the alae of the nose and the cheeks sometimes fail to discharge and the secretions collect, forming a semi-solid mass which we call a "black-head." These

sometimes become invaded by pus-producing organisms, and boils or pimples may result.

The secreting portions of the sebaceous glands are sac-like masses of epithelial cells. The outer, more or less columnar cells rest upon a basement membrane adjoining the corium. They produce by mitosis the inner, polyhedral, secreting cells of the gland. These cells become swollen with droplets of a fatty material, undergo physiologic degeneration, and as they are crowded toward the duct of the gland, disintegrate to form the secretion. Other cells move up from the reproducing layer to take their place.

Sudoriferous Glands (Figs. 6–1, 2, 5). These are the **sweat glands** that are in the form of simple coiled tubules extending down into the reticular layer of the dermis or in some cases into the subcutaneous tissue. They are found in most skin areas, being absent only on the margins of the lips, on the concave surface of the external ear, the skin of the nipple and portions of the skin of the genital organs. As many as 90 per square centimeter are found on the leg, 400 per square centimeter on the palms and soles, and even greater numbers on the fingertips where they open on the friction ridges.

The coiled part (*glomerulus*) of the gland contains the simple columnar secreting cells and is embedded in a fibrous capsule. Some spindle-shaped **myoepithelial** cells lie between the epithelium and a basement membrane that lines the capsule. About one-fourth of the coiled part of the tubule is duct and leads into the relatively straight or slightly coiled part of the duct system that empties on the skin surface. The innermost part of the duct is without myoepithelial cells but contains two or three layers of low columnar epithelium. As the duct reaches the epidermis its cells blend into the epidermal cells and its basement membrane with that of the corium. Only the lumen is left to twist its way to the surface of the skin. The diameter of the glomerular part of the gland is from 0.1 to 0.5 mm., that of the duct about half as much.

In the skin of the anal region, the scrotum, labia majora, and axilla special sudoriferous glands occur. They are larger, their glomerular parts measuring from 1 to 5 mm. in diameter and they extend into the subcutaneous tissue. Their ducts open into hair follicles rather than on the skin surface. Their secretions are different from sweat and contain part of the secreting cells. Hence they are apocrine glands. Also these glands become enlarged at puberty and in the female enlarge with each menstrual cycle and diminish again after the menstrual period.

Other apocrine glands believed to be related to the sweat glands are the compound tubular glands of the areola of the breasts, the relatively simple **ciliary glands** of the eyelids and the **ceruminous** (*wax*) **glands** of the external auditory meatus.

The secretion of the sweat glands is a thin, watery solution containing mainly sodium chloride, and some sulfates, phosphates, and urea. The amount of sweat secreted depends upon the temper-ature and humidity of the surrounding air, the amount of muscular work, and the emotional state of the individual. About 500 cubic centimeters of sweat are produced on the average each day. Under conditions of exercise or high temperature the amount of perspiration may be 2000 to 3000 cubic centimeters.

Mammary Glands. These are paired glands of the integument which are rudimentary in the male, larger in the female. They are closely associated functionally with the reproductive system in the female and hence will be considered with that system.

BLOOD AND LYMPHATIC VESSELS

Arteries come to the skin through the subcutaneous tissue. They form a **cutaneous plexus** at the interface of the corium and subcutaneous layers from which branches go to the sweat and sebaceous glands, the deep part of the hair follicles, and the fat tissue. Another plexus forms at the level of the papillary layer and is called the **papillary plexus.** It sends vessels to the capillary tufts of the dermal papillae, to the sebaceous glands and the upper parts of the hair follicles.

There are similar plexuses on the venous side and an additional one of the larger subcutaneous veins which receives branches from all parts of the skin.

The lymphatics begin as blind outgrowths or capillary networks in the dermal papillae. They join into a dense, flat mesh-work of lymphatic capillaries in the papillary layer. From this point the lymphatics form a deeper network at the boundaries of the corium and subcutaneous layers. The lymph leaves the skin area through lymphatics that travel with the arteries and veins.

NERVES AND NERVE ENDINGS

The integument is very well supplied with nerves and nerve endings. Some of the nerves are **afferent somatics** for general sensations of pain, pressure, touch, heat and cold. Others are **efferent** (*motor and secretory*) **autonomic** (*sympathetic*) fibers which go to the smooth muscle of blood vessels, to the myoepithelial cells of glands and the arrector pili muscles of the hair follicles.

Afferent or sensory nerve endings (*receptors*) of the skin are varied and unequally distributed. Some are merely naked nerve endings which extend between the cells of the basal layer of the epidermis, others terminate in plexuses around the hair follicles. **Encapsulated tactile corpuscles** (Meissner's) that mediate touch are found in the dermal papillae and **Pacinian corpuscles,** sensitive to pressure, are found in the dermis and subcutaneous tissue (Fig. 6–1).

QUESTIONS

1. What are the two major layers of the integument and how are they related to the subcutaneous tissue?
2. What kind of tissue makes up the epidermis? the dermis? the subcutaneous layer?
3. Give two other names that are commonly applied to the subcutaneous layer.
4. Name the accessory structures of the skin.
5. Compare the epidermis of thin and thick skin and state where each is found.
6. What are friction ridges, flexion lines and creases, "finger prints," calluses, and blisters?
7. What is the source and the distribution of pigment in the skin?
8. Discuss growth and regeneration of the epidermis.
9. Since there are no blood vessels in the epidermis, how does it receive food and oxygen and rid itself of waste materials?
10. Compare the structure of the papillary and reticular layers of the dermis.
11. Describe a fingernail and tell how it grows.
12. Compare sebaceous and sudoriferous glands as to structure, distribution, and functions.
13. Describe a hair and its follicle in terms of development, structure, and relationship to sebaceous glands and arrector pili muscles.
14. Describe the blood and lymphatic vessels of the skin.
15. Name some of the skin receptors.
16. Make a list of the functions of the skin and underline those which are most vital and important.

Chapter 7

General Osteology and Arthrology

GENERAL OSTEOLOGY

Definition and Functions. The skeletal system may be narrowly defined as the **bones** and related **cartilages** of the body, or more broadly to include the **joints** or **articulations**. Since the articulations hold the bones in meaningful relationships for action we will include them in our definition of the skeletal system. The study of the bones is called **osteology,** that of the joints is **arthrology.**

The human skeleton is an **endo-skeleton.** It lies within the soft tissues of the body and is a dynamic, living structure capable of growth, adaptation, and regeneration. In this way it differs from the non-living exoskeletons of arthropods, such as insects, which have to be shed and replaced at each growth period of the animal.

A chief function of the skeletal system is that of giving **support** for the soft tissues of the body. It gives direct attachment to most of the skeletal muscles and together they give the body its basic form. An artist seeking to produce a likeness of the living human figure should be well versed in the anatomy of these two systems.

The skeleton is also a basis for **movement.** Though its role in movement is passive, the bones do serve as levers and the joints as fulcra upon which the muscles act (see Chapter 11, p. 188). The bones at the articulations, by their shapes and relationships, are a factor in determining the kind and extent of movement.

Protection is given to many of the vital organs by the skeleton The brain is housed securely within the cranial cavity of the skull, the spinal cord within a canal formed by the vertebral column. The heart, lungs and major blood vessels lie within the rib cage, while the urinary bladder, uterus and related organs are protected by the bony pelvis.

Finally, **blood cell formation** (*hemopoiesis*) takes place in the red marrow in the proximal epiphyses of the humerus and femur, in the vertebrae, ribs, sternum and in the diploë of the skull bones

Classification of Bones. The 206

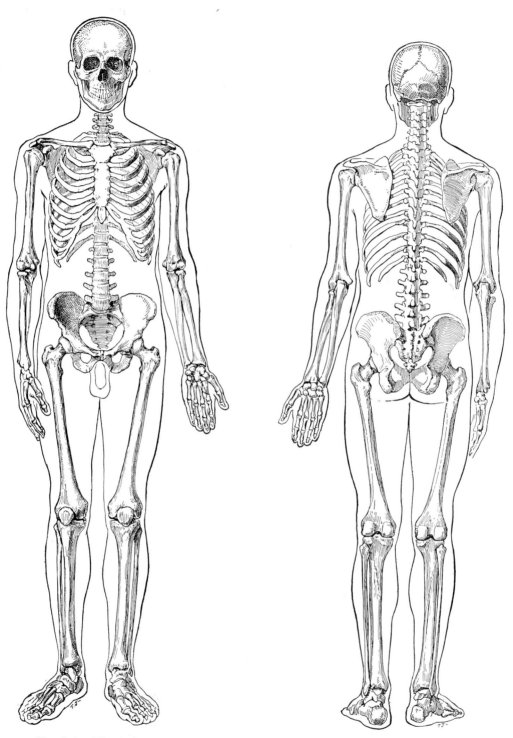

Fig. 7–1. The skeleton as projected on the surface of the body viewed from in front and from behind. (Eycleshymer and Jones.)

bones, which are the organs of the skeletal system, may be arranged in the following groups (Fig 7–1).

Axial skeleton (80 bones)
Skull	29 bones
Vertebral column	26
Thorax	
Sternum	1
Ribs	24
Appendicular skeleton (126 bones)	
Pectoral (*shoulder*) girdle . .	4
Bones of the upper limbs . .	60
Pelvic (*hip*) girdle	2
Bones of the lower limbs . .	60
	206 bones

If we count the two sesamoid bones which occur under the head of the first metatarsal of each foot, it brings the total to 210.

Bones may also be classified according to their shapes as **long, short, flat** and **irregular** (Fig 7–2) The meanings of these terms are quite evident from the following examples **Long** bones are those of the limbs such as the femur and humerus The **short** bones are those of the wrist (*carpus*) and ankle (*tarsus*) regions **Flat** bones make up the vault of the skull and also the ribs and sternum are so classified **Irregular** bones constitute the vertebral column and the pectoral and pelvic girdles and some are found in the floor of the skull and in the facial region.

Surface Features. Bone surfaces present a variety of features which suggest their relationship to other tissues and organs of the body These are projections ranging from ridges, to irregular roughened surfaces, to well-defined knobs or processes. These may serve to strengthen the bone, articulate it to neighboring bones, or for attachments of fibrous membranes, ligaments, or muscles. Depres-

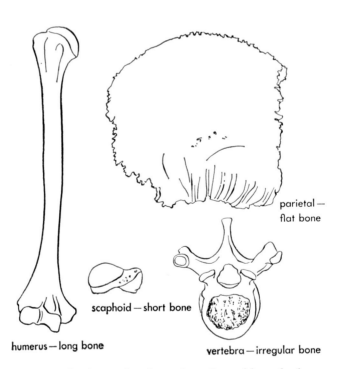

parietal — flat bone

scaphoid — short bone

vertebra — irregular bone

humerus — long bone

FIG. 7–2.   Sketches to show long, short, flat and irregular bones.

sions, some shallow and some deep, serve similar functions. Perforations are present also which allow the passage of blood vessels and nerves.

Special terms are assigned to these surface features and they are an important part of the vocabulary necessary for describing bones. An attempt is made in the following list to define each term concisely but some of the projections are difficult to differentiate clearly.

process—a broad designation for any prominence or prolongation

spine—an abrupt or pointed projection— the spine of the scapula

trochanter—a large, usually blunt process —the trochanters of the femur

tubercle—a smaller, rounded eminence— the tubercle of the humerus

tuberosity—a large, often rough, eminence —the tuberosity of the ischium

hamulus—a projection shaped like a hook —the hamulus on the sphenoid bone

line—a slight ridge—the linea aspera of the femur

crest—a prominent ridge—the crest of the ilium

head—a large, rounded articular end of a bone, often set off from the shaft by a neck—the head of the femur

condyle—a rounded articular prominence of a bone—occipital condyles

epicondyle—a projection above a condyle—the epicondyles of the femur

facet—a smooth, flat, or nearly flat, articular surface—facets of vertebrae for articulation of ribs

fovea—a shallow depression—the fovea capitis on the head of the femur

fossa—a deeper depression—the olecranon fossa of the humerus

sulcus—a groove—the sagittal sulcus in the roof of the skull

foramen—a hole—the foramina of the skull for passage of nerves and blood vessels

meatus—a canal—the external auditory meatus of the temporal bone

fissure—a narrow, cleft-like passage—the pterygoid fissure of the sphenoid

Structure of Bone (Fig. 7–3). A longitudinal section of a dried, long bone, such as the femur, demonstrates two types of bony substance, namely, dense, ivory-like **compact** tissue and light, spongy **cancellous** tissue. The relative amounts of these substances vary in different bones of the body and within parts of the same bone, depending upon their particular requirements for strength or lightness. The compact tissue forms a thin layer around the enlarged ends of **epiphyses** of the femur but becomes much thicker in the shaft or **diaphysis.** Cancellous tissue fills the epiphyses, whereas the diaphysis is hollow. Since a given amount of material arranged in a hollow cylinder gives greater strength than when arranged in a solid structure of the same length, this kind of arrangement meets well the requirements of the long bones of the limbs which must carry great weight or bear forceful pulls of muscles and often sustain considerable direct or indirect violence from the environment. The shafts of many long bones are strengthened by ridges for muscular attachment. Also the long bones are curved in their length and their ridges and grooves tend to spiral or curve, which gives elasticity to overcome shocks and blows and to make fracture less likely.

The enlarged ends of long bones are also outstanding examples of adaptation. They give a wide base for attachment of muscles and insure steadier joint action. Yet they are light as well as strong and elastic because they are composed largely of cancellous bone in which the lamellae and trabeculae are so arranged as to best withstand the stresses and strains which these bones sustain.

In **short** and **irregular** bones, the inner substance is cancellous, covered externally by thin compact tissue. Again, the arrangement is based upon good

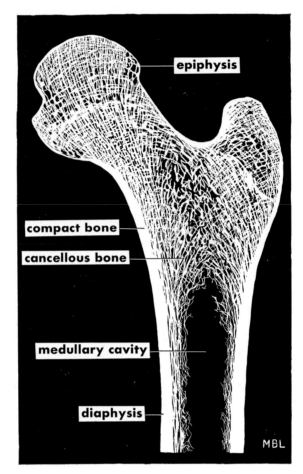

FIG. 7–3.   Longitudinal section of proximal end of femur.

engineering and the bones are both light and strong. They, too, are reinforced by local thickening where extra strength is needed (Fig. 7–4).

The **flat** bones have outer and inner coverings or **tables** of compact tissue between which is a cancellous layer, the **diploë** (Fig. 7–5). This kind of structure as seen in the bones of the cranial vault helps to protect the underlying brain from trauma by their strength and elasticity.

In the living or fresh state, bone tissue has a pinkish color externally and is red within because it is permeated with a rich blood supply. The cavities within the long bones, the **medullary** or **marrow** cavities, are filled with a **yellow marrow** consisting mostly of fat with a few primitive blood cells. A **red marrow** is found in the proximal ends of humerus and femur, in the short and flat bones, and in the bodies of vertebrae. Red marrow is associated with cancellous tissue and diploë and consists of primitive blood cells both immature and mature, and of macrophages, fat cells, and giant cells. Both kinds of marrow are supported by a stroma of reticular fibers accompanied by primitive and phagocytic reticular cells. Red bone marrow is the site of manufacture of red corpuscles and granular leukocytes.

Living bone, except where it is covered by articular cartilage, is encased in a fibrous membrane, the **periosteum**. It

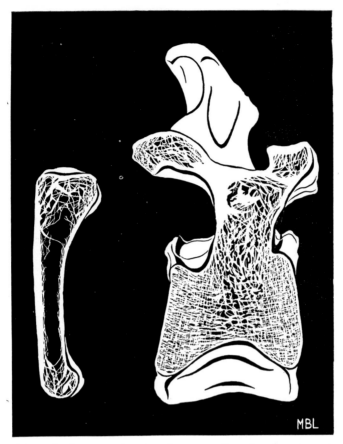

FIG. 7–4.   Long and irregular bones sectioned to show cancellous interiors covered externally by thin layer of compact bone. Bones reinforced by extra thickenings where extra strength is needed.

is composed of two layer, the outer one being dense and fibrous and containing blood vessels, the inner one looser with fewer white and more elastic fibers. Some of the white fibers are continuous with those in the matrix of the bone and give a strong attachment. They are known as **Sharpey's** or **perforating**

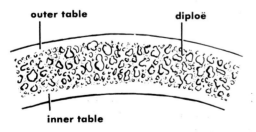

FIG. 7–5.   Section of a flat bone.

fibers. Where the tendons or ligaments attach to the bones their fibers are also continuous with those of the periosteum. This "continuity" of the connective tissues is everywhere apparent.

The periosteum of young bones is thicker and more vascular than that of the adult and the inner layer is osteogenic and contains many **osteoblasts** or bone-forming cells. In the adult, under normal conditions the **osteogenic** capacity of the periosteum is lost, but osteoblasts again appear in case of bone injury and become active in the regeneration process.

A membrane, the **endosteum,** lines the marrow cavities of bones and can be found even in the walls of the spaces in cancellous bone. It is thinner and less well defined than the periosteum and is considered to have both hematopoietic and osteogenic potencies.

Blood Supply. The blood suppy of bone is very rich. It consists of two sets of arteries, the **periosteal** and **medullary** or **nutrient.** The **periosteal arteries** are derived from a dense network of vessels in the outer layer of the periosteum. They penetrate into the compact bone through minute passages called **Volkmann's canals.** These canals, in turn, connect with the **Haversian canals** which were discussed and illustrated in Chapter 5 on tissues (pp. 63–64 and Fig. 5–8). Larger and less numerous arteries penetrate the compact tissue of the epiphyses to supply the many spaces and lamellae of the subjacent cancellous bone. A **medullary** or **nutrient** artery usually enters near the middle of the shaft of a long bone through an obliquely placed canal in the compact layer and upon reaching the medullary cavity sends branches toward both ends of the bone. These vessels ramify in the marrow, give off twigs to adjoining canals and **anastomose** (*interconnect*) freely with the vessels of the cancellous parts of the bone. Usually, one or two veins accompany the medullary artery; other veins, large and small, emerge from the cancellous tissues at the extremities of long bones through large and small foramina in the compact layer. Many tiny veins pass out through Volkmann's canals into the periosteum. Short, flat, and irregular-shaped bones are all similarly supplied with blood. The blood, by this system of gross and minute vessels, brings food and oxygen to all of the osteocytes, buried as they are in the matrix, and to the marrow which is so active in the formation of blood cells. The blood, as it leaves bone, not only carries away the waste products of metabolism but may also have gained many newly formed cells (Fig. 7–6). **Lymphatic** vessels are also found in the periosteum and, according to some, pene-

trate the bone and travel in the Haversian canals. **Nerves** are abundant in the periosteum and may accompany nutrient arteries into the bone itself.

Chemical Composition. About two-thirds of the weight of bone is inorganic material, mostly calcium phosphate and calcium carbonate. The remaining one-third is organic, consisting of cells such as osteoblasts and fibroblasts, the matrix, and collagenous fibers. If a bone, a long bone, for example, is immersed in a dilute mineral acid to remove the inorganic material, it will become so flexible that it can be tied into a knot. Yet, it retains its original shape and the arrangement of Haversian canals, lacunae, lamellae, and canaliculi have not been altered. Conversely, if the organic material is removed by calcination (*heat*), the bone will retain its original form but will be white and brittle and only a slight blow will cause it to crumble.

Development and Growth of Cartilage and Bone (Fig. 7–6). You should recall at this time that connective tissues including **cartilage** and **bone** develop from mesenchyme. Mesenchyme arises largely from the mesodermal somites and the somatic and splanchnic mesoderm (Fig. 4–12). Cartilage formation is evident at about the fifth week of embryonic life when the mesenchyme cells in a given region proliferate and enlarge and ultimately become cartilage cells. Matrix material is laid down by these cells and they become more widely spaced. A connective tissue sheath develops around the growing cartilage, the **perichondrium,** and its inner cells undergo mitosis to produce new cells. These transform into cartilage cells (*chondrocytes*) and become buried in lacunae within their own matrix. The cartilage with which we are now most concerned is **hyaline** which has fine masked fibrils within its homogeneous matrix. It precedes the development of most of the bones of the body.

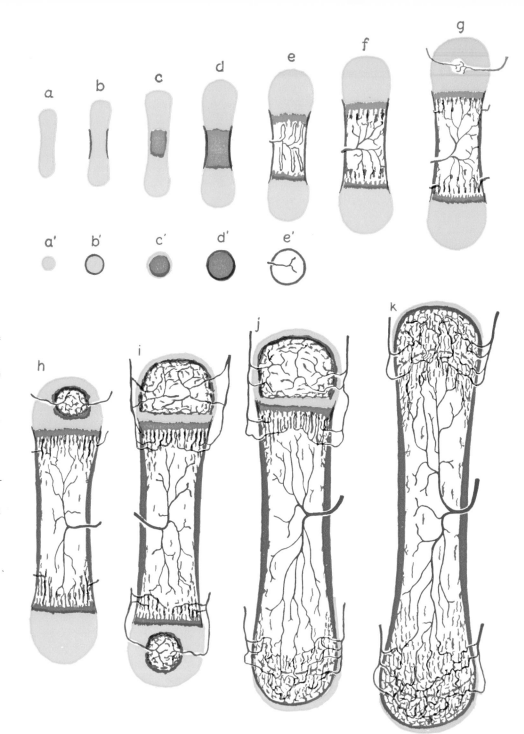

Fig. 7–6. Diagram of the development of a typical long bone as shown in longitudinal sections. *Pale green*, cartilage; *purple*, calcified cartilage; *blue*, bone; *red*, arteries. *a′, b′, c′, d′, e′*, cross-sections through the centers of *a, b, c, d, e*, respectively. *a*, cartilage model, appearance of the periosteal bone collar; *b*, before the development of calcified cartilage; *c*, or after it, *d; e*, vascular mesenchyme has entered the calcified cartilage matrix and divided it into two zones of ossification, *f; g*, blood vessels and mesenchyme enter upper epiphyseal cartilage; *h*, epiphyseal ossification center develops and grows larger; *i*, ossification center develops in lower epiphyseal cartilage; *j*, the lower and, *k*, the upper epiphyseal cartilages disappear as the bone ceases to grow in length, and the bone marrow cavity is continuous throughout the length of the bone. After the disappearance of the cartilage plates at the zones of ossification, the blood vessels of the diaphysis, metaphysis, and epiphysis intercommunicate. (Bloom and Fawcett, *A Textbook of Histology*, courtesy of W. B. Saunders Co.)

Bones are of two types on the basis of their means of development: (1) **intramembranous bones** which form directly in a fibrous membrane; and (2) **endochondral bones** which form by the replacement of hyaline cartilage. In other ways the manner of bone formation (*osteogenesis*) is the same and the resulting bone is identical histologically. In each case specialized mesenchymal cells, the osteoblasts, are the bone formers. They lay down a matrix of homogeneous ground substance and fibrillae called **osteoid tissue** (*ossein*). Osteoblasts and blood vessels become imprisoned in the lacunae, canaliculi, Haversian, and Volkmann's canals, in the bone matrix (Fig. 5–8). Lime salts are now deposited in this matrix through the action of the enzyme phosphatase, a process called **calcification.** The process depends upon an adequate supply of calcium, phosphorus, vitamin D, and a hormone from the parathyroid glands. In the absence of any of these materials, **rickets** may develop in which there is an overproduction of osteoid tissue, a lack of calcification, and deformities of the skeleton such as bowlegs and the appearance of bead-like enlargements at the sternal ends of the ribs called rachitic rosary. Throughout life there is a delicate balance between blood calcium and that stored in the skeletal system. To keep the blood calcium at a proper level the supply in the bones is called upon, or excesses in the blood are stored in the bone. If the parathyroids, which play an important role in the maintenance of this balance, become overactive, excessive withdrawal of calcium from the bones takes place and the osteoid substance is once again soft and pliable. This condition is called **osteomalacia.** Hormones of the hypophysis (*pituitary*) and the thyroid glands also influence bone development and growth.

Intramembranous Ossification. Intramembranous bones (*membrane bones*) in the human skeleton are those flat bones of the cranial vault and some bones of the face. **Intramembranous ossification** begins at the more highly vascularized areas in the primitive connective tissue membrane, most often at the center of the future bone. Here, **osteoblasts** appear and begin to lay down the osteoid material in the form of spicules. As the spicules form, the osteoblasts take up their positions upon them in somewhat the manner of an epithelium. In this position they continue to lay down material causing the spicules to broaden and lengthen and join with adjacent ones to form a meshwork of **trabeculae.** This primary center of ossification spreads peripherally; calcification takes place as new matrix is formed, and soon a periosteum forms around the entire primordium (*developing structure*). Osteoblasts appear on the inner surface of the periosteum and start to lay down osteoid substance. This is sometimes referred to as **periosteal ossification** and results in the laying down of the compact bone of the **outer** and **inner tables.** The spongy mass between the tables is the diploë. As these bones grow, a considerable amount of remodeling takes place within them through the action of bone-destroying cells, the osteoclasts, or perhaps by other agents. Again the element of uncertainty enters our story of the human body. We know that bone is resorbed and that **osteoclasts** are present in these areas. We are not sure that these cells are directly responsible. However, as a result of these processes, the bone structure is arranged and modified in accordance with the stresses encountered in postnatal life, although the general shape and arrangement are hereditary.

Intramembranous ossification is incomplete at birth so that considerable membrane is left around the bones. These membranous areas are the **fontanels.** They make it possible for the head of the fetus to be molded somewhat to the shape of the birth canal and hence ease the process of delivery.

Endochondral Ossification (Fig. 7–6). Most of the bones of the body are preceded by a cartilage model which takes on somewhat the shape of the definitive (*future*) bone. In endochondral ossification, the cartilage is first destroyed, and secondly, its place is taken by the bony tissue. For this reason the resulting bone is often called a **cartilage** bone or, perhaps even better, a **replacement** bone. Once the cartilage is removed, ossification is essentially like that in membrane bone. Bone is formed both inside the cartilage model where the cartilage has been removed and under the **perichondrium** which we call the **periosteum** as soon as some bone is laid down.

The development of a long bone serves well to illustrate how endochondral ossification takes place. There are usually at least three ossification centers in long bones, a primary one in the center of the cartilage of the future shaft or diaphysis and a secondary center in each end or epiphysis. With a few exceptions, as the wrists and ankle bones, the primary centers appear shortly after the second month of fetal life, while the secondary centers in the epiphyses, again with few exceptions, do not appear until after birth. The diaphyses and epiphyses are separated for some time after birth by **growth** or **epiphyseal cartilages** at which points the bones continue to grow in length (Fig. 7-8). Finally, these cartilages yield to ossification and epiphyseal lines mark the point of union of diaphysis and epiphysis. The diameter of the bone is increased by periosteal ossification and the marrow cavity is enlarged, probably by the action of the osteoclasts. It is not until the twenty-fifth year that the skeleton completes its ossification and growth.

The first evidence of endochondral ossification is when the cartilage cells in a primary center begin to multiply, enlarge and arrange themselves in rows. Lime deposits also appear in their matrix. Next the cartilage cells and some of the matrix are destroyed and primordial marrow cavities are thus formed. **Primary marrow tissue,** derived from the inner layer of the perichondrium, begins to grow into the cartilage and marrow spaces giving rise to **osteoblasts** and **vascular marrow.** Osteoblasts take up their positions on spicules of eroding calcified cartilage and deposit osteoid substances. A cancellous tissue thus forms with primary marrow spaces. While these changes go on at the primary ossification center, new cartilage cells are formed at each end of the cartilage to increase its length and the bone-forming process itself progresses toward the epiphyses. Meanwhile, the inner layer of cells (*osteoblasts*) of the perichondrium lay down a **compact** bone matrix around the periphery of the cartilage. This also grows gradually toward the epiphyses of the bone. Finally, the trabeculae of the spongy endochondral bone of the interior are dissolved and the primary marrow spaces coalesce to form the marrow cavity.

Bone formation in the secondary centers of the epiphyses follows essentially the same pattern as above except that they remain cancellous. Also, a layer of smooth articular cartilage will remain on the surface of those epiphyses which become a part of an articulation. Other secondary centers of ossification, as the trochanters of the femur, afford processes for the attachment of muscles and become fully ossified.

Skeletal Disorders. Since bones are vital tissues, they are subject to a variety of disorders ranging from congenital defects, malnutrition, and hormone imbalance to degenerative and inflammatory diseases and injuries. Some of these have been mentioned in our previous discussions; others will receive some comment as we discuss the individual parts of the skeleton. All of them should be more understandable if you have grasped the full meaning of our study of bone. When a bone is fractured and has been properly set the regenerative processes or "knitting" of the bone are very much the same as the primary ossification process we have just described.

Osteomyelitis is an acute inflammation of the bone (*osteitis*) and bone marrow (*myelitis*). Though the inflammation involves only the soft parts of the bone, it results in a softening of the hard structure through the dissolving away of the calcium.

Osteomyelitis is most common among young people in whom bone growth is taking place. It usually affects the area around the epiphyseal cartilages where the blood supply is rich and where a twist of the bone or a blow may cause the vessels to rupture and form clots. Here the causative organisms, usually staphylococci, settle down and multiply rapidly. Staphylococci generally enter the body through the skin where they may cause a boil and then, travelling through the blood stream, secondarily attack the bones. The condition is highly damaging to bone and if untreated may extend into the joints and into the general circulation. It then spreads to other organs, as the lungs, where multiple abscesses may appear.

Fortunately, the introduction of antibiotics has simplified the treatment of osteomyelitis. If it is diagnosed early enough, it can be treated medically, and widespread bacterial infection avoided. Surgery used to be the principal means of treatment and very often many operations were necessary to arrest the progress of the infection.

Similar to osteomyelitis is tuberculosis of the bones and joints. It, however, is chronic from the beginning, rather than acute. Usually the primary infection is in the lungs or lymph glands from which it metastasizes to reach the bones and joints.

GENERAL ARTHROLOGY

Definition and Function. The bones serve as the levers upon which most of the skeletal muscles of the body act in producing movement. The joints or articulations are the areas where adjacent bones or cartilages join. The movable joints serve as the centers or fulcra around which movement takes place. The study of the articulations is called **arthrology.**

Classification of Articulations (Table 7–1). Articulations fall into three cate-

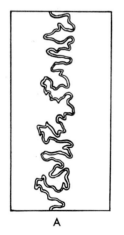

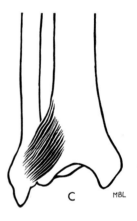

Fig. 7–7.   Types of fibrous articulations.
A and *B*, Suture joints—synarthroses.   *C*, Syndesmosis.

gories on the basis of movement: **immovable, slightly movable,** and **freely movable.** These functional differences are reflected in the structural details of the articulations. On a structural basis we recognize **fibrous, cartilaginous,** and **synovial** joints (Figs. 7–7, 8, 9). Recall that during embryological development mesenchyme pushed out into the growing limb buds and that centrally it condensed to give rise to cartilage and ultimately to bone. In between the adjacent ends of the developing bones the mesenchyme persists as the **primitive joint plate.** It is the subsequent history of the joint plate that determines the type of articulation. Also in the intramembranous bones of the vault of the skull it is the fate of the intermediate membranous area which is significant in joint formation.

Fibrous Joints (Fig. 7–7). The fibrous joints are those in which the primitive joint plate develops into a fibrous tissue. If the amount of fibrous tissue is minimal, you have a **suture,** as is well illustrated among the flat bones of the skull. The edges of the articulatory bones may be dentate, serrate, beveled, or grooved for close fitting. The fibrous tissue joining them is continuous with the fibers of the periosteum. You will recall that at birth, areas of membrane are still apparent

and these are the **fontanels.** The fontanels are nearly closed by the end of the eighteenth month of life. At about twenty-five years of age, the sutures begin to fuse over with bone, starting on the inside of the skull. This continues throughout life and it is only in old age that the sutures become obliterated. The condition is called **synostosis.** The suture joints are immovable and are called **synarthroses.**

Articulations with larger amounts of fibrous tissue between the bones are called **syndesmosis.** The joint between the distal ends of the tibia and fibula, the **tibiofibular joint,** is an example. The fibers are short, holding the bones in close apposition and allowing only a minimal movement. In other syndesmoses the fibers are longer and considerable movement is possible as in the fibrous membrane between the borders of the radius and ulna or the tibia and fibula.

Cartilaginous Joints (Fig. 7–8). In these joints the mesenchyme of the primitive joint plates become cartilaginous. They are of two types, one known as a **synchondrosis,** the other a **symphysis.** Examples of synchondroses are the growth or epiphyseal cartilages (*hyaline*) which are apparent between the diaphyses and epiphyses of bones and which enable

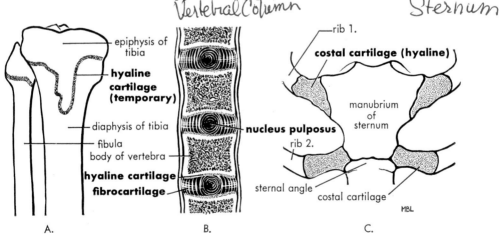

FIG. 7–8. Types of cartilaginous articulations.
A, Temporary synchondrosis; *B,* symphysis; *C,* permanent synchondrosis.

them to grow. When growth is complete ossification takes place in the epiphyseal cartilages and the epiphyseal joints are eliminated. The joints between the ribs and costal cartilages and sternum are permanent cartilaginous joints.

Symphyses occur between the bodies of adjacent vertebrae and at the point where the pubic bones articulate anteriorly. The articulating bones at a symphysis have the adjacent surfaces covered with hyaline cartilage and these in turn are joined to a pad of fibrous tissue or fibrocartilage. The fibrocartilage pads are fairly thick and compressible, and slight movement occurs at these joints. Between the vertebrae the pads are called the intervertebral discs. These, too, are examples of synarthroses (sometimes called amphiarthroses).

Synovial Joints. Also called **diarthroses,** the synovial joints are distinguished by the development of one or more **synovial** or **articular cavities** between the bones and by their freedom of movement. Most of the joints of the body are of this type and though they vary in details they are all built on the same general plan (Fig. 7–9). Some of the individual variations in synovial joints will be considered when the bones of the appendicular skeleton are described.

In synovial joints the ends of the articulating bones are covered with cartilage which is, with few exceptions, hyaline cartilage. Its free surface lacks a **perichondrium** and the superficial chondrocytes are flattened and arranged in rows. The deeper part of the cartilage is calcified. The joint is surrounded by a **capsule** of dense fibrous connective tissue whose fibers are continuous with those of the periosteum of the bones. The capsule serves to protect and in some cases to strengthen the joint. Inside of the joint the capsule is lined with a **synovial membrane** of variable character. It may be dense fibrous, areolar or adipose tissue and is continuous with the perichondrium at the edges of the articular cartilages. The synovial membrane is often thrown into folds called the **synovial villi** which may in turn have **secondary villi.** Some of the villi are highly vascular and may be covered with flattened connective tissue cells. The synovial membrane produces a **synovial fluid** which lubricates the joint, insuring smooth action. In some synovial joints, as the knee, articular discs of dense fibrous connective tissue or fibrocartilage, continuous with the synovial membrane, are wedged into the articular cavity dividing it into two.

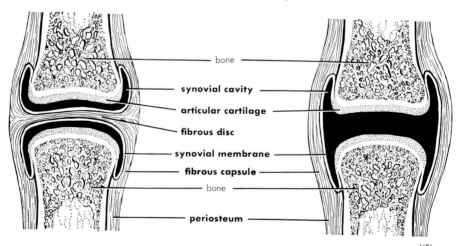

bone

synovial cavity

articular cartilage

fibrous disc

synovial membrane

fibrous capsule

bone

periosteum

MBL

FIG. 7–9.   Types of synovial articulations.

Outside of the joint capsule, a variety of ligaments further bind and strengthen the joint. The arrangement of the ligaments, as well as the shapes of the ends of the articulatory bones, are factors which are important in determining the freedom and direction of movement of the joint. Muscles also contribute to the strengthening of a joint and serve as its first line of protection, as well as being the active agents in its movement.

Many varieties of synovial joints occur in the body, based upon the kind of movement they permit. These will be named and described very briefly and as you continue your study of the skeleton you should attempt to identify them and consider their action.

1. **Gliding** (*arthrodia*) joints are those in which the articulating surfaces are small and flat, or slightly concave, and one slides over the other. They are found between the articular processes of vertebrae and in many of the joints between the carpal bones of the wrist and the tarsal bones of the ankle region.

2. **Hinge** (*ginglymus*) joints allow movement in only one axis at right angles to the bones, as at the elbow joint and in the joints between the bones of the fingers. The joint capsules are lax on the bending surfaces, but strong collateral ligaments bind the bones on the margins of the joint. Only flexion and extension are possible at these joints.

3. **Pivot** (*trochoid*) joints allow movement around a single axis—in this case the longitudinal axis of the bone. The articulation of the first cervical vertebra upon a process of the second, around which it rotates, and the one between the proximal ends of radius and ulna, are good examples.

4. **Condyloid** (*ellipsoid*) joints are biaxial, allowing movement in two directions at right angles to each other—flexion—extension and abduction—adduction. Circumduction is also possible. The wrist joint is an example.

5. **Saddle** joints are also biaxial, the articulating surfaces being reciprocally concavoconvex. The carpometacarpal articulation of the thumb is a good example.

6. **Ball and socket** articulations are "universal" joints. They allow movement in many axes with one common center. The shoulder and hip joints are examples.

Movements in Articulations. Special terms are used to describe the effect of muscle contraction on the joints of the

TABLE 7–1. CLASSIFICATION OF ARTICULATIONS

Joints	*Examples*
Fibrous—joint cavity absent	
Suture	Bones of skull
Immovable	
Syndesmosis	Distal ends of tibia and fibula;
Slightly to quite freely movable	borders of radius and ulna; laminae of vertebrae
Cartilaginous—joint cavity absent	
Synchondrosis	Epiphyseal cartilages of long
Slightly movable	bones; ribs and costal cartilages
Symphysis	Pubic bones; between bodies of
Slightly movable	vertebrae
Synovial—joint cavity present	
Freely movable	Mandible; extremities; articulating processes of vertebrae

body. These terms are related to the body as seen in the **anatomical position** (p. 12). In this position most body parts are extended (Fig. 2–1).

1. **Flexion** is a movement which results in a decrease in the angle between two bones as when you bend the elbow or the knee joints. Since the foot, in the anatomical position, is already flexed, any further bending is called **dorsiflexion.** The term dorsiflexion also applies to the over-extended hand.

2. **Extension** involves an increase in the angle between two bones at an articulation. Straightening the arm at the elbow, or the leg at the knee are examples. Extending the foot is often called **plantarflexion.**

3. **Abduction** refers to movement away from the central axis of the body or from a plane wholly within a part such as the hands or feet. Hence when we move the legs laterally, lift the arm from the side, or separate the fingers or toes we are abducting.

4. **Adduction** is the opposite of abduction, the parts being moved toward the body or the midline of a part.

5. **Rotation** is the movement of a bone around an axis, either its own or that of another bone. Turning the anterior face of the humerus toward the midplane of the body is **medial rotation,** away from the midplane is **lateral rotation.** In these movements the humerus turns on its own axis. The rotation of the atlas vertebra on the dens of the axis is a case of one bone turning on the axis of another. The radius, as it rotates around the ulna, moves in an axis which is not quite parallel to the long axis of the bone.

6. **Supination** is the lateral rotation of the forearm which brings the palm of the hand into the forward (*upward*) position. In this position the radius and ulna are parallel.

7. **Pronation** is the medial rotation of the forearm which brings the palm of the hand into the backward (*downward*) posi-

tion. The radius, in this case, lies diagonally across the ulna.

8. **Eversion** is the rotation of the foot so that the sole turns outward.

9. **Inversion** is the rotation of the foot which turns the sole inward.

10. **Circumduction** is an action involving flexion, extension, abduction, adduction and rotation as when the arm is moved to circumscribe a cone.

Bursae and Tendon Sheaths (Figs. 11–7, 8). Closely associated with joints are connective tissue spaces, the **bursae.** They are lined with synovial membrane and they contain a fluid similar to the synovial fluid. They are located under tendons and muscles and between muscles where they permit an easy movement of one part upon another. Some lie between the skin and bony prominences, as over the knee cap or at the elbow, and allow the skin to move with less friction over the underlying parts. The **tendon** or **vaginal** sheaths which surround tendons as they pass through osteofibrous canals and over grooves of bones are also lined with synovial membrane. They reduce friction and are well illustrated by the tendons of the flexor muscles of the fingers.

Blood and Nerve Supply of Joints (Fig. 7–6). Networks of **vessels** around the articulations give rise to **arteries** which supply the ligaments and capsules of joints. Branches also form networks of fine vessels in the synovial membranes. **Veins** follow a pattern similar to that of the arteries, and **lymphatic** networks are present in the capsules.

The **nerves** to joints are numerous and are derived as branches from the nerves supplying the muscles which move the joints and also from those to the overlying skin. **Proprioception** (*kinesthetic*), those sensations which enable us to know the positions and movements of joints and muscles, is mediated by nerves which have their receptive endings in the joint capsules. Impulses traveling over these nerves go to the spinal cord and brain, some reaching the conscious centers of the cortex. Others result in reflex control of joint muscles through the spinal cord. The capsules and ligaments also have pain fibers which are particularly sensitive to stretching and twisting of the joint. Pain is poorly localized and may be referred to other joints, skin areas or muscles. Visceral disturbances may also accompany joint pain. Other nerves are also present, both sensory and motor, for the control of the blood vessels around the joints.

Disorders of the Joints. So many people ask, "What is arthritis?" It comes from **arthron,** meaning joint, and **itis**, inflammation. We may define arthritis, therefore, as an inflammation of a joint. This condition is marked by local tenderness and pain, local redness and swelling, and local elevation of temperature. It may be the result of infection or trauma and may leave the joint stiff. Stiffness is sometimes due to fibrous connective tissue, cartilage, or bone joining the articulating surfaces or to spasms of the muscles surrounding the joint.

Chronic or rheumatoid arthritis is one of the most common crippling diseases.

Dislocations of joints are common and may be caused by falling on a joint or by a blow to it or less commonly by violent muscular contractions. Deformity of the joint, painful movement, and swelling are the local symptoms. The ligaments are sprained or may even be torn in severe cases. Blood vessels are often ruptured and nerves may be compressed. Sometimes the bones are fractured and may produce an open wound. The proper treatment is to immobilize the part and to seek medical aid. One should not try to relocate the bones (*reduction*) as further injury could easily be done to blood vessels and nerves.

The most common injury to joints is **sprain** in which ligaments are stretched beyond their normal limits. Pain is usually very severe, especially when the joint is moved or touched. There may be discoloration due to ruptured blood vessels and local elevation of temperature. The movement of the joint may be impaired. If severe, and broken bones are suspected, a physician should be consulted.

QUESTIONS

1. Compare the endoskeleton of man with the exoskeleton of invertebrates such as insects.
2. What are the functions of the endoskeleton of man?
3. Most books state that there are 206 bones in the human skeleton, a few say there are 210. How do you account for this difference?
4. Define hamulus, epicondyle, facet, fovea, meatus, fissure, and sulcus.
5. Discuss the relationship of cancellous and compact bone tissue as found in the epiphyses and diaphyses of long bones.
6. What is diploë?
7. What is the distribution of white and red bone marrow in long and in flat bones?
8. Discuss the periosteum and its functions in the bones of young and old individuals.
9. Use the following words in sentences to demonstrate your understanding of their meaning: osteoblasts, osteogenic, osteocyte, and osteoclasts.
10. Under certain conditions a long bone could be tied into a knot. Explain.
11. Discuss the blood supply of a bone.
12. State in general terms the steps in the formation of bone tissue.
13. How do intramembranous and endochrondral bones differ?
14. How do bones grow in length and diameter?

15. Compare osteomyelitis and tuberculosis of bone as to causative factors, progress of the infections and symptoms.
16. Define the following terms relative to joints: syndesmosis, symphysis, synchondrosis, diarthrosis, synovial, and synarthrosis.
17. Describe, using diagrams, single cavity and double cavity synovial joints.
18. Give examples of condyloid, trochoid (*pivot*), ginglymus (*hinge*), and saddle joints.
19. Define pronation, abduction, eversion, flexion, and circumduction.
20. Why are tendon sheaths important?
21. What is a bursa?
22. What is the importance of the proprioceptive senses?
23. Discuss some of the disorders of joints.

"What is the use of a book without pictures?"

—Alice in Wonderland

Chapter 8

The Axial Skeleton

HAVING considered the general features of the skeleton and its articulations, we turn now to a closer study of individual bones and some of the more important joints.

The axial skeleton, you will recall, consists of the skull, vertebral column and thorax.

THE SKULL

Classification of Bones. The bones of the skull are placed in two groups, those forming the walls of the cranial cavity which houses the brain, and those of the facial region, supporting the eyes and entry ways to the respiratory and digestive systems.

> Bones of the cranium (8)
>> Single bones:
>>> Frontal
>>> Occipital
>>> Ethmoid
>>> Sphenoid
>> Paired bones:
>>> Temporal
>>> Parietal

8

Facial bones (14)
    Single bones:
        Mandible
        Vomer
    Paired bones:
        Maxillae
        Zygomatic
        Nasal
        Lacrimal
        Palatine
        Inferior nasal conchae

Included within the middle ear cavity of each temporal bone are three small **ear ossicles,** the **malleus, incus** and **stapes.** The **hyoid bone,** which is in the base of the tongue, is usually listed among the skull bones, bringing the total number to twenty-nine.

Study of the illustrations in Figures 8–1 to 6 will enable you to identify and learn the relative positions of most of the bones of the skull. The diagrams should be followed closely as you read the following description.

The Anterior View of the Skull. This view shows the deep **orbital fossae** which house the eyes; the **nasal fossae** for the nose; a prominent frontal bone, the base for the forehead; and the upper and lower jaws bearing the teeth (Fig. 8–1). The anterior margins of the orbital fossae are formed superiorly by the frontal bone, laterally by the **zygomatics,** inferiorly and medially by the **maxillae.** Going deeper into the fossae on the medial side are the small, thin and grooved **lacrimal bones** which, with the

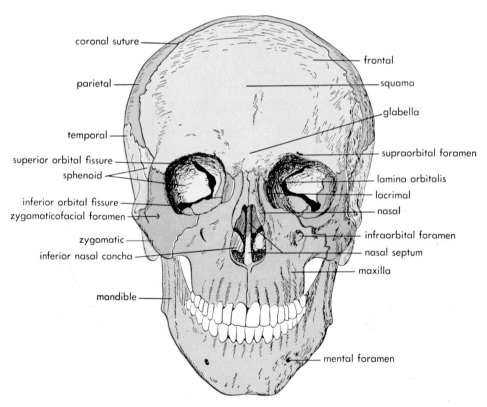

Fig. 8–1.   Anterior view of the skull.

maxillae, form the walls of the nasolacrimal canals connecting with the
nasal fossae. Beyond the lacrimals on
the medial sides of the orbital fossae are
the **laminae orbitalis** of the **ethmoid**
bones and behind these a part of the
sphenoid bone with the round **optic
foramina** and the large **superior orbital
fissures.** The frontal bone forms the
roof of the orbital fossae, the maxillae
most of the floor, and the **greater wings**
of the **sphenoid** form the lateral posterior
walls. Bounded by the maxillae, the
greater wings of the sphenoid, and the
zygomatics in the floor of the orbits are
the large **inferior orbital fissures.**

The **frontal** bone, from the anterior
view, presents a vertical portion, the
squama and a horizontal or **orbital** portion forming the roofs of the orbits.
Where these two join is the **supraorbital
margin** in which there may be on each
side a **notch** or **foramen, the supraorbital
foramen.** On either side of the squama
is a **superciliary arch** and between these
medially a smooth area, the **glabella.**
The **frontal sinuses** lie medially behind
the superciliary arches. They drain into
the nasal fossae.

Infraorbital foramina of the maxillae
and the **mental foramina** of the mandible should be noted.

The openings into the **nasal fossae** are
bounded superiorly by the sharp edges of
the **nasal** bones which form the bridge
of the nose. Laterally and inferiorly they
are bounded by the **maxillae.** Separating the two nasal fossae medially is the
vertical plate formed by the **ethmoid** and
the **vomer** which is called the **nasal
septum.** In the living specimen the
nasal septum is extended forward by
hyaline cartilage as are also the anterior
margins of the maxillae. Projecting from
the lateral walls into the fossae are the
shelf-life **inferior nasal conchae** and
above these the **superior** and **middle
nasal conchae** which are parts of the
lateral masses of the ethmoid. The

floor of the nasal fossa is formed by the
horizontal plates of the **maxillae** and
palatines.

The Lateral View of the Skull.
Laterally, the skull may be divided into
the relatively smooth, dome-shaped cranial portion and the anterior facial portion with its irregularly shaped bones
(Fig. 8–2). The cranial portion is marked
by three main sutures, the **coronal** between the frontal and parietal bones, the
squamosal between the **squamous** part
of the temporal bone and the parietal,
and the **lambdoidal** between the parietal
and occipital bones. Along the sutures,
particularly the lambdoidal, small pieces
of bone, called **sutural** or **Wormian**
bones may sometimes be seen. At the
posterior inferior angle of the cranium, a
large rounded process, the **mastoid** of the
temporal, justs downward and just in
front of it is an opening, the **external
acoustic meatus.** An elongated slender
projection of the temporal bone, the
styloid process, lies below and medial to
the external auditory meatus. It is often
broken off in the preparation and handling
of skulls. Important muscles and ligaments attach to it in the living specimen.
In front of the external acoustic meatus
is the **mandibular fossa** for the articulation of the **condyloid process** of the
mandible and above it is the root of the
zygomatic process of the temporal bone.
Anteriorly, the zygomatic process articulates by an oblique suture with the
zygomatic bone and together they form
the **zygomatic arch.** A straight line
drawn along the superior border of the
zygomatic arch and extending posteriorly
would mark the inferior limits of the
large but shallow **temporal fossa** in
which the fan-shaped **Temporalis** muscle
is located. The temporal fossa is bounded
above and behind by the inconspicuous
temporal lines which curve from the
zygomatic process of the frontal bone
upward and backward over the frontal
and parietal bones and then forward

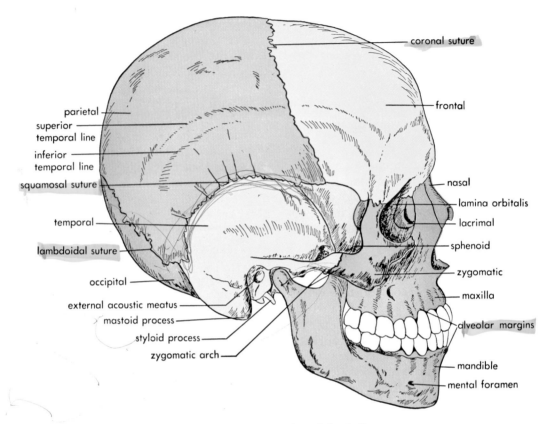

FIG. 8–2. Lateral view of the skull.

above the mastoid to the posterior root of the zygomatic process of the temporal bone. Below the superior border of the zygomatic arch and medial to it is the deep **infratemporal fossa.** The remaining facial portions of the lateral view of the skull are the maxilla and the mandible with their **alveolar** margins containing the teeth. The mandible has a prominent **coronoid process** on the medial side of which the temporal muscle inserts. The maxilla not only supports teeth but forms part of the floor and margin of the orbit and the lateral wall of the nasal fossa. Internally, it has a large cavity, the **maxillary sinus,** which communicates with the nasal fossa. The nasal bones and the prominent **superciliary**

arches of the frontal complete the profile of the facial bones.

The Skull from Below. In this region we find a complex arrangement of bones, their processes and foramina. It is more easily seen if the mandible is removed. Since most of the features are illustrated and labeled in Figure 8–3, the following description will be brief.

Anteriorly, this view is marked by the concave **hard palate** formed by the maxillae and palatine bones, and the teeth borne on the **alveolar** margins of the maxillae. Posterior to and above the hard palate are the large **choanae** separated from one another by the **vomer,** limited superiorly by the body of the sphenoid bone and laterally by the

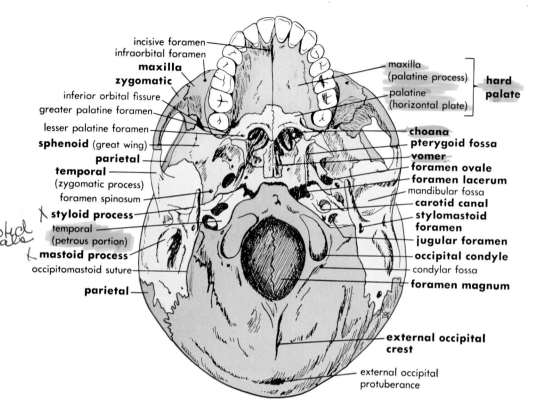

incisive foramen
infraorbital foramen
maxilla
zygomatic
inferior orbital fissure
greater palatine foramen
lesser palatine foramen
sphenoid (great wing)
parietal
temporal
(zygomatic process)
foramen spinosum
styloid process
temporal
(petrous portion)
mastoid process
occipitomastoid suture
parietal

maxilla
(palatine process) } **hard**
palatine   **palate**
(horizontal plate)
choana
pterygoid fossa
vomer
foramen ovale
foramen lacerum
mandibular fossa
carotid canal
stylomastoid
foramen
jugular foramen
occipital condyle
condylar fossa
foramen magnum

external occipital
crest
external occipital
protuberance

FIG. 8–3.   Inferior view of the skull.

medial pterygoid plate of the sphenoid bone. Lateral to the **medial pterygoid plate** is a prominent **pterygoid fossa** and its lateral wall is formed by the **lateral pterygoid plate**. The pterygoid fossa is the point of origin of the **Pterygoideus internus** muscle. Between the **lateral pterygoid** plates and the zygomatic arches are the **infratemporal fossae** described above, which in life are occupied by the muscles of mastication.

The area immediately behind the **choanae** and in front of the base of the **occipital bone** is occupied by the **pharynx** in the live specimen. Laterally, the **petrous portions** of the **temporal** bones push in toward the midline and with the sphenoid and occipital bones form the

irregular boundaries of the large **foramen lacerum**. Posterolaterally in the petrous temporal are the **carotid canals** which carry the internal carotid arteries. Just behind the carotid canals are the large **jugular foramina** bounded by the petrous temporals and the occipital bone. Lateral to the jugular foramina are the **styloid processes** and anterior to them at the bases of the zygomatic processes of the temporals are the **mandibular fossae** for articulation of the mandible. Behind the base of the occipital is the large **foramen magnum** for the passage of the spinal cord to the brain and lateral to it the prominent **occipital condyles** for articulation of the skull with the vertebral column. Posterior to and above the foramen magnum on the midline of the occipital bone is the **external occipital protuberance.** The obvious **mastoid processes** of the temporal bone should be noted laterally.

The Cranial Cavity. Looking inside of the skull and at the floor of the cranial cavity one sees three distinct fossae arranged like steps (Figs. 8–4, 5). The uppermost is the **anterior cranial fossa** which lies above the orbits and receives the frontal lobes of the brain; the middle step is the **middle cranial fossa** which lies behind the orbits and receives the temporal lobes of the brain, the lowest step is the **posterior cranial fossa** housing the medulla and cerebellum. The walls of each of these fossae are marked by depressions for brain convolutions and by grooves for blood vessels. The many foramina accommodate the passage of cranial nerves and also of blood vessels.

The **anterior cranial fossa** is formed by the orbital plates of the frontal bone, the cribriform plate of the ethmoid and the small wings and part of the body of the sphenoid. On the midline anteriorly is the **frontal crest** and behind it an abrupt projection, the **crista galli** of the ethmoid, both of which afford attachment to the falx cerebri, a membranous partition between the two halves of the cerebrum of the brain. On either side of the base of the crista galli is an **olfactory groove,** the floor of which is the **cribriform plate** of the ethmoid. The grooves support the olfactory bulbs and the cribriform plates have numerous foramina for

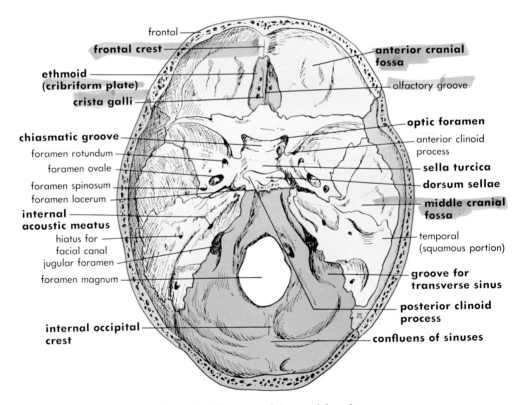

frontal
frontal crest
ethmoid (cribriform plate)
crista galli
chiasmatic groove
foramen rotundum
foramen ovale
foramen spinosum
foramen lacerum
internal acoustic meatus
hiatus for facial canal
jugular foramen
foramen magnum
internal occipital crest

anterior cranial fossa
olfactory groove
optic foramen
anterior clinoid process
sella turcica
dorsum sellae
middle cranial fossa
temporal (squamous portion)
groove for transverse sinus
posterior clinoid process
confluens of sinuses

FIG. 8–4. The floor of the cranial cavity.

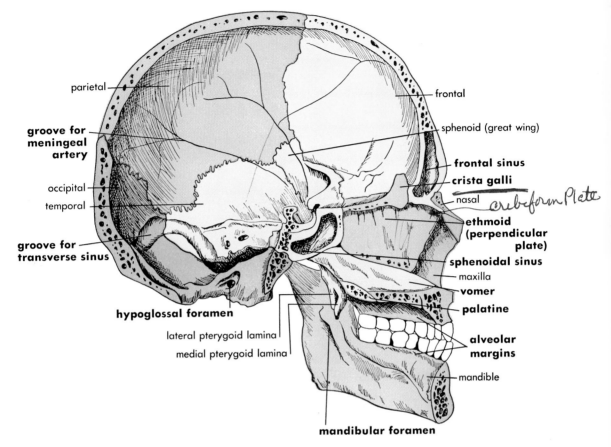

parietal

groove for
meningeal
artery

occipital

temporal

groove for
transverse sinus

frontal

sphenoid (great wing)

frontal sinus

crista galli

nasal *cribeform Plate*

ethmoid
(perpendicular
plate)

sphenoidal sinus

maxilla

vomer

palatine

hypoglossal foramen

lateral pterygoid lamina

medial pterygoid lamina

alveolar
margins

mandible

mandibular foramen

FIG. 8–5. Median section of th e skull.

the passage of the olfactory nerves (Fig. 19–4). The anterior fossa is limited posteriorly by the anterior margins of the **chiasmatic groove** and the posterior margins of the **small wings** and **anterior clinoid processes** of the sphenoid (Fig. 8–4).

The **middle cranial fossa** is narrow and shallow medially and is deep and much wider laterally. Its narrow medial portion is marked by a shallow chiasmatic groove anteriorly, each end of which leads into an **optic foramen** for the passage of the **optic nerves** and **ophthalmic arteries** to the orbit. Behind this is a deeper depression, the **sella turcica** (*Turk's saddle*), which lodges the **hypophysis** or **pituitary gland,** an important gland of the

endocrine system. The anterior wall of the sella turcica is marked by the **tuberculum sellae** with its **middle clinoid processes,** and posteriorly it is bounded by the **dorsum sellae** with its **posterior clinoid processes.** At each side of the sella turcica there is a shallow carotid groove coming up from the **foramen lacerum.** It lodges the **internal carotid artery** and the **cavernous sinus.** Lateral to the carotid grooves are the deep parts of the middle cranial fossae. They are limited laterally by the squamae of the temporal bones, the sphenoidal angles of the parietal and the great wings of the sphenoid; posteriorly by the superior angles of the petrous portions of the temporal bones. Anteriorly and

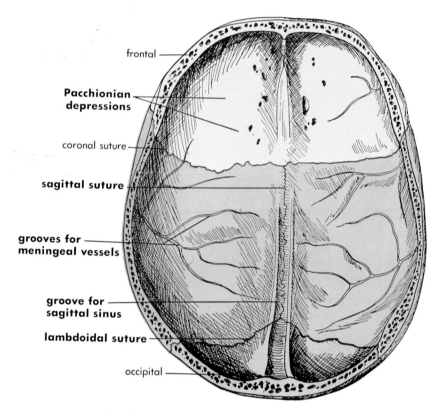

frontal

Pacchionian depressions

coronal suture

sagittal suture

grooves for meningeal vessels

groove for sagittal sinus

lambdoidal suture

occipital

FIG. 8–6.   Internal view of the "cap" of skull.

medially between the small and great wings of the sphenoid are the **superior orbital fissures** that carry the **oculomotor** (III), **trochlear** (IV), **ophthalmic division of the trigeminal** (V), and the **abducens** (VI) nerves into the orbits. Behind the medial ends of the superior orbital fissures are the **foramina rotunda** carrying the **maxillary nerves** and behind and lateral to these the **foramina ovale** for the passage of the **mandibular nerve.** The **foramina spinosae** are the last ones in these curved rows of foramina and they are for the passage of the **middle meningeal arteries** and **recurrent branches** of the mandibular nerves. The anterior surfaces of the **petrous portions** of the temporal bones, housing the **internal** and **middle ear** structures, complete this fossa.

The **posterior cranial fossa,** the largest and deepest of the three, is bounded anteriorly by the dorsum sellae of the sphenoid, the base of the **occipital,** and the petrous portions of the temporals. Laterally, the mastoid temporals, mastoid angles of the parietals, and the occipital limit the fossa. The occipital completes the posterior wall of the fossa. The large **foramen magnum** occupies the center of the floor of this fossa and an **internal occipital crest** extends from it along the midline to give attachment to the falx cerebri. A groove beside the crest serves for the passage of the **occipital sinus** and similar grooves for the **transverse sinuses** go laterally, then turn downward to end at the **jugular foramina.** In the mastoid temporal part of this groove there is a **mastoid foramen** and just before the groove reaches the jugular foramen there is a **condyloid canal.** Neither the mastoid foramen nor condyloid canal are found consistently. The large opening on the posterior surface of the petrous temporal is the **internal acoustic meatus** for the **facial** and **vestibulocochlear nerves** and the **auditory artery.**

The inside surface of the **cap of the skull,** like the fossae just described, shows depressions for the convolutions of the cerebrum and numerous grooves for the **meningeal vessels** (Fig. 8–6). A prominent longitudinal groove on the midline is for the **sagittal sinus.** This groove narrows in front and broadens posteriorly, continuing downward on the occipital to join similar grooves in the posterior cranial fossa, for the occipital and transverse sinuses. In the parietal region of the sagittal sinus, the **parietal foramina** may often be found. The edges of the groove for the sagittal sinus afford the attachments of the falx cerebri. Several depressions along either side of the groove mark the position of **arachnoid granulations** (*Pacchionian bodies*). These granulations form a part of the drainage system for **cerebrospinal fluid** into the venous sinuses. The skull cap is crossed in front by the **coronal suture** and behind by the **lambdoidal.** The **sagittal suture** lies in the medial plane and joins the two parietal bones.

Table 8–1 will serve to review the principal foramina of the human skull and some of the nerves and blood vessels they carry.

Development of the Skull (Fig. 8–7). In the section on general osteology the development of intramembranous and endochondral bones was described. In the formation and growth of the skull the best example of the intimate relationship of these two kinds of bones is seen. The roof and sidewalls of the cranial cavity are made up of intramembranous bones, as are the bones of the face (*dermatocranium*). The bones in the floor of the cranial cavity are endochondral (*chondro-*

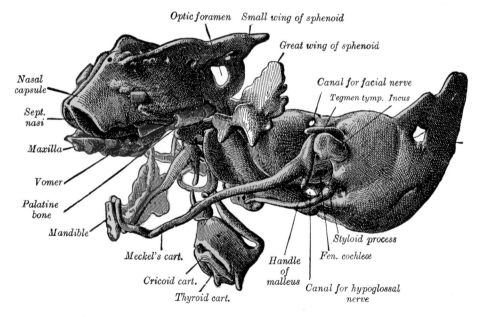

Fig. 8–7. Model of the chondrocranium of a human embryo, 8 cm. long. Certain of the membrane bones of the right side are represented in yellow. (Hertwig.)

TABLE 8-1

FORAMINA OF THE FACIAL BONES

Foramina	Bones Involved	Structures Passing Through
1. Incisive (Stensen) (foramina of Scarpa sometimes present)	Horizontal part of maxillae, in back of incisor teeth	Anterior branches of descending palatine vessels (Stensen); nasopalatine nerves (Scarpa)
2. Greater palatine (lesser palatine)	Palatine bones—at posterior angle of hard palate	Posterior branches of descending palatine vessels; anterior palatine nerves
3. Supraorbital (notch sometimes)	Frontal	Supraorbital nerves and vessels
4. Infraorbital	Maxilla	Infraorbital nerves and vessels
5. Zygomaticofacial	Zygomatic	Zygomaticofacial nerve
6. Mental	Mandible—lateral surface	Mental nerves and vessels
7. Mandibular	Mandible—medial surface	Inferior alveolar vessels and nerve
8. Lacrimal	Lacrimal	Tear duct

FORAMINA OF CRANIAL BONES

Foramina	Bones Involved	Structures Passing Through
1. Olfactory	Cribriform plate of ethmoid	Olfactory nerves (I)
2. Optic	Sphenoid	Optic nerves (II)
3. Superior orbital fissure	Sphenoid	Oculomotor (III), trochlear (IV), ophthalmic of trigeminal (V), abducens (VI) nerves
4. Inferior orbital fissure	Sphenoid, maxilla, palatine, zygomatic	Maxillary nerve (V), infraorbital vessels
5. Rotundum	Sphenoid	Maxillary nerve (V)
6. Ovale	Sphenoid	Mandibular nerve (V)
7. Spinosum	Sphenoid	Middle meningeal vessels
8. Lacerum	Sphenoid, temporal, occipital	Meningeal branch of the ascending pharyngeal artery, internal carotid artery
9. (Internal acoustic meatus)	Petrous portion of temporal	Facial (VII) and vestibulocochlear (VIII) nerves, internal auditory artery
10. Jugular	Petrous temporal and occipital	Glossopharyngeal (IX), vagus (X) and accessory (XI) nerves, internal jugular vein
11. (Hypoglossal canal)	Occipital bone	Hypoglossal nerve (XII)
12. (Carotid canal)	Petrous temporal	Internal carotid artery
13. Stylomastoid	Temporal—between mastoid and styloid processes	Facial nerve (VII)
14. (Condyloid canal)	Occipital	Vein to transverse sinus
15. Foramen magnum	Occipital	Medulla oblongata and its membranes; accessory nerves; vertebral arteries
16. Mastoid	Mastoid portion of temporal	An emissary vein

cranium). In some of the more complex bones, as the temporal and sphenoid, the completed bones are derived from both intramembranous and endochondral components. These "bone complexes" will be described later in greater detail. A series of paired cartilaginous visceral arches (*splanchnocranium*) which are associated with the mouth and pharynx also make contributions to the skull and to the larynx. The first visceral arch called **Meckel's cartilage** occupies the position of the future lower jaw and extends upward into the tympanic cavity. The mandible forms around part of Meckel's cartilage and this part gradually disappears. The remaining portions, in the tympanic cavity, give rise to the middle ear bones (*ossicles*), the **malleus** and **incus.** The third ear ossicle, the **stapes,** is derived from the second visceral arch as are also the styloid processes of the temporal bones and the **lesser cornua** of the hyoid. The third visceral arch forms the remainder of the hyoid bone. The other arches contribute to the cartilages of the voice box, the **larynx.** The skull therefore has developed from three sources, the **dermatocranium, chondrocranium and splanchnocranium.**

The Skull from Birth to Old Age (Figs. 8–8 to 10). The skull at birth is large in proportion to the rest of the skeleton. Its facial portion equals about one-eighth that of the cranium in size, while in the adult it is one-half. This small size of the face is due to the undeveloped condition of the mandibles and maxillae, the lack of teeth, and the small size of the nasal cavities and maxillary sinuses. Superciliary arches and mastoid processes are not developed and many of the bones such as the temporals, mandible, sphenoid, occipital, and frontal consist of more than one piece. A prominent **frontal suture** separates the two halves of the frontal bone. Unossified membrane remains at the angles of the parietal bones and these constitute the

"soft spots" or **fontanels,** of which there are six. The largest of these is the **anterior** or **frontal fontanel** which lies at the junction of the sagittal, coronal, and frontal sutures. It is about 4 cm. long by 2.5 cm. in diameter. It usually closes about eighteen months after birth. The **posterior** or **occipital fontanel** is at the junction of lambdoidal and sagittal sutures and is small and triangular in shape. The two lateral pairs of fontanels are the **anterolateral** or **sphenoidal** and **posterolateral** or **mastoidal.** They are small and of irregular shape. Posterior and lateral fontanels are usually closed by the end of the second month after birth.

The presence of the fontanels makes possible the molding of the head and therefore eases the process of childbirth and also allows for the expansion of the cranium to accommodate the rapidly growing brain. Premature closure of fontanels and sutures may be seen in case of retarded brain growth or **microcephalus.** Late or nonclosure may occur in individuals with **rickets** or untreated **cretinism,** or in cases of increased intracranial pressure, as in **hydrocephalus.**

A common malformation of the face is **cleft palate** (Fig. 4–22). In the severest and most complete cases, it involves the soft and hard palates, the alveoli of the maxillae and even the lip may be cleft— **hare-lip.** All degrees of severity occur— the slightest being just a cleft in the soft palate. The condition is due to the failure of the bilateral components of the face to join on the midline, or, in the case of the maxillae, to join the more median **premaxillae.** The cleft in the alveolus would, therefore, be on one side of the midline, or on both sides, and this would also be true for the lip which overlies this area.

In the first seven years, skull growth is rapid. The face enlarges as the teeth appear and the orbital fossae reach almost adult size. Growth then slows

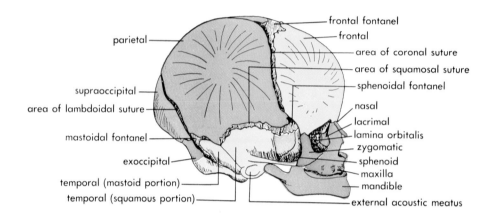

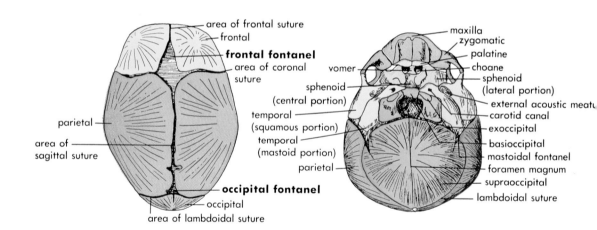

FIG. 8–8.   Superior surface of the skull at birth.

FIG. 8–9.   Lateral view of the skull at birth.

FIG. 8–10.   Inferior surface of the skull at birth.

until puberty when a second active period results in an over-all increase in the size of the skull, but especially in the facial region, due in part to the enlargement of the air sinuses.

The closure of the sutures begins at about the twenty-second year, though there is some individual variation. It is first apparent in the sagittal and sphenofrontal sutures, followed in the twenty-fourth year in the coronal, and in the twenty-sixth in the lambdoidal and mastooccipital. Closure is rapid between the twenty-sixth and thirtieth years, then slows, to again accelerate in old age.

In a very old skull the maxillae and mandible become reduced in size with the loss of teeth and subsequent absorption of the alveolar processes. The vertical measurement of the face is also reduced and the angle of the mandible is altered.

THE BONES OF THE SKULL

One should, having reached this point in the description of the skull, know well the positions of most of the bones and some of their most conspicuous features. Reference to Figures 8–1 to 8 will refresh your memory.

The following descriptions of individual bones will be brief, with emphasis on major features of structure, relationships, and development.

Frontal

The **frontal bone,** consisting of two main parts, the vertical portion or **squama** and the horizontal portion or **orbital,** forms the support for the forehead and part of the roofs of the orbits, respectively (Figs. 8–1, 2, 4). The **frontal sinuses** lie between the outer and inner tables of this bone in the region above the nose. The frontal sinuses, like others that will be mentioned, are lined by mucous membrane and each communicates with the corresponding nasal cavity by means of a duct (Fig. 14–2). Because of these connections the sinuses are subject to infection by any organisms which may involve the mucous membranes of the nose. The common cold, for example, should always be considered as a possible forerunner of a serious sinus infection and treated accordingly.

The frontal bone ossifies from two primary centers, one for each half, and from a few minor secondary centers. The bone, at birth, is in two halves, separated by the frontal suture. This suture usually disappears by the eighth year, except possibly at its lower end, or may, in a few individuals, persist throughout life.

Occipital

The **occipital bone** forms the lower and back part of the wall of the cranial cavity (Figs. 8–11, 12). Through its foramen magnum, the cranial cavity and vertebral canal are connected. The bone may be divided into three main parts in relationship to the foramen magnum; behind and above it the **squama,** to either side the **lateral portions,** and in front of it the thick **basilar part.**

On the external surface of the squama a ridge extends backward along the median line from the rim of the foramen magnum, the **median nuchal line.** It ends in a prominence, the **external occipital protuberance.** Extending laterally from the median nuchal line are the **inferior nuchal lines** and from the external occipital protuberance the **superior** and **highest nuchal lines.** The internal surface of the squama is divided into four fossae by a **cruciate** (*cross*) **eminence** and at the crossing of the four arms of the eminence an **internal occipital protuberance.** Paralleling the arms of the cruciate eminence are grooves or sulci for the superior sagittal, transverse, and occipital sinuses. The lower of the four fossae accommodate the

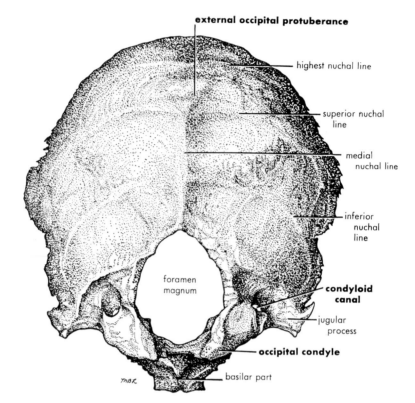

external occipital protuberance

highest nuchal line

superior nuchal line

medial nuchal line

inferior nuchal line

foramen magnum

condyloid canal

jugular process

occipital condyle

basilar part

FIG. 8-11.   Occipital bone—outer surface.

cerebellar hemispheres of the brain, the upper two the occipital lobes of the cerebrum.

The lateral portions of the occipital are marked externally by the large **occipital condyles** for the articulation of the skull with the atlas (*first cervical vertebra*). Behind each condyle is a depression, the **condyloid fossa,** the floor of which may sometimes be perforated by a **condyloid canal.** Anterior to each condyle is a foramen, sometimes divided into two by a thin plate of bone, the **hypoglossal foramen** for the passage of the twelfth cranial or **hypoglossal nerve.** Lateral to each condyle is a **jugular process.**

The **basilar part** of the occipital extends forward and upward from the foramen magnum. It is grooved internally to receive the medulla oblongata of the brain.

The upper part of the squama ossifies in membrane and usually from four centers. Sometimes it fails to join with the rest of the occipital, in which case it is called the **interparietal bone.** The remainder of the occipital ossifies in cartilage, two centers for the squama, two each for the lateral portions, and one or two for the basilar part. By the sixth year the various parts have united into one bone.

ETHMOID

The **ethmoid bone** is one of the lightest and most delicate of the skull (Figs. 8–4, 5, 13). Located in the front and base of the skull it contributes to the wall of the cranial cavity and to the nasal and orbital fossae. Its four parts are the **horizontal** or **cribriform plate,** the **perpendicular plate,** and the two labyrinth-like **lateral masses.**

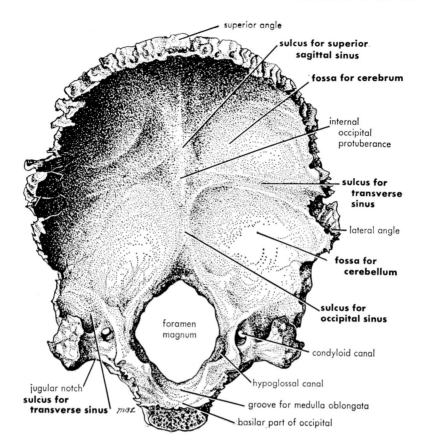

superior angle

sulcus for superior sagittal sinus

fossa for cerebrum

internal occipital protuberance

sulcus for transverse sinus

lateral angle

fossa for cerebellum

sulcus for occipital sinus

condyloid canal

foramen magnum

hypoglossal canal

jugular notch
sulcus for transverse sinus

groove for medulla oblongata

basilar part of occipital

FIG. 8–12.   Occipital bone—inner surface.

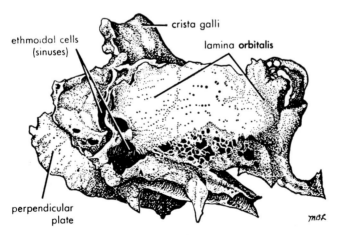

crista galli

lamina orbitalis

ethmoidal cells (sinuses)

perpendicular plate

FIG. 8–13.   Ethmoid bone from the left side.

The **cribriform plate** fits into a notch in the frontal bone and forms the roof of the nasal fossae as well as a part of the floor of the cranial cavity. The cribriform plate is narrow and grooved on each side of the midline to receive the olfactory bulbs and is perforated with small foramina for the passage of the olfactory nerves. Projecting from the midline of the cribriform plate into the cranial cavity is the prominent **crista galli.**

The **perpendicular plate** descends from the underside of the cribriform plate to form a part of the septum between the two nasal fossae. The septum is completed by the vomer and by hyaline cartilage.

The **lateral masses** are thin walled and contain many air spaces, the **ethmoidal sinuses.** The lateral surface of each mass consists of a smooth, thin plate, the **lamina orbitalis,** which forms a part of the medial wall of the orbit. The medial surface of each mass is a thin scroll-like bone, the lower part of which is called the **middle nasal concha.** The

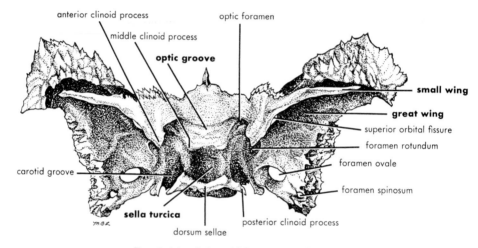

FIG. 8–14. Sphenoid bone—superior view.

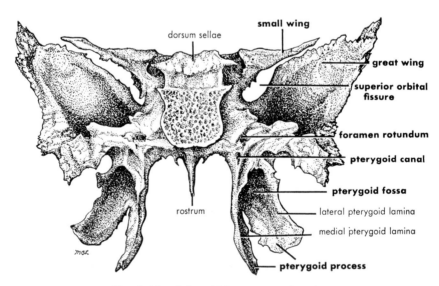

FIG. 8–15. Sphenoid bone—posterior view.

back part of this surface is divided by a narrow fissure, the **superior meatus** of the nose, and the thin, curved plate above it is the **superior nasal concha.**

The ethmoid ossifies in the cartilage of the nasal capsules from three centers, one for the perpendicular plate and one for each of the lateral masses.

Sphenoid

The **sphenoid bone** is in the base of the skull in front of the occipital and temporal bones (Figs. 8–14, 15). It is often described as resembling a bat with its wings expanded and its feet dangling below. The bone is an extremely complicated one, consisting of a **body, great wings, small wings,** and **pterygoid processes.** Since the bone has been quite fully described in the description of the cranial cavity and the lower surface of the skull, it will not be repeated here. It should be mentioned, however, that the body of this bone contains the large **sphenoid sinuses** which are separated from each other by a thin septum.

The embryological development of the sphenoid is most interesting. The body comes from two parts, one in front, the **presphenoid,** with which the small wings are associated, the other the **postsphenoid,** associated with the great wings and the pterygoid processes. The presphenoid and postsphenoid do not join until about the eighth month of intrauterine life. The small wings develop, each from one ossific center, and are sometimes called the **orbitosphenoids,** while the great wings may be called the **alisphenoids.** While most of the sphenoid is formed in cartilage, some parts, such as the pterygoid processes and the orbital plate, are intramembranous. The sphenoid bone reminds us of the skulls of many of the lower vertebrates, the cat, for example, where these componentss, the presphenoids, postsphenoids, orbitosphenoids, and alisphenoids remain more or less distinct throughout life. It is a good example of the concept of **recapitulation;** namely, that you see in the developmen

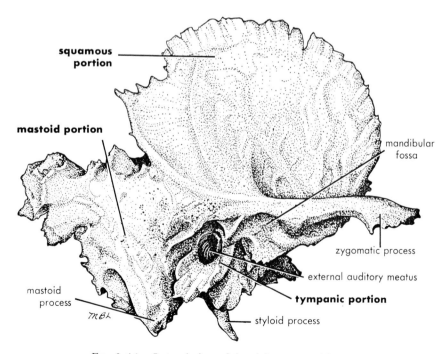

squamous portion

mastoid portion

mandibular fossa

zygomatic process

external auditory meatus

mastoid process

tympanic portion

styloid process

FIG. 8–16.   Lateral view of the right temporal bone.

of the individual (*ontogeny*), a brief résumé of the evolutionary history (*phylogeny*) of a species.

TEMPORAL

The **temporal bones,** like the sphenoid, are extremely complex and diverse in the origin and functions of their many parts (Figs. 8–2, 4, 16, 17). They have been described in relationship to the other skull parts in earlier sections of this chapter. Each temporal bone consists of four parts, a **squamous** portion forming a large part of the temporal fossa and with its zygomatic process, ossified in membrane; a **tympanic** portion, also intramembranous in origin, which forms a large part of the wall of the external auditory meatus and the tympanic cavity; a **mastoid** portion with its prominent mastoid process containing air cells or sinuses located behind the external auditory meatus and ossified in cartilage; and finally a **petrous** portion, also formed in cartilage. The latter contributes to the floor of the cranial cavity and houses the vital ear structures. An elongated process of the temporal, the **styloid,** is derived from the second visceral or hyoid arch. Inside of the tympanic or middle ear cavity are located the three ear ossicles, the **malleus, incus** and **stapes,** which are derived also from cartilaginous visceral arches, the first and second. The ear structures will be described in greater detail in the chapter on sense organs (p. 573).

PARIETAL

The **parietal bones** are almost square in outline and form a large part of the side walls and roof of the cranial cavity. They meet each other in the sagittal suture. With the frontals they form the coronal, with the occipital the lambdoidal, and with the temporal the squamosal sutures (Fig. 8–2).

The parietals are ossified in membrane, each from a single center in the region of the **parietal eminence.** Ossification

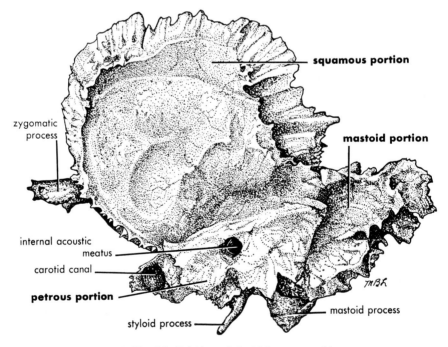

FIG. 8–17.   Medial view of the right temporal bone.

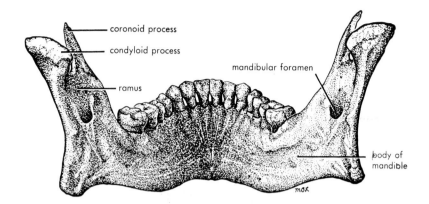

coronoid process

condyloid process

mandibular foramen

ramus

body of mandible

Fig. 8–18. Posterior view of the mandible.

progresses radially from the center, hence the corners or angles are the last to ossify. It is here we find the **fontanels** described earlier (pp. 105–107 and Figs. 8–8, 9).

MANDIBLE

The **mandible** consists of a horizontal horseshoe-shaped **body** and two perpendicular **rami** (Figs. 8–2, 18). The **body** has an **alveolar border** with sixteen cavities for the reception of the teeth. **Mental foramina** for the passage of mental vessels and nerves are on the external surface below the second premolar teeth. A faint ridge in the median line marks the **symphysis** where the two pieces of the body were joined earlier in life. On the internal surface of the body in the region of the symphysis are **mental spines** for muscle attachment.

The rami are quadrilateral in shape. The lateral surface is flat; the medial has an oblique **mandibular foramen** for the passage of the **inferior alveolar vessels** and **nerve.** Since all the lower teeth on one side are innervated by this nerve, the dentist needs to make only one injection to render the whole half of the jaw insensitive. There is no single point on the maxilla where such an injection can be accomplished. The upper border of the ramus has two prominent processes,

the thin, triangular **coronoid process** in front and the thick **condyloid process** in back. Between them is the **mandibular notch.** The coronoid process gives insertion to the temporalis muscle; the condyloid process articulates with the temporal bone at the **mandibular fossa.**

The **temporomandibular joint** is the only synovial joint of the skull. It is provided with an **articular disc** which fits between the condyloid process and the mandibular fossa dividing the joint into two cavities, each with a synovial membrane. The whole is enclosed in a thin, loose **articular capsule** attached to the circumference of the mandibular fossa and articular tubercle above and to the neck of the condyle of the mandible below. It is given further support by ligaments. The structure is such as to allow the opening and closing of the jaws, and the protrusion and lateral movements of the mandible.

The mandible is for the most part an intramembranous bone derived from the fibrous membrane covering the outer surface of Meckel's cartilages. Only a small part of each cartilage, below and behind the incisor teeth, becomes ossified and contributes to the mandible. The fate of the other portions of Meckel's cartilage has already been described (p. 105).

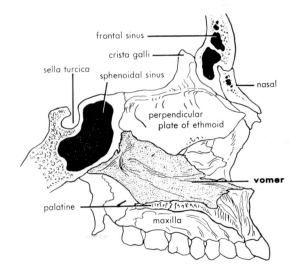

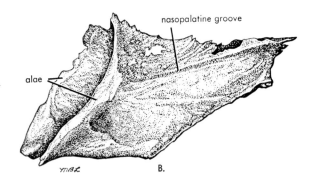

Fig. 8–19. *A*, Position and relations of the vomer. *B*, The vomer.

Vomer

The **vomer** forms the back and lower part of the nasal septum (Figs. 8–5, 19). The anterior part of the bone is a thin plate; its superior portion is thicker and has a furrow with horizontal wings (*alae*) extending out from it. This part fits against the sphenoid bone. Inferiorly, it joins with the maxillae and palatine bones. Its anterior border slopes downward and forward. The upper part of this border articulates with the perpendicular plate of the ethmoid, while its lower part is grooved to receive the inferior margin of the cartilage of the nasal septum.

The vomer ossifies in the membrane which covers the posteroinferior part of the cartilaginous septum of the nose of the fetus. The vomer therefore consists of two lamellae with intervening cartilage. The cartilage is gradually absorbed and by puberty the lamellae are almost completely joined into a median plate.

Maxillae

The **maxillae** are paired bones of irregular shape which together constitute the upper jaw (Figs. 8–1, 2, 5, 20). They, along with the palatine bones, also form the hard palate in the roof of the mouth,

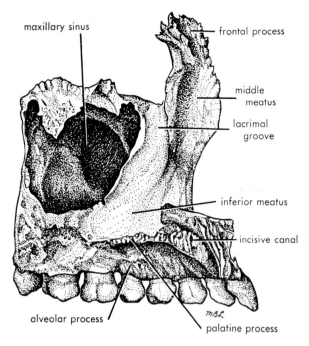

maxillary sinus

frontal process

middle meatus

lacrimal groove

inferior meatus

incisive canal

alveolar process

palatine process

Fig. 8–20. The left maxilla, medial surface.

and at the same time, they form the floor of the nasal fossae (Fig. 8–3). Each bone consists of a **body,** in which is located the large **maxillary sinus,** and four processes, the alveolar, bearing teeth, the **zygomatic** and **frontal** forming parts of the rims of the orbital and nasal fossae, and the **palatine** by which the two maxillae are joined. The maxillae are ossified in membrane and at first each has a **premaxillary** portion in front bearing the incisor teeth. The premaxillary remains as a separate bone in many vertebrate animals but becomes a part of the maxilla in man.

Zygomatic

The **zygomatic bones** form the prominences of the cheecks and are sometimes called the malar or cheek bones (Fig. 8–2). They have four processes by which they articulate to four bones—the maxillary, frontosphenoid, orbital and zygomatic (*temporal*). They form a part of the

floor and lateral walls of the orbits and contribute to the boundaries of the temporal and infratemporal fossae. They ossify in membrane.

Nasal

The **nasal bones** are small and articulate on the midline to form the "bridge" of the nose (Fig. 8–1, 21A). They vary considerably in shape and size among individuals. They articulate with the frontal, ethmoid, and maxillary bones. They ossify in the membrane above the cartilaginous nasal capsules.

Lacrimal

The **lacrimal bones** are the smallest and most delicate bones of the face (Figs. 8–2, 21). They are located on the medial sides of the orbital fossae. They have a sharp vertical ridge on their lateral surfaces, in front of which is a longitudinal groove. The maxillae con-

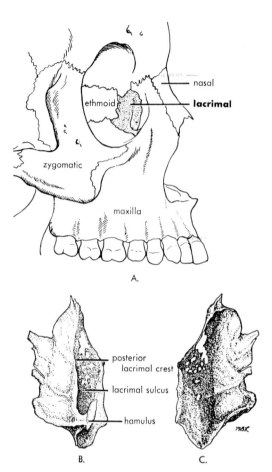

A.

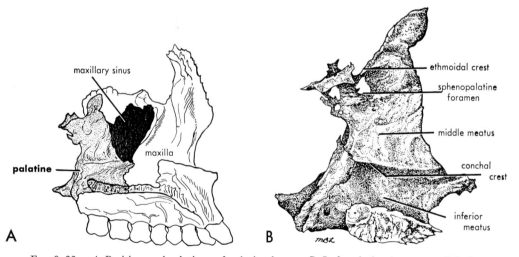

B.   C.

FIG. 8–21.   *A*, Position and relations of lacrimal bone.   *B*, Right lacrimal bone—lateral view. *C*, Right lacrimal bone—medial view.

tribute to the anterior side of this groove to form the **lacrimal sulcus** which lodges the **lacrimal sac** and its **duct**. The lacrimal sulcus leads into the **lacrimal canal** which goes into the lateral wall of the nasal fossa. The lacrimals are ossified in membrane.

PALATINE

The **palatine bones** are L-shaped, consisting of a horizontal plate which forms the posterior portion of the hard palate, and a vertical part in the posterior lateral wall of the nasal fossae (Figs. 8–3, 22). A small part of the bone contributes to the medial portion of the back of the orbit and just below the orbital surface is the **sphenopalatine foramen** for the passage of the sphenopalatine vessels and the superior nasal and nasopalatine nerves. The nasal surface of the bone has a horizontal ridge, the **conchal crest,** for articulation with the inferior nasal concha, and above this a less conspicuous ethmoid crest, for articulation with the **middle nasal concha** of the ethmoid bone. These crests mark the boundaries between the **superior, middle,** and **inferior meatuses** which appear as shallow depressions on the nasal surface of this bone. The palatine is an intramembranous bone.

FIG. 8–22.   *A*, Position and relations of palatine bone.   *B*, Left palatine bone—medial view.

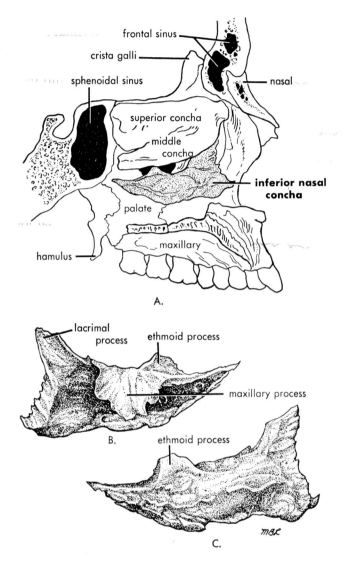

FIG. 8–23.   *A*, Position and relations of inferior nasal concha.   *B*, Left inferior nasal concha—lateral
surface.   *C*, Left inferior nasal concha—medial surface.

INFERIOR NASAL CONCHAE

The **inferior nasal conchae** are scroll-like bones formed in the lateral walls of the cartilaginous nasal capsules (Fig. 8–23). They protrude into the nasal fossae like a shelf beneath which is the inferior meatus and above them the middle meatus. They form a part of the walls of the nasolacrimal canals and of the maxillary sinuses.

HYOID

The **hyoid bone** is horseshoe-shaped and lies just above the thyroid cartilage or "Adam's apple" where it can be palpated by pressing hard with the fingers. It is sometimes called the "tongue bone" (Fig. 8–24). It is connected to the styloid processes of the temporal by the **stylohyoid ligaments.** It consists of a **body** and two **greater**

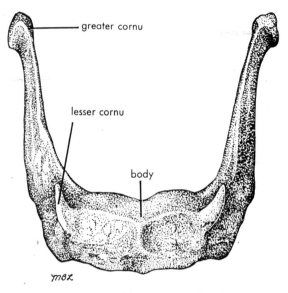

greater cornu

lesser cornu

body

max

Fig. 8–24.   Hyoid bone—anterior view.

and two **lesser cornua** or horns. The body serves for the attachment of many muscles. The greater cornua extend backward from the lateral borders of the body and the lesser cornua attach at the angle of junction between the body and greater cornua. The bone ossifies in the cartilaginous second and third visceral arches. It belongs, therefore, more to the splanchnocranium than to the skull proper.

THE VERTEBRAL COLUMN

Definition and Function (Fig. 8–25). The **vertebral column** is composed of a series of bony units, the vertebrae, between which are pads of **fibrocartilage, the intervertebral discs.** It is located in the mid-dorsal region and serves as the chief axial support of the body and is the key to the posture of the trunk. The skull rests on its superior end; the thorax, and through it the forelimb, is supported by it; and it gives direct attachment to the pelvic girdle. It gives protection to the spinal cord of the central nervous system which passes through a **vertebral canal** formed by the vertebrae.

Openings between adjacent vertebrae, the **intervertebral foramina,** allow for the passage of the paired **spinal nerves** which carry nervous impulses to and from the spinal cord. The vertebral column is also the point of origin and insertion of a great many **skeletal muscles,** some of which produce movements of the column itself; others move parts of the body in relationship to the column.

The vertebral column consists of 33 vertebrae, grouped and named according to the regions of the body they occupy. They are the 7 **cervical vertebrae,** 12 **thoracic,** 5 **lumbar,** 5 **sacral,** and 4 **coccygeal.** These numbers vary in some individuals, though seldom in the cervical region. The vertebrae of the first three regions remain distinct throughout life, those of the sacral and coccygeal regions fuse; the five sacrals to form the **sacrum,** the four coccygeals to form the **coccyx.** On this basis the adult has 26 distinct bones in the vertebral column.

Characteristics of a Vertebra (Fig. 8–26). A vertebra consists of two parts, the anteriorly placed **body** and the posterior **vertebral** or **neural arch.**

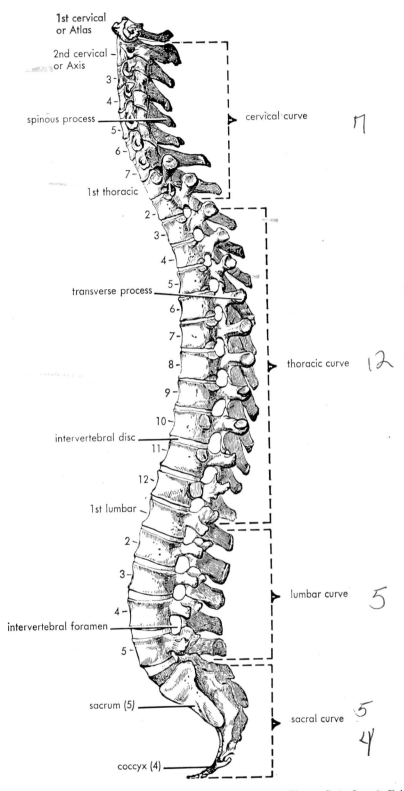

FIG. 8–25.   Lateral view of the vertebral column.   (Gray, *Anatomy of the Human Body*, Lea & Febiger.)

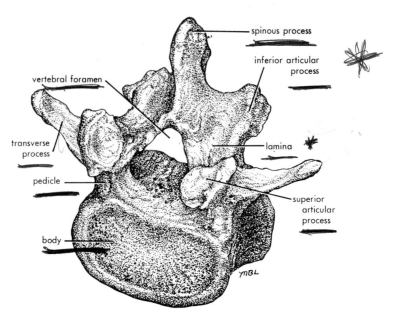

Fig. 8–26.   Lumbar vertebra—superior and posterior view.

They form the walls of the **vertebral foramen.**

The **body** is the thickest part of the vertebra. Its superior and inferior surfaces are flattened and rough, giving attachment to the intervertebral discs. Anteriorly, the body is convex from side to side and concave in the superior-inferior direction. It has a few small foramina for the passage of nutrient vessels. Posteriorly, the body is nearly flat with a slight concavity in the central region in which are one or more large irregular apertures for the passage of the basivertebral veins from the body of the vertebra.

The **vertebral arch** consists of two **pedicles** and two **laminae** from which arise **seven processes.** The **pedicles** are two thick, short structures which arise from the posterolateral sides of the body. Extending posteriorly and medially from the ends of the pedicles to join on the midline are the broad, flat **laminae.** Projecting posteriorly and inferiorly from the point of junction of the laminae is the **spinous process.** Where the lamina and pedicle meet on each side a process ex-tends laterally, the **transverse process.** The **articular processes** are four in number, two **superior** and two **inferior,** and they also arise from the point of junction of the pedicles and laminae. Their articular surfaces are covered with a layer of hyaline cartilage. The spinous and transverse processes serve for the attachment of muscles and ligaments, the articular process for the joining of adjacent vertebrae.

Regional Differences (Figs. 8–27 to 31). While it is true that all vertebrae are built on the same general plan, it is nevertheless possible to assign any single vertebra to its proper region. This is so because of the different demands placed upon vertebrae at different levels in the vertebral column. The **bodies,** for example, are the weight-bearing parts of the vertebrae, and as the column descends to the lumbar region, each vertebra has to carry more weight, hence the bodies become more and more massive. The intervertebral discs increase correspondingly in size and are thickest in the lumbar region.

The **vertebral canal** varies only slightly

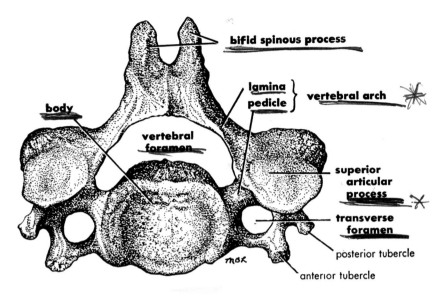

FIG. 8–27. A cervical vertebra—superior view.

in size from cervical to lumbar regions, as the structure it contains, the spinal cord, is approximately the same size throughout. However, where the cord does increase slightly in diameter, we see larger vertebral foramina to accommodate it. Therefore, the cervical and lumbar vertebrae have somewhat larger and triangular vertebral foramina for the cervical and lumbar enlargements of the cord, while the thoracic canal is smaller and rounder.

The **spinous processes** show a considerable regional differentiation. Those in the cervical region are short and tend to be bifid, *i.e.*, to have a double tip; the thoracic spines are long and slender and project downward more than posteriorly to overlap like shingles; the lumbar spines are massive and square and project posteriorly. The types of spines of each region only gradually change over from one type to another, hence we find intermediate forms.

Transverse processes also vary. Those in the cervical region are most distinctive, as they have a **transverse foramen** for the passage of the vertebral arteries and veins. The anterior part of this foramen is formed by a small rib element which thus becomes a part of the vertebra. Thoracic vertebrae can be identified by the articular facets or demifacets for ribs on most of their transverse processes, as well as on their bodies. In the lumbar vertebrae, the transverse processes have neither foramina nor articulating facets. The processes come straight out to the sides and are quite long and slender.

Superior and **inferior articular processes** differ considerably in detail up and down the vertebral column. The most obvious differences are the greater distances between right and left processes in the cervical region; and in the lumbar region, the way the superior processes turn inward to grasp the outward-turning inferior processes of the next vertebra. This arrangement makes rotation impossible in the lumbar part of the vertebral column.

Specialized Vertebrae (Figs. 8–28, 31). A few of the vertebrae have become highly modified, and while most of them possess the basic vertebral structures, their special adaptations are of sufficient importance to warrant attention.

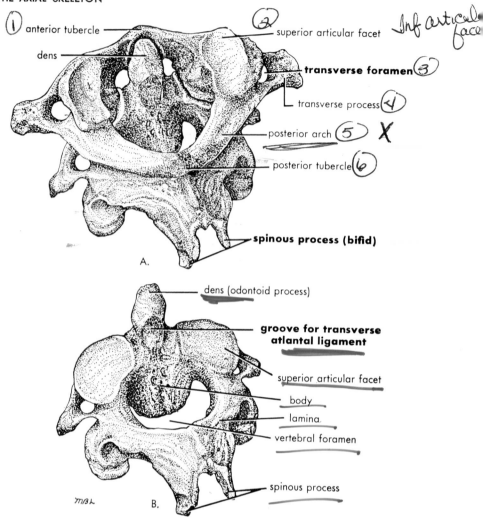

① anterior tubercle

dens

② superior articular facet *Inf articul face*

transverse foramen ③

transverse process ④

posterior arch ⑤ **X**

posterior tubercle ⑥

spinous process (bifid)

A.

dens (odontoid process)

groove for transverse atlantal ligament

superior articular facet

body

lamina

vertebral foramen

spinous process

mBL   B.

FIG. 8–28.   *A,* Atlas and axis in articulated position.   *B,* The axis from behind and above.

The **first cervical vertebra,** the **atlas,** is a wide, ring-like bone which supports the skull. It has lost its body and consists of two **lateral masses** joined anteriorly and posteriorly by arches. The superior surfaces of the lateral masses have elliptical, concave facets which articulate with the **occipital condyles** of the skull. It is at this articulation that we perform the nodding or "yes" movements of the head. The inferior surface of the lateral masses has circular, flat **facets** for articulation with the second cervical vertebra, the **axis.** The **transverse processes** are wide and their **transverse foramina** are large. On the medial side of each lateral mass, a short distance behind the anterior arch, is a small tubercle for the attachment of the **transverse atlantal ligament.** This ligament divides the vertebral foramen into unequal parts—a small anterior compartment that receives the **dens** or **odontoid process** of the second cervical vertebra, a larger posterior compartment that transmits the spinal cord and its covering membranes.

The anterior arch forms about one-fifth of the ring of the atlas. On its convex anterior surface is the **anterior tubercle** for muscle attachment; on its

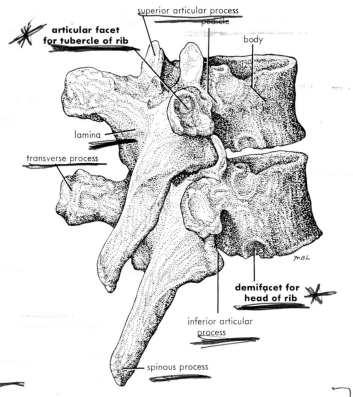

articular facet
for tubercle of rib

superior articular process
pedicle

body

lamina

transverse process

demifacet for
head of rib

inferior articular
process

spinous process

FIG. 8–29. Two thoracic vertebrae—intervertebral disc removed.

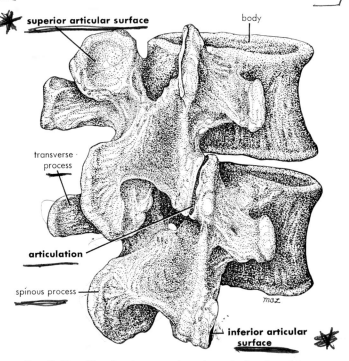

superior articular surface

body

transverse
process

articulation

spinous process

inferior articular
surface

FIG. 8–30. Two lumbar vertebrae in articulated position.

articulates ilium

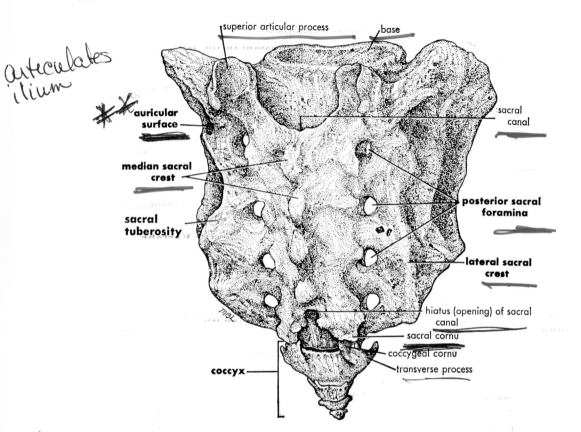

superior articular process · base

auricular surface · sacral canal

median sacral crest

posterior sacral foramina

sacral tuberosity

lateral sacral crest

hiatus (opening) of sacral canal

sacral cornu

coccygeal cornu

coccyx · transverse process

FIG. 8–31.   Posterior view of sacrum and coccyx.

concave posterior surface is a smooth oval facet for articulation with the dens of the second cervical vertebra. The **posterior arch** forms about two-fifths of the ring of the atlas. It has on its posterior surface a **posterior tubercle,** which represents a rudimentary **spinous process** and serves for muscle attachment.

The second cervical vertebra or **axis** is easily distinguished by its prominent **dens** (*odontoid*) which projects upward from its body. The dens is, in reality, the **body of the atlas,** as embryological studies demonstrate. By fitting into the anterior compartment of the vertebral foramen of the atlas, where it is held by the transverse atlantal ligament, it serves as a pivot around which the atlas turns. This mechanism serves as the basis for the rotation or "no" movement of the head.

The **body** of the axis is prolonged downward anteriorly, where it overlaps the superior part of the body of the third cervical vertebra. The transverse processes are very small and the spinous process is large, strong and bifid.

The **seventh cervical** or **vertebra prominens** is distinctive because of its long and nearly horizontal **spinous process.** This process is not bifid, but has a tubercle to which the lower end of the ligamentum nuchae is attached. The transverse foramina vary, sometimes being small, occasionally double, or even absent. The usual arrangement is for the vertebral arteries and veins to pass in front of, rather than through, the foramen, though there are many variations.

The **sacrum** is a large triangular bone made up of five vertebrae and their intervertebral discs which have fused together (Fig. 8–31). The lines of fusion can easily be seen on the concave **anterior surface.** Lateral to these lines of fusion are the four pairs of **anterior sacral foramina.** They give passage to the anterior divisions of the sacral nerves. Lateral to the foramina, the sacrum is grooved for the sacral nerves and the intervening elevated areas give origin for the Piriformis muscles.

The **posterior surface** of the sacrum is convex, and marked by prominent crests and grooves. The **middle sacral crest** is on the midline and the three or four tubercles extending from it represent reduced spinous processes. Lateral to the middle sacral crest is the **sacral groove.** The **posterior sacral foramina** are obvious and transmit the posterior divisions of the sacral nerves. Near the apex are the **sacral cornua** which represent the elongated inferior articular processes of the fifth sacral vertebra.

The **lateral surface** of the sacrum is broad above but little more than a ridge below. The upper part has a large, ear-shaped area, the **auricular surface** for the articulation of the ilium. Posterior to it is a rough surface with three deep impressions, the **sacral tuberosity,** for the attachment of the sacroiliac ligament.

The base of the sacrum is broad and is directed upward and forward to its articulation with the fifth lumbar vertebra by an intervertebral fibrocartilage. Its superior articular processes resemble those of the lumbar vertebrae. Lateral to the body are large, triangular surfaces, the **alae,** which support the Psoas major and the lumbosacral trunk. The posterior part of each ala represents the transverse process, the anterior parts the costal processes.

The **coccyx** consists of three to five rudimentary vertebrae (Fig. 8–31). The first is the largest and has forward-projecting articular processes, the **coccygeal cornua,** which join the sacral cornua to

complete on either side, the foramina for transmission of the posterior branches of the fifth sacral nerves. The remaining coccygeal vertebrae diminish in size and consist mostly of reduced bodies with small tubercles representing articular processes. The last three are usually fused into one bone, while the first may or may not fuse with them. The coccyx gives attachment to Coccygeal, Gluteus maximus, and External sphincter ani muscles, and to the sacrotuberous and sacrospinous ligaments.

Curvatures of the Vertebral Column. When viewed from the anterior or posterior sides the vertebral column is almost straight, except for a slight **lateral** curve to the right side.

When viewed from the side the column presents four normal curves named for the regions involved, **cervical, thoracic, lumbar** and **sacral** or **pelvic** (Figs. 8–25, 32). The cervical curve is slightly convex forward, and extends between the second cervical and the second thoracic vertebrae. The thoracic curve is concave forward and extends through the remainder of the thoracic region. The lumbar curve is convex forward and is more pronounced in the female than in the male. It ends at the angle formed by the articulation of the fifth lumbar and the sacrum. The sacral or pelvic curve is concave forward and downward and ends at the apex of the coccyx.

In the fetus the vertebral column has but one curve, concave anteriorly (Fig. 8–32). Since the thoracic and pelvic curves of the adult are in the same direction as the fetal curve, they are often called the **primary curves.** The cervical and lumbar curves being in the opposite direction and acquired after birth are called **secondary curves.** The cervical curve develops around the third or fourth month after birth as the child begins to hold his head up and at nine months to sit upright. When the child begins to stand and walk, at from twelve

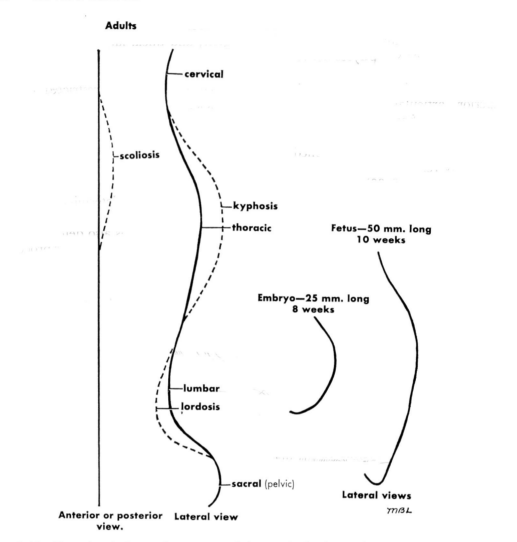

Adults

cervical

scoliosis

kyphosis

thoracic

Fetus—50 mm. long
10 weeks

Embryo—25 mm. long
8 weeks

lumbar

lordosis

sacral (pelvic)

Lateral views
mßL

Anterior or posterior   Lateral view
view.

FIG. 8–32.   Normal and abnormal curvatures of the vertebral column of man.   Normal curves are shown by solid lines; abnormal curves by broken lines.

to eighteen months, the lumbar curve gradually develops.

Abnormal curves of the vertebral column are not uncommon (Fig. 8–32). An abnormal lateral curvature, or **scoliosis,** may be the result of muscular paralysis on one side of the body, disease of the vertebrae, muscular imbalance or poor posture.   An excessive posterior convexity, a **kyphosis** or **"hump-back,"** is commonly, though not always, due to tuberculosis in one or more of the vertebral bodies.   The diseased bodies are weakened, eaten away and crushed by the superimposed weight of the body. Muscular imbalance and poor posture may also contribute to kyphosis.   An exaggerated lumbar curve is called **lordosis.**   While it may be the result of causes already mentioned above, it is believed by some to be due to the wearing of excessively high heels and therefore may be more common in women.   It is a frequent cause of backache.   Rickets is often a cause of these abnormalities.

Articulations and Movements of the

Vertebral Column (Figs. 8–25, 28, 33, 34, 35). The principal articulations of the spine are the **symphyses** between the bodies of adjacent vertebrae, and the **synovial joints,** between the inferior and superior articular processes. Special articulations between the atlas and occipital bones, and between the atlas and axis have been mentioned and their bony relationships and movements described (Fig. 8–33). The thoracic vertebrae articulate to the ribs by synovial joints, and the sacrum with the ilia, at the **sacroiliac joints.**

The **symphyses** consist of compressible **fibrocartilaginous discs** between the bodies of the vertebrae. The discs are thin in the cervical region; larger in sur-face area, but thinner in the thoracic region; and thick, large, and resilient in the lumbar region. For these reasons movement is greatest in the cervical and lumbar regions, more restricted in the thoracic. Also, the rib attachments and the long, sloping and overlapping spinous processes further restrict movement in the thoracic region.

The amount of movement at the **synovial joints** between articular processes is, of course, determined to some degree by limitations on movement at the symphyses. It is also determined by the relationships of the articular processes themselves. Those in the cervical region are directed, in general, upward and downward; those in the thoracic region

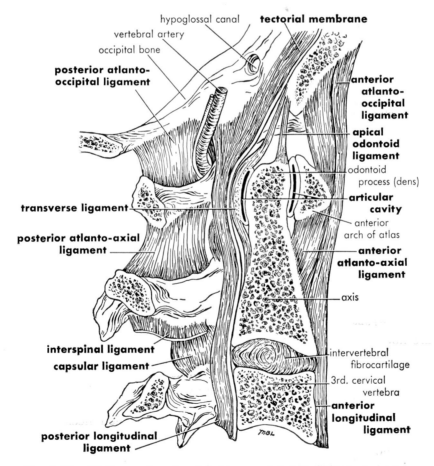

FIG. 8–33.   Median section of occipital bone and first three cervical vertebrae to show ligamentous attachments.

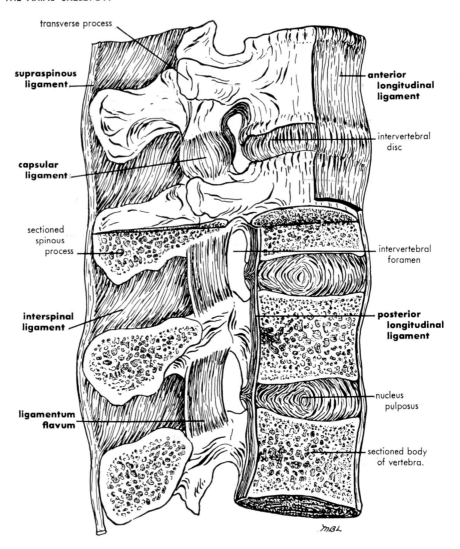

FIG. 8–34. Ligaments of the vertebral column as seen in lumbar region. The lower two and part of another vertebra are shown in median section.

anteriorly and posteriorly; in the lumbar region the surfaces look medialward and lateralward. Consequently, in the neck region movement is relatively free in many directions—flexion, extension, abduction, adduction and rotation. It makes possible a wide range of head movements. In the thoracic region, all of these movements are possible but on a restricted scale. In the lumbar area, because of the way in which the articular processes grasp one another, rotation is almost impossible. Yet, because of the very thick discs, flexion and extension are much more extensive than in any other part of the column. Abduction and adduction also take place. While movements between any two vertebrae are not great, they do add up to a considerable movement over the entire spine.

The **ligaments,** in binding together and strengthening the units of the backbone, also limit its movement (Fig. 8–34). The **anterior and posterior longitudinal**

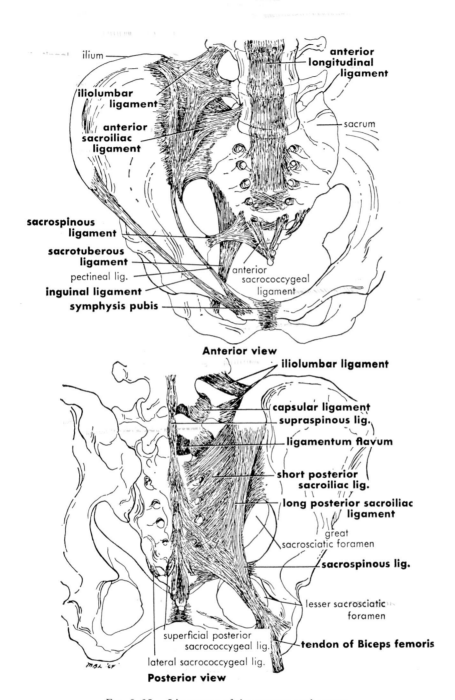

FIG. 8–35.   Ligaments of the sacrum and coccyx.

ligaments run the length of the backbone in front and back. **Lateral ligaments** connect the bodies of adjacent vertebrae. There are also restraining ligaments running between the arches. The elastic **ligamenta flava** stretch between adjacent laminae. They not only help to bring the column back into the extended position after it has been flexed, but they serve to complete the wall of the vertebral canal posteriorly and to protect the spinal cord. **Interspinous ligaments** run between adjacent spinous processes, and **supraspinous ligaments** connect over a series of spinous processes. In the cervical region, the supraspinous ligament enlarges to form the **ligamentum nuchae,** an elastic ligament of special importance in lower animals, where it aids in holding up their heads.

The **sacroiliac joint** is a dual structure, one part **synovial,** the other **fibrous.** The auricular surfaces of the sacrum and corresponding surfaces on the ilium are covered with articular cartilage and fit very closely together, allowing only a minimum of movement. A small synovial cavity is present. In later years of life the two surfaces may actually adhere. Above and behind the auricular surface of the ilium is a broad, roughened area, the **tuberosity of the ilium.** Between this and the sacral tuberosity is a considerable cleft which is filled by a large mass of fibers constituting the strong **interosseous** and **posterior sacroiliac ligaments.** In addition, an **iliolumbar ligament** extends from the iliac crest to the transverse process of the fifth lumbar vertebra (Fig. 8–35). There are also some secondary ligaments.

It is the fibrous part of the sacroiliac joint that receives most of the weight of the body and transmits it to the hip bone. It is under constant stress except when the individual is in the recumbent position. The synovial joint, by itself, would be too weak to maintain its own integrity. Because of the great weight that the sacroiliac joint carries, it is quite subject to strain, especially in pregnant women. The resulting pain may be intense, usually radiating down the back of the thigh.

Development of the Vertebral Column. The notochord, listed earlier as one of the diagnostic characteristics of chordates, forms a central axis around which the vertebral column develops. The vertebral column develops from mesodermal material, the **sclerotomes,** which are derived from the medial side of the primitive somites (see Figs. 4–12, 13). These sclerotomes form into a **sclerotogenous layer** along both sides of the notochord and neural tubes. Proliferation of the cells of this layer results in surrounding the notochord and neural tube with membranous material, the **membranous vertebral column** (Fig. 8–36). Segmentation is still apparent in this material. There follows a complicated modification and shifting of materials in this membranous column. There then develop two pairs of cartilages, one in the membranous material lateral to the notochord, the other lateral to the neural tube. The former extends around the notochord to form the cartilaginous body of the vertebra, the latter around the neural tube to form the neural arch. A spinous process develops where the two sides of the arch meet on the midposterior line. Separate cartilaginous centers appear to produce costal processes, while transverse processes grow out from the vertebral arch behind the costal processes (Fig. 8–36). Intervertebral fibrocartilages lie between the developing bodies of the vertebrae. The notochord, surrounded by the developing bodies of the vertebrae, ultimately disappears, while that in the region of the fibrocartilages remains as a **nucleus pulposus** within them (see Fig. 1–2).

In general, ossification of the cartilaginous vertebrae is from three primary centers, one for the body and one each

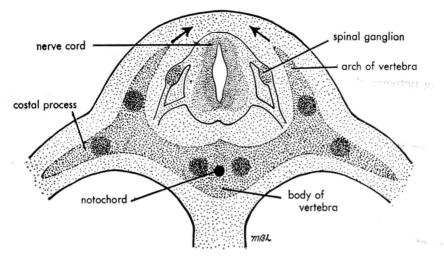

FIG. 8–36. Mesenchyme stage in the development of a vertebra. Six centers of cartilage formation will appear in the areas shown in close stipple. Bone will replace cartilage. (Modified from Dodds.)

for each side of the neural arch and transverse process. At birth the vertebrae are in three parts, a body and the two halves of the arch. During the first year, the two halves of the arch unite behind, but it is not until the sixth year that the arches have all united to the bodies. About the sixteenth year, five secondary centers of ossification appear, one at the end of each transverse process and at the end of the spinous process, and one each on the lower and upper surfaces of the body. These do not join the rest of the vertebra until about the twenty-fifth year of life.

There are, of course, exceptions to the above description in some of the special vertebrae of the column. Consideration of these is beyond the scope of this book.

Anomalies. Sometimes the laminae of the vertebral arches do not meet and join on the midline. This leaves a cleft through which the meninges (*membranes*) and often the spinal cord may protrude. It may be limited to the lumbosacral region, which is most common, to the thoracic and cervical areas, or may involve the whole column. It is known as **spina bifida.** The fifth lumbar is sometimes fused to the sacrum (*sacraliza-*

tion) and the costal processes of certain vertebrae may develop into **supernumerary** ribs.

THE THORAX

Scope and Functions. The skeleton of the **thorax** or **chest** is composed of the bony **sternum, ribs,** and the **costal cartilages.** The thoracic vertebrae complete the thorax wall behind. Its shape is conical with the small end, the **inlet,** above and the broad **outlet** below (Fig. 8–37). It is flattened from front to back and has its longest measurement posteriorly. In transverse section it is kidney-shaped, because the ribs swing backward from the thoracic vertebrae before they turn forward again. Viewed laterally, the thoracic "cage" slants forward and downward. The lower opening of the thorax is closed by the dome-shaped diaphragm, which forms its floor. Anteriorly, the costal cartilages of ribs seven through ten turn upward to the sternum and form the **subcostal angle.** In the living individual the spaces between the ribs are closed with the external and internal intercostal muscles.

This rather substantial thoracic struc-

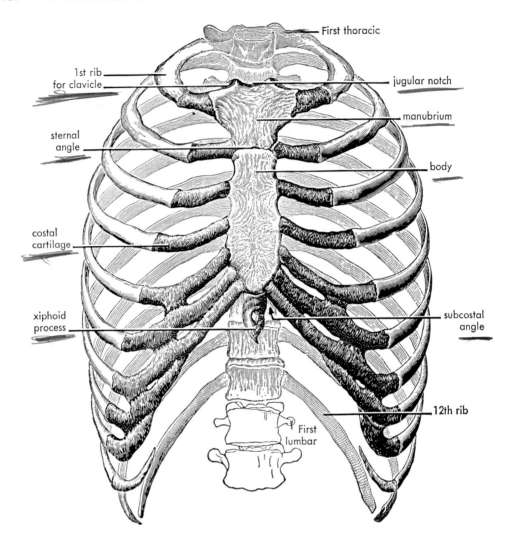

First thoracic

1st rib
for clavicle

jugular notch

manubrium

sternal
angle

body

costal
cartilage

xiphoid
process

subcostal
angle

12th rib

First
lumbar

FIG. 8–37.   The thorax from in front.   (Spalteholz.)

ture (1) houses and gives protection to vital organs of the circulatory and respiratory system such as the heart and major blood vessels and the lungs. (2) It gives support to the shoulder girdle of the appendicular skeleton and attachment to many extrinsic muscles of the upper limb and the vertebral column and skull. (3) It also plays an important role in the process of breathing.

The **Sternum** (Figs. 8–37, 38).   The **sternum** or breast bone lies in the middle region of the anterior wall of the thorax. It is flat and elongated and consists of

three parts, the superior **manubrium,** the middle portion or **body** (*gladiolus*), and the inferior **xiphoid** (*ensiform*) **process.**   The manubrium is broad above and narrows below to join the body. Its superior border has a shallow notch medially called the **jugular** or **presternal** notch.   Lateral to the notch on each side is an oval articular surface for the sternal end of the clavicle of the shoulder girdle.   The lateral border has a depression above for attachment of the **costal cartilage** of the first rib and below a small **demifacet,** which, with a similar one on

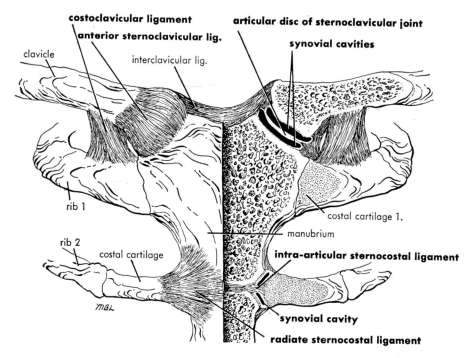

Fig. 8–38.   Anterior view of sternoclavicular and sternocostal articulations.   The left side is shown in frontal section to reveal the synovial cavities.

the body (*gladiolus*), forms a facet for the attachment of the second costal cartilage. Where the manubrium and body join they form a slight transverse elevation, the **sternal angle.** This is easily palpated and is an important surface feature or point of reference by which the position of soft structures within the chest can be located.

The **body** of the sternum is broadest at its lower end. Its lateral borders have demifacets at the top and bottom and between these, four facets on each side. These serve for the attachment of costal cartilages two through seven.

The **xiphoid process** is small and varies in form; it may be broad, bifid, pointed, curved, perforated or deflected to one side. With the body it forms a part of the facet for the seventh rib cartilage and at its inferior end gives attachment to the **linea alba** which marks the midline of the abdomen. The xiphoid process also is a landmark for

measurements of chest girth in weight analysis or the excursion of the chest in breathing.

The Ribs and Costal Cartilages (Figs. 8–37, 38).   There are twelve pairs of ribs and costal cartilages. The first seven pairs, the "**true**" ribs, articulate through their costal cartilage to the sternum. The last five pairs are called "**false**" ribs, since their costal cartilages do not articulate directly to the sternum. Of the "false" ribs, numbers eight, nine and ten have their cartilages articulating to the next cartilage above them, while the cartilages of eleven and twelve are small and end free in the body wall. These last two pairs of "false" ribs are for this reason often called "**floating**" or **vertebral ribs.** The ribs increase in length from the first to the seventh, then decrease to the twelfth.

A typical rib, such as number six, has a **head** divided by an **interarticular crest** into two facets for articulation with the demifacets of two adjacent thoracic vertebrae (Figs. 8–39, 40). The intervening interarticular crest attaches to the intervertebral disc by an **interarticular** ligament. Next to the head is the short neck of the rib which ends at an eminence, the **tubercle.** The tubercle is divided into an articular and a non-articular portion. The articular portion is the lower and more medial of the two and articulates with a facet on the transverse process of a corresponding thoracic vertebra. The non-articular portion is for the attachment of a ligament.

The next and major part of the rib is the body or shaft which, a short distance from the tubercle, bends rather sharply to form the **angle.** The superior border of the body of the rib is rounded, the inferior border is sharper. The external surface is, in general, convex and smooth. The internal surface is smooth and is marked by a **costal groove** for the intercostal vessels and nerves. The anterior extremity of the rib is flattened and has on its end an oval, concave depression for the reception of the costal cartilage.

As one would expect, some of the ribs

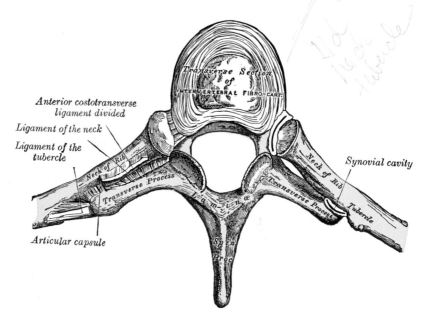

FIG. 8–39.   Costotransverse articulation.   (Seen from above.)
(Gray, *Anatomy of the Human Body*, Lea & Febiger.)

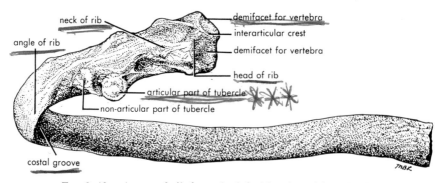

FIG. 8–40.   A central rib from the left side; viewed from behind.

show variations from the typical rib. The first two and the last three require some special comment (Figs. 8–41, 42).

The **first, tenth, eleventh** and **twelfth** ribs have only a single articular facet on their heads for articulation with the thoracic vertebrae. The **eleventh** and **twelfth** have no necks or tubercles. The **twelfth** may be without an angle and costal groove and may sometimes be even shorter than the first rib. The **first** rib is ordinarily the shortest rib and most curved though it lacks the angle. It is broad and flat and its surfaces are superior and inferior rather than anterior and posterior. The superior surface has

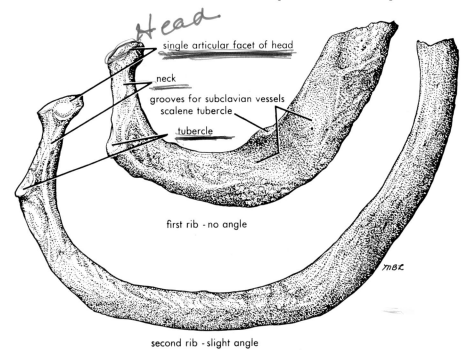

Head

single articular facet of head

neck

grooves for subclavian vessels
scalene tubercle

tubercle

first rib - no angle

second rib - slight angle

Fig. 8–41.   Drawings of first and second ribs showing special features.

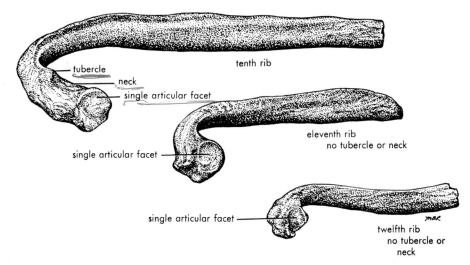

tubercle

neck

single articular facet

tenth rib

single articular facet

eleventh rib
no tubercle or neck

single articular facet

twelfth rib
no tubercle or
neck

Fig. 8–42.   The last three ribs—all have angles.

two shallow grooves for the subclavian artery and vein. Between the grooves is a slight ridge ending on the anterior border in the **scalene tubercle.** The inferior surface lacks the costal groove. The second rib is transitional in character between the first and the typical ribs.

The costal cartilages are of the hyaline variety and give considerable elasticity to the thoracic cage. In old age they usually show some superficial ossification.

Movements of the Thorax. The most vital of the movements of the thorax are those involved in breathing. The inspiration of air depends upon the active contractions of a variety of muscles which increase all the diameters of the thorax. The ribs are so articulated to the thoracic vertebrae by synovial joints and to the sternum through their costal cartilages that when they are elevated they swing laterally and forward so that the transverse diameter of the thorax is increased markedly and the anteroposterior diameter to a lesser extent. At the same time the contraction of the diaphragm increases the vertical diameter of the thorax. As a result of this increase in volume of the thorax and the decrease in intrathoracic pressure, air rushes into the lungs. Expiration is passive and takes place as the muscles of inspiration are relaxed. The elastic nature of the hyaline cartilage aids by recoil in expiration as do the elastic fibers in the lungs themselves. Expiration may be made an active process if one so chooses by contractions of the muscles of the abdominal wall. This places pressure upon the viscera which in turn forces the diaphragm upward.

Development of the Thorax. The ribs are formed from the costal processes of the vertebrae which extend out between the muscle plates. They later become separated from the vertebrae by the development of a joint between them, the **costocentral** joint. It is only in the thoracic region that ribs normally develop as separate bones. The costal processes in the cervical region form the anterolateral boundary of the transverse foramen. In the lumbar region they contribute to the transverse process and in the sacral region to the lateral masses of that bone. The coccygeal vertebrae and fifth sacral are without costal processes. Ossification in most ribs takes place in four centers, a primary center for the body and three secondary centers, one each for the head, articulating, and non-articulating tubercles. The eleventh and twelfth ribs, lacking tubercles, have only two centers of ossification. While ossification begins in the body of the rib early in fetal life, the epiphyses for the head and tubercles do not appear until the sixteenth to twentieth year and do not unite to the body of the rib until the twenty-fifth year. The costal cartilages are essentially unossified parts of the ribs.

The sternum is formed from longitudinal bars, the **sternal plates,** formed by the joining of the anterior ends of the ribs. These bars unite opposite the first seven pairs of ribs to form the manubrium and body while the xiphoid grows at the the lower end of the plate. Ossification takes place in six centers, one for the manubrium, four for the body and one for the xiphoid process. Union of the various centers of the body starts at puberty and is completed by the twenty-fifth year. The xiphoid process may join the body by the age of thirty but more commonly not until forty. In some cases it does not join at all. The manubrium is separated from the body by an intervening cartilage and only in later life does it join to the body by bone.

Both the sternum and ribs are composed of a highly vascular cancellous tissue covered by a thin layer of compact bone. They are important centers of blood manufacture.

Applied Anatomy. The sternum, like other bones, is subject to fracture, disease, and deformities. The site of the fracture is most often at the sternal angle and may be caused by a blow or exaggerated movements in the thoracic region. If these parts are displaced to any great extent posteriorly, injury might be done to the trachea, vessels, heart or lungs. Rickets, during the development of an individual, often results in deformities of the thorax, such as **"pigeon breast"** when the sternum pushes too far forward, or **"rachitic rosary,"** in which bead-like enlargements occur at the junctions of the ribs and their costal cartilages. These are clearly visible externally on a lean individual. Tuberculosis often results in a flat-chested appearance as the body becomes emaciated. Sometimes the paired sternal plates fail to join and the sternum is **cleft** or **bifid,** or a partial fusion might leave a foramen, a **perforated sternum.** Ribs are also frequently fractured if not by direct violence, by coughing or sneezing. The fourth to the eighth ribs are the most frequently fractured. The fractured ends may show little if any displacement or in severe cases they may puncture the pleural cavities, the lungs, or blood vessels. Fracture of a rib may be accompanied by localized pain often intensified by the breathing movements. The chest is usually taped to immobilize the ribs.

Extra ribs are not uncommon, especially on the seventh cervical vertebra where they may interfere, in a serious and painful way, with the brachial plexus of nerves, or with the subclavian vessels. A rib on the first lumbar is quite frequent and it may cause pressure on spinal nerves in that area. The appearance of extra ribs is not surprising, knowing what we do about their embryology and comparative anatomy.

QUESTIONS

1. Name the bones which form the anterior margin of the orbital fossa.
2. What bones form the walls of the nasal fossa?
3. What bones constitute the hard palate? The nasal septum?
4. Of what bone are the following a part? sella turcica; posterior clinoid process; foramen ovale.
5. What structure is found in the sella turcica in the live individual?
6. What part of the temporal bone houses the ear ossicles; the organ of Corti; the semicircular canals?
7. Name and locate five sutures of the skull.
8. What is the most common type of articulation to be found in the adult skull? What other type is present?
9. Name the important foramina in the mandible.
10. What constitutes the zygomatic arch?
11. Name the sinuses of the skull. Where do they drain?
12. How could one insert a probe into the middle ear (*tympanic*) cavity without puncturing the tympanic membrane?
13. Of what use are the foramina in the cribriform plate?
14. Name the cranial fossae in the floor of the skull.
15. Name the nerves which pass through the superior orbital fissure.
16. All of the teeth on one side of the mandible can be anesthetized by one injection. Explain.
17. Distinguish between chondrocranium and dermatocranium.
18. What contribution does Meckel's cartilage make to the adult skull?
19. Compare some of the important features of the skull of the newborn with those of the adult skull.
20. Name one bone of the skull which in its development receives contributions from the chondrocranium, dermatocranium, and splanchnocranium.

21. Name the intramembranous bones of the cranial wall.
22. Name the parts of a typical vertebra.
23. What are the primary curves of the vertebral column?
24. Name those abnormal curvatures which often occur in the human spine.
25. Describe the special structural features of the atlas and axis.
26. What are the special demands which are served by the sacrum?
27. What two general types of joints are found in the vertebral column?
28. Give an example, from the vertebral column, of how movement may be limited by articulatory bones.
29. Name the components of the thorax.
30. Name the parts of the sternum.
31. What structures articulate with the manubrium and body of the sternum?
32. Describe a typical rib.
33. Describe the movements of the thorax as they contribute to breathing.

"A little picture is worth a million words"

—Chinese Proverb

Chapter 9

The Appendicular Skeleton

DEFINITION AND FUNCTIONS

THE appendicular skeleton consists of the two **girdles, pectoral** (*shoulder*) and **pelvic** (*hip*), and the paired limbs which they support. The **pectoral girdle** attaches to the axial skeleton by way of the sternum only, making it relatively unstable and weak, but highly versatile in terms of movements. The instability is not a problem because in man the **upper limb** (*forelimb*) is no longer used in locomotion nor to support the body.

139

Rather, it is freed to serve as one of man's most important tools for manipulating the external environment and fashioning it to his needs and desires. The upper limb and its supporting girdle constitute a complex mechanism, the actions of which can be understood only by a careful analysis of each of its parts and the relationships which they bear to the axial skeleton. These actions range from the highly complicated and coordinated movements of the wrist and fingers to the less complex but important movements of forearm, arm, and shoulder.

The **pelvic girdle,** in contrast to the pectoral, is constructed for strength and stability. It attaches directly to the sacrum by a slightly movable joint which must bear the weight of the superimposed body parts. The girdle in turn transfers the weight to the lower limb (*hindlimb*) at the femoral joint. The lower limb, consisting of thigh, leg, and foot, is supported and given stability by strong bones and joints and powerful muscles. More delicate muscles provide for the intricate movements of the foot. Since the body must balance on the two lower limbs in standing, or on each one alternately in ambulation, the demands for strength, stability, motility, and balance are evident (Fig. 7–1).

PECTORAL GIRDLE

The pectoral girdle is superimposed upon the thorax. It consists of two pairs of bones, the anterior **clavicles** (*collar bones*) and the posterior **scapulae** (*shoulder blades*).

CLAVICLES

The **clavicles** are double-curved bones which articulate medially with the manubrium of the sternum and laterally with the scapulae (Fig. 7–1). The medial ends are rounded, the lateral ones are broad and flat. The inferior surface has

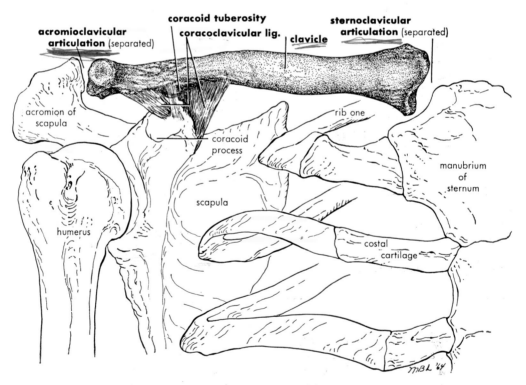

FIG. 9–1. The right clavicle, showing its relationship to the shoulder and thorax.

Clavicle

a prominence called the **coracoid tuberosity** near the lateral end for the attachment of the **coracoclavicular ligament** (Fig. 9–1). The clavicle is subcutaneous and hence very vulnerable to blows on the shoulder. Also, because it is the only bone of the pectoral girdle which is joined to the axial skeleton, it gets the full force of falls on the outstretched upper limb. For these reasons, it is one of the most frequently broken bones of the body.

The clavicle, by projecting laterally and articulating with the scapula, holds the shoulder out and in a position which allows the free swinging of the arms at the sides. When the clavicle is broken, the whole shoulder collapses. The **sternoclavicular** attachment is a synovial joint with an **articular disc** separating two cavities. The clavicle is also bound securely to the costal cartilage of the first rib by the **costoclavicular ligament.** It allows free movement to the extent that

the lateral end of the clavicle and the attached scapula can be swung forward and backward and upward and downward and also in intermediate directions. The coracoclavicular ligament forms a strong fibrous joint between the coracoid process of the scapula and the clavicle. The **acromioclavicular** joint between the acromion process of the scapula and the clavicle allows only a slight gliding movement.

The clavicle is the first of the bones to begin ossification, two primary centers, one for the medial and one for the lateral parts, appearing during the fifth to sixth week of fetal life. A secondary center appears at the sternal end at the eighteenth or twentieth year and joins with the rest of the bone at about age twenty-five.

SCAPULAE

The **scapulae** are made up of flat, thin, triangular bodies which fit over the

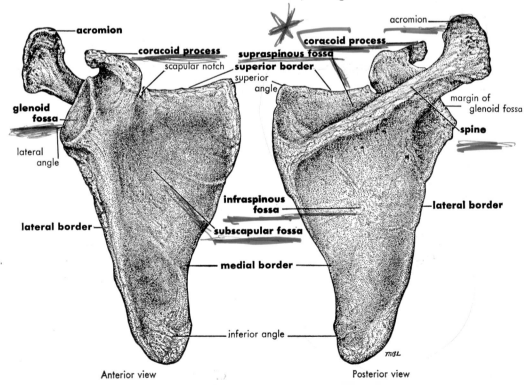

Anterior view　　　　　　Posterior view

FIG. 9–2. The right scapula.

posterior wall of the thorax between ribs two to seven (Fig. 9–2). They have thin, vertical **medial borders** which lie about two inches lateral to the vertebral column. Inferiorly, there is a sharp angle, the **inferior angle,** which continues into the **axillary** or **lateral border.** The lateral border is thick and heavy in order not to buckle under the action of heavy muscles upon it. It is directed laterally and upward to the **lateral angle** which is marked by a shallow depression about 3.5 by 2.3 cm. in size, the **glenoid fossa,** for articulation with the humerus of the arm. This region is the thickest and strongest part of the scapula. Running medially and almost horizontally from the upper part of the **glenoid fossa** is the **superior border.** It is thick at the glenoid fossa just medial to which it gives off a strong **coracoid process** which is bent forward and in the direction of the arm. The remainder of the superior border is thin and ends medially in the **superior angle** which is continuous with the medial border. The coracoid process, as stated previously, is joined to the clavicle by the coracoclavicular ligaments and also serves for attachments of muscles. It has an interesting evolutionary history, since it is believed to be homologous to a coracoid bone which is present as a separate component of the pectoral girdle of some vertebrate animals.

Arising from the body of the scapula posteriorly and about one-third of the way down the medial border is a **spine** which becomes higher laterally and ends in a free, flat **acromion process** to which the clavicle attaches. The free end of the acromion process is the most lateral bony prominence of the shoulder. The depression above the spine is the **supraspinous fossa,** that below it the **infraspinous fossa** which serve for the fleshy attachments of muscles which have similar names. On the anterior face of the body of the scapula is a shallow **subscapular fossa,** for attachment of the Subscapularis muscle.

It is important to remember that the scapulae have no articulation with the axial skeleton and that they "float" in a complex of muscles. For this reason, they are very versatile in their movements and seldom fractured. They are oriented in such a way that their glenoid fossae are directed as much forward as laterally, which is a logical arrangement for the attachment of the arm, which works largely in front of the body rather than behind it.

The spine, acromion, coracoid process and the borders and angles of the scapula, except for the superior angle and border, can be easily palpated and give important landmarks for determining the positions of soft parts in the area.

The scapula ossifies from seven centers, sometimes more. There is one center for the body, one each for the medial border and inferior angle, and two each for the acromion and the coracoid process. The body begins to ossify about the second fetal month and the spine grows up from its dorsal surface about the third month. Ossification in the remaining centers and at the glenoid fossa begins after birth. Ossification is usually completed by the twentieth year and the various parts fully joined by the twenty-fifth year.

UPPER LIMB

The bones of the upper limb consist of the **humerus** in the **arm,** the **radius** and **ulna** in the **forearm,** the eight **carpals** in the **wrist,** the five **metacarpals** in the **palm,** and the fourteen **phalanges** in the five **digits** (Fig. 7–1).

HUMERUS

The **humerus,** the longest and largest bone of the upper limb, articulates proximally with the glenoid fossa of the scapula and distally with the radius and ulna (Fig. 9–3). Its proximal end consists of a large rounded **head,** which joins to the **body** by a constricted area, the **anatomical neck;** and two prominences, the lateral **greater** and the anterior **lesser tubercles.** Between the tubercles is a deep **intertubercular** (*bicipital*) **groove.** The narrow part of the body of the humerus, distal to the tubercles,

is called the **surgical neck** because it is frequently the location of fractures.

The **body** of the humerus is cylindrical in its proximal one-half but becomes somewhat flattened anteroposteriorly and widened at the distal end. Midway down the lateral side is a roughened surface, the **deltoid tuberosity,** for the insertion of the Deltoideus muscle.

The distal or lower end of the humerus

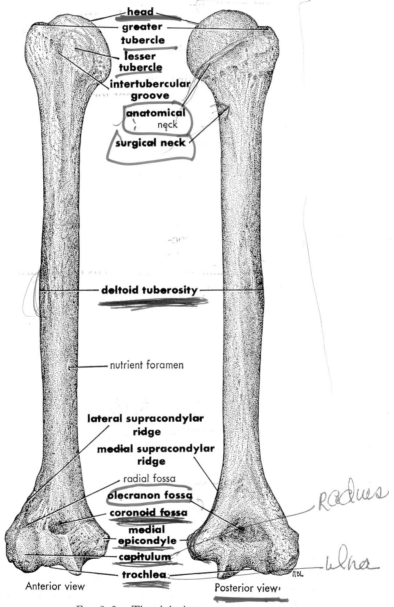

head
greater tubercle
lesser tubercle
intertubercular groove
anatomical neck
surgical neck
deltoid tuberosity
nutrient foramen
lateral supracondylar ridge
medial supracondylar ridge
radial fossa
olecranon fossa
coronoid fossa
medial epicondyle
capitulum
trochlea
Anterior view
Posterior view

FIG. 9–3. The right humerus.

curves forward and has prominent medial and lateral processes, the **medial** and **lateral epicondyles.** Extending proximally from these epicondyles are the **medial** and **lateral supracondylar ridges** which contribute to the width of the lower end of the body or shaft of the humerus. Lateral to the medial epicondyle is a pulley-like surface, the **trochlea,** which fits into a deep, rounded notch on the proximal end of the ulna to form the hinge-like elbow joint. Immediately above the trochlea, on the front of the bone, is a coronoid fossa and on the posterior side an olecranon fossa, for accommodation of corresponding processes of the ulna. Lateral to the trochlea is a small, rounded eminence, the **capitulum,** for articulation in the cup-shaped depression on the head of the radius. Above the capitulum, on the anterior surface of the humerus, is a shallow depression, the **radial fossa,** for receiving the anterior border of the head of the radius when the forearm is flexed.

The humerus ossifies from eight centers,

one for the body, which appears about the eighth week of fetal life and by birth has formed most of the length of the bone. The other seven centers are in the head, the greater and lesser tubercles, the trochlea, the capitulum, and the two epicondyles, and ossification starts in them after birth. By the sixth year the centers for the head and tubercles have joined and together form a single epiphysis which fuses with the body about the twentieth year. At the distal end of the bone, the various centers have joined to the body by the eighteenth year.

SHOULDER JOINT

The **shoulder joint** is a ball-and-socket joint which allows extremely free movement (Fig. 9-4). It is so loosely constructed that it gives little stability and is very frequently dislocated. Its movements should be clearly distinguished from those of the clavicle and scapula, though their movements commonly occur together and are related.

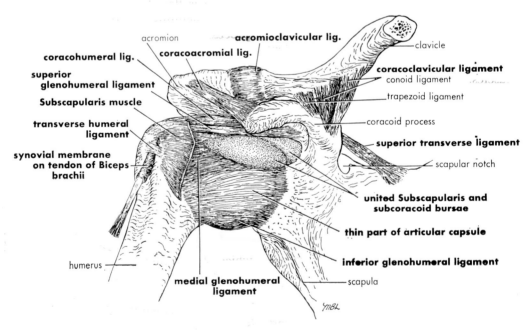

FIG. 9-4. Anterior view of the right shoulder, acromioclavicular joint, and the ligaments of the scapula.

The articular surfaces of this joint are the pear-shaped, shallow glenoid fossa of the scapula and the considerably larger head of the humerus. Their surfaces are covered with articular cartilage. There is relatively little area of contact between these two bones at any time, and always a considerable part of the head of the humerus is in contact with the joint capsule. Though the shallow glenoid fossa does have a rim of fibrocartilage, the **glenoid labrum,** around its periphery to deepen it, it still remains a weak, insecure joint.

The **articular capsule** attaches to the rim of the bony glenoid fossa and to the anatomical neck of the humerus. It is extremely loose and allows the articulating surfaces of the bones to be separated by as much as 2.5 cm. The most prominent of the restraining ligaments is the **coracohumeral ligament** which extends from the coracoid process to the greater tubercle of the humerus to strengthen the upper part of the capsule. The **glenohumeral ligaments** add some strength to the anterior part of the capsule.

Lacking support from capsule and ligaments the shoulder joint is dependent for its integrity upon the surrounding muscles. The Supraspinatus above, the long head of the Triceps brachii below, the tendons of the Teres minor and Infraspinatus behind, and the Subscapularis in front help to strengthen the joint. The long tendon of the Biceps brachii, from its origin on the superior border of the glenoid fossa, passes inside the capsule of the joint.

The shoulder joint is protected above and anteriorly by the coracoid process and above by the acromion of the scapula and by a **coracoacromial ligament.** Beneath many of the muscles of the shoulder joint are fluid-filled spaces, the bursae, which help to cut down friction around moving parts. Some of them, such as the one under the Subscapularis muscle, connect with the synovial cavity of the joint (Fig. 11–35).

The shoulder joint is capable of a wide range of movement: flexion, extension, abduction, adduction, rotation, and circumduction. These will be given further consideration in Chapter 11 on the muscular system.

Ulna

The ulna is the longer of the two bones of the forearm. It lies medial and parallel to the radius. It is thick at its proximal end where it forms a major part of the elbow joint. It diminishes in size, distally, to a small end which articulates only with the radius (Fig. 9–5).

The proximal end of the ulna is marked by a prominent **olecranon process** which, when the forearm is extended, fits into the olecranon fossa at the distal end of the humerus. It forms the tip of the elbow. Distal to the olecranon on the anterior surface of the ulna is another process, the **coronoid,** which, when the forearm is flexed, fits into the coronoid fossa of the humerus. Between these two processes, and formed by them, is the deep **semilunar notch** which fits over the trochlea of the humerus and with it constitutes the hinge-type elbow joint. On the lateral side of the coronoid process is a shallow depression, the **radial notch,** to receive the circumference of the circular head of the radius. Below the coronoid process on the anterior surface is a roughened area, the **tuberosity** of the **ulna,** for the insertion of the Brachialis muscle.

The shaft of the ulna has a sharp lateral border, the **interosseous border,** which faces a similar sharp medial border of the radius and the two are connected in life by an **interosseous membrane.** The small distal end of the ulna is disc-like and from its posterior part there is a small, blunt projection, the **styloid process.**

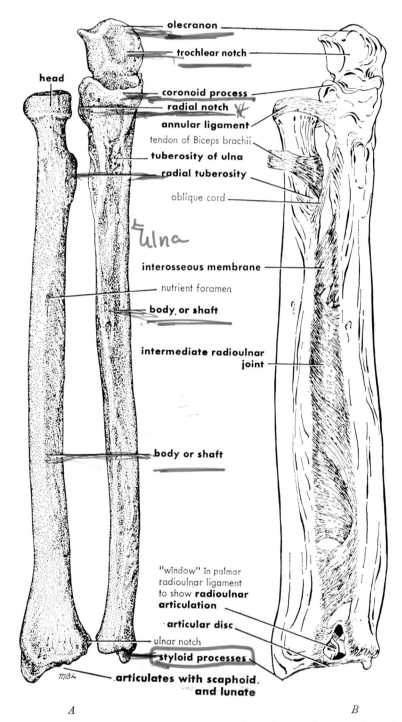

head

olecranon

trochlear notch

coronoid process

radial notch

annular ligament

tendon of Biceps brachii

tuberosity of ulna

radial tuberosity

oblique cord

Ulna

Radius →

interosseous membrane

nutrient foramen

body, or shaft

intermediate radioulnar joint

body or shaft

"window" in palmar radioulnar ligament to show **radioulnar articulation**

articular disc

ulnar notch

styloid processes

articulates with scaphoid, and lunate

A   *B*

FIG. 9–5.   *A,* Anterior view of right radius and ulna.   *B,* Anterior view of proximal, intermediate, and distal radioulnar articulations.

RADIUS

The radius, the lateral bone of the forearm, in contrast to the ulna, is about twice as wide at the distal as at the proximal end. It plays a minor role in the elbow joint, but it alone articulates with the proximal part of the wrist.

The proximal end of the radius is a circular disc with a slight concavity for articulation with the capitulum of the humerus. Its circumference in part fits into the radial notch of the ulna and the remainder into a circular **annular ligament** which holds it in position, yet allows it to turn freely. Below the proximal end and on the medial side is a roughened projection, the **radial tuberosity,** for the insertion of the tendon of the Biceps brachii muscle.

The shaft of the radius shows a con-

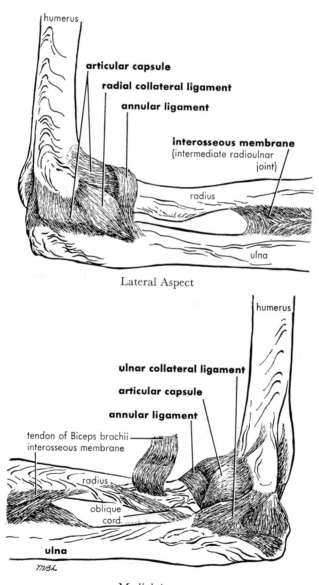

Lateral Aspect

Medial Aspect

FIG. 9–6. The right elbow joint.

vexity along its lateral border, while its medial border is sharp and, as indicated above, is connected to the ulna by the interosseous membrane. The posterior surface immediately above the lower end is convex and marked by a number of vertical grooves for the passage of tendons. The corresponding anterior surface is smooth.

The lower end of the radius shows medially a concave **ulnar notch** into which the head of the ulna fits. On the opposite side is a prominent **styloid process.** Between the ulnar notch and the styloid process is a large, concave surface for articulation with the scaphoid and lunate bones of the wrist. The radius and ulna each ossify from three centers, one for the body and one each for the proximal and distal ends.

Elbow Joint

The elbow joint properly consists of the semilunar notch of the ulna clasping the trochlea, and the shallow end of the radius riding upon the capitulum of the humerus. It is a hinge-type, synovial joint. However, since the articulation between the head of the radius and the radial notch of the ulna (**proximal radioulnar joint**) are included in the same articular capsule as the above and have a common joint cavity, we can consider them together.

As the radial notch of the ulna can accommodate only about one-fourth of the circumference of the head of the radius at one time, the remainder of the circumference is embraced and held in place by a circular **annular ligament** (Fig. 9–5). Within this annular ligament the head of the radius rotates freely, which enables one to pronate and supinate the hand. The diameter of the annular ligament is less below than above to keep the radius from slipping out of it distally. This accident sometimes happens, more commonly in children than adults, when too great a pull is exerted on the hand.

The articular capsule of the elbow joint is thin and loose anteriorly and posteriorly, which permits freedom of movement. Laterally and medially, however, it is thick and strong and forms the **ulnar** and **radial collateral ligaments** (Fig. 9–6).

Intermediate Radioulnar Joint

An intermediate radioulnar joint has already been mentioned in the form of the interosseous membrane which extends between the shafts of the radius and ulna. This membrane, while holding radius and ulna together, does not impede the movement of the radius over the ulna. It also increases the area of attachment both anteriorly and posteriorly for the numerous muscles originating in the forearm (Fig. 9–5).

Distal Radioulnar Joint

The distal radioulnar joint is a pivot joint between the head of the ulna and the ulnar notch on the radius. It is held together by **palmar** and **dorsal radioulnar ligaments.** An **articular disc,** triangular in shape, lies transversely under the head of the ulna, attached to its styloid process and to the lower margin of the ulnar notch of the radius. This disc excludes the ulna from participation in the wrist joint (Figs. 9–5, 8 and 9).

The movements in the joints just described may be summarized as follows: At the elbow joint, between humerus and radius and ulna, flexion and extension are the only significant movements possible. Between the radius and ulna, at their proximal and distal articulations, rotation takes place, resulting in pronation and supination of the hands. Pronation carries the distal part of the radius anterior to the ulna and to its medial side. Supination returns it to the anatomical position.

Skeleton of Hand

The skeleton of the hand consists of the bones of the **carpus or wrist;** the **metacarpus or palm;** and the **phalanges** or bones of the **digits** (Fig. 9–7).

Carpus

The carpus consists of eight bones arranged in two transverse rows of four each. The proximal row from lateral to medial side is made up of the **scaphoid, lunate, triangular** and **pisiform;** those of the distal row, in the same order, are called the **trapezium, trapezoid, capitate,** and **hamate.**

The carpal bones are small and irregular in shape, as can be seen by reference to Figure 9–7. Their names are more or less suggestive of their appearance. The smallest is the pisiform which rests on the front of the triangular, rather than at its side, and can be easily palpated when the wrist is flexed. The bones are closely fitted together and bound by ligaments but they do have synovial cavities and some movement is allowed between them. The greatest movement, however, is between the proximal and distal rows, at the **transverse intercarpal** or **midcarpal joint** (Fig. 9–9). The rows of bones are also placed in such a manner as to form a transverse arch with its concavity anterior. In life, this concavity is bridged by a ligament, the **flexor retinaculum,** under which pass the numerous flexor tendons to the wrist and fingers. Due to the irregularity of carpal bones, fractures and other injuries are very serious and often may not be restored to 100 per cent function.

Metacarpus

The metacarpal bones are five in number and have no special individual

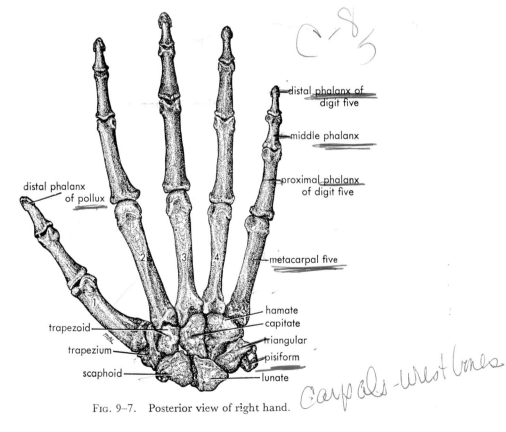

distal phalanx of digit five

middle phalanx

proximal phalanx of digit five

distal phalanx of pollux

metacarpal five

hamate
capitate
triangular
pisiform
lunate

trapezoid
trapezium
scaphoid

Fig. 9–7.   Posterior view of right hand.

Carpals - Wrist bones

names, but are numbered from the lateral to the medial side. Each one has a **base** proximally, a **shaft,** and distal extremity or **head.** The four medial metacarpals articulate at their bases with each other and with adjacent carpal bones and are closely bound together by ligaments. The first metacarpal articulates at its base only with the trapezium and hence is freely movable. The heads of the metacarpals articulate with the digits and form the "knuckles."

Phalanges

The phalanges are fourteen in number, three for each of the digits two to five, and two in the thumb or digit one. They are called proximal, middle, and distal phalanges. The proximal one is the largest, the distal the smallest. Each consists of a body or shaft and two extremities, the base and head. The bases of the proximal phalanges have oval, concave, articular surfaces, while those of the middle and distal rows have a double concavity separated by a shallow groove. The articular surfaces also extend farther on to the anterior than on to the dorsal surfaces. The distal phalanges, of course, lack articulating surfaces on their heads but instead have horseshoe-shaped elevations on their anterior or palmar surfaces which support the sensitive pulp of the fingertips (Fig. 9–7).

Ossification of the carpus takes place entirely after birth from one center in each bone. The metacarpals each have two centers of ossification, one for the body and one for the head, except for metacarpal one, which has a center in the base instead of the head. In the phalanges, the pattern is that of the first metacarpal, one center for the body and one for the base. This fact has caused some anatomists to consider the first metacarpal a proximal phalanx. Ossification begins in the bodies of metacarpals

and phalanges about the eighth week of fetal life, but not until about the third year in the heads of the metacarpals and the bases of the phalanges. The union of the heads and bodies of metacarpals and the bases and bodies of phalanges does not take place until the eighteenth to twentieth years.

JOINTS OF THE HAND

The joints of the hand are many and complex (Figs. 9–8, 9, 10). The **radiocarpal** or **wrist joint** is of the **condyloid** type and allows flexion, extension, abduction, adduction and circumduction of the hand. It is composed of the inferior concave surface of the end of the radius and the accompanying articular disc at the distal end of the ulna, which fit over the convexity formed by the scaphoid, lunate and triangular bones of the carpus. Its articular capsule completely surrounds the joint and is made up of the **palmar radiocarpal, dorsal radiocarpal, ulnar collateral** and **radial collateral** ligaments (Figs. 9–8, 9).

The joints within the carpus are illustrated in Figure 9–9. They have a continuous single joint cavity which includes the bases of the metacarpal bones. Only the joint between pisiform and triangular and between the first metacarpal and the carpus are excluded. The bones of each row of carpals are bound together by **palmar, dorsal** and **interosseous ligaments** which allow only a minimal amount of gliding motion. They are arthrodial joints.

The articulations between the carpals of the proximal and distal rows are connected by **dorsal, palmar** and **collateral ligaments.** They constitute collectively the **midcarpal joint.** More freedom of movement is allowed here than between the individual carpals of each row. The principal movements are flexion and extension, though slight rotation is possible. The midcarpal joint working in

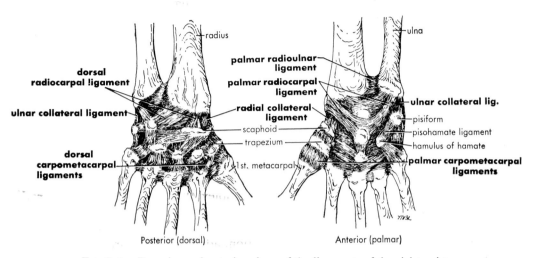

Fig. 9–8.   Posterior and anterior views of the ligaments of the right wrist.

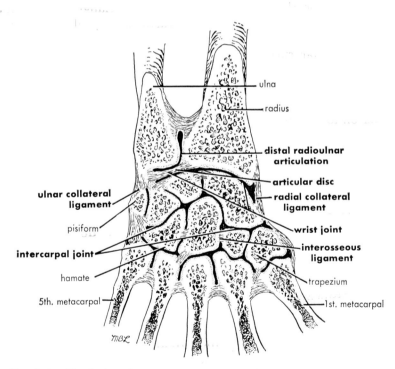

Fig. 9–9.   Vertical section through the wrist showing the articulations and their synovial cavities.

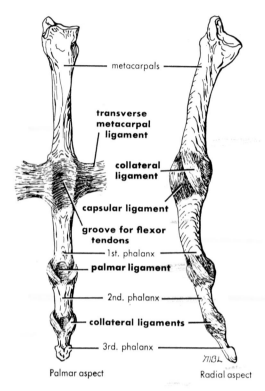

metacarpals

transverse
metacarpal
ligament

collateral
ligament

capsular ligament

groove for flexor
tendons

1st. phalanx

palmar ligament

2nd. phalanx

collateral ligaments

3rd. phalanx

Palmar aspect

Radial aspect

FIG. 9–10.   Metacarpophalangeal and interphalangeal articulations
of the third digit of the right hand.

conjunction with the wrist joint increases the range of extension and flexion of the hand.

The **carpometacarpal joints** of metacarpals two through five are of the arthrodial type and allow a minimum of gliding movement. They are bound by **dorsal, palmar** and **interosseous ligaments.** The articulation of the metacarpal of the thumb (*first digit*) with the trapezium of the carpus is saddle-shaped and enjoys great freedom of movement. Also, the first metacarpal is set in such a position that the thumb faces medially and can be used with the remaining digits or opposite them. It is said to be **opposable.** The joint is covered by a thick but loose capsule and is lined by its own synovial membrane.

The **metacarpophalangeal joints** are condyloid, the rounded heads of the metacarpals fitting into shallow concavi-ties on the bases of the proximal phalanges. The thumb is again an exception and its articulation with the first metacarpal is hinge-like (*ginglymoid*). Each articulation is provided with a **palmar** and two **collateral ligaments.** The collateral ligaments are strong, rounded cords placed on the sides of the joints, while the palmar ligaments are dense, thick, fibrocartilages fitting in the intervals between them and connected to them. The palmar ligaments are loosely united to the metacarpals but securely attached to the bases of the first phalanges. On each side they blend with the transverse metacarpal ligament which binds the metacarpals together except for metacarpal one. The palmar ligaments present grooves anteriorly for the passage of flexor tendons (Fig. 9–10).

The movements of the metacarpophalangeal joints are flexion and exten-

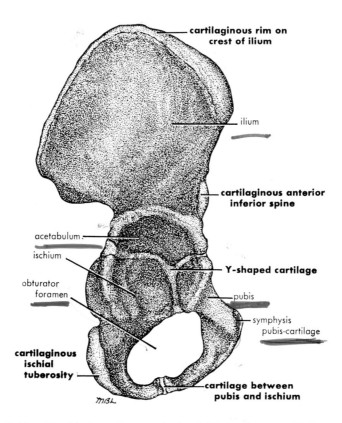

cartilaginous rim on
crest of ilium

ilium

cartilaginous anterior
inferior spine

acetabulum

ischium

Y-shaped cartilage

obturator
foramen

pubis

symphysis
pubis-cartilage

cartilaginous
ischial
tuberosity

cartilage between
pubis and ischium

MBL

FIG. 9–11.   The hip bone (os coxae) of a child showing unossified areas.
Its origin from three bones is clear.

sion and, to a limited extent, abduction
and adduction.   Abduction and adduction are limited by the collateral ligaments, and prevented entirely when the
fingers are flexed.   Flexion and extension
only are possible between the thumb and
its metacarpal.

The **interphalangeal** articulations are
hinge joints and each has a palmar and
two collateral ligaments.   The extensor
muscle tendons serve as posterior ligaments.   They are capable of flexion and
extension only and the latter is limited
by the palmar and collateral ligaments.

PELVIC GIRDLE (Figs. 9–11 to 14)

Return now to the beginning of this
chapter and review the material given
under **"Definition and Function"** in

order to see the following in proper perspective.

The pelvic girdle, in the adult, consists
of the **os coxae** or **innominate** bones and
the sacrum and coccyx.   The os coxae
articulate with the sacrum posteriorly
and with each other anteriorly at the
symphysis pubis.   They form a deep
acetabulum for articulation of the inferior limb (Fig. 7–1).   In the young
subject each hip bone consists of three
bones; a large, expanded, upper **ilium,**
which supports the flank and contributes
about two-fifths of the wall of the
acetabulum; the **ischium,** inferior and
posterior, contributing about two-fifths of
the wall of the acetabulum; and the
anterior and inferior **pubis,** contributing
the remaining one-fifth of the acetabular
wall.   The ischium with the pubis forms a

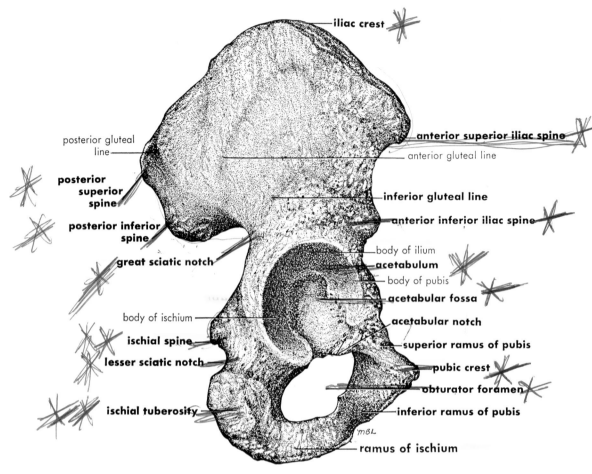

FIG. 9–12.   Right os coxae—external surface.

large aperture, the **obturator foramen.** In the adult these three bones become completely fused.

ILIUM

The ilium is the superior, broad and expanded portion of the os coxae. Its superior border is called the **iliac crest** and is the part you feel when you place your "hands upon your hips." The iliac crest ends anteriorly in the **anterior superior iliac spine,** below which is a notch, and below it the **anterior inferior iliac spine.** At the posterior end of the iliac crest is the **posterior superior iliac spine** separated by a shallow notch from the **posterior inferior iliac spine.** The

deep notch below the posterior inferior iliac spine is the **greater sciatic notch,** the lower part of which belongs to the ischium. The greater sciatic notch is, in the living subject, converted into a foramen by the **sacrospinous ligament.**

The external surface of the ilium is marked by three arched lines, the **posterior, anterior,** and **inferior gluteal lines.** On the surfaces enclosed by these lines are the origins of the three gluteal muscles. The internal surface of the ilium has a large, smooth and concave surface anteriorly, called the **iliac fossa** which gives origin to the large iliacus muscle, and behind the fossa, a roughened surface which is divided into two parts. One part, the **auricular surface** (*ear-*

shaped), is for articulation with a similar-shaped surface on the sacrum, and is, in the living state, covered with articular cartilage. The other portion, above and behind, is elevated and rough for the attachment of the **posterior sacral ligaments** and is called the **iliac tuberosity.** Below the iliac fossa is a smooth, rounded border which extends downward, forward, and medialward, the **arcuate** or **iliopectineal line.** This line marks the inferior boundary of the **false pelvis** (see page 157). The remainder of the ilium contributes to the acetabulum.

Ischium

The ischium contributes to the lower and back part of the acetabulum. Behind the acetabulum is a triangular **ischial spine** forming the inferior border of the greater sciatic notch mentioned above. Below the ischial spine is the **lesser sciatic notch** which is bounded inferiorly by the prominent **ischial tuberosity** on which the body rests when in the sitting position. The remaining portion of the ischium, the **ramus,** is thin and flattened and extends forward and slightly upward to join the inferior ramus of the pubis.

Pubis

The pubis forms the inferior and anterior portion of the hip bone. It joins with its fellow of the opposite side at the **symphysis pubis.** From here an **inferior ramus** extends downward and backward

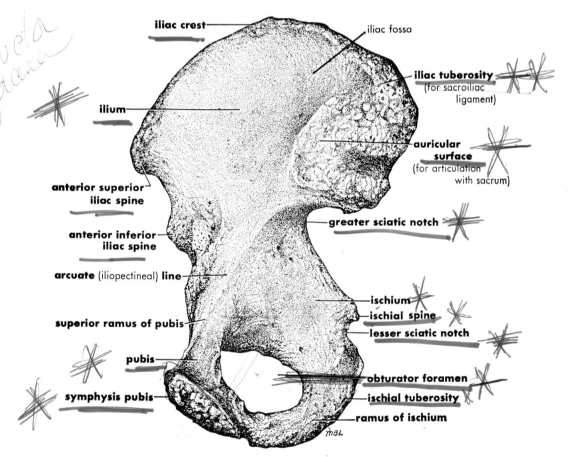

Fig. 9–13.   Right os coxae—internal surface.

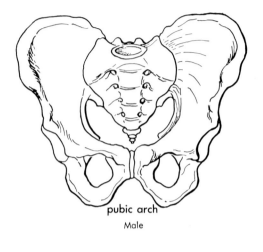

pubic arch

Male

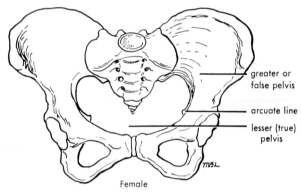

greater or
false pelvis

arcuate line

lesser (true)
pelvis

Female

FIG. 9–14.   The male and female pelves.

to join the ramus of the ischium. A **superior ramus** extends upward from the symphysis to end in the **body** of the pubis which contributes the remaining portion of the acetabulum.

The **acetabulum,** as has been indicated, is formed from contributions of the ilium, ischium, and pubis. It is deep and cup-shaped and is directed lateralward, downward, and forward. Its rim has a deep notch below, the **acetabular notch,** which is continuous with a circular non-articular depression, the **acetabular fossa.** The acetabular notch is converted, by a **transverse ligament,** into a foramen for the passage of nutrient vessels into the hip joint. The rest of the surface of the acetabulum is articular, the **lunate surface.** The acetabulum is also somewhat

deepened and its orifice narrowed by the presence of a fibrocartilaginous lip around its rim, the **acetabular labrum.**

The **obturator foramen,** the large opening between the ischium and pubis, is closed by a fibrous membrane, in the living specimen, except for a passage at its superior border. Through this passage obturator nerves and vessels pass out of the pelvis.

The ossification of the hip bone takes place from three primary centers, one for each of the constituent bones, and from five secondary centers. The secondary centers are one each for the crest of the ilium, the anterior inferior spine, the tuberosity of the ischium, the pubic symphysis and for the Y-shaped area at the bottom of the acetabulum (Fig. 9–11).

THE BONY PELVIS

The pelvis, meaning basin, is composed of the two os coxae laterally and anteriorly and the sacrum and coccyx posteriorly. The pelvis is divided into two parts by a plane passing through the promontory of the sacrum, the arcuate (*iliopectineal*) lines, and upper margin of the symphysis pubis. The circumference of this plane is the **pelvic brim.** The part of the pelvis above the pelvic brim is the **greater** or **false pelvis,** that below it the **lesser** or **true pelvis** (Fig. 9–14).

The **false pelvis** is bounded on either side by the ilium but has no bony component in front. The expanded ilia, with their iliac fossae, give support to the intestines and transmit some of their weight to the abdominal wall.

The **true pelvis** is located below the pelvic brim. It has more complete bony walls than the false pelvis, consisting of a part of the ilium, the ischium and pubis and the sacrum and coccyx. Its **superior circumference,** marked by the pelvic brim is called the **inlet;** its **lower circumference** marked by the tip of the coccyx and the tuberosities and spines of the ischia, is the **outlet.** The true pelvis houses the rectum behind and the urinary bladder in front. Between these organs, in the female, are the uterus and vagina.

When the body is in the anatomical position (Chapter 2), the pelvis is in an oblique position in reference to the trunk. The plane of the inlet of the true pelvis makes an angle of about 60° with the horizontal. The anterior spine of the ilium is in the same vertical plane with the top of the symphysis pubis (Fig. 7–1).

MALE AND FEMALE PELVES (Fig. 9–14)

The female pelvis is built on lighter lines than that of the male and shows definite adaptations for child bearing. The bones are more delicate, muscular impressions are slight, and the pelvis is shallow. The ilia are less sloped and the distance between the anterior superior spines is greater. This accounts for the greater prominence of the hips in a woman. The inlet of the true pelvis of the female is larger than in the male and is almost circular, rather than heart-shaped. Its cavity is shallower and wider; the sacrum is shorter, wider, and less curved above. The pelvic outlet is also wider than in the male; the coccyx is more movable; the spines and tuberosities of the ischia are farther apart or everted; the sciatic notches are wider and shallower; the pubic symphysis is shallower and the pubic arch wider and more rounded. The pubic arch is a distinct acute angle in the male. Finally, in the female the obturator foramina are smaller and more triangular and the acetabula are smaller and turn more forward.

JOINTS OF THE PELVIS (Fig. 8–35, p. 129)

The joints of the pelvis are the **sacroiliac,** the **symphysis pubis** and the **sacrococcygeal symphysis.**

The **sacroiliac joint** has been described in Chapter 8, page 130 and should be reviewed at this time. Recall that the upper part of the joint is fibrous, the lower synovial. The synovial portion, however, is so closely joined by irregular interlocking surfaces that almost no movement is allowed. It does give resilience to the pelvis since it comes under the influence of weight transmitted to it by the sacrum.

Two ligaments related to the sacroiliac joints which are of special importance in keeping the sacrum from tilting beyond the normal under influence of superimposed weight are the **sacrospinous** and **sacrotuberous ligaments.** These ligaments attach to the lower sacrum and coccyx, the former extending to the ischial spine, the latter to the ischial tuberosity. Besides preventing tilt they convert the greater and lesser sciatic notches into foramina and complete the wall of the true pelvis posteriorly (Fig. 8–35).

The **symphysis pubis** is between the two pubic bones anteriorly. Each pubic bone is covered with hyaline cartilage and these in turn are united by a pad of fibrocartilage to form the symphysis joint. The bones are held securely together by **superior and arcuate pubic ligaments,** by crossing fibers from the fibrocartilaginous pad, and by fibers of the External oblique aponeuroses and tendons of the Recti abdominis muscles. Movement in this joint is slight but does increase some in the female, especially during pregnancy, when the ligamentous structures of the pelvis become softened. This aids in the passage of the fetus at birth.

The **sacrococcygeal symphysis** is homologous to those between the bodies of other vertebrae and has similar ligaments. Like the symphysis pubis and the sacro-iliac joint it is often more freely movable during pregnancy. In such cases it may have a synovial membrane. At an advanced age this joint becomes fused over. The joints between the remaining coccygeal vertebrae are symphyses. In the male these bones become fused at an early age, but only in later years in the female.

THE LOWER LIMB

The lower limb consists of the **femur** located in the **thigh,** the **patella,** a sesamoid bone at the knee, the **tibia** and **fibula** in the **leg,** seven **tarsals** in the ankle region, five **metatarsals** in the foot, and fourteen phalanges in the five **digits** (Fig. 7–1).

FEMUR (Fig. 9–15)

The femur is the longest bone of the body, about 18 inches in length. It consists of a relatively uncomplicated **shaft,** and **superior** and **inferior extremities**

which are highly specialized for the hip and knee joints. The upper extremities of the two femurs are separated by the diameter of the true pelvis and the shafts slope downward and medially to bring the knee joints near the line of gravity of the body. Since the pelvis is broader in the female, the angle of inclination of the femurs is greater than in the male, though it varies in different individuals of either sex.

The **superior** or **proximal extremity** of the femur has a rounded **head** medially to fit into the acetabulum. Below and behind the center of the head is a depression, the **fovea capitis** for the attachment of the **ligamentum teres.** The remainder of the head is covered, in the living state, by smooth, articular cartilage. The head is extended medially, upwards and slightly forward by the **neck** which makes an angle of about 125° with the main body of the femur. This angle varies with age, sex, and stature of the individual.

Distal to the neck of the femur are two large processes, the **greater** and **lesser trochanters** which afford attachment and leverage to muscles. The **greater trochanter** projects upward and lateralward and has on its medial surface a depression, the **trochanteric fossa.** Inferior, medial and posterior to the great trochanter is the **lesser trochanter.** Connecting the two trochanters posteriorly is a prominent **intertrochanteric crest,** while anteriorly they are joined by a less prominent **intertrochanteric line.**

The **shaft** has as its most prominent feature, the **linea aspera,** which extends longitudinally along the posterior surface of its middle third and consists of lateral and medial lips and a roughened, narrow intermediate line. The lateral lip of the linea aspera extends upward toward the base of the greater trochanter as the **gluteal tuberosity.** The medial lip is continuous with the lesser trochanter superiorly while between these is a faint **pectineal line.** Inferiorly, the linea

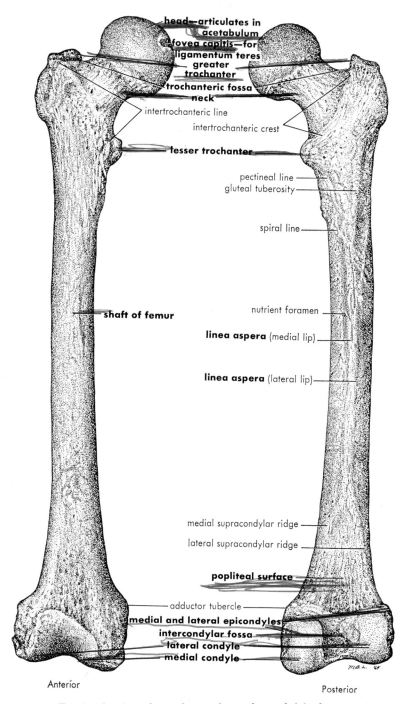

head—articulates in acetabulum
fovea capitis—for ligamentum teres
greater trochanter
trochanteric fossa
neck
intertrochanteric line
intertrochanteric crest
lesser trochanter
pectineal line
gluteal tuberosity
spiral line
shaft of femur
nutrient foramen
linea aspera (medial lip)
linea aspera (lateral lip)
medial supracondylar ridge
lateral supracondylar ridge
popliteal surface
adductor tubercle
medial and lateral epicondyles
intercondylar fossa
lateral condyle
medial condyle

Anterior

Posterior

FIG. 9–15.   Anterior and posterior surfaces of right femur.

aspera divides into two supracondylar ridges enclosing between them a triangular area, the **popliteal surface.** The lateral of these two ridges extends downward to the lateral condyle of the femur; the medial ends in a small tubercle above the medial condyle, the **adductor tubercle.** A **nutrient canal** perforates the linea aspera just below its center. The remaining borders and surfaces of the shaft are relatively smooth. The shaft, almost circular in cross-section through much of its length, is flattened and broad at its lower end.

The **inferior** or **distal extremity** of the femur is marked by the prominent **lateral** and **medial condyles** which rest upon the tibia. They project considerably on the posterior side of the femur, leaving a deep notch between them, the **intercondylar fossa.** In front, they are less prominent and a shallow depression separates them, the **patellar surface,** which in life receives the **patella** or **knee-cap.** Each condyle has an elevation upon it, the **lateral** and **medial epicondyles,** the latter being the more prominent. They give attachment to the collateral ligaments of the knee joint.

Ossification of the femur starts as early as the seventh week, being preceded only by the clavicle among the long bones. There are five centers, one each for the body, the head, the greater and lesser trochanters, and the distal extremity. The epiphyses begin ossification later than the shaft; the lesser trochanter not until the thirteenth or fourteenth year. They join the body about the eighteenth year except for the distal extremity which does not join until the twentieth year.

THE HIP JOINT (*coxal articulation*)
(Fig. 9–16)

This joint is of the **ball-and-socket** (*enarthrodial*) type being formed by the head of the femur fitting into the acetabulum of the coxal (*hip*) bone. Relative to the shoulder joint, the articular surfaces are more congruent. The acetabulum is very deep, and with its **acetabular labrum** and **transverse ligament** (see page 156) virtually grasps the head of the femur. The articular cartilage of this joint is of special interest. On the head of the femur it is thicker in the center than on the margins, while in the acetabulum it is thickest above and thins gradually below. It is horseshoe-shaped in appearance with the open part below at the acetabular fossa. The acetabular fossa is, in life, filled with a mass of fatty tissue and covered with synovial membrane. The **ligamentum teres** attaches medially to the acetabular fossa, with some fibers coming from the transverse ligament (Fig. 9–16). It passes laterally to the fovea capitis of the head of the femur. This ligament, since it remains loose in all positions of the joint, plays little part in holding the articular surfaces together. It appears to serve more for carrying nutrient vessels to the head of the femur and possibly for synovial fluid in the joint.

The joint is completely invested by an **articular capsule** which attaches to the rim of the acetabulum, and on the femoral side of the intertrochanteric line and parts of the base of the neck and adjacent areas. It therefore encloses not only the head of the femur, but the neck as well (Fig. 9–16). It is lined with a **synovial membrane** which extends from the rim of the cartilaginous surface of the head of the femur, over the part of the neck enclosed within the joint and is then reflected onto the inside of the articular capsule. It covers both surfaces of the acetabular labrum, the mass of fat in the acetabular fossa, and ensheathes the ligamentum teres to its point of attachment with the head of the femur. It may sometimes communicate with a bursa. The articular capsule is made up of longitudinal and circular fibers and is strongest and thickest at the upper and

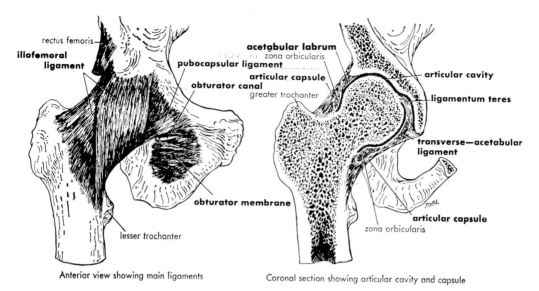

rectus femoris

iliofemoral
ligament

acetabular labrum
zona orbicularis

pubocapsular ligament

articular capsule

obturator canal

greater trochanter

articular cavity

ligamentum teres

transverse—acetabular
ligament

obturator membrane

articular capsule
zona orbicularis

lesser trochanter

Anterior view showing main ligaments

Coronal section showing articular cavity and capsule

FIG. 9–16.   The right hip joint.

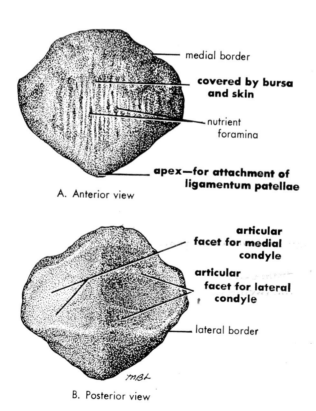

medial border

covered by bursa
and skin

nutrient
foramina

apex—for attachment of
ligamentum patellae

A. Anterior view

articular
facet for medial
condyle

articular
facet for lateral
condyle

lateral border

B. Posterior view

FIG. 9–17.   The right patella.

anterior part of the joint, thinner and looser behind and below. It is supported externally by accessory ligaments of which the **iliofemoral** is the most important and is anteriorly placed. The **pubocapsular ligament** is on the inferior aspect of the capsule, the **ischiocapsular** (*ischiofemoral*) on the posterior surface.

Movements of the Hip Joint. The movements of flexion, extension, abduction, adduction, rotation, and circumduction of the thigh are all possible in this ball-and-socket joint. The length and angle of the neck of the femur relative to the body have marked influence upon these movements and the direction they take. Also, the ligaments and articular capsule place certain limitations upon the movements at the hip. The strong iliofemoral ligament, for example, is put on stretch by any attempt to extend the thigh beyond a straight line with the trunk of the body (*hyperextension*). It ensures us, when in the erect position, against falling over backward at the hip joints and does so without muscular effort or fatigue.

Patella (Fig. 9–17)

The patella or knee-cap is a sesamoid bone formed in the tendon of the Quadriceps femoris muscle. It is triangular in outline and has its apex inferiorly. It is subcutaneous and its shape can readily be palpated. It lies in front of the knee joint which it protects and it also increases the leverage of the strong Quadriceps femoris muscle. Its apex is held firmly to the tibial tuberosity by the **ligamentum patellae,** while its posterior surface has two **articular facets, lateral** and **medial,** separated by a vertical ridge. These facets ride on the lateral and medial condyles of the femur respectively, while the vertical ridge fits into the groove between the condyles. The anterior surface of the patella is slightly convex and perforated with nutrient foramina. Some fibers of the Quadriceps femoris tendon continue over this surface and are continuous with the ligamentum patellae. A bursa lies between this surface of the patella and the skin.

The patella is usually ossified from a single center, which appears in the second or third year, and is completed by puberty. The bone is densely cancellous internally, covered by a thin, compact layer.

Tibia (Fig. 9–18)

The tibia or shin-bone is the large, massive long bone on the medial side of the leg and can be readily palpated anteriorly and medially. It receives the weight transmitted to it by the femur and directs that weight to the foot.

The upper end of the bone is expanded, and has on its surface two large, almost flat **condyles** which articulate with the condyles of the femur, and support peripherally the **menisci** (*cartilages*) of the knee joint. The lateral condyle is almost circular in outline; the medial one is oval. Between the condyles are **anterior** and **posterior intercondylar fossae** and nearer the posterior than anterior side of the bone an **intercondylar eminence** with two tubercles on either side is present. On the anterior surface of the proximal end of the bone is a large oblong elevation, the **tuberosity** of the tibia, to which the ligamentum patellae, mentioned above, attaches. On the posterior side of the lateral condyle is a circular, flat **articular facet** for articulation with the head of the fibula.

From the expanded upper-end, the shaft tapers downward to expand again, though to a lesser degree, at the distal end of the bone. Projecting downward on the medial side of the lower end is the **medial malleolus** which with the expanded undersurface of the tibia forms an articulation for the talus of the foot.

The tibia presents, on the lateral side of its distal end, a triangular concave surface for articulation with the **fibula,** which projects downward beyond the tibia, joining into the articulation with the talus.

FIBULA (Fig. 9–18)

The fibula is a long, slender, twisted bone which articulates on the lateral side of the tibia. Its upper end is expanded into a rounded head which has an articular surface medially for articulation with the tibia. Its lower end is triangular and extends beyond the tibia as the **lateral malleolus.** It completes the ankle joint laterally as stated above, and is easily palpated.

The tibia and fibula each ossify from three centers; one for the body and one for each end. The centers in the bodies appear about the seventh or eighth week of fetal life, while those in the ends appear usually after birth. They join the bodies of their respective bones between the eighteenth and the twenty-fifth year.

THE KNEE JOINT (Fig. 9–19)

The knee joint, the largest and most complicated in the body, is, in essence, three articulations in one. Indeed, in some lower mammals there are in this joint three distinct synovial cavities, in others three cavities connected only by small openings. In man, the synovial cavities are not entirely separate. There is, however, some separation of the joints in the form of the **cruciate ligaments** which pass between the two condyles of the femur and the corresponding condyles of the tibia. The infrapatellar synovial fold also contributes to this separation. The cruciate ligaments might be interpreted as the collateral ligaments of the lateral and medial joints of the knee. The third articulation is that between the patella and the femur.

Following the above line of reasoning

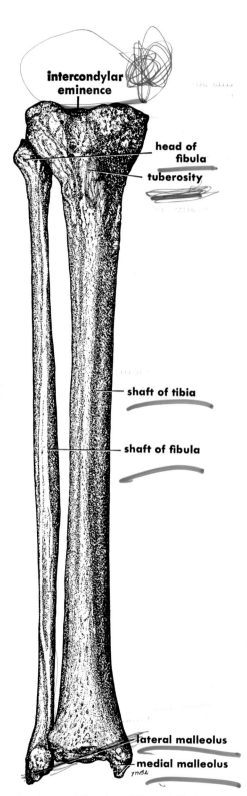

intercondylar eminence

head of fibula

tuberosity

shaft of tibia

shaft of fibula

lateral malleolus

medial malleolus

FIG. 9–18.   The right tibia and fibula—anterior view.

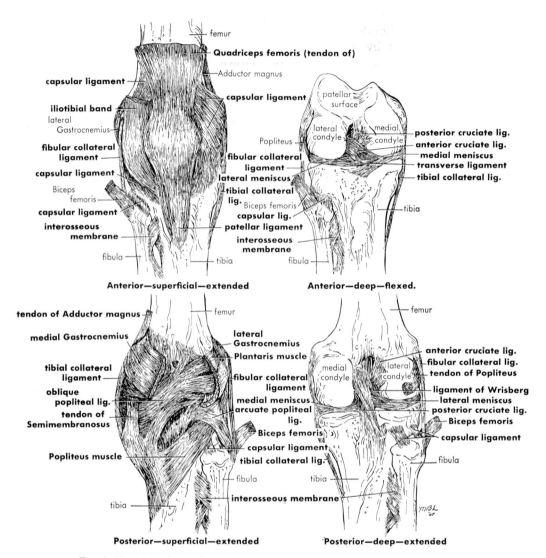

FIG. 9–19.   Anterior and posterior views of the right knee joint showing superficial and deep ligaments and related muscles.

the knee joint, though usually considered a ginglymus or hinge joint, would be considered two condyloid joints, lateral and medial, between femur and tibia; and one arthrodial joint, not entirely typical, between the patella and the femur. While the chief movement at the knee is flexion and extension, as one would expect in a hinge joint, some rotation does take place under certain conditions, and a gliding movement takes place in the patellofemoral articulation.

Another remarkable thing about the knee joint is that it combines a wide range of movement in one direction, with a great weight-bearing capacity, and considerable stability. The superior end of the tibia is the largest weight-bearing surface of the skeleton. Its two articulating condyles are deepened in life by the crescent-shaped **semilunar cartilages** or **menisci**. The menisci are thickened and convex on their peripheral borders, and are thin, concave, and free on their opposite borders. They are connected anteriorly and peripherally by a **transverse ligament,** and by a part of the capsule of the knee joint, to the head of the tibia. These menisci lend some stability to the joint. Additional stability is given by the strong **anterior** and **posterior cruciate ligaments** which connect the tibia and femur inside of the joint and cross each other like the letter X. The anterior cruciate ligament extends from the front of the intercondylar eminence of the tibia upward and backward to the medial side of the lateral condyle of the femur; the posterior one, from the posterior intercondylar fossa of the tibia, upward and forward to the lateral side of the medial condyle of the femur (Fig. 9–19).

Finally, the outer **fibrous capsule** of the knee joint is a complicated one compounded of thin, but strong fibrous membrane, special ligaments, and strong expansions of the muscle tendons which pass over the joint. As already mentioned, the patella in the tendon of the Quadriceps femoris muscle, and the ligamentum patellae make up a large part of the front and sides of the joint. Also, expansions of fibrous tissue from the **fascia lata** and its **iliotibial bands** fill in anteriorly and laterally. **Tibial** and **fibular collateral ligaments** give strength and stability to the medial and lateral sides of the knee joint respectively. **Oblique popliteal** and **arcuate popliteal ligaments** give additional support laterally and posteriorly. Reference to Figure 9–19 will help you to visualize these complicated structures. The integrity of the joint is due not so much to the bony interrelations as it is to external and internal ligaments, muscle tendons, and menisci. Numerous bursae are located in relationship to this joint and to its complicated synovial cavity. Injuries to such a complicated structure are often difficult to remedy without some loss of function and discomfort.

Movements of the Knee Joint. As stated before, flexion and extension are the primary movements of the knee joint. However, during flexion, the axis around which movement takes place shifts backward, and during extension it shifts forward. This is different than in a more typical hinge joint, such as the elbow, where the axis does not shift. Also, as full extension is reached the articulation is rotated inward and the joint is "locked" with the ligaments taut. This gives the joint great stability in the extended position. As flexion is initiated, there is an external rotation of the femur which "unlocks" the joint. In the flexed position there is slight inward and outward rotation, the movements taking place mostly between the tibia and the menisci.

TIBIOFIBULAR ARTICULATIONS
(Figs. 9–19, 22, 23, 24)

There are three articulations between the tibia and fibula: 1. the **superior tibiofibular joint,** 2. the **interosseous membrane** between their shafts, and 3. the **inferior tibiofibular joint.** The former is an arthrodial joint which allows a gliding movement between the articulating surfaces. The **interosseous membrane** forms a fibrous joint, the membrane serving to separate the muscles on the front of the leg from those on the back. The **inferior tibiofibular joint** is a syndesmosis and the two bones are firmly held together by a strong **interosseous ligament,** by **anterior** and **posterior inferior tibiofibular ligaments,** and an **inferior transverse ligament.** Movements between tibia and fibula are slight and are dependent mostly on flexion and extension of the ankle joint which displace the fibula upward and downward.

TARSUS (*Ankle*) (Figs. 9–20, 21)

The tarsus is made up of seven bones: the **talus, calcaneus, navicular, cuboid,** and three **cuneiforms,** which occupy about one-half of the foot region. The most prominent and proximal of these are the **talus**, which articulates with the leg to form the ankle joint, and below the talus, the **calcaneus,** the largest tarsal, which forms the heel and serves as a strong lever for the muscles of the calf. The **navicular** is located in front of the talus on the medial side, the **cuboid** in front of the calcaneus on the lateral side of the tarsus. The first, second, and third **cuneiforms** complete the tarsus anteriorly and with the cuboid serve as points of articulation with the **metatarsals.**

METATARSUS (Figs. 9–20, 21)

The **metatarsals** and **phalanges** form the skeleton of the anterior half of the foot. There are five metatarsals which are numbered from the medial side. The **first** is the stoutest and shortest and supports the proximal phalanx of the hallux or great toe. The remaining four metatarsals are large and more slender and taper from proximal to distal ends where they attach to the remaining toes. All five are concave below and convex above with their wedge-shaped proximal ends articulating with the distal row of tarsals and with each other. The distal ends have convex articulating surfaces which extend farther backward below than above and are grooved on their plantar surfaces for passage of the flexor tendons. Their sides have tubercles for attachment of ligaments.

PHALANGES (Figs. 9–20, 21)

The phalanges are fourteen in number as is the case in the hand. There are two in the great toe and three in each of the remaining toes. They are shorter than the phalanges of the hand and those of the proximal row especially, quite compressed laterally, concave below and

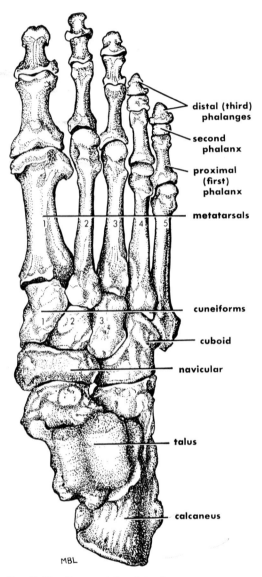

FIG. 9–20.   Bones of the right foot, superior view.

convex above. The distal row, the **ungual phalanges,** resemble those of the fingers, but are smaller and flatter.

Sesamoid bones are found in the foot. The most frequent ones are a pair on the plantar surface of the first metatarsophalangeal joint.

The bones of the foot are all ossified in cartilage. Each tarsal has one center of ossification, except the calcaneus which has an additional one at its posterior

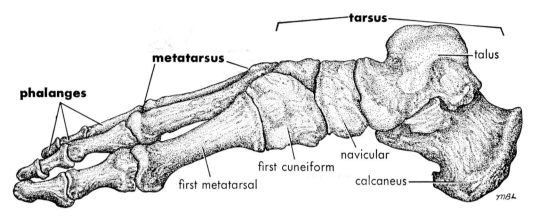

FIG. 9–21. Medial aspect of right foot showing medial longitudinal arch.

extremity. The metatarsals have two centers, one for the head and one for the body except number one which has a center in the base rather than in the head. The phalanges have one center for each body and one for each base. Ossification and the union of the various centers is completed at different ages, but all are joined by the twentieth year.

ARCHES OF THE FOOT (Fig. 9–21)

There are two **longitudinal** and one **transverse** arch in the human foot. The longitudinal are the **medial** which is high, the **lateral** which is low and less complicated. The medial longitudinal arch is formed by the calcaneus, talus, navicular, three cuneiforms, and the three medial metatarsals. The talus is the keystone of the arch, the calcaneus is the posterior pillar, while the other elements form the anterior pillar. The posterior pillar carries much of the weight of the body. The lateral longitudinal arch consists of the calcaneus as the posterior pillar, the cuboid as the keystone, and the two lateral metatarsals as the anterior pillars.

The **transverse arch** or metatarsal arch is well shown at the level of the cuboid and cuneiforms. The middle cuneiform is the keystone of this arch.

The arches are supported primarily by ligaments with the added help of a few muscles such as the Tibialis posterior and the Peroneus longus. This complex of structures spreads the superimposed weight of the erect body about equally between the calcaneus or heel and the heads of the metatarsals. In walking, the weight is first carried by the calcaneus and then shifted forward along the lateral side of the foot to the heads of the metatarsals. As one comes up on the front part of the foot, the main thrust is centered at the head of metatarsal one, which is enlarged to carry this extra burden, about 50 per cent of the weight. The arches, the character of the various joints, the ligaments, tendons, and muscles give to the foot not only strength and stability but its characteristic mobility and resilience.

THE ANKLE JOINT (Figs. 9–22, 23, 24)

This is a ginglymus (*hinge*) joint formed by the lower end of the tibia and its medial malleolus, the lateral malleolus of the fibula, and the talus. The tibia rests upon the convex upper surface of the talus and the malleoli grasp its sides. This arrangement allows dorsiflexion and plantar flexion of the foot but prevents any side movements. Since the upper

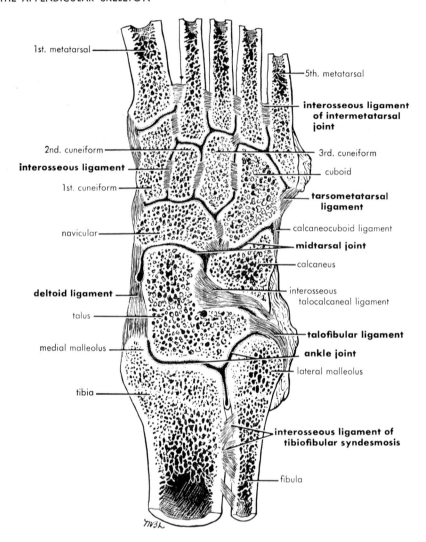

FIG. 9-22.   Articulations of right ankle and foot.   Foot is held in position of plantar flexion.
Articulations are cut in oblique plane.   Synovial cavities are in black.

articular surface of the talus is broader in front than in back, the lateral malleolus moves outward in dorsiflexion of the foot and the grasp of both malleoli on the talus is increased, giving added security to the joint. In this movement also, there is a slight gliding movement at the tibiofibular joint, and a twisting of the fibula. In plantar flexion, the ankle joint is least secure, since the narrower part of the talus is then grasped less tightly by the malleoli.

In addition to those ligaments described above which secure the inferior tibiofibular joint, the following are most important in supporting the ankle joint. The **articular capsule** or ligament is of variable thickness and is supported by the following collateral ligaments which can best be understood by reference to Figures 9-22, 23, and 24. The large, strong and triangular **deltoid ligament** attaches the medial malleolus firmly to the navicular, talus, and calcaneus. Laterally, the **anterior** and **posterior talofibular** and the **calcaneofibular liga-**

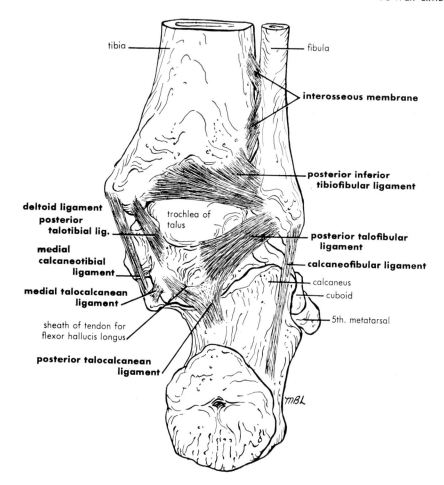

FIG. 9–23.   Posterior view of right ankle joint.

ments secure the fibula to the talus and calcaneus, respectively. So strong are the ligaments of the ankle that when the ankle is twisted—such as in a skiing accident—the medial malleolus and the lower end of the fibula may fracture before the ligaments are injured. This is called Pott's fracture, named for the London surgeon who suffered such an accident and published a description of it.

INTERTARSAL JOINTS

These are the joints between individual tarsal bones and are of the arthrodial type. They are too numerous and too complex to describe here in detail, but reference to Figures 9–22, 24, and 25 will give a general impression of them. They should be compared to those of the carpal region of the hand. As you would expect, the tarsus has stronger ligaments than the carpus, since they must sustain the weight of the body. Among the more prominent of the ligaments are the **long** and **short plantars** and the **plantar calcaneonavicular** (Fig. 9–24). The latter connects the calcaneus and the plantar surface of the navicular and supports the head of the talus, and thus forms a part of the articular cavity. This part of the ligament has a fibrocartilaginous facet and a synovial membrane on which this portion of the head of the talus rests. This ligament also contains numerous elastic fibers giving elasticity to the longitudinal arch and spring to the foot and therefore is often referred to as the **"spring" ligament.** The plantar calcaneonavicular ligament is also supported by the tendon of the Tibialis posterior muscle which inserts on its plantar surface and on neighboring tarsal and metatarsal bones (Fig. 9–25). A weakening of this ligament and muscle allows the talus to move downward under the weight of the body and the foot flattens, expands, and turns laterally— the condition called **flat-foot.**

The joint between the talus and navicular and the calcaneus and cuboid is called the **midtarsal joint** and is a plane at which movements of abduction, adduction, inversion and eversion of the foot take place in cooperation with these movements at the talocalcaneal joint (Fig. 9–22).

Tarsometatarsal Joints
(Figs. 9–22, 24, 25)

These are also arthrodial joints and the bones involved are the three cuneiforms, the cuboid and the bases of the metatarsals. The first cuneiform projects beyond

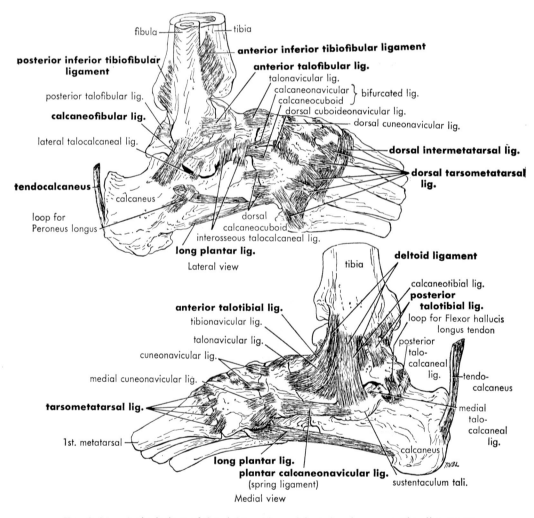

Fig. 9–24. Articulations of the right ankle and foot showing supporting ligaments. Lateral and medial views.

the others and receives the base of the first metatarsal. The second metatarsal is wedged in between the first and third cuneiforms and articulates with the second at its base; the third articulates with the third cuneiform; the fourth and fifth with the cuboid. **Dorsal, plantar, and interosseous ligaments** connect these bones and movement is limited to a slight gliding of one bone on another.

INTERMETATARSAL JOINTS
(Figs. 9–22, 24, 25)

These are arthrodial joints formed by the adjacent sides of the metatarsal bones. They are held by **dorsal** and **plantar**

intermetatarsal ligaments and by **interosseous ligaments** except between the first and second metatarsal where only the interosseous ligament is present. Only a slight gliding movement is allowed at these articulations.

The distal ends or heads of the metatarsal bones are all joined by the transverse metatarsal ligament. This differs from the corresponding ligament of the metacarpus which excluded the first metacarpal.

METATARSOPHALANGEAL JOINTS

The rounded heads of the metatarsal bones fit into the corresponding shallow

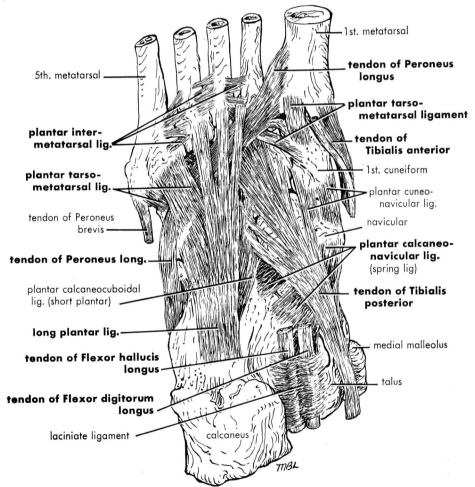

FIG. 9–25. Ligaments and tendons of the plantar surface of the right foot.

cavities at the proximal ends of the first or proximal phalanges to form **condyloid joints.** The movements of flexion, extension, abduction and adduction are permitted.

The joints are held by thick, dense fibrous **plantar ligaments** which blend with the **transverse metatarsal ligament** and are grooved on their plantar surfaces for the passage of the flexor tendons. **Collateral ligaments** are strong, rounded cords which attach along the sides of each of the metatarsophalangeal joints are are connected to the plantar ligament of each joint.

Extensor muscle tendons take the place of dorsal ligaments on the dorsal sides of the joints.

INTERPHALANGEAL JOINTS

Interphalangeal joints are essentially like the metatarsophalangeal joints in the arrangements of their ligaments— the **plantars** and **collaterals.** However, the articulating surfaces of the bones are such as to allow only flexion and extension and they are, therefore, **ginglymoid** joints.

QUESTIONS

1. Compare the pectoral and pelvic girdles as to strength, function, and versatility of action.
2. Why is the clavicle more frequently fractured than the scapula?
3. What functions does the clavicle serve in the skeletal system?
4. To what bones does the scapula articulate?
5. What is the strongest part of the scapula?
6. What special interest does the coracoid process have phylogenetically?
7. Compare the shoulder and hip joints as to strength, function, and versatility.
8. What parts of the scapula can you easily palpate?
9. Describe the bony arrangements which are involved in the elbow joints.
10. Describe the articulation of the radius and ulna at their proximal ends. What movement takes place at this articulation?
11. Describe briefly the ossification of the humerus.
12. What bones form the skeleton of the hand?
13. What types of joints are each of the following: the wrist joint; those between the carpals; carpometacarpal; between the first metacarpal and thumb; metacarpophalangeal; interphalangeal?
14. Name the bones of the pelvic girdle and the joints in which they are involved.
15. Distinguish between the "true" and "false" pelves.
16. Compare the bony pelves of male and female.
17. Describe the ossification of the hip bone.
18. Name the most prominent anatomical features of the femur.
19. What ligament of the hip joint keeps the body, when in the anatomical position, from over-extending (*falling backward at hip*)?
20. What kind of a bone is the patella?
21. Describe the knee joint, indicating some of its weaknesses and strengths.
22. What types of joints are represented by the three tibiofibular articulations?
23. What is the largest bone of the tarsus?
24. Name the arches of the foot. What serves as the keystone for each?
25. Where do we commonly find sesamoid bones in the foot?
26. What is the importance of the plantar calcaneonavicular ligament?
27. Describe the movements within the foot.
28. What is Pott's fracture?

29. What parts of the lower limb are most palpable? Why are these palpable areas important to know?
30. The hand and foot are similar in their skeletal and joint structures, but serve quite different functions. How do they differ in structure to serve their diverse functions?

"Strength does not come from physical capacity. It comes from an indomitable will."

—Gandhi

Chapter 10

General Myology

DEFINITION AND FUNCTIONS

WE have given considerable attention to osteology and arthrology, the study of the skeleton and the joints, respectively. In our discussions frequent reference has been made to the essentially passive nature of bones and joints and to their dependence upon the muscles for **action. Myology** is the study of muscle tissue and organs. The muscular system is made up of the voluntary skeletal muscle organs of the body. Each of these organs is a complex of **striated** (*skeletal*) **muscle tissue,** a connective tissue framework, and a rich supply of blood vessels, lymphatics, and nerves. The skeletal muscles make up about 40 per cent of the body weight. They produce all the voluntary movements, hence are one of the most important systems of the body. They are essential to the performance of a great variety of body functions.

The skeletal muscles maintain the posture of the body by their constant state of partial contraction (*tonus*), and give stability to the articulations. They enable us to inspire air containing the vital oxygen. They enable us to obtain and chew our food and to swallow it, thus passing it along to the involuntary organs of the digestive system. They are involved in defecation, the elimination of the debris left over after digestion and absorption have taken place. They make the emptying of the urinary bladder a voluntary act.

The voluntary muscle tissue is used in articulate speech by means of which we communicate with others. The facial muscles are the basis for communication of a different sort, as when we smile, frown, show fear, worry, indifference or other emotions. We may even whistle, wink, or kiss as additional means of expression by voluntary muscles. And, they give to the body its characteristic form.

Voluntary muscles play a role in the adjustment of sense organs for most efficient action, sometimes to the extent of saving one from destruction. The quick movements of the eyes, the turning of the head for better advantage in hearing, the sniffing with the nose.

Locomotion is one of the most obvious functions of this system. We creep, walk, and run. But beyond all of this, among our most important and unique tools are our hands—miracles of quick and coordinated action—capable of doing a vast number of things from swinging a club to fingering the keys of a piano. Among the great events of man's evolution was the attainment of the bipedal condition by which the hands were freed from locomotion for other uses. This and his highly developed brain have enabled him to gain vast control over his external environment and to alter it as no other species can.

While the muscles contribute to the functioning of many of the body systems, they are equally dependent upon them. Their food and oxygen are made available through the digestive, respiratory, and circulatory systems and the waste products of their metabolism are excreted by the kidneys. The skin not only protects the muscles but serves as a means of dissipating the heat of muscular contraction. The nervous system with its sense organs initiates, controls, and coordinates muscular activity.

We are reminded again of the **oneness** of the human body. Its trillions of cells, its "institutions" of tissues, organs, and systems have their own identities and functions, but their survival depends equally upon their relationship to the whole organism. It is important to remember this as we proceed to analyze the structure of the human body.

KINDS OF MUSCLE TISSUES
(Fig. 10–1)

While the **muscular system** includes only the voluntary skeletal muscles, it is appropriate at this time to describe and compare the three kinds of muscle tissue found in the human body. These are the **smooth, cardiac** and **skeletal** muscle tissues. They are all specialized in the remarkable physiological property of **contractility,** the capacity to change form. They are all composed of elongated cells, called **muscle fibers,** which in turn contain fine fibrils, the **myofibrils,** within their cytoplasm (*sarco-*

13

plasm). This gives to the muscle fibers a faint longitudinal striation.

Smooth Muscle. Smooth muscle is so called because it, unlike cardiac and skeletal, has no transverse striations as seen under the microscope. It is involuntary, being under the control of the autonomic nervous system.

Smooth muscle cells or fibers are spindle-shaped and have an elongated

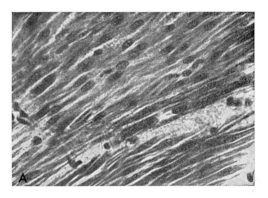

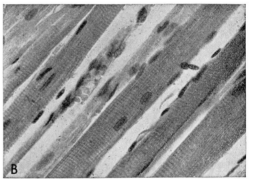

intercalated disc

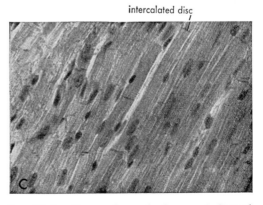

FIG. 10–1. Types of muscle tissue. *A*, Smooth muscle. *B*, Skeletal muscle. *C*, Cardiac muscle.

oval nucleus in the center of the cell (Fig. 10–1A). The myofibrils (*sarcostyles*) nearly fill the cytoplasm (*sarcoplasm*) and are less distinct than in the other muscle tissues. The muscle fibers are about 6 microns in diameter and vary in length from 15 to 500 microns. Their length varies under certain conditions, as in the walls of the uterus during pregnancy where they may become eight times their original length. This property of **extensibility** is accompanied by one of **elasticity** which enables the fibers to return to their original condition. Smooth muscle is also highly **irritable,** enabling it to respond to stimulation and it is in a state of **tonus,** or partial contraction—a condition of tension or readiness to contract.

Smooth muscle fibers are connected by white, elastic and reticular fibers and form sheets or layers of tissues which constitute parts of the walls of the hollow viscera and vessels of the body. When circularly arranged around the organs, their contraction results in decreasing the size of the cavity (*lumen*) of the hollow organ. When arranged longitudinally, their contraction shortens the organ involved. The fibers are usually so closely packed that individual cell boundaries cannot be seen. The main or larger layers of smooth muscle in the walls of organs are separated by connective tissue septa which allow for the passage of nerves and blood vessels. Neither blood nor nerve supply is as rich to smooth muscles as it is to the voluntary muscles. Most fibers are without nerves but the smooth muscle is highly conductive, and denervation does not cause the atrophy (*degeneration*) of the fibers as it does in voluntary muscle. In a few locations, as in the iris of the eye and some glands, smooth muscle cells are scattered among epithelial cells reminding us of the primitive contractile elements of lower animals (*Coelenterata*). These primitive cells are called **myoepithelial.**

Smooth muscle contractions are slow, sustained, and may produce great force as in the "labor" of childbirth. It fatigues slowly, and therefore can function over long periods of time, as in digestion.

It should be apparent that smooth muscle contributes to the functioning of many body systems. Contractions of blood vessels are important in the regulation of blood flow and blood pressure. Contractions of the walls of digestive organs manipulate and move the contained food during digestion and absorption. Urine is passed through the ureters to the urinary bladder; the urinary bladder contracts to expel urine, and the uterus forces the fetus into the birth canal. In all of these functions, smooth muscle is at work. Even the contraction of the iris of the eye and the erection of the hair are smooth muscle functions.

Cardiac Muscle (Fig. 10–1C). Cardiac muscle, as the name suggests, is the contractile tissue of the heart wall. Like skeletal muscle its fibers are striated and it has a sarcolemma, though it is not prominent. The single nucleus is centrally located, as in smooth muscle, and is oval in shape. Features peculiarly its own are its **branching, anastomosing fibers** which produce a continuous network, and lines or bands oriented transversely to the long axis of the fibers, the **intercalated disks.** These stain darkly with various dyes. They appear rather late in development of cardiac muscle and increase in number with age. Their function is not definitely known.

Cardiac muscle tissue is similar to smooth muscle functionally, since it exhibits considerable tonus, is difficult to fatigue, is highly "automatic," and receives its nerve supply from the autonomic or involuntary nervous system.

Skeletal Muscle (Figs. 10–1B, 2). Skeletal muscle is the flesh, the "red meat" of the body. It attaches to and moves the skeleton. Its fibers are long, straight, seldom-branching cylinders. They range

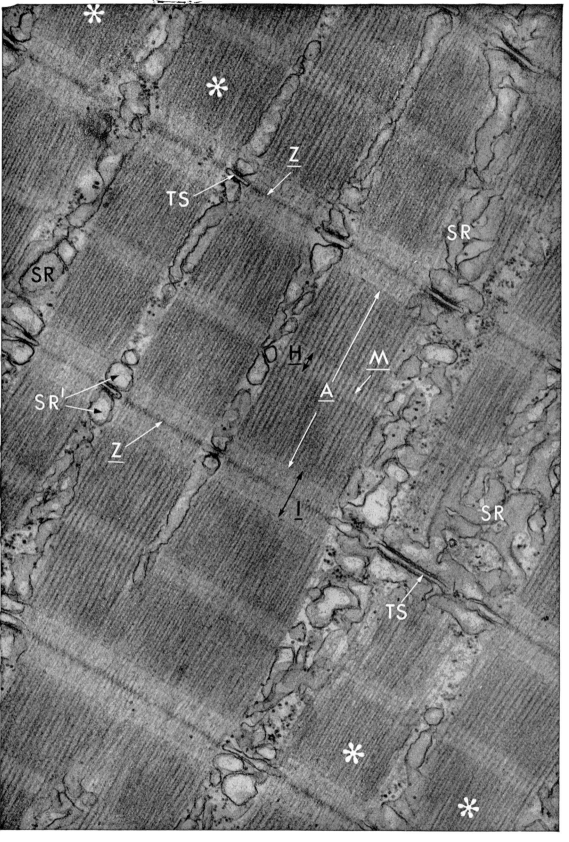

FIG. 10–2. Skeletal muscle and the sarcoplasmic reticulum. (Porter and Bonneville:
Fine Structure of Cells and Tissues. Ed. 3, Philadelphia, Lea & Febiger, 1968.)

from 10 to 100 microns in diameter and up to 4 cm. in length which places them among the largest cells of the body. The thickest cells are found in muscles of the appendages, the thinnest in the very active muscles, like those moving the eyes. Skeletal muscle cells or fibers are enclosed in a special membrane, the **sarcolemma.** It consists of the plasma membrane of the cell plus basement membrane material and associated reticular fibers. The cytoplasm of a muscle cell is called **sarcoplasm.** Longitudinally arranged **myofibrils** in the sarcoplasm show alternating light and dark bands or discs. They are so arranged that the light and dark bands of all the myofibrils line up across the entire cell and appear to form continuous bands, the **striations.** For this reason skeletal muscle is often called **striated muscle tissue.**

Electron micrographs reveal in more detail the structure of myofibrils. Figure 10–2 shows a number of myofibrils (*) from a relaxed fiber of a skeletal muscle magnified 29,000 times. Note that the myofibrils themselves are also constructed of myofilaments of two types which relate to the alternating light and dark bands (*striations*).

The **dark band** is made up of filaments of the protein **myosin** and is called the **anisotropic band** (*A-band*). It is bisected by a narrow somewhat paler band, the H-band, through the center of which an M-line is frequently seen.

The **light band** of the myofibril is called the **isotopic band** (*I-band*). It is made up only of thin filaments of **actin** which at their ends intermingle with the myosin filaments of the A-band reaching about to the edges of the H-band. The I-band is crossed by a prominent, dense line, the Z-line. The segment between two Z-lines is the histological and functional unit of a muscle fiber. It is called a **sarcomere.** It includes all of an A-band and the half of each I-band adjacent to the A. When the muscle contracts the

thin actin filaments slide between the myosin filaments with the subsequent partial or complete disappearance of the I- and H-bands. The A-band remains constant in length. This explanation of contraction does not involve the shortening of myosin or actin myofilaments but merely a change in their relationship.

The skeletal muscle micrograph also shows the interfibrillar **sarcoplasm** with its prominent **endoplasmic** (*sarcoplasmic*) **reticulum** (*SR*). It is made up of anastomosing tubules and vesicles forming lace-like sleeves around the myofibrils and at the Z-line forming peculiar dilated sacs which lie close to a transverse tubular system called the **T-system** (*TS*). The T-system is continuous with the sarcolemma of the muscle fiber. This T-system may occur, in more rapidly contracting muscles, at the junction of A- and I-bands and be more extensive. It is interpreted as the structure which enables myofibrils at the center of a muscle fiber to contract simultaneously with those at the surface which are adjacent to the excitatory action potential.

Skeletal muscle fibers have **numerous** small, ovoid **nuclei,** most of which lie **peripherally,** close to the sarcolemma. This multinuclear condition may be due to the fusion of many myoblasts (*embryonic cells*) in the formation of a muscle cell, to repeated divisions of the original single nucleus, or to a combination of these.

Skeletal muscle fibers differ from smooth muscle by responding more rapidly to stimulation, by fatiguing more readily, and by being less extensible. They are under the control of the voluntary nervous system, and therefore responsive to the will of the individual. They are often called **voluntary muscle.**

EMBRYOLOGICAL DEVELOPMENT OF MUSCLE

Two processes are involved in the embryological development of the muscles of the body. One of these is **myogenesis,** the development of the muscle tissues, the other **morphogenesis** (*organogenesis*), the formation of the muscles as organs.

Myogenesis. With the exception of the **muscles of the iris** of the eye and the **Arrector pili** muscles, which are ectodermal, all muscle tissues are derived from formative **myoblasts** of mesodermal origin. As early as the fifth week of embryonic life mesenchyme cells appear around the epithelial esophagus and begin to differentiate into **smooth muscle** fibers. Similarly, wherever smooth muscle is to be formed, in the walls of the hollow viscera or of blood and lymphatic vessels, the same process takes place.

The cardiac muscle tissue develops from the splanchnic mesoderm, through mesenchyme, which invests the primitive heart tubes.

Skeletal muscle tissue has a double origin, part of it coming from the paired somites (Fig. 4–13), the other from mesenchyme around the branchial arches.

The portion of the somite which is left, after giving rise to the **sclerotome** medially and possibly a **dermatome** laterally, is the **myotome.** The sclerotome gives rise to vertebrae, the dermatome to the dermis of the skin. Myoblast cells differentiate in the myotome, arrange themselves in the long axis of the body and develop into skeletal muscle fibers. Those muscles which are associated with the primitive branchial arches come from mesenchyme of the lateral mesodermal plates, possibly by specialization of the smooth muscle of the primitive gut. In any case, both types of skeletal muscle tissue are identical histologically, although those from the myotomes are called **metameric** or **somatic** while those from the mesodermal plates are known as the **branchiomeric** or **special visceral** musculature.

Morphogenesis of Skeletal Muscles (Fig. 10–3). The myotomes appear early in development and by the fifth week some are already beginning to dif-

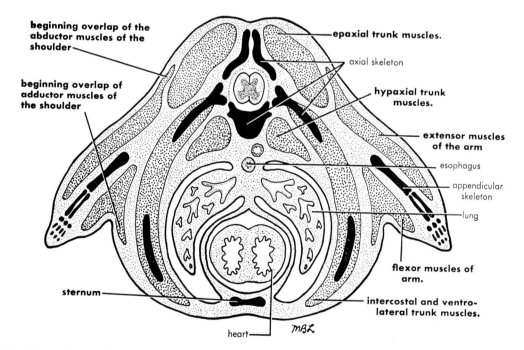

FIG. 10–3. Schematic representation of some of the major developing muscle masses showing their relationship to the axial and appendicular skeleton. (Modified from Patten.)

ferentiate into muscle organs. By the eighth week the definitive muscles of the fetus are well defined and are soon capable of coordinated movements. While no attempt will be made to describe morphogenesis in detail, the following statements may help to visualize the general process.

1. The muscle fibers undergo many shifts in direction from the original craniocaudal orientation in the myotomes.

2. There is a migration of myotome material from the original position, some moving dorsally to the midline, the **epaxial** mass, which is innervated by the dorsal branches of the spinal nerves; some moves to the ventral line, the **hypaxial** mass, and is innervated by the ventral branches of the spinal nerves. Migration from these masses takes place in many directions. The diaphragm, a muscle in the thoracic region, comes from cervical myotomes, as does the Latissimus dorsi which extends back to the lumbar region. Branchiomeric muscles also migrate during their differentiation. The facial muscles, for example, derive from the mesenchyme of the second branchial arch and from there move forward to the face.

3. Fusion of parts of successive myotomes to form individual muscles is of regular occurrence. The Rectus abdominis is a good example.

4. Myotome and branchial arch muscle primordia also undergo considerable longitudinal and tangential splitting to form subdivisions and layers. The Trapezius and Sternothyroid and the abdominal muscles are examples of such splitting.

5. Fascia, ligaments and aponeuroses are in some cases the result of degeneration of all or parts of myotomes.

There is still some debate as to the origin of the muscles of the limbs. Studies in comparative anatomy suggest that they form by migration from the ventral (*hypaxial*) parts of the myotomes adjacent to the limbs. However, in birds and mammals such migrations are not apparent and some believe that they derive from an undifferentiated mesenchyme in the limb buds. The fact that limb muscles receive their nerve supply from the ventral rami of spinal nerves suggests, but does not prove, a myotomic origin. In either case the limb muscles do differentiate into flexor (*ventral*) and extensor (*dorsal*) muscle groups (Fig. 10–3).

Morphogenesis of muscles of the head region is a complex and unresolved problem. Probably tongue and extrinsic eye muscles are derived from myotomes, though they are innervated by cranial nerves.

Muscles of the face, jaws, palatine, and pharyngeal regions are branchiomeric muscles. They are innervated by fibers which are different from those of the myotomic muscles. They are **special visceral fibers** of cranial nerves V, VII, IX, X and XI, as opposed to the **somatic motor fibers** of the metameric musculature.

GROSS STRUCTURE OF SKELETAL MUSCLES AS ORGANS
(Fig. 10–4)

Muscle organs consist of two parts, one predominantly of muscle fibers, the **belly;** the other of fibrous connective tissue, the **tendons.** The tendons attach, in most cases, to the skeleton. The one which attaches to the less movable or proximal structure is the **tendon of origin,** that attaching to the more movable or distal part is the **tendon of insertion.** The origin of a muscle is often attached directly to the periosteum of a bone without any intervening tendon, while the insertion is usually tendinous. Some muscles have two or more origins, as the Biceps brachii, and insert on a common tendon. This serves to concentrate the full force of the muscle on a single spot; or a muscle may have a

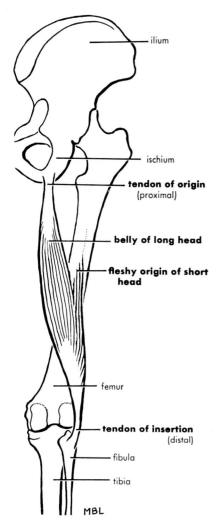

- ilium
- ischium
- tendon of origin (proximal)
- belly of long head
- fleshy origin of short head
- femur
- tendon of insertion (distal)
- fibula
- tibia
- MBL

FIG. 10–4. Diagram to show tendons of origin and insertion and the belly of a muscle—the biceps femoris.

The latter are called **aponeuroses** and are well illustrated by the abdominal muscles (Fig. 11–27). In some cases the origin and insertion of a muscle are interchangeable as is true with the Rectus abdominis. Its origin is considered to be on the pubis yet when one, lying on his back, elevates his thighs and legs the greater movement is at the pubic ends of these muscles. Bones develop in the tendons around some of the joints and are known collectively as **sesamoid** bones. The **patella** or knee-cap is the best known.

MICROSCOPIC STRUCTURE OF SKELETAL MUSCLE ORGANS
(Fig. 10–5)

Careful examination of cross-sections of the belly of a skeletal muscle will reveal the intimate association of muscle fibers with connective tissues. Each muscle fiber has a thin sheath-like covering of connective tissue, the **endomysium**, containing some reticular fibers, fibroblasts, and fixed macrophages. The latter are phagocytic cells and play an important role in case of muscle inflammation. Individual muscle fibers are collected into bundles, the **fasciculi,** which in turn are covered with a somewhat thicker layer of loose connective tissue, the **perimysium.** The perimysium sends connective tissue partitions, the **trabeculae,** into the bundles to partially subdivide them. A number of bundles make up the total belly of a muscle. The belly is covered externally by a loose connective tissue, the **epimysium** or **deep fascia,** which is continuous with the perimysium of the bundles.

The association of the belly with the tendons of the muscle is a very close and strong one. The **sarcolemma** of the muscle fiber fuses with the collagenous bundles of the tendon as do the collagenous fibers of the **perimysium.** The collagenous bundles of the muscle or the

number of tendons of insertion as those to the digits, the Extensor or Flexor digitorum, spreading the action over numerous joints. Tendons are also advantageous in making it possible to keep the weight and bulk of the strong muscles of the limbs nearer to the center of the body. The feet and particularly the hands, for this reason, remain relatively delicate in construction yet have great versatility of action. Tendons vary in shape from short and stout to long and slender and some form broad sheets.

tendons are also directly continuous with those of the periosteum of the bone to make a firm attachment.

Blood and Lymph Vessels of Skeletal Muscles (Fig. 10–5). The connective tissue framework of the belly of the muscle serves to carry, support and protect the blood and lymph vessels. Skeletal muscles, in consideration of their active role in movement and as generators of heat, are richly supplied with blood.

Arteries and veins and in many cases lymphatics enter the muscle together along the connective tissue trabeculae between the fasciculi. Here they break up into capillaries which form extensive networks in the endomysium surrounding individual muscle fibers.

When muscles are active, the contained vessels are squeezed so that each muscle serves as a pump to speed the flow of blood and lymph. This results in not only an increased flow of blood to the heart, but an increase in heart action to speed the entire circulation. Other changes in the body such as constriction of the minute vessels to the viscera and dilation of those to the skeletal muscles make for a greater flow of blood to the muscles. In this way their increased need for oxygen and nourishment is met, and their waste products are more quickly removed.

Nerve Supply of Skeletal Muscles (Fig. 10–5). Skeletal muscles are provided with nerves containing both motor and sensory fibers. The nerves travel along the intermuscular septa and usually enter a muscle with the blood vessels. They branch out through the connective tissue framework of the muscle to reach the muscle fibers.

The **sensory nerve fibers** (*afferent*) of skeletal muscles come from groups of specialized muscle fibers and carry impulses to the central nervous system. They are stimulated by muscular contractions and by stretching and are responsible for providing data indicating the status of a muscle. They carry impulses involving reflex and other activity of the muscle, such as maintenance of tonus and posture (see pp. 566–569 and Figs. 19–1F, 3 and 4).

Motor or efferent neurons to muscles are of at least two types: those known as alpha motor neurons ending in motor end-plates or myoneural junctions on the regular muscle fibers (Figs. 10–5, 6) and neurons with fibers of smaller diameter, the gamma motor neurons, ending in motor end-plates or similar structures on the smaller muscle fibers of muscle spindles (see pp. 566–569 and Figs. 19–1F, 3 and 4).

Motor Units (Fig. 10–6). Each terminal fiber of a motor nerve branches to supply several to a hundred or more skeletal muscle fibers. These are called **motor units.** A given skeletal muscle contains a certain number of these motor units. Each motor unit if it responds at all to a stimulus will give a maximal response. This is an example of the **all-or-none law** of the physiologists. To obtain a maximal response from an entire skeletal muscle, therefore, the stimulus must be of sufficient strength to activate all the motor units of that muscle. Weaker stimuli to a muscle, theoretically, bring fewer motor units into action; each unit, however, responds at its maximum. In this way we can grade our muscular responses to the difficulty of the task to be performed, *i.e.*, to the lifting of a light or a heavy object.

Motor units and their actions give us a basis for understanding phenomena such as muscle tonus and the maintenance of posture.

Muscle Tonus. Muscle tonus is a sustained partial contraction of muscle. Since the cutting of motor nerves to a muscle or the sensory nerves from a muscle causes it to lose tonus, we judge tonus to be of a reflex nature. Any influence which places muscles on the stretch, such as the persistent force of gravity, stimu-

tendon *of origin*

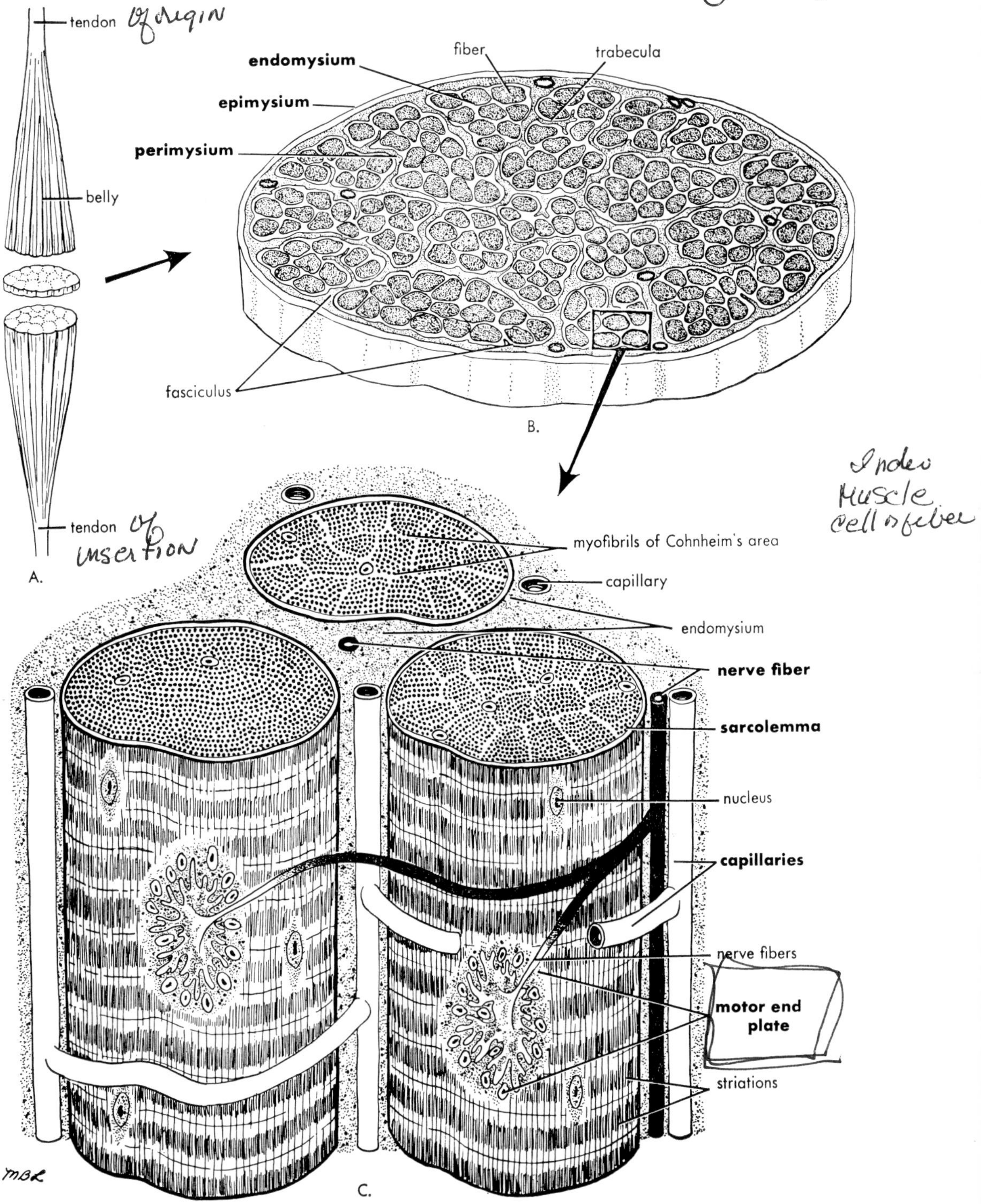

endomysium   fiber   trabecula

epimysium

perimysium

belly

fasciculus

B.

tendon *of insertion*

A.

myofibrils of Cohnheim's area

capillary

endomysium

nerve fiber

sarcolemma

nucleus

capillaries

nerve fibers

motor end plate

striations

Index Muscle Cell of fiber

MBL

C.

FIG. 10–5. Schematic diagram to show the structure of the belly of a skeletal muscle. *A*, The entire muscle—sectioned. *B*, A cross-section of the belly. *C*, Detail of three fibers and related structures.

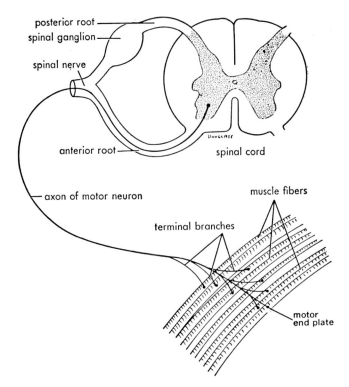

posterior root

spinal ganglion

spinal nerve

anterior root

spinal cord

axon of motor neuron

muscle fibers

terminal branches

motor end plate

FIG. 10–6.   A motor unit.

lates the receptors (*muscle spindles*) of the stretched muscles. The resulting reflex response counteracts the influence of gravity. Thus, while the mandible tends to drop, the head to fall forward, or the knees to buckle due to gravity, they are constantly prevented from doing so by the reflex tone of the anti-gravity muscles. The normal posture of the body is thus maintained. In maintaining tonus it is believed that relatively few motor units are active at any one time and that fatigue is prevented because the motor units "take turns" in this action.

QUESTIONS

1. Define myology.
2. What constitutes the muscular system?
3. What are some of the functions of the muscular system?
4. Construct a table in which you compare the three kinds of muscle tissues in terms of their structural and functional characteristics.
5. What is myoepithelium?
6. Describe briefly the embryological development of the muscle tissues.
7. What are the two kinds of skeletal muscles on the basis of origin?
8. Make a labeled diagram to show the microscopic structure of a skeletal muscle.
9. Distinguish, by definition, between the origin and insertion of a skeletal muscle.
10. Define motor end-plates, myoneural junctions, kinesthetic sense (*proprioception*), tonus, and sarcolemma.

Chapter 11

The Skeletal Muscles and Fasciae

AN APPROACH

In the preceding chapter kinds of muscle tissues and their functions have been considered. Skeletal muscles as organs were described as gross entities and in terms of their microscopic structure and relationship to the body systems. It is the purpose of this chapter to elaborate upon the skeletal muscles; to consider the mechanics of their action; how they move the bones of the body at the articulations; and how body movements

185

represent group action rather than actions of individual muscles.

While most of the muscles of the body will be briefly described and/or illustrated in this chapter, special attention will be brought to those which the author considers of special importance or interest. These will be printed in bold-face type for easy recognition. No student in elementary human anatomy should be expected to master but a few of the muscles. It is often helpful, however, to be able to make reference to other muscles, and it is to fulfill this need that so many are included. An alphabetized table of the muscles giving origin, insertion, innervation, and action is placed at the end of the chapter for easy reference and study (Table 11-1, p. 260). The majority of the muscles are illustrated by schematized drawings. O is used to indicate origin, I is used for insertion.

MECHANICS OF MUSCLE ACTION

As an initial goal each student should try to learn the origin and insertion of each of the major muscles of the body. This, of necessity, would include some knowledge of the joint or joints over which the muscle passed and, on this basis, an understanding of the action produced. All actions will be described from the anatomical position unless otherwise stated.

A second level of effort would be to consider the origins, insertions, and actions of other muscles operating over the same articulation, to see how each muscle of the group participated in a given movement of the joint. In making such an analysis one soon learns that muscles are not always directly involved in the production of movement or external work (*isotonic action*). They may exercise either a restraining or holding action in which considerable tension is developed within the muscle (*isometric action*), or they may be involved in negative work (*eccentric action*) in which a muscle is exerting tension while being lengthened by an outside force such as gravity. The fact is that in a normal functioning body the brain initiates movements such as the flexion of the knee or the extending of the elbow and is not concerned with individual muscles. In one kind of movement of a part, a given muscle would function in a certain way, in another kind of movement of the same part the same muscle would have a different role.

Roles of Muscles. To designate the roles of muscles in different actions of the body, a system of nomenclature was developed. If a muscle is the primary agent in the production of a desired movement, it is called a **prime mover.** If the same muscle serves to counteract or slow the action of another prime mover, it is called an **antagonist.** If it serves to steady a movement, or to eliminate an undesired movement of a joint, it is said to be a **synergist.** In movements which involve considerable muscle effort, the body, or some part of it, may have to assume a position or "fix" in order for the movement to be carried out efficiently. The muscles which maintain the position of the body or part while not actually taking part in the action are called **fixators.**

Examples should help to clarify the points just made. The Biceps brachii is a flexor of the elbow joint and in this movement would be considered a **prime mover** (Fig. 11-1). When it relaxes, however, gravity tends to cause the elbow to extend, but this movement is slowed by slight contraction or holding by the Biceps brachii. In this situation the Biceps brachii is serving as an **antagonist** to gravity, or to the Triceps if it were to contract to aid extension. It should be noted that muscles in general are arranged in opposing positions on the various parts of the body, flexors-exten-

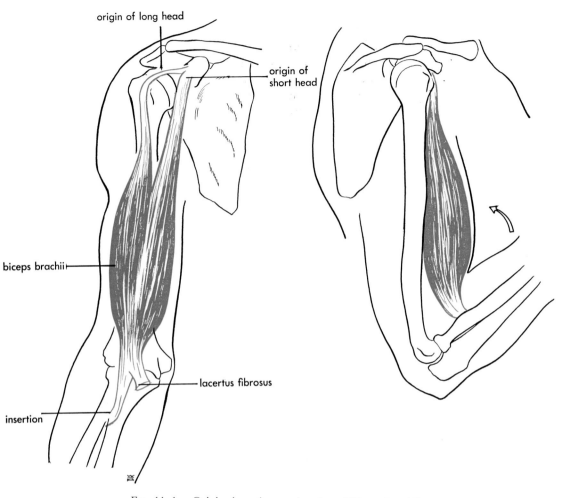

origin of long head

origin of short head

biceps brachii

lacertus fibrosus

insertion

FIG. 11–1.   Origin, insertion, and action of Biceps brachii.

sors, abductors-adductors and so on. Further, when one clenches the fist, the flexors of the fingers and thumb contract, and if this action were unopposed, the wrist would bend too, and interfere with the fist clenching. Flexion of the wrist in this situation is opposed by the extensors of the wrist which aid the fist-clenching operation. The extensors are acting as **synergists.** Finally, in movements of the elbow joint, involved in making a fist and threatening someone, it is feasible to have the shoulder fixed in an appropriate position to facilitate the movement. The muscles contributing in this way are **fixators.**

In all of this interaction or cooperation among muscles in the performance of a given movement, coordination is essential. This **coordination** comes from that part of the brain called the cerebellum. Interference with the nerve pathways to any one muscle of a group may seriously affect the action of the whole "team" or may stop the action through muscle paralysis.

The relationship of the Biceps brachii to the skeleton illustrates three other situations in muscle mechanics (Fig. 11–1). First, the Biceps brachii, because it inserts in part on the tuberosity of the radius, rotates the radius to supinate the hand as it also flexes the elbow joint. Here we have a double action on the part of one muscle. Secondly, the Biceps brachii has its origins, one, the short head, on the coracoid process, the other, the long head, on the superior rim of the glenoid fossa of the scapula. It, therefore, crosses another articulation—the shoulder joint. The whole muscle contracting together may flex and rotate the shoulder joint inward. It also tends to pull the head of the humerus closer to the glenoid fossa and thus to secure the shoulder joint. Thirdly, a muscle with two origins may contract in such a way as to use these origins as separate muscles. When the long head of the Biceps acts alone, it is an abductor of the humerus, the short head may adduct.

Levers. Whether recognizing it or not, almost everyone has had some experience with levers. The wheelbarrow, the beer can opener, the crowbar, the seesaw, the scales in the laboratory, the use of golf clubs or tennis rackets are all examples. In each of these there are certain common factors; a rigid bar is involved, which moves around a point or **fulcrum** when a **force** is applied. The result is the overcoming of a resistance or **weight.** In the case of the wheelbarrow, the beer can opener and the crowbar, a great resistance is overcome with relatively little force. In the seesaw and the laboratory scales, a balance is established between weight and power around a fulcrum, while with the golf clubs and tennis racket, a wide range of motion and speed are achieved at the expense of force.

In studying the actions of the skeletal, articular, and muscular systems we are seeing leverage in use. The bones individually or collectively serve as rigid bars or **levers,** though some of them have peculiar shapes. The articulations are the fulcra around which movement takes place. The force is exerted by the skeletal muscles. When the Biceps brachii

by its contraction pulls upon the radius and causes the elbow joint to bend we are seeing leverage in action. When one raises the arm, extends forearm and hand laterally (*abduction*), the shoulder joint is serving as **fulcrum,** the whole upper extremity as **lever** and the Deltoid as the **force.** The weight is in the arm, forearm, and the hand. It is true that some of the movements of the body are difficult to analyze in terms of leverage, but others are quite obvious.

Three **classes of levers** are recognized and are designated by number.

Class I—(Fig. 11–2). The fulcrum lies between the force point and the weight to be lifted. The seesaw or the crowbar is a good example, when one end is put under the object to be moved and a point in between is supported by a block while force is exerted on the other end. In the human body the head tipping backward on the atlas would be an example. The face might be considered the **weight,** the joint between occipital condyles and atlas, the **fulcrum,** and Splenius capitis, one of the muscles, to exert the **force.**

Class II—(Fig. 11–3). The weight is between the fulcrum and the point at which pull is exerted. The wheelbarrow is a good example. By this means, heavier loads can be carried or moved with less force exerted. It is controversial as to whether this class of leverage is represented in the human body. Some interpret raising the body on the toes as an example, in which case the point of contact of the toes with the ground would be the fulcrum, the weight at the ankle joint, and the pull exerted on the calcaneus by the Gastrocnemius and Soleus muscles through the tendon of Achilles. Another example may be the opening of the

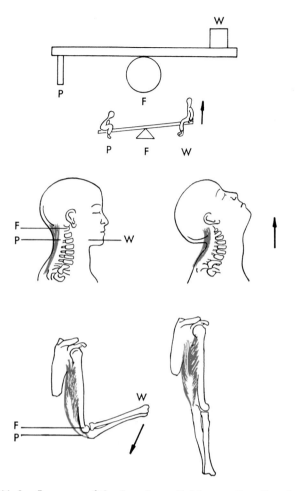

Fig. 11–2.   Leverage of the first class—*F*, fulcrum; *P*, pull or force;
W, weight or resistance.   (Modified from Wells.)

mouth against resistance, as when the teeth are stuck together with taffy.

Class III—(Fig. 11–4).   The force is exerted between the fulcrum and the weight.   The use of golf clubs and the tennis racket are examples where considerable power is required to initiate the movement, but a wide range of action is accomplished.   Flexing the arm at the elbow joint is a good example where the elbow is the fulcrum, the radius and/or ulna the levers, the pull exerted by Biceps and Brachialis muscles, and the weight is at the hand.   Class III levers are the most common in the musculoskeletal system.

The portion of a lever between the fulcrum and the power point is called the **power arm;** that between the fulcrum and the weight to be lifted is the **weight arm.**   A lever whose power arm is longer than the weight arm, whether it be of class one or two, requires less force to lift the weight.   Range and speed are sacrificed.   When the weight arm is longer, however, in either a first- or third-class lever, range of action and speed are favored at the sacrifice of power.   Refer

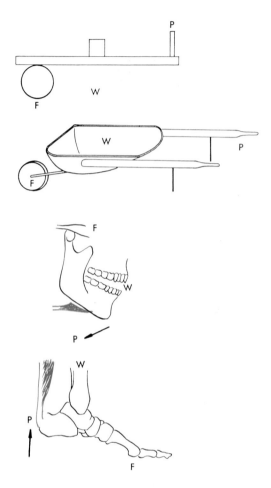

FIG. 11–3. Leverage of the second class—*F*, fulcrum; *P*, pull or force; *W*, weight or resistance.

to Figures 11–2, 3, 4 and try to visualize what influence changing the relative lengths of power and weight arms would have on the actions illustrated.

The **angle of pull** of a muscle on a lever is another factor that needs consideration (Fig. 11–5). The optimum angle is 90 degrees, for then the full force of the muscle is applied directly to the bone. At greater or lesser angles than 90 degrees some of the muscle energy is dissipated. A survey of the muscles of man shows that most of them pull at

something other than a 90-degree angle. Many pull at very small angles, but while power may thus be sacrificed, the range and speed of action are great, which serve well the needs of the organism.

Fiber Direction and Muscle Action. Some muscles have their fiber bundles (*fasciculi*) all running parallel to each other and to the long axis of the muscle. Such a muscle can shorten by about one-third to one-half of the length of its belly because each fiber can shorten by this much. Other muscles have their fasciculi leading into the side of a tendon which runs the entire length of a muscle. Because this arrangement gives the appearance of a feather, it is called **pennate.** If the fasciculi all come in from one side of the tendon, the muscle is **unipennate,** as in the Semimembranosus; if they come in from two sides to the central tendon it is **bipennate,** as in the Rectus femoris; and if they come in to converge on many tendons they are **multipennate** as in the Deltoideus. In a few muscles as the Temporalis, the fibers converge from a broad area into a common tendinous point and are said to be **radiate** (Fig. 11–6).

Pennate muscles of the various types give greater power because the force of a great many fibers can be brought to bear upon the tendon. They produce, however, a shorter range of motion than those with their fasciculi running on the longitudinal axis of the muscle.

One should notice also that a number of the muscles that we study have their tendons operating over pulleys which change the angle at which the force is applied. They, in some cases, enable the muscle to produce a movement which it could not otherwise perform. The Gracilis muscle, for example, has its tendon of insertion passing over the bulging medial condyle of the knee just before it attaches to the tibia and this changes its angle of pull (Fig. 11–5). The tendon of insertion of the Peroneus longus passes behind the

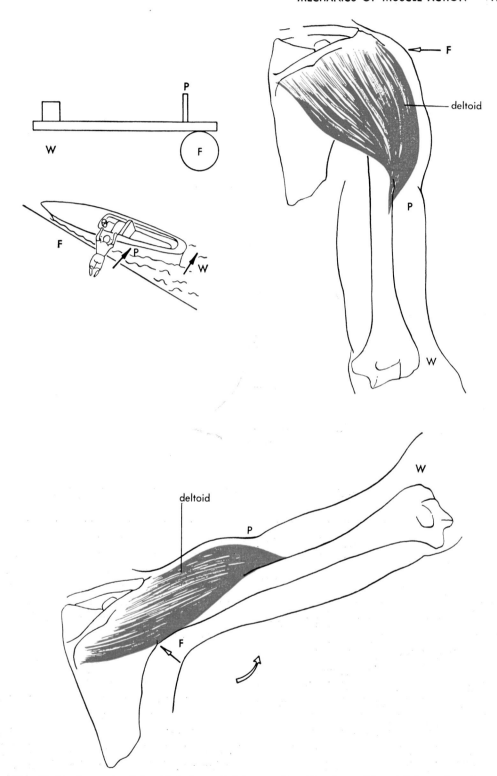

Fig. 11–4. Leverage of the third class—*F*, fulcrum; *P*, pull or force; *W*, weight or resistance. (Modified from Wells.)

14

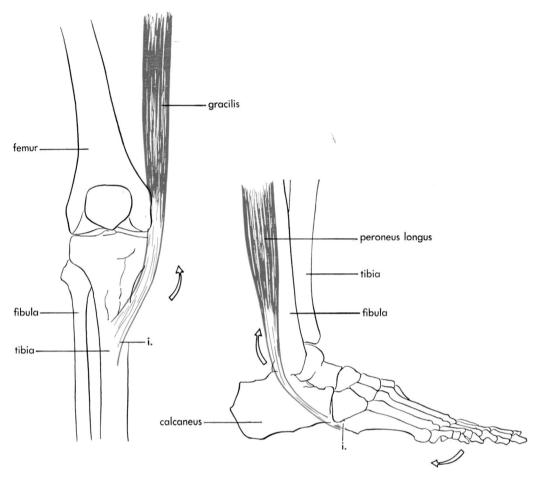

FIG. 11–5. Pulleys change the angle at which the force of a muscle is applied.

lateral malleolus of the fibula and then inserts on the underside of the first cuneiform and the base of the first metatarsal. It plantar-flexes the foot at the ankle. If this tendon passed in front of the lateral malleolus, its pull would be in front of the ankle joint and it would dorsiflex the foot (Fig. 11–5).

Gravity. The force of gravity has an important constant influence upon the action of muscles. Indeed, its influence upon the evolution of life from the water to less buoyant air, and upon the rise of man as a bipedal organism has been great. It is an important factor in the early life of each of us as we first attempt to lift our weight from the substratum, to creep, to stand, and finally to walk.

It remains throughout life a function of our muscles to maintain the posture of the body, and to provide the power for a multitude of other motor and locomotor activities against, or in cooperation with, the force of gravity. Gravity must always be a consideration in our study of muscle action.

FASCIAE

Thus far in the book the term fasciae has been used in reference to certain connective tissues with no effort to define it. In *Gray's Anatomy*, by Goss (1966), fasciae are defined as "The dissectable, fibrous connective tissues of the body, other than the specifically organized structures, tendons, aponeuroses, and ligaments." The fasciae enclose or invest various structures of the body, and receive special names. They vary in thickness and strength, and in the relative amounts of fat, collagenous and elastic fibers, and tissue fluid which they contain.

They constitute a structural and functional system of the body.

The fasciae fall into three categories; the **superficial, deep,** and **subserous.** The superficial fascia, already briefly described in Chapter 6, is found between the skin and the deep fascia over the

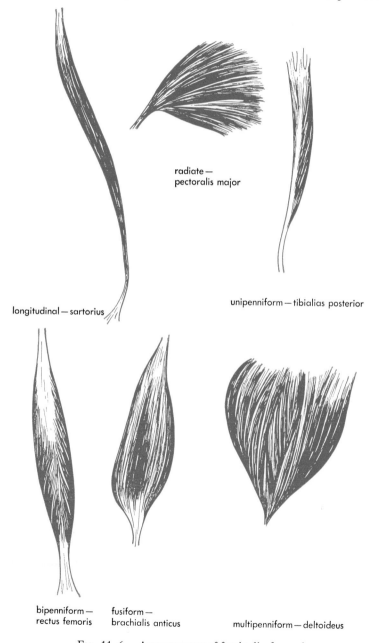

radiate —
pectoralis major

longitudinal — sartorius

unipenniform — tibialias posterior

bipenniform —
rectus femoris

fusiform —
brachialis anticus

multipenniform — deltoideus

FIG. 11–6.   Arrangement of fasciculi of muscles.

entire body. It is composed of two layers, the outer of which is the **panniculus adiposus,** which usually contains fat and varies considerably in thickness. The inner layer is thin, membranous, and quite elastic. The two layers can be separated by careful dissection and between them lie the superficial arteries, veins, lymphatics, nerves, mammary glands, and the facial muscles.

The superficial and deep fasciae are easily separated by blunt dissection in most parts of the body, but adhere over the bony prominences.

The **deep fascia** is the most extensive of the three subdivisions, the one which invests the body wall and appendages and penetrates among the structures which constitute them. It holds muscles and other structures together or separates them into independently functioning groups. The deep fascia forms a continuous system of membranes throughout, membranes which undergo much splitting, rejoining, and fusing with such structures as periosteum, perichondrium,

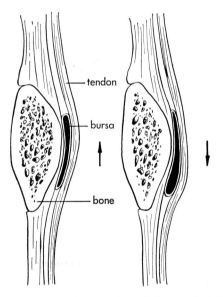

Fig. 11–7. A bursa. In a living subject the bursa would be collapsed and its inner walls moistened by synovial fluid. Arrows indicate directions of movement of tendon. (Modified from Basmajian.)

and aponeurosis. The membranes, therefore, vary in thickness and strength, depending upon the functions they serve.

The deep fascia, though continuous, can be divided into three parts. The **outer investing layer** lies just under the superficial fascia and covers the trunk, neck, part of the head, and the limbs. The **internal investing layer** is an extensive sheet which lies on the inside of the body wall in the trunk region where it forms a part of the walls of the body cavities, and is covered internally by the subserous fascia and the serous membranes. The **intermediate membranes** connect the outer and internal investing layers from which they are derived by splitting. They form the compartments around muscles and other structures of the body wall.

The deep fasciae are most complex and some of them will be described in connection with our study of the body systems, particularly the skeletal muscles. Bursae and tendon sheaths, described below, are special modifications of the deep fascia which facilitate movements between parts.

The **subserous fascia** is located between the internal investing layer of the deep fascia and the serous membranes lining the body cavities. A cleft of variable thickness separates subserous and deep fascia, allowing a sliding motion between them. The subserous fascia is thin in some areas, thick in others, such as around the kidneys where considerable fat is located (Fig. 15–4). The parietal layer of the subserous fascia is continuous with that of the visceral layer where the serous membranes are reflected over organs and at the mesenteries.

Bursae and Tendon Sheaths. These structures have been briefly described in Chapter 7, page 92 and should be reviewed at this time. Figures 11–7, 8 show some of the details of structure of tendon sheaths and bursae. Note that the tendon is held to the bone, often in a groove in a

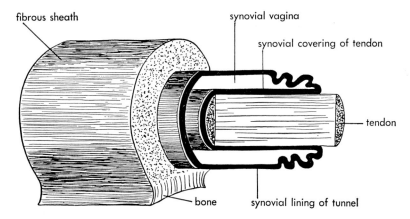

fibrous sheath

synovial vagina

synovial covering of tendon

tendon

bone

synovial lining of tunnel

FIG. 11–8.   Scheme of tendon synovial sheath.   In the living subject the tendon fits snugly in its tunnel.

bone, by a fibrous sheath. This fibrous sheath is lined by a synovial membrane which doubles back at the end of the tunnel to form a covering on the tendon. The synovial membrane secretes a small amount of fluid which lubricates the membrane surfaces and allows them to move freely one over the other. Such sheaths are most common at the distal ends of the limbs. They make it possible for muscles, having their insertions at some distance out on the extremity from their bellies, to act with ease and efficiency, yet their tendons cannot pull away from the underlying skeletal components. (See also Figures 11–48 and 49, pp. 238–241.)

CLASSIFICATION OF MUSCLES

The classification of muscles will follow the same plan as was used for the classification of the parts of the skeleton, as follows:

A. Muscles of the Axial Skeleton
   1. Muscles of the head and neck
   2. Muscles of the vertebral column
   3. Muscles of the thorax
   4. Muscles of the abdomen
   5. Muscles of the pelvis
   6. Muscles of the perineum

B. Muscles of the Upper Extremity
   1. Muscles connecting the axial skeleton and the shoulder girdle
   2. Muscles connecting the axial skeleton and the arm
   3. Muscles connecting the shoulder girdle and the arm
   4. Muscles and fascia of the arm
   5. Muscles and fascia of the forearm
   6. Intrinsic muscles of the hand
      a. Fascia of the hand
      b. Practical considerations

C. Muscles of the Lower Extremity
   1. Muscles of the hip
   2. Muscles of the thigh
   3. Muscles of the leg
   4. Intrinsic muscles of the foot

Names of Muscles. Unfortunately, there is no one common criterion for the naming of individual muscles. Rather, many criteria have been used, among them, shape, size, location, action, origin, insertion, and others. But, the names are descriptive and this is of some help in learning the muscles. The shape of the muscle is the criterion for naming in the Trapezius and the Rhomboideus. Size relationships are indicated in Gluteus maximus, Gluteus medius, and Gluteus minimus; location in Rectus abdominis and Intercostals; size, action, and location in Extensor digitorum longus; origins and insertions in Sternocleidomastoideus and Coracobrachialis. One who has labored through the naming of the bones and their parts, wondering why it was important, will now be able to apply that knowledge in the learning of muscles. Understanding comes with increasing knowledge of facts, and understanding is one of our goals. Aristotle said, "Let us first understand the facts, and then we may seek the cause."

MUSCLES OF THE AXIAL SKELETON

MUSCLES OF THE HEAD AND NECK

Facial Muscles. These are the muscles of facial expression by which man willingly, or sometimes unwillingly, reveals his deeper feelings—his emotions. They are essential tools of the actor who attempts to turn the words of a play into living experience and the characters into real personalities. Indeed, as Shakespeare once said, "all the world is a stage . . ." These muscles serve other functions too, some of which will be mentioned as the following individual muscles are briefly described.

> **Orbicularis oculi**
> Corrugator
> Procerus
> **Orbicularis oris**
> Dilator naris anterior
> Dilator naris posterior
> **Buccinator**
> **Epicranius**
> **Platysma**

Study of Figure 11–9 will help you to realize the number of facial muscles

which we have and the variations in their shapes, sizes, and actions. They have their origin in the superficial fascia, or some of them, on the underlying bones and they generally insert into the skin. They are **cutaneous** muscles and are supplied by the seventh cranial or **facial nerve.** Some of them are composed of pale and scattered bundles of fibers, others are more distinct. Many tend to merge one into another.

Conspicuous among the facial muscles are the sphincter types, **Orbicularis oculi** and **Orbicularis oris. Orbicularis oculi** arises from the maxillary and frontal bones on the medial side of the orbit and from a short fibrous band, the **medial palpebral ligament.** It passes around the circumference of the orbit and spreads up onto the frontal bone and downward to the cheek. A thinner part, the palpebral portion, passes in the eyelids to insert laterally into a raphe. The palpebral portion is used in gentle closing of the eyes as in sleeping or blinking. The whole muscle is involved in winking and in any stronger contraction of the muscle. Such strong contractions cause the skin of the forehead and cheek to be drawn toward the medial side of the orbit and to form folds radiating out at the lateral angles of the eye. These become permanent in old age and are called "crow's feet." The strong contraction of the muscle is also used in emergency to protect the eye and it also compresses the lacrimal sac, helping in the flow of tears. The vertical folds which we sometimes see above the medial sides of the orbits are caused by the Corrugator muscle—the "frowning" or "sorrow" muscle. It is closely associated with the Orbicularis oculi.

The **Orbicularis oris** encircles the mouth. Its contraction closes the lips and pushes them forward and therefore is sometimes called the "kissing" muscle. It is directly involved with neighboring muscles of the lips and cheeks, some of

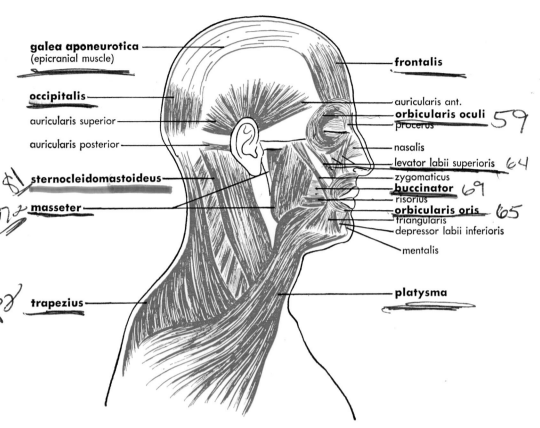

galea aponeurotica
(epicranial muscle)

occipitalis

auricularis superior

auricularis posterior

sternocleidomastoideus

masseter

trapezius

frontalis

auricularis ant.
orbicularis oculi
procerus

nasalis
levator labii superioris
zygomaticus
buccinator
risorius
orbicularis oris
triangularis
depressor labii inferioris

mentalis

platysma

Fig. 11–9.   Some muscles of the head and neck.

whose fibers become a part of it. These muscles, depending on their position, raise the upper lip, depress the lower one, draw out the corners of the mouth as in grinning, as others draw the corners upward as in smiling, or downward as in a grimace.

The muscles of the nose are mostly small and inconspicuous. The Procerus muscles produce transverse wrinkles over the bridge of the nose by drawing down the medial angle of the eyebrows. Anterior and Posterior dilator muscles enlarge the external nares and resist the tendency for them to be closed by atmospheric pressure. They also give expression to emotions, e.g., their strong contractions in anger.

The **Buccinator** is the principal muscle of the cheek. Its origin is from the outer surface of the alveolar processes of the

mandible and maxilla opposite the molar teeth. It inserts into the corner of the mouth, its fibers blending with those of the Orbicularis oris. It compresses the cheek so that in mastication it helps to hold the food between the teeth. The word Buccinator, from the Latin, means trumpeter and this muscle is sometimes called the "trumpeter" muscle because it compresses the cheeks and forces air out of the mouth as is necessary in playing a trumpet.

Under the thick skin of the scalp is the **Epicranius** or Occipitofrontalis muscle. It consists of a broad, tendinous layer over the upper part of the cranium, the **galea aponeurotica,** and two flat, thin muscle bellies, the one over the frontal bone, the **Frontalis,** the other over the occipital bone, the **Occipitalis.** The Occipitalis has its origin on the occipital bone and inserts into the galea aponeu-rotica. Its contraction draws the scalp backward. The Frontalis originates in the galea aponeurotica, having no attachment to bone, and inserts into the fibers of the Procerus, Corrugator, and Orbicularis oculi muscles. It raises the eyebrows and wrinkles the skin of the forehead transversely. This is sometimes called the expression of surprise.

One other muscle, the **Platysma,** should be mentioned, although it could as well be described with the muscles of the neck. It arises from the fascia cover-ing the superior parts of the Pectoralis major and Deltoideus muscles, crosses the clavicle, and goes obliquely up the side of the neck to join fibers of its counterpart from the opposite side behind the symphysis of the mandible. Some of its fibers go to the mandible, others to skin areas of the face intermingling with muscles around the corners and lower parts of the mouth. Its action draws the angle of the mouth and lower lip downward. Under some circumstances it may lower the mandible. In strong contractions it may pull the skin upward from

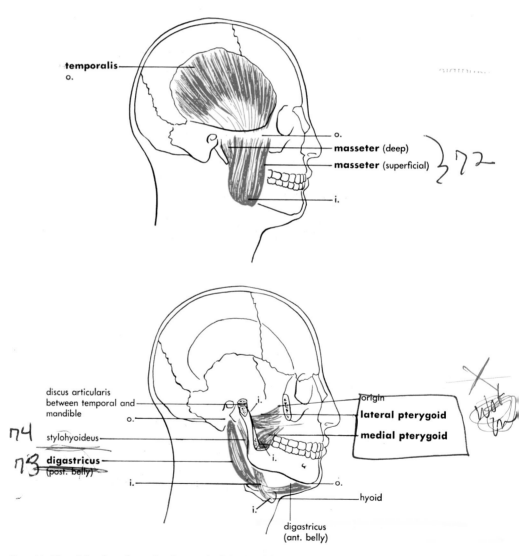

FIG. 11–10.   Muscles of mastication and of the hyoid.   Origins indicated by *O*, insertions by *I*.

the region of the clavicle and increase the width of the neck.

Muscles of Mastication. The following four pairs of muscles are used in biting and chewing. Their origins and insertions should be noted carefully on Figures 11–10, 11.

> **Temporalis**
> **Masseter**
> **Pterygoideus medialis**
> (*Internal pterygoid*)
> **Pterygoideus lateralis**
> (*External pterygoid*)

The **Temporalis** muscle is an extensive radiate muscle on the side of the skull, arising from the temporal fossa and inserted into the coronoid process of the mandible. Its action closes the jaws and

it is a powerful chewing muscle. Its contractions can be seen or palpated on the side of the head.

The **Masseter,** which comes from the Greek meaning "chewer," is a thick, quadrangular muscle extending between the zygomatic arch and the external surface of the ramus of the mandible. Its outline can easily be felt when you "clench the teeth."

The **Pterygoideus medialis** inserts on the medial surface of the ramus of the mandible opposite the Masseter and originates on the medial surface of the lateral pterygoid plate of the sphenoid. The Pterygoideus medialis and the Masseter suspend the angle of the mandible as in a sling, the **mandibular sling,** and in their contractions move the mandible

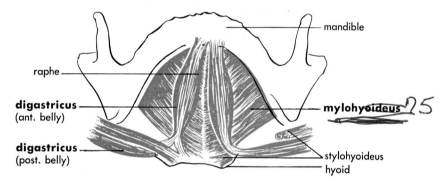

FIG. 11–11.   Muscles of the floor of the mouth and hyoid.

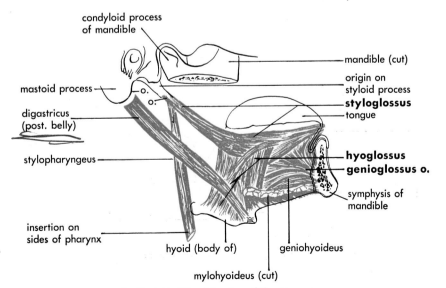

FIG. 11–12.   Muscles of the tongue.

in relationship to the maxilla, using the temporomandibular joint as a guide. In addition to bringing the jaws together in biting and chewing, the Pterygoideus medialis pulls the mandible to the opposite side, as in grinding movements.

The **Pterygoideus lateralis** takes an entirely different direction than the other muscles of mastication, its fibers being at nearly right angles to theirs. It arises from the lateral surface of the lateral pterygoid plate, opposite the Pterygoideus medialis, and inserts into the neck of the condyle of the mandible and into the disc of the temporomandibular joint. Its action, therefore, is to pull the jaw forward, as in protruding the chin. It also opens the mouth, usually with the help of muscles of the floor of the mouth (Fig. 11–11). The Pterygoideus lateralis,

acting alternately with its fellow of the opposite side, produces side-to-side movements of the jaw, as in chewing and grinding.

The muscles of mastication are innervated by the mandibular branch of the trigeminal or fifth cranial nerve.

Muscles of the Tongue (Fig. 11–12). The tongue is primarily a muscular organ covered with mucous membrane which lifts up from the floor of the mouth. Its muscles may be divided into two types, those which lie entirely within the tongue, the **intrinsic muscles,** and those which reach the tongue from some neighboring part, called the **extrinsic muscles.**

The **intrinsic muscles** are arranged in the longitudinal, vertical, and transverse planes of the tongue and by their con-

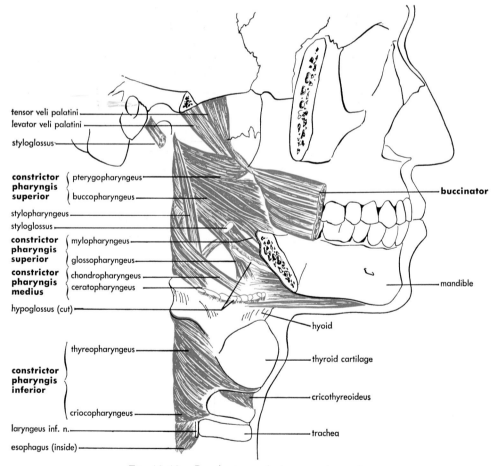

FIG. 11–13. Buccinator and pharyngeal muscles.

tractions alter the shape of the tongue, as in articulate speech, whistling, and manipulating food.

The **extrinsic muscles** are three in number (Fig. 11–12).

Styloglossus
Hyoglossus
Genioglossus

The origins and insertions of the extrinsic tongue muscles are clearly shown in Figure 11–12 and further indicated in their names. Together with the intrinsic muscles they make the tongue one of the most versatile organs of the body as far as movement is concerned. We can, at will, change its form, or it can be retracted, protracted, elevated, or depressed.

The Glossopalatinus (Palatoglossus) extends from the palate to the tongue, but is usually considered a palate muscle

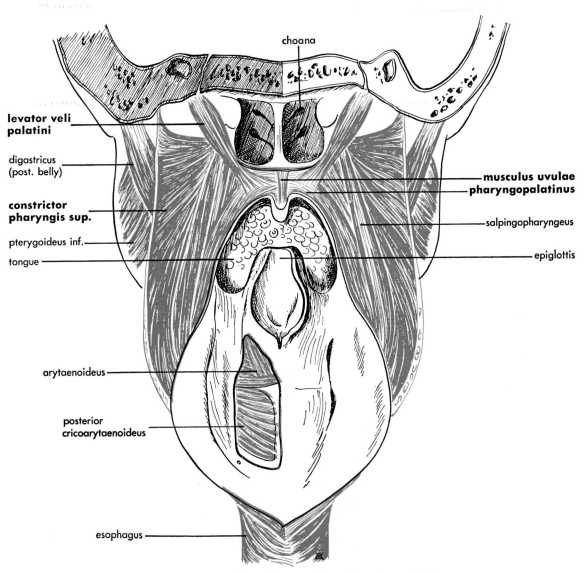

FIG. 11–14.   Posterior view of the muscles of the palate and pharynx.

because it develops with that group and its nerve supply is from the spinal accessory (11th cranial) rather than the **hypoglossal** or twelfth cranial which supplies the tongue muscles.

Muscles of the Pharynx and Palate. The muscles of the pharynx are (Figs. 11–13, 14):

> **Constrictor inferior**
> **Constrictor medius**
> **Constrictor superior**
> **Stylopharyngeus**
> Salpingopharyngeus
> **Pharyngopalatinus**

The muscles of the pharynx are arranged in two layers, an outer circular and an inner longitudinal. The outer circular layer is the more prominent, and consists of three large, paired muscles, the **Inferior, Middle,** and **Superior constrictors.** The other three muscles constitute the longitudinal layer. Study these muscles carefully on Figures 11–12, 13. Notice that the Constrictors all have narrow origins anteriorly from which they widen and spread backward and medialward to insert with the muscle of the opposite side, into the medial posterior raphe of the pharynx. The posterior parts of the Constrictors overlap posteriorly; the Inferior over the Medius; the Medius over the Superior. Notice too, that at its origin the Constrictor superior is almost continuous with the Buccinator, the pterygomandibular raphe marking the point where their fibers interlace. The origin of the Constrictor medius is associated with the hyoid bone, that of the Constrictor inferior with the thyroid and cricoid cartilages of the larynx.

The **Stylopharyngeus** muscle has a narrow origin on the styloid process of the temporal bone. It passes into the pharynx in the space above the Middle constrictor and spreads out over the inner surface of the Middle and Inferior constrictors. Its lowermost fibers reach

the posterior border of the thyroid cartilage on which they insert.

The Salpingopharyngeus has its origin on the inferior part of the auditory tube and passes downward to blend with the Pharyngopalatinus. The Pharyngopalatinus will be described below with the palatine muscles which are as follows:

> Levator veli palatini
> Tensor veli palatini
> Musculus uvulae
> **Glossopalatinus**
> **Pharyngopalatinus**

Close study of Figure 11–14 will help you to place these muscles and to realize their close relationships to the pharyngeal muscles in position and actions. The Levator palati and Tensor palati arise from the base of the skull lateral to the cranial attachment of the pharynx and descend to become a part of the soft palate. Between these two muscles superiorly is the opening of the auditory tube which is, as you may recall, homologous to the first gill cleft.

The **Glossopalatine** muscle descends from the soft palate to the tongue, the **Pharyngopalatine** from the palate to the pharynx. They are covered with mucous membrane and stand out prominently, forming between them the fossa which lodges the palatine tonsil. These structures can be seen easily through the open mouth. They form the **glossopalatine** and **pharyngopalatine arches** (Fig. 13–4).

The muscles of the mouth, tongue, pharynx, and palate work together in the complex function of deglutition or swallowing, a process which will receive further attention in Chapter 13, page 359, on the Digestive System.

Cervical fasciae (*fascia colli*). The neck is a connecting structure between the head and the thorax and therefore a very complex and important part of the body. Its structures are highly compartmentalized by the cervical fascia which

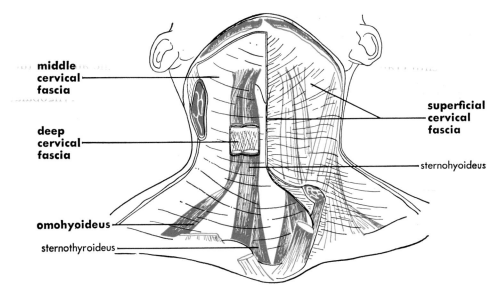

middle
cervical
fascia

deep
cervical
fascia

superficial
cervical
fascia

sternohyoideus

omohyoideus

sternothyroideus

FIG. 11–15.   The cervical fasciae.

is illustrated in Figure 11–15. Study of this illustration will convey a general impression of the fascia without a detailed description which would go beyond the purpose of this book. The cervical fascia extends upward to the zygomatic arch and downward to the clavicle. It encloses the parotid salivary gland, and when this gland is infected, as in the mumps, the fascia is so restrictive that the swelling of the gland creates great pressure, which in turn causes the pain characteristic of this condition.

Cervical Triangles. Removal of the Platysma, a large, superficial cervical muscle which was described above, reveals two roughly triangular areas on each side of the neck, the cervical triangles. The dividing line between them is the Sternocleidomastoideus muscle which runs an oblique course along the side of the neck (Fig. 11–16). The **anterior triangle** lies between the midline anteriorly and the Sternocleidomastoideus laterally. The mandible marks its superior boundary. The **posterior triangle** is between the adjacent borders of the Sternocleidomastoideus and the Trape-

zius muscles, with the clavicle forming its inferior border. Various muscles, nerves, and blood vessels are located in relationship to these triangular areas.

Lateral Cervical Muscles.   These muscles are two in number:

Sternocleidomastoideus
Trapezius

The Trapezius will be considered later with the muscles of the back and shoulder.

The **Sternocleidomastoideus,** as its name suggests, has its origin by two heads, one on the manubrium of the sternum, the other on the clavicle; its insertion is on the mastoid process of the temporal bone. These muscles contracting together flex the cervical vertebral column, and the chin is slightly elevated as the head moves forward. Acting singly, they flex the vertebral column laterally, bringing the head closer to the shoulder of the same side while the chin is elevated and turned to the opposite side (Fig. 11–16).

Suprahyoid Muscles. These muscles lie in the floor of the mouth above the hyoid bone to which they have attachments.

Digastricus
Stylohyoideus
Mylohyoideus
Geniohyoideus

Figures 11–11, 12, and 17*A* show the relationships of these muscles. Note especially the **Digastricus** with its anterior and posterior bellies separated by an intermediate tendon which perforates the Stylohyoideus muscle and is held to the hyoid bone by a fibrous loop.

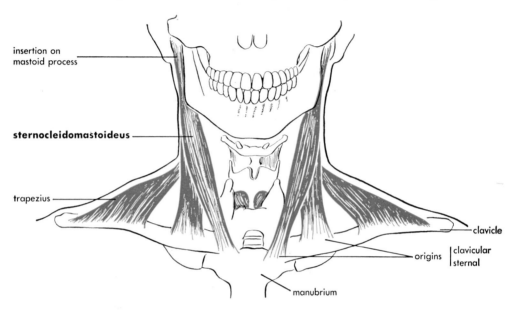

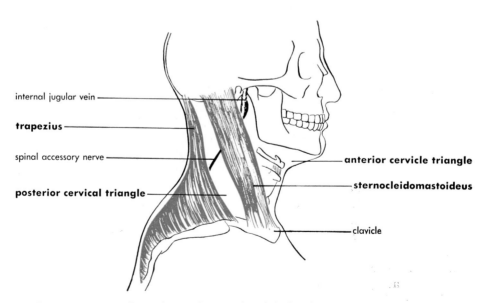

FIG. 11–16.   Anterior and posterior muscles of the head, neck and cervical triangles.

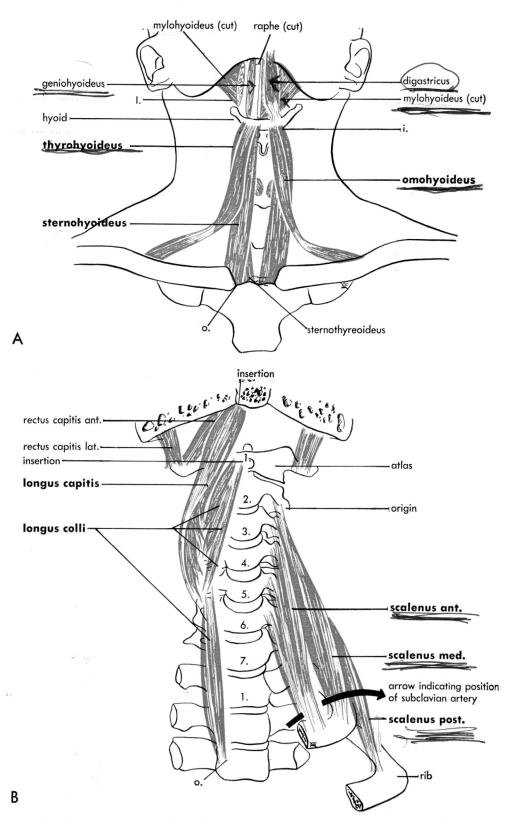

FIG. 11–17. Anterior and lateral muscles acting upon structures of the head and neck.
A, Suprahyoid and infrahyoid ("strap") muscles. *B*, Anterior and lateral vertebral muscles.

The combined actions of the suprahyoid muscles are to elevate the hyoid bone and with it the base of the tongue and larynx as in deglutition; when the hyoid bone is fixed by the infrahyoid muscles, they lower the mandible. The anterior belly of the Digastricus contracting alone pulls the hyoid forward; when the posterior belly contracts alone it pulls the hyoid backward.

Infrahyoid Muscles. The relationships of these muscles, their origins, and insertions are shown in Figure 11–17A. They are sometimes called the "strap muscles."

> Sternohyoideus
> Sternothyroideus
> Thyrohyoideus
> **Omohyoideus**

The **Omohyoideus,** like the Digastricus above, is a two-bellied muscle with an intermediate tendon. This tendon is held in position by a long process of the deep fascia which sheathes it and attaches it to the clavicle and first rib. This accounts for the angular form of the muscle as it passes from its origin on the superior border of the scapula to its insertion on the hyoid bone. It is this muscle, too, which divides the cervical triangles into upper and lower triangles (Fig. 11–15).

The infrahyoid muscles depress the hyoid and larynx after they have been elevated by the suprahyoid and other muscles in the process of deglutition. The Omohyoid also pulls the hyoid backward and to the side and tenses the cervical fascia.

Anterior Vertebral Muscles (Fig. 11–17B).

> **Longus colli**
> **Longus capitis**
> **Rectus capitis anterior**
> **Rectus capitis lateralis**

These muscles flex the head when acting together. The Longus colli slightly rotates the cervical vertebral column, thus turning the head to the side. The Rectus capitis lateralis, acting on one side, bends the head laterally.

Lateral Vertebral Muscles (Fig. 11–17B).

> **Scalenus anterior**
> **Scalenus posterior**
> **Scalenus medius**

Since these muscles arise from the transverse processes of cervical vertebrae and insert on the first and second ribs, they have a two-fold function. They elevate the first two ribs and thus aid in inspiration of air into the lungs, and they flex the cervical spine. If the muscles on only one side contract, they bend the vertebral column to that side.

Posterior Vertebral Muscles. Turning now to the posterior side of the head and neck we find muscles which oppose the flexing action of the anterior and lateral muscles by extending the head and neck. They also, like those anteriorly and laterally, rotate the head and move it laterally. They are separated on the midline of the neck by the supraspinous ligaments which form a prominent partition, the **ligamentum nuchae.** In lower animals, as the horse, this elastic ligament holds up the head.

Some muscles in this area, like those anterior and lateral muscles of the neck described above, classify also as muscles of the vertebral column. Only a few will be considered at this point; others will be mentioned later under that category.

Some prominent posterior muscles of the head and neck are the following which should be studied in Figures 11–18, 22, 23.

> **Splenius capitis**
> Semispinalis capitis
> Longissimus capitis
> **Splenius cervicis**
> Semispinalis cervicis
> Longissimus cervicis

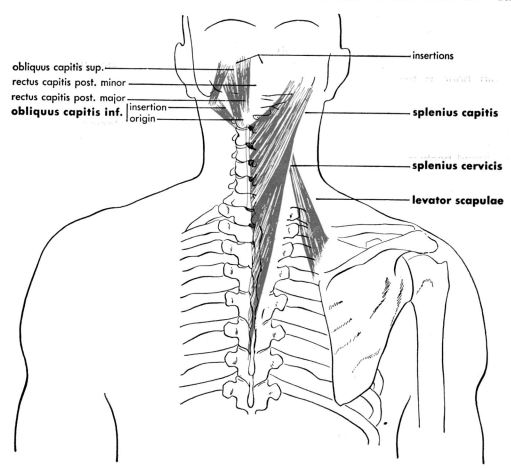

obliquus capitis sup.
rectus capitis post. minor
rectus capitis post. major
obliquus capitis inf. |insertion
                          |origin

insertions

splenius capitis

splenius cervicis

levator scapulae

FIG. 11–18.   Posterior and lateral vertebral and suboccipital muscles.

Smaller, though important, muscles of this area are the suboccipital muscles, specifically the

> Rectus capitis posterior major
> **Obliquus capitis inferior**
> Rectus capitis posterior minor
> Obliquus capitis superior

Notice especially that the **Obliquus capitis inferior** muscles originate on the spinous process of the axis (*second cervical*) and insert on the transverse processes of the atlas. These muscles, contracting alternately, rotate the atlas on the dens of the axis and thus account for the "no" movements of the head.

15

MUSCLES OF THE VERTEBRAL COLUMN

The muscles of the vertebral column constitute two groups, the prevertebral and postvertebral. These muscles contribute not only to the movements of the vertebral column and its adjacent parts, but are involved in maintaining its integrity and balance. They also give the column stability in its weight-supporting function and when it is involved in movements, such as those of the upper and lower extremities.

The **prevertebral muscles** are found primarily in the cervical and lumbar regions of the spine. They are less

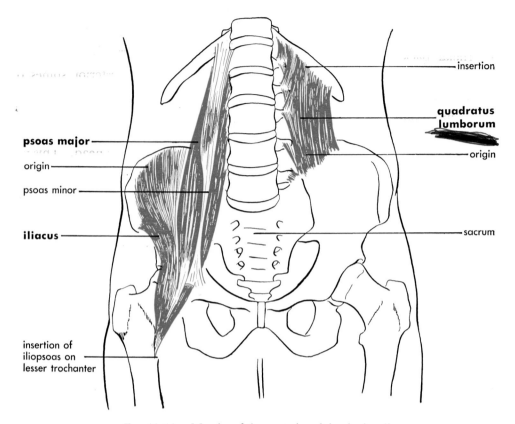

FIG. 11–19.   Muscles of the posterior abdominal wall.

numerous than the postvertebral muscles and are better defined in form and function. Those in the cervical region have been described under the section on head and neck muscles. It leaves only those of the lumbar region for consideration, namely the:

Psoas major	**Diaphragm**
Psoas minor	**Iliacus**

The Psoas muscles are shown in Figure 11–19. They belong as much to the lower extremity as to the vertebral column. The **Psoas major** is large and strong and flexes the lumbar spine and bends it laterally as well as flexing and rotating the thigh. It opposes and balances the action of the postvertebral muscles in the lumbar region. The Psoas minor is often absent and at best is a very small muscle which flexes the pelvis and the lumbar spine. Closely associated with the Psoas major is the **Iliacus** muscle, so close, in fact, that they are often considered one muscle, the **Iliopsoas.**

The **Diaphragm,** though chiefly a muscle of respiration, does have important attachments, in the form of strong tendons, the **crura,** to the bodies of the first three lumbar vertebrae, and by the **lateral** and **medial lumbocostal arches** to the transverse process and bodies of the first two lumbar vertebrae (Fig. 11–20). The diaphragm inserts on a central tendon. Further consideration of the Diaphragm will be given under the muscles of the thorax and in the chapter on the respiratory system.

The **postvertebral muscles** and muscles of the back are numerous, and varied in structure and function. The more superficial back muscles are concerned with the movements of the upper extremity, another deeper group with the function of breathing. Discussion of all these will be postponed until those functions are considered.

The deep back muscles, those of the posterior vertebral column, are our immediate concern. They are separated from the more superficial muscles by the **thoracolumbar** (*lumbodorsal*) and the **nuchal fasciae** (Figs. 11–21, 15). These fascial coverings attach medially to the ligamentum nuchae, the spinous processes and supraspinous ligaments of the

in the aponeurosis of origin of the Transversus abdominis muscle (Fig. 11–21).

The muscles in this group extend from the sacrum and posterior spines of the ilium to the skull. They constitute a serially arranged complex of muscles which in a functional sense could be looked upon as one great extensor of the vertebral column and head. This muscle mass can be divided into a superficial stratum whose fasciculi turn mostly laterally as they ascend, the **transverso-costal group,** and a deeper stratum whose fasciculi turn medially as they ascend, the **transverso-spinal group.** These groups may be subdivided as follows (Figs. 11–22, 23):

Transverso-costal Muscles

Splenius capitis ⎫
Splenius cervicis ⎭ studied under muscles of head and neck (Fig. 11–18).

Erector spinae (*Sacrospinalis*)

Lateral Column	**Intermediate Column**	**Medial Column**
Iliocostalis	**Longissimus**	**Spinalis**
I. lumborum	L. thoracis	S. thoracis
I. thoracis	L. cervicis	S. cervicis
I. cervicis	L. capitis	S. capitis

Transverso-spinal Muscles

Semispinalis   Rotatores   Interspinales
Multifidus   Intertransversarii

whole vertebral column and to the median sacral crest. Laterally, they extend to the angles of the ribs beyond the Iliocostalis muscle and in the lumbar region to the aponeurosis of origin of the Transversus abdominis muscle. The **nuchal fascia** is a part of the cervical fascia described above and covers the Splenius muscle of the neck region. The **thoracolumbar fascia** has a posterior layer which extends laterally over the Erector spinae muscle and an anterior layer covering the same muscle anteriorly and ending medially in the transverse processes of the lumbar vertebrae. The two layers meet laterally in the lumbar region

The position and extent of each of the above muscles are pretty well indicated in their names, and with the help of the illustrations, their general functions become clear. In addition to extending the vertebral column, some of them bend it laterally, a few rotate it. Others move the head in relationship to the neck and trunk.

Muscles of the Thorax

These muscles may be defined as those which have their insertions on the rib or thoracic cage and by elevating or depressing the ribs cause breathing. Included, of course, is the diaphragm, which inserts on its own central tendon (Fig. 11–20). These muscles are overlain and therefore obscured by large superficial muscles which, functionally, belong to the appendicular musculature. Examples are the Latissimus dorsi, Trapezius, and Pectoralis major. The thoracic muscles are as follows. (Figs. 11–21 to 24):

Outer Wall of Thoracic Cage

External intercostals
Levatores costarum
Serratus posterior superior
Serratus posterior inferior

Inner Wall of Thoracic Cage

Internal intercostals
Subcostals
Transverse thoracic

Floor of Thoracic Cage

Diaphragm

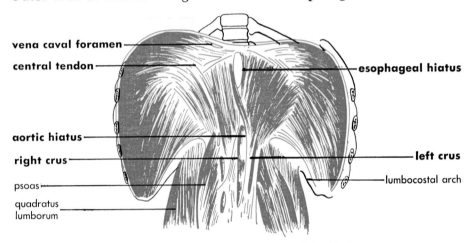

Fig. 11–20.   View of the inferior surface of the diaphragm.

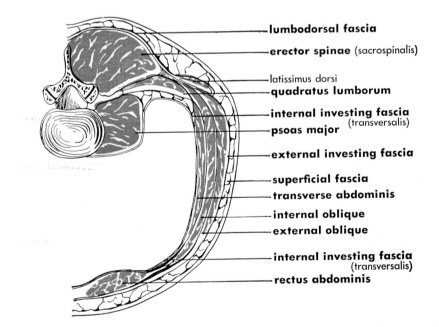

Fig. 11–21.   Fasciae and muscles of the back and abdomen.   Diagrammatic cross-section.

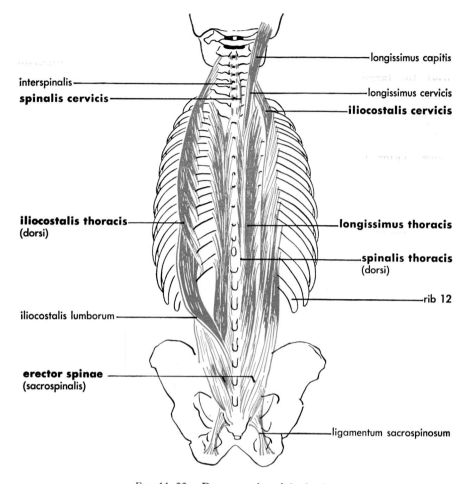

interspinalis

spinalis cervicis

iliocostalis thoracis
(dorsi)

iliocostalis lumborum

erector spinae
(sacrospinalis)

longissimus capitis

longissimus cervicis

iliocostalis cervicis

longissimus thoracis

spinalis thoracis
(dorsi)

rib 12

ligamentum sacrospinosum

Fig. 11–22.   Deep muscles of the back.

Deep Fascia. The thoracic cage proper, which is composed of the ribs, costal cartilages, sternum and the intercostal muscles is covered externally by the **external intercostal fascia** and internally by the **endothoracic fascia** and **subserous fascia.**

The **external intercostal fascia** is continuous superiorly with the fascia of the Scalenus muscle, inferiorly with a fascia between the Internal and External abdominal muscles. Posteriorly, this fascia joins the thoracolumbar (*lumbodorsal*) fascia which, as described above, has posterior and anterior layers embracing the Erector spinae muscles.

The **subserous fascia** lies between the endothoracic fascia and the pleural membranes and furnishes the connective tissue investment for the mediastinal structures.

Muscles of Outer Wall of Thorax. The **External intercostal** muscles are eleven in number on each side and are the superficial muscles of the intercostal spaces. They have their origin from the inferior border of each rib and insert into the superior border of the next rib below. In the posterior wall of the thorax their fibers run obliquely downward and lateralward; in the anterior wall downward, forward, and medialward. They extend from the tubercles

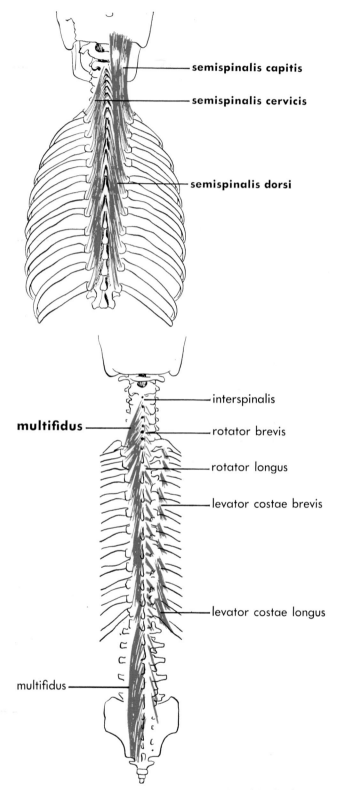

FIG. 11–23.   Some of the deep muscles of the back.

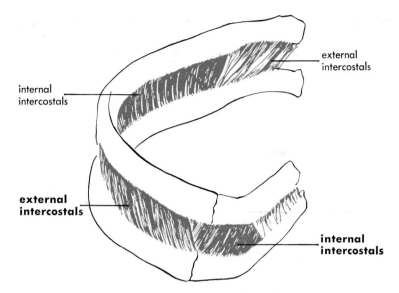

external
intercostals

internal
intercostals

**external
intercostals**

**internal
intercostals**

Fig. 11–24.   The intercostal muscles.

of the ribs posteriorly to the costal cartilages anteriorly and from this point they
continue to the sternum as thin **anterior
intercostal membranes.** Their action
may be to draw adjacent ribs together,
or when the first ribs are fixed by the
Scaleni muscles, they elevate the ribs at
their costovertebral articulations and thus
increase the volume of the thoracic
cavity. According to some, the External
intercostals in conjunction with the
Internal intercostals help to keep the
intercostal spaces constant during breathing. They may prevent the blowing out
or sucking in of the intercostal spaces with
the changes in intrathoracic pressures.

The Levatores costarum muscles, twelve
in number, belong as much to the back
as the thorax. They originate on the
transverse processes of the seventh cervical and the upper eleven thoracic vertebrae and insert between the tubercle and
angle of the next rib below. A few of
the lower muscles of this group send some
fasciculi to insert two ribs below their
origin. Their functions are to help raise
the ribs, thus increasing the size of the
thoracic cavity and to extend, bend
laterally, or rotate the vertebral column.

The Serratus posterior superior and
inferior are both thin quadrilateral muscles. The Superior lies at the junction of
the neck and back, originating from the
ligamentum nuchae and the spinous
processes of the seventh cervical and the
first two or three thoracic vertebrae. Its
four muscular bellies insert into the
upper border of ribs two through five a
little lateral to their angles. The Inferior
lies at the junction of the thoracic and
lumbar regions, and through the thoracolumbar fascia, attaches to the last two
thoracic and the first two lumber vertebrae at their spinous processes. It extends lateralward and upward to insert
by its four bellies into the lower borders
of the last four ribs. The Superior
elevates the upper ribs and thus increases
the diameters of the thoracic cage and
raises the sternum, contributing to the
inspiration of air in breathing. The
Inferior pulls outward on the lower ribs,
and depresses them, aiding in expiration.
**Muscles of the Inner Wall of the
Thorax.** The **Internal intercostal** muscles, also eleven in number on each side,
lie just inside of the External intercostals,
separated from them by a thin layer of

fascia. They extend to the sternum anteriorly between the true ribs and their cartilages, and to the ends of the cartilages of the false ribs. Posteriorly, their fibers end at the angles of the ribs, but their aponeuroses, the **posterior intercostal membranes,** extend to the vertebral column. They run from the superior borders of the ribs and costal cartilages into the inferior borders of the ribs and costal cartilages above. Their fibers are obliquely oriented and in the opposite direction of those of the External intercostals. They probably function with the External intercostals in breathing.

The Subcostals originate on the anterior surfaces of the ribs near their angles, and extending slightly mediad, insert between the angles and necks of the second or third ribs below. Their contraction depresses the ribs in expiration.

The Transverse thoracic originates on the posterior surface of the body and xiphoid process of the sternum and inserts into the posterior surfaces of the cartilages of the second to the sixth ribs. The lower fibers of this muscle are horizontal and continuous with those of the Transverse abdominis muscle. Its upper fibers range in direction from oblique to vertical. The function is to depress the ribs in expiration.

Muscle of the Floor of the Thorax. The **Diaphragm,** mentioned before under the muscles of the posterior vertebral column, is a dome-shaped structure separating the thoracic and abdominal cavities (Fig. 11–20). Its convex, superior surface forms the floor of the thoracic cage; its concave inferior surface, the roof of the abdominal cavity. It has its origins from the circumference of the thoracic outlet at three points: one from the posterior side of the xiphoid process of the sternum, one from the costal cartilages and adjacent surfaces of the lower six pairs of ribs, and one from the lumbocostal arches and crura of the lumbar vertebrae. Anatom-

ically, its insertion is on its own thin but strong central tendon.

The Diaphragm is pierced by three large openings, one, just anterior to the first lumbar vertebra for the passage of the aorta, azygos vein, and thoracic duct, one above, in front of and to the left of the aortic opening for the passage of the esophagus, vagus nerves, and small esophageal blood vessels, and one, the highest, at the level of the eighth and ninth thoracic vertebrae, for the passage of the vena cava and some branches of the phrenic nerve. Smaller apertures carry the greater and lesser splanchnic nerves and the hemiazygos veins. The ganglionated sympathetic nerve trunks enter the abdomen through apertures behind the Diaphragm.

The contraction of the Diaphragm increases the vertical diameter of the thoracic cavity and thus reduces intrathoracic pressure, which results in inspiration. It also reduces the size of the abdominal cavity and thus increases pressure, which may be used to assist in defecation, vomiting, urination or childbirth.

Muscles of the Abdomen
(Figs. 11-19, 25 to 28)

The anterior and lateral abdominal walls, unlike other parts of the trunk region, receive no direct support or protection from skeletal structures. The abdominal muscles with their extensive, flat bellies and aponeuroses are arranged in reinforcing layers connecting the lower rim of the thoracic cage and a part of the vertebral column with the pelvic girdle. On the midline of the abdomen, from the xiphoid process to the symphysis pubis, is a tough band which serves as a point of attachment for many of the abdominal muscles. It is the **linea alba.** The direction which the fibers of the various abdominal muscles take is different in each, ranging from oblique, to vertical, to transverse. This arrangement allows the muscles to function in various combinations. It also strengthens the wall and helps to stabilize the trunk. Other general and collective functions of these muscles are to hold the abdominal viscera in position, to compress the viscera and thus aid in forced expiration, and by the same means to assist in defecation, urination, vomiting, or childbirth. By pulling on the thoracic cage they may assist in flexion of the thoracic and lumbar spine. When the subject is in the supine position, they aid in flexing the thighs by stabilizing the vertebral column, rib cage, and pelvis. They, of course, have no attachments on the femurs. They are serving here as fixators, rather than prime movers.

The abdominal muscles may be placed into two groups: the **anterolateral** and the **posterior muscles.**

the midline, over the lower end of the linea alba, it forms a strong ligament, the **suspensory ligament of the penis,** which attaches the dorsal side of the penis to the symphysis pubis. Both the superficial and deep fasciae have complex involvements with the structures of the perineal region such as the penis, labia, scrotum, and spermatic cord. Some of these will be considered later in this section and under the reproductive system.

Anterolateral Muscles. The **Obliquus externus abdominis** is a broad muscle which is, in a developmental sense, a continuation of the External intercostals, and like it, has its fibers extending downward, forward, and medialward (Fig. 11–25). It has its origin by finger-like slips from the external surfaces of the lower eight ribs and its insertion by a broad, thin and tough aponeurosis into

Anterolateral Muscles

Obliquus externus abdominis	Transversus abdominis
Obliquus internus abdominis	Rectus abdominis
Cremaster	Pyramidalis

Posterior Muscles

Psoas major	Iliacus
Psoas minor	Quadratus lumborum

Superficial Fascia (Fig. 11–21). The superficial fascia of the anterior abdominal wall is soft and movable. Below the umbilicus its two layers are quite distinct and can be easily separated by blunt dissection. The outer one, the **panniculus adiposus,** tends to store fat, which, in obese individuals, may be several centimeters thick. The inner layer of the superficial fascia is membranous with many elastic fibers.

Deep Fascia (Fig. 11–21). The **outer investing layer** of the deep fascia is apparent on the muscular part of the external oblique, but while present, is difficult to separate from the aponeurosis. Below, it attaches to the inguinal ligament and at

the linea alba, where it fuses with its counterpart from the opposite side. The most posterior fleshy fibers run almost vertically and insert on the anterior half of the iliac crest.

Below and laterally, the aponeurosis of the External oblique folds inward and forms a strong ligamentous band, the **inguinal ligament,** extending from the anterior superior spine of the ilium to the pubic tubercle (Fig. 11–25). It lies in the groin between the anterior abdominal wall and the anterior surface of the thigh. Just above the medial end of the inguinal ligament is a triangular split in the aponeurosis which in the male gives passage to the spermatic cord and in the female to the round ligament of the uterus. Although triangular in shape, this opening is called the **subcutaneous inguinal ring** and is the outer opening of the **inguinal canal.** The fibers forming and supporting the boundaries of the inguinal ring are the **crura** and are named for their position relative to the opening, the **superior, inferior,** and **intercrural** fibers. Study of Figure 11–25 will help to clarify these structures.

The **Obliquus internus abdominis** lies beneath the External oblique and its fibers run obliquely upward and medialward, similar in direction to those of the Internal intercostals. It has its origin from the lateral half of the inguinal ligament, the anterior two-thirds of the iliac crest, and from the thoracolumbar fascia. It inserts into the inferior borders or cartilages of the lower three or four ribs by fleshy attachments and by a broad aponeurosis into the linea alba from the sternum to the symphysis pubis. The aponeurosis, when it reaches the lateral border of the Rectus abdominis in its upper two-thirds, splits and sends a layer in front and one behind this muscle to form a sheath. In the lower third, the aponeurosis does not split, but goes anterior to the Rectus where it fuses with the aponeurosis of the Transversus abdominis to form the **falx inguinalis** (conjoined tendon). This is attached to the crest of the pubis and strengthens the abdominal wall medial to the subcutaneous inguinal ring (Fig. 11–26).

The Internal oblique gives off some muscle fibers, at a point near the middle of the inguinal ligament, which come to lie within the inguinal canal. These fibers pass out of the subcutaneous inguinal

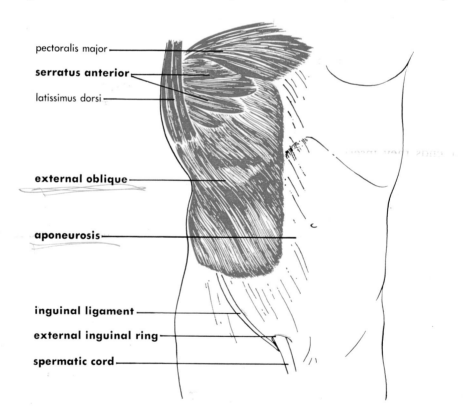

pectoralis major
serratus anterior
latissimus dorsi

external oblique

aponeurosis

inguinal ligament
external inguinal ring
spermatic cord

FIG. 11–25. Muscles of the abdominal and chest wall. (Outer layer.)

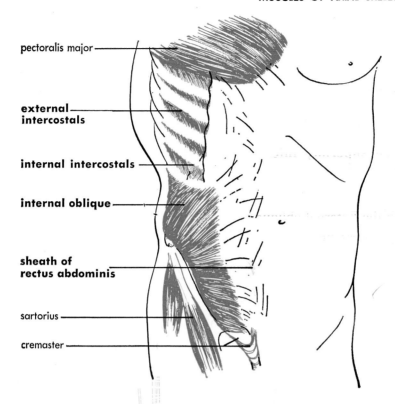

pectoralis major

external
intercostals

internal intercostals

internal oblique

sheath of
rectus abdominis

sartorius

cremaster

FIG. 11–26.   Muscles of the abdominal and chest wall.   External oblique removed.

ring and form a series of loops, the longest of which reaches to the testis.   At their other ends they insert into the tubercle and crest of the pubis.   This is called the Cremaster muscle and its contraction draws the testis upward toward the subcutaneous inguinal ring (Fig. 11–26).

The **Transversus abdominis** is the innermost and thinnest of the abdominal muscles.   It has its origin by fleshy fibers from the inguinal ligament and iliac crest, from the thoracodorsal fascia, and the inner surfaces of the lower six ribs and their cartilages.   Its fibers, except the lowermost, pass horizontally forward to an aponeurosis which fuses with that of the Internal oblique then joins the aponeurosis of the opposite side to insert into the linea alba.   The aponeurosis of the Transversus contributes to the sheath of the Rectus abdominis.   The upper part of the aponeurosis passes entirely behind the Rectus; below the umbilicus for a short distance some fibers pass in front and some behind the Rectus, while below they pass entirely in front of it (Figs. 11–27, 28).   Some of the lower portion, as stated above, joins with the Internal oblique aponeurosis to form the falx inguinalis.

The **Rectus abdominis** muscles are vertically oriented and lie one to either side of the midline.   Their vertical fibers are interrupted along their course by three transverse fibrous bands called **tendinous inscriptions,** believed to be remnants of the old myosepta which separated muscle segments or myotomes during embryological development.   The Rectus muscles arise by two tendons, the lateral from the crest of the pubis, the medial, interlacing with the one from the

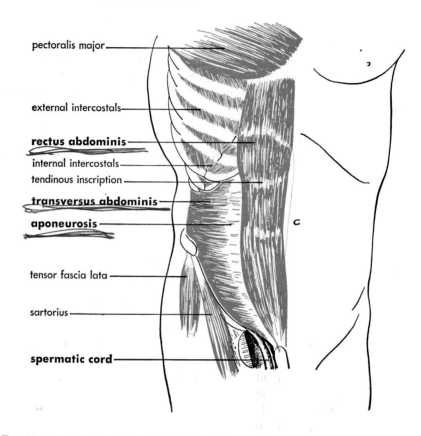

pectoralis major

external intercostals

rectus abdominis

internal intercostals

tendinous inscription

transversus abdominis

aponeurosis

tensor fascia lata

sartorius

spermatic cord

FIG. 11–27.   Muscles of the abdominal and chest wall—external and internal oblique muscles removed.

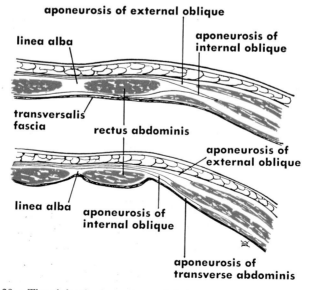

aponeurosis of external oblique

linea alba

aponeurosis of internal oblique

transversalis fascia

rectus abdominis

aponeurosis of external oblique

linea alba

aponeurosis of internal oblique

aponeurosis of transverse abdominis

FIG. 11–28.   The abdominal muscles and their aponeuroses and related fasciae.

opposite side, from the ligaments on the front of the symphysis pubis. The Rectus inserts by three unequal slips into the xiphoid process and the cartilages of ribs five to seven (Fig. 11–27).

The Pyramidalis is a small and triangular muscle at the inferior end of the Rectus abdominis and is enclosed in its sheath. It arises in front of the pubis and inserts into the linea alba, and probably acts to tense that structure. It is inconstant in occurrence.

Posterior Muscles. The Psoas major and minor and the Iliacus muscles will be described in conjunction with the muscles of the lower limb.

The **Quadratus lumborum** is a quadrilateral muscle originating on the crest of the ilium and transverse processes of the lower lumbar vertebrae and inserting on the inferior border of the twelfth rib and the transverse processes of the upper lumbar vertebrae. It is an important lateral flexor of the spine and fixes the lower ribs in forced expiration. It may, under certain conditions, elevate the pelvis laterally or extend the spine.

The Quadratus lumborum with the Psoas muscles complete the muscular posterior wall of the abdominal cavity (Fig. 11–19).

Transversalis Fascia. The term **transversalis fascia** was at one time applied only to the deep fascia on the inner side of the Transversus abdominis muscle. Now it is applied to all of the **internal investing layer** of deep fascia of the abdomen and even of the pelvis. It is a highly complicated structure of great interest and importance to the surgeon. We are more concerned with an interruption in this fascia, the abdominal (internal) **inguinal ring**, where the spermatic cords, or in the female the round ligament, penetrate the anterior abdominal wall carrying with them a sleeve-like investment of the transversalis fascia, the **internal spermatic fascia,** as they pass through the inguinal canal. The **femoral sheath** around the femoral vessels in the upper thigh is another outgrowth of the transversalis fascia.

Subserous Fascia. This fascia lies between the transversalis fascia and the

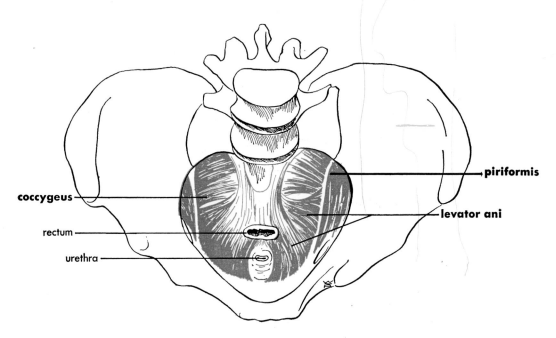

coccygeus

rectum

urethra

piriformis

levator ani

Fig. 11–29.   Muscles of the superior surface of the pelvic floor.

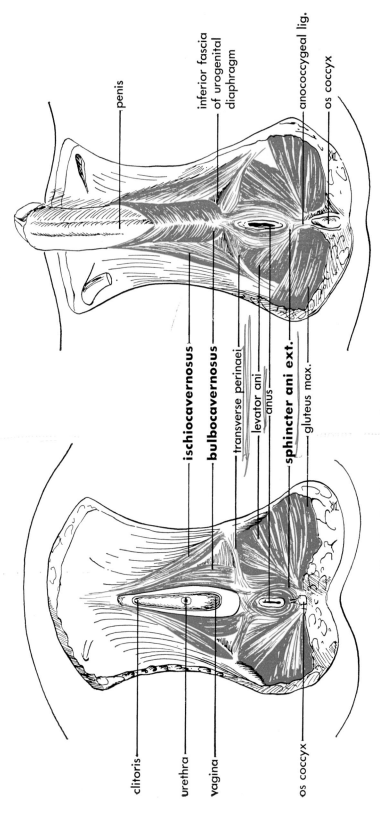

Fig. 11–30.   Muscles of the female and male perineum.   (Modified from Gray.)

peritoneum which lines the abdominal and pelvic cavities. It was described in an earlier section of this chapter. Figures 11–21, 28 give a quick review of the layers present in the walls of the abdominal cavity.

MUSCLES OF THE PELVIS

The muscles of the pelvis are the:

Levator ani   Coccygeus

The Piriformes and Obturator internus originate within the pelvis and form a part of its wall, but belong functionally with the muscles of the lower limb with which they will be described.

The Pelvic Diaphragm (Figs. 11–29, 30). The Levator ani and Coccygeus, together with their internal and external fasciae, constitute the **pelvic diaphragm.** It is the floor of the abdominopelvic cavity and as such gives support to the pelvic viscera. The urethra, vagina, and anal canal open through it and the underlying perineal structures.

The **Levator ani** is a broad, thin muscle forming much of the funnel-shaped floor of the pelvis. It has its origin from the inner surface of the superior ramus of the pubis, from the spine of the ischium and a fascia connecting these two lines along the lateral pelvic wall. Its more posterior fibers insert into the coccyx; the more anterior ones unite in the midline behind and in front of the anus (Figs. 11–29, 30).

The Levator ani supports and raises the floor of the pelvis and by drawing the anus toward the pubis constricts it, thus aiding in the regulation of defecation.

The Coccygeus, a triangular muscle, is behind the Levator ani where it arises by its apex from the spine of the ischium and is inserted into the coccyx and sacrum. It draws the coccyx forward and helps to complete and support the pelvic floor.

MUSCLES OF THE PERINEUM

The perineum is the region of the outlet of the pelvis. It contains the structures between the symphysis pubis in front and the coccyx behind. Laterally, it is limited by the thighs. The urethral, vaginal, and anal orifices are in the perineum. An imaginary line drawn between the ischial tuberosities and passing just in front of the anus divides it into urogenital and anal triangles. The muscles and fasciae of the urogenital triangle form the urogenital diaphragm which, with the pelvic diaphragm described above, helps to support the pelvic floor.

The principal muscles of the perineum in the male and female are shown in Figures 11–29, 30.

MUSCLES OF THE UPPER LIMB

One should refer back at this time to Chapter 9, page 139, and review the skeletal and articular structures of the upper limb and the movements of which they are capable. Their capability for movement, of course, is dependent upon the muscles with which we are now concerned.

MUSCLES CONNECTING THE AXIAL SKELETON AND THE SHOULDER GIRDLE

These muscles may be divided into posterior and anterior groups as follows:

Posterior (Fig. 11–31)
  Trapezius
  Rhomboideus major
  Rhomboideus minor
  Levator scapulae
Anterior (Figs. 11–32, 33)
  Pectoralis major
  Pectoralis minor
  Serratus anterior
  Subclavius

Posterior Muscles. The **Trapezius** is a large, flat, triangular muscle which, with the one on the opposite side, forms an irregular, four-sided figure, a trapezium. The Trapezius originates on the occipital bone of the skull, the ligamentum nuchae, and the spines of the seventh cervical and all of the thoracic vertebrae. It inserts onto the clavicle, acromion and spine of the scapula. Its fibers form three groups which in some mammals, as the cat, form separate muscles—the clavo-, acromio-, and spinotrapezius, respectively. The disposition of these three groups of fibers is a clue to the functions of the muscle. The upper fibers working alone elevate and rotate

the scapulae and elevate the tips of the shoulders, as when one "shrugs" his shoulders; the middle fibers adduct the scapulae, as when we stand at attention; the lower fibers pull the scapulae downward and depress the tips of the shoulders. Working together, all groups of fibers pull the scapulae backward toward the vertebral column (retraction). With the scapulae fixed, the contraction of the upper fibers on the two sides draws the head backward, or one side acting alone draws the head to that side and turns the face to the opposite side.

The **Rhomboideus major** and **minor**

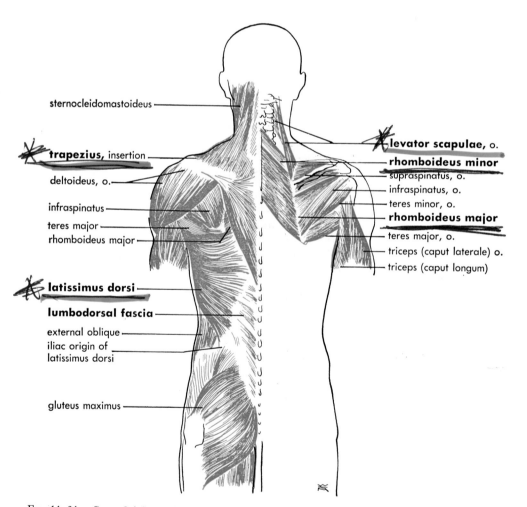

FIG. 11–31.   Superficial muscles of the back and those connecting axial skeleton to shoulder girdle (posterior view).

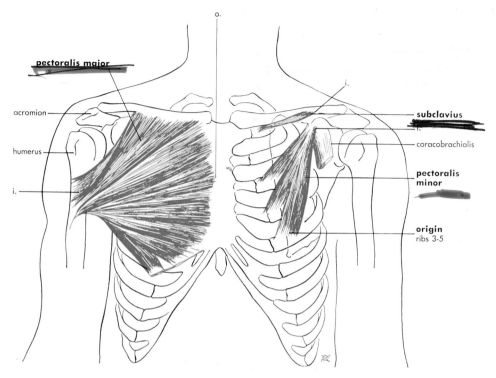

Fig. 11–32. Anterior muscles to shoulder girdle and arm.

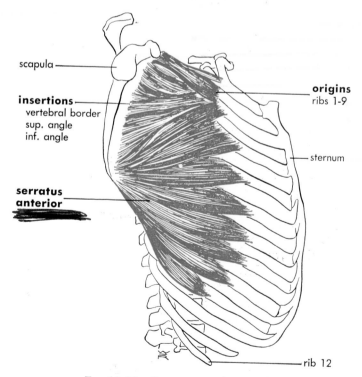

Fig. 11–33. Serratus anterior muscle.

16

lie underneath the Trapezius and have their origins from the spinous processes of the vertebrae, from the seventh cervical to the fifth thoracic. They insert into the medial border of the scapula. Since their fibers run laterally and inferiorly from their origins, these muscles adduct the scapula and elevate the inferior angle, also rotating it slightly to depress the lateral angle.

The **Levator scapulae** also lies beneath the Trapezius on the side and back of the neck. From its origin on the transverse processes of the first four cervical vertebrae its fibers pass downward and lateralward to insert on the medial border of the scapula above the spine. Its contraction elevates the scapula, tends to draw it medialward and rotates it to lower the lateral angle. If the scapula is fixed, it bends the neck laterally and rotates it slightly.

Anterior Muscles (Figs. 11–32, 33). The **Pectoralis major** is the largest and most superficial muscle of the anterior chest wall. It is thick, flat, and fan-shaped and consists of three portions: one originating on the clavicle, another on the sternum and costal cartilages of the second to the sixth ribs, and another on the aponeurosis of the External oblique abdominal muscle. The fibers of all three portions converge to a flat, wide tendon which inserts into the lateral lip of the intertubercular groove of the humerus.

The primary action of the Pectoralis major is on the humerus, and this depends on the position of the humerus at the time of Pectoralis action. If the humerus is pendant, the Pectoralis pulls it forward (*flexes*), rotates it medially, and the clavicular portion tends also to elevate and adduct it. When the humerus is at the horizontal or vertical position, the muscle brings it forward and downward. If the humerus is fixed in the pendant position, however, the Pectoralis major depresses the shoulder and may aid in forced inspiration by elevating the ribs.

The **Pectoralis minor** is a flat, thin triangular muscle which lies beneath the Pectoralis major. It arises from the sternal ends of the second to the fifth ribs and inserts into the coracoid process of the scapula. When the ribs are fixed, the contraction of the Pectoralis minor draws the scapula forward and downward, rotating it so as to depress the lateral angle. When the scapula is fixed it elevates the ribs and thus aids in forced inspiration.

The **Serratus anterior** is a broad and flat muscle on the anterolateral wall of the thorax which arises by fleshy digitations from the outer surfaces of the upper nine ribs and inserts into the superior and inferior angles and medial border of the scapula. Its action is to abduct the scapula and to elevate the point of the shoulder, as when the arm is fully flexed and abducted. It may aid in forced inspiration by elevating the ribs when the scapula is fixed. Full abduction of the humerus is possible only because of the scapular abduction by the Serratus anterior.

The Subclavius muscle, as the name suggests, lies beneath the clavicle. It extends laterally from its origin on the first rib and its costal cartilage to insert into the subclavian groove of the clavicle. It draws the shoulder forward and downward. By fixation of the clavicle it elevates the first rib and thus aids in forced inspiration.

MUSCLES CONNECTING THE AXIAL SKELETON AND ARM (Fig. 11–31)

Most muscles of the arm originate on the shoulder with the exception of the:

Latissimus dorsi   Pectoralis major

The **Latissimus dorsi,** as the name suggests, is the widest muscle of the back. It has its origin from the thoracolumbar fascia (Fig. 11–32), the crest of the ilium, the lower three or four ribs, and as it

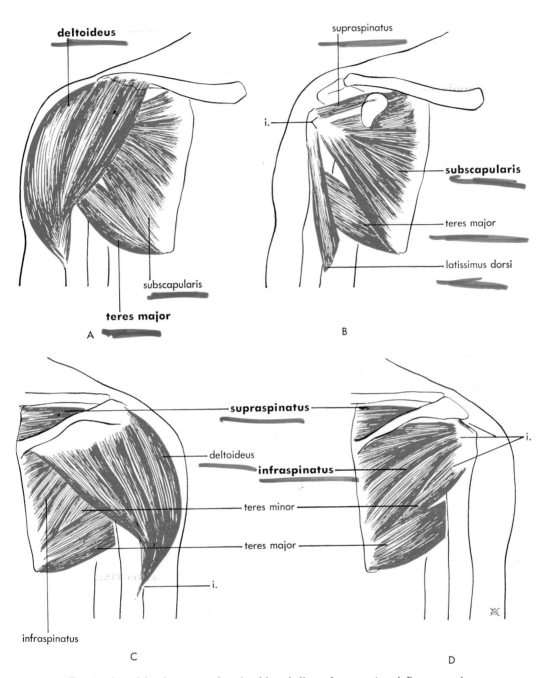

Fig. 11–34.   Muscles connecting shoulder girdle and arm.   *A* and *B* are anterior;
C and *D* are posterior aspects.

passes over the inferior angle of the scapula it receives a few fasciculi from that bone. From this broad origin it passes laterally and superiorly, its upper fibers almost horizontal, to the axilla of the arm where its fasciculi converge into a flat, narrow tendon which spirals almost 180 degrees before inserting into the bottom of the intertubercular groove of the humerus.

The Latissimus dorsi extends, adducts, and rotates the arm medialward. It also draws the shoulder downward and backward. It is used extensively in swimming and in paddling a canoe.

The Pectoralis major has already been considered in the previous section and its action in reference to the arm should be reviewed at this time.

MUSCLES CONNECTING THE SHOULDER GIRDLE AND THE ARM (Fig. 11–34)

Deltoideus   **Teres major**
Subscapularis   Teres minor
Supraspinatus   Coracobrachialis
Infraspinatus

The **Deltoideus** is a large, thick triangular, and coarse-textured muscle which forms the roundness of the shoulder. Like the Trapezius, with which it almost appears continuous at the scapula and clavicle, it has three parts which in the cat are separate muscles. Its anterior fibers arise from the lateral third of the anterior border and upper surface of the clavicle; its middle fibers from the lateral border and upper surface of the acromion; its posterior fibers from the posterior border of the spine of the scapula. The middle portion of the muscle is the strongest and its fibers are arranged in a complex bipennate fashion. The fibers of all portions converge into a thick tendon which inserts into the deltoid tuberosity on the middle lateral surface of the humerus. A large bursa is found beneath the Deltoideus, the subacromial bursa (Fig. 11–35).

The muscle, working as a whole, is an abductor of the humerus. The anterior fibers flex and rotate the humerus medially; the posterior fibers extend and rotate it laterally.

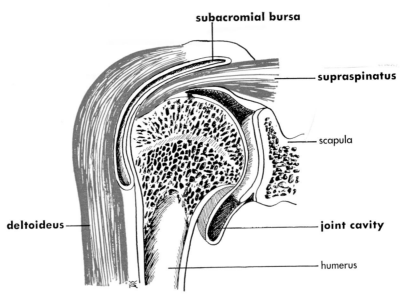

FIG. 11–35. Section through shoulder joint to show joint cavity and bursa.

The Subscapularis, Supraspinatus, Infraspinatus and Teres minor, all of them deep muscles of the shoulder, are closely associated in strengthening and stabilizing the shoulder joint. They also serve as the chief rotators of the arm, and constitute what is sometimes referred to as the "rotator cuff" (Fig. 11–34).

The **Subscapularis,** as its name suggests, arises in the subscapular fossa where it has a multipennate origin. It converges laterally into a broad tendon which passes over the front of the fibrous capsule of the shoulder joint which it reinforces. It inserts on the lesser tubercle of the humerus. Beneath this tendon is a bursa which communicates with the shoulder joint (Fig. 11–34B). The Subscapularis is a medial rotator of the arm. Its superior fibers weakly abduct while its inferior ones weakly adduct the humerus.

The **Supraspinatus** arises from the supraspinous fossa of the scapula which it completely fills. Its tendon passes over the superior part of the fibrous capsule of the shoulder joint, with which it adheres, and inserts on the highest point of the greater tubercle of the humerus. The Supraspinatus is an important abductor of the humerus. It initiates abduction, and works with the Deltoideus. The Deltoideus is quite handicapped without it, since its pull is at first directly upward along the line of the humerus, not in a direction to initiate abduction.

The **Infraspinatus** has a multipennate origin from the infraspinous fossa of the scapula. Closely associated with it along its lateral border is the small Teres minor. These muscles both insert on the greater tubercle of the humerus, the Infraspinatus just above the Teres minor. Their tendons reinforce the shoulder joint cap-

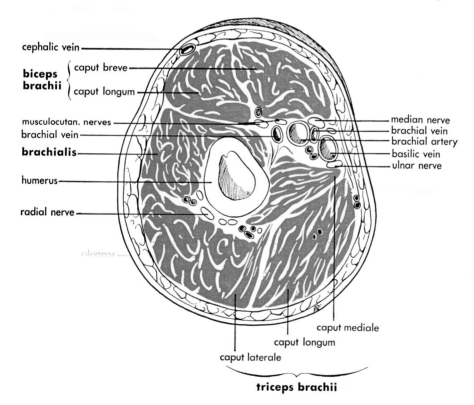

FIG. 11–36. Cross-section through the middle of upper arm.

sule.   They are lateral rotators of the humerus.

The **Teres major** arises from the inferior angle and lower part of the axillary border of the scapula and inserts on the crest of the lesser tubercle of the humerus.   Its insertion is just behind that of the Latissimus dorsi with which it is closely associated embryologically and functionally.   It adducts, extends, and rotates the humerus medially.

The Coracobrachialis is a small muscle having a common origin with the short head of the Biceps brachii, on the tip of the coracoid process of the scapula.   It inserts by a flat tendon half-way down the medial border of the shaft of the humerus. It is perforated by the musculocutaneous nerve.   Its function is to flex and adduct the humerus.

Muscles of the Arm (Figs. 11–37 to 39)

Biceps brachii	Coracobrachialis
Brachialis	**Triceps brachii**

Brachial Fascia.   The arm is covered by an investing layer of deep fascia which is continuous above with the deltoid, axillary, and pectoral fascia.   Below

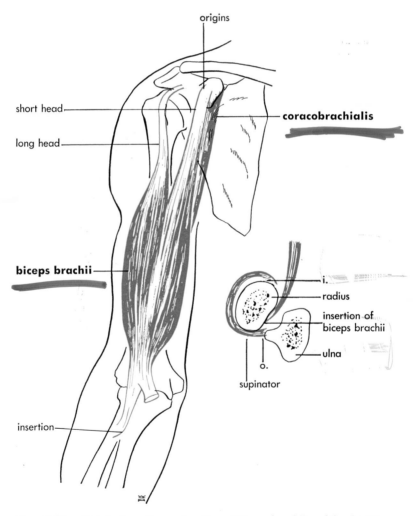

Fig. 11–37.   Origin, insertion and action of Biceps brachii and Supinator.

it attaches to the epicondyles of the humerus and the olecranon and is continuous with the antebrachial fascia of the forearm. Medial and lateral intermuscular septa extend from the investing layer to the bone dividing the arm into flexor and extensor compartments (Fig. 11–36).

The **Biceps brachii** is a long, spindle-shaped muscle originating by two heads; a short head from the coracoid process of the scapula and a long head from the supraglenoid tubercle. The tendon of the long head lies within the capsule of the shoulder joint which provides a special synovial sheath for it. It emerges from the capsule and descends in the intertubercular groove. The two bellies

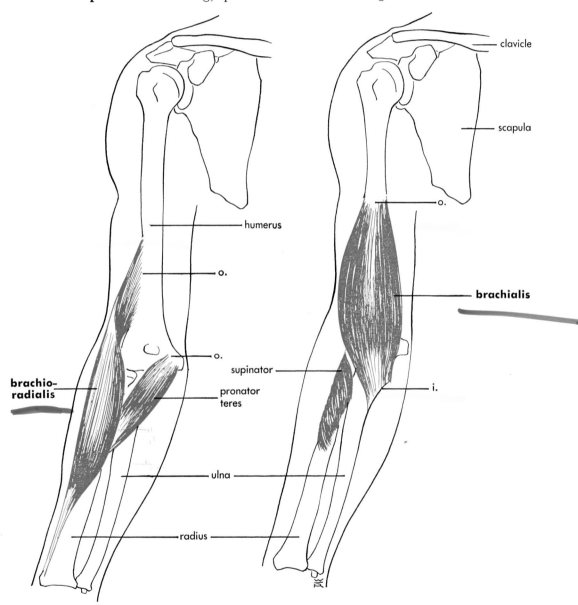

Fig. 11–38.   Muscles extending from the arm to the forearm on the anterior aspect of the upper limb.

can be easily separated through most of their length, but at the inferior end they become one. The tendon of insertion is flattened and fastens into the radial tuberosity. The Biceps brachii is a flexor of the arm and forearm and also a supinator of the forearm and hand (Fig. 11–37).

The **Brachialis** has its origin from the lower half of the anterior surface of the humerus and its insertion on the tuberosity and coronoid process of the ulna. It is an important flexor of the forearm (Fig. 11–38).

The Coracobrachialis was considered in the previous section. It is the smallest muscle of this group.

The **Triceps brachii** is a large muscle occupying the whole posterior surface of the humerus. As its name suggests, it has

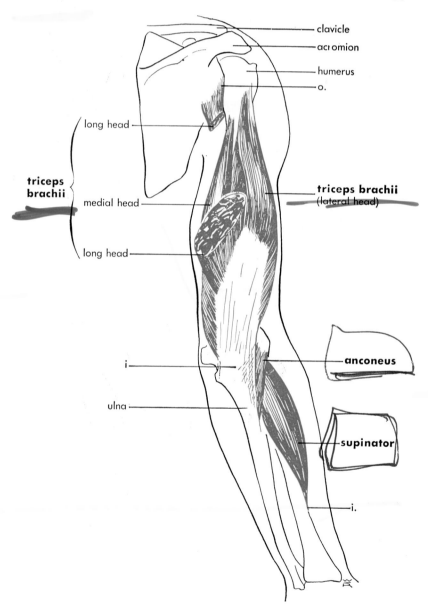

Fig. 11–39.   Muscles running from arm to forearm on the posterior side.

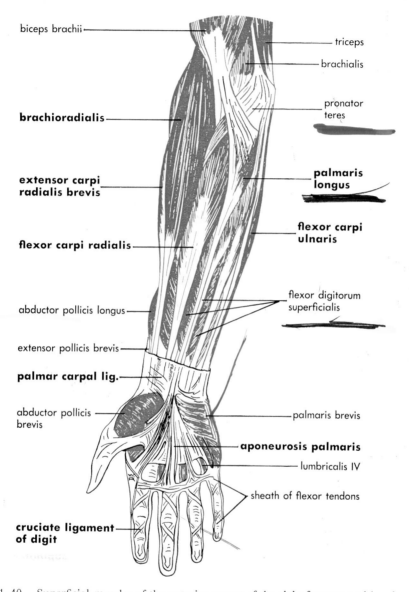

FIG. 11–40.   Superficial muscles of the anterior aspect of the right forearm and hand.

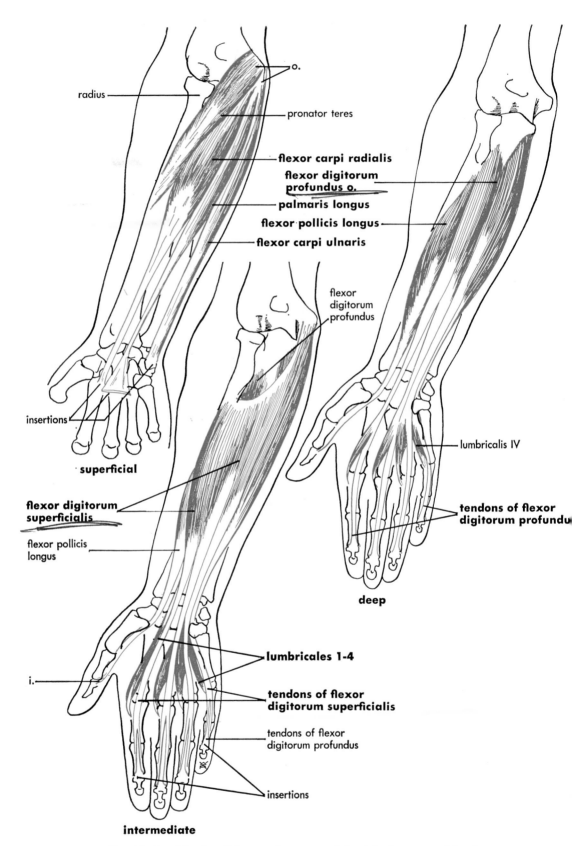

radius

o.

pronator teres

flexor carpi radialis

flexor digitorum profundus o.

palmaris longus

flexor pollicis longus

flexor carpi ulnaris

insertions

superficial

flexor digitorum superficialis

flexor pollicis longus

flexor digitorum profundus

flexor digitorum superficialis

lumbricalis IV

tendons of flexor digitorum profundu

deep

i.

lumbricales 1-4

tendons of flexor digitorum superficialis

tendons of flexor digitorum profundus

insertions

intermediate

FIG. 11–41. Muscles of the anterior aspect of the right forearm and hand.

232

three heads—long, lateral, and medial. The long head arises from the infraglenoid tuberosity of the scapula, the lateral from the posterolateral surface of the proximal end of the humerus, and the medial from the lower two-thirds of the posteromedial surface (Fig. 11–39). They all insert by a common tendon into the olecranon process of the ulna. It is the important extensor of the forearm.

MUSCLES OF THE FOREARM

The muscles in this group are many and varied. Fortunately, their names often include words descriptive of their chief functions and sometimes indicate their origins or insertions. These muscles arise mostly from the distal end of the humerus, a few from the radius and ulna. Their bellies form the fullness of the proximal end of the forearm. Distally they become tendinous and many pass into the hand. They may be divided into two groups on the basis of position and function, a palmar (*anterior*) or flexor and pronator group, and a posterior or extensor and supinator group. Each of these groups may, in turn, be divided into superficial and deep muscles.

Rather than describe these muscles individually they are presented adequately in the illustrations and in Table 11–1. These sources should be studied carefully.

The muscles may be organized as follows:

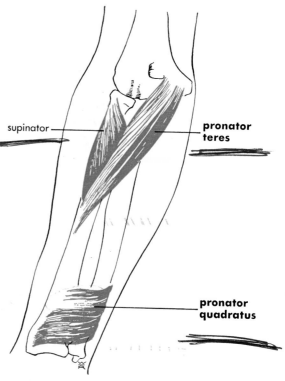

Fig. 11–42.   Pronator and Supinator muscles of the right forearm.

Palmar (**Flexor-Pronator Group**) (Fig. 11–40)

Superficial (Figs. 11–41, 42)

Pronator teres	Palmaris longus
Flexor carpi radialis	Flexor carpi ulnaris

Intermediate (Fig. 11–41)
Flexor digitorum superficialis (sublimis)

Deep (Figs. 11–41, 42, 43)

Flexor digitorum profundus	Flexor pollicis longus
Pronator quadratus	

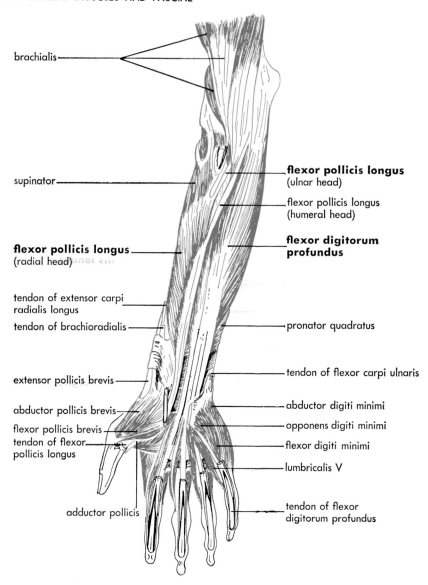

brachialis

supinator

flexor pollicis longus
(ulnar head)

flexor pollicis longus
(humeral head)

flexor digitorum
profundus

flexor pollicis longus
(radial head)

tendon of extensor carpi
radialis longus

tendon of brachioradialis

pronator quadratus

extensor pollicis brevis

tendon of flexor carpi ulnaris

abductor pollicis brevis

abductor digiti minimi

flexor pollicis brevis

opponens digiti minimi

tendon of flexor
pollicis longus

flexor digiti minimi

lumbricalis V

adductor pollicis

tendon of flexor
digitorum profundus

FIG. 11–43.   Some deep muscles of the anterior aspect of the right forearm and hand.

Posterior (Extensor-Supinator) Group (Figs. 11–44 to 46)

Superficial (Figs. 11–44, 46)

(Brachioradialis) *
Extensor carpi radialis longus
Extensor carpi radialis brevis
Extensor digitorum communis

Extensor digiti minimi
Extensor carpi ulnaris
Anconeus

* This muscle on the basis of its location and nerve supply is placed with the extensor group though
it is actually a flexor of the forearm.

Deep (Figs. 11–45, 46)

Supinator   Extensor pollicis longus
Abductor pollicis longus   Extensor indicis
Extensor pollicis brevis

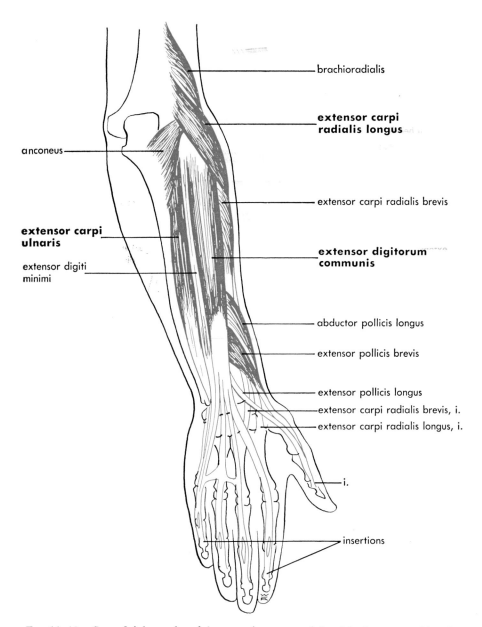

FIG. 11–44. Superficial muscles of the posterior aspect of the right forearm and hand.

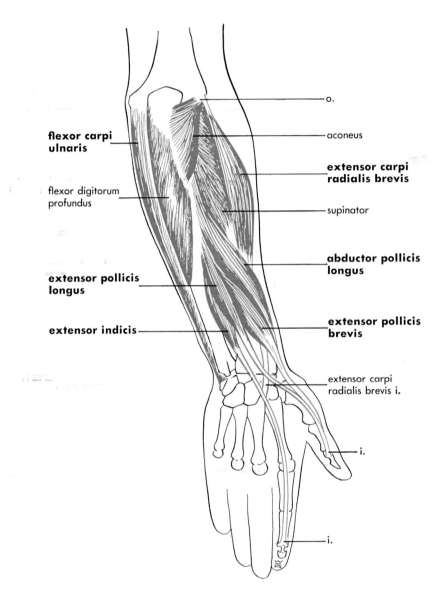

flexor carpi
ulnaris

flexor digitorum
profundus

extensor pollicis
longus

extensor indicis

o.

aconeus

extensor carpi
radialis brevis

supinator

abductor pollicis
longus

extensor pollicis
brevis

extensor carpi
radialis brevis i.

i.

i.

FIG. 11–45. Deep muscles of the posterior aspect of the right forearm and hand.

Antebrachial fascia (Fig. 11–47). This is a continuation of the brachial fascia which forms a strong investment of the forearm. It is attached below to the distal parts of the radius and ulna and is continuous with the fascia of the hand. Through most of its length it is attached to the dorsal border of the ulna, thus closing off the palmar or flexor from the posterior or extensor compartments of the forearm. These compartments are further separated by the interosseous membrane between the radius and ulna. Intermuscular septa extend deeply from the investing fascia toward the bones to further compartmentalize the muscles of the forearm. At the distal end of the forearm the fascia is thickened to form palmar and dorsal carpal ligaments, which serve to bind down the long

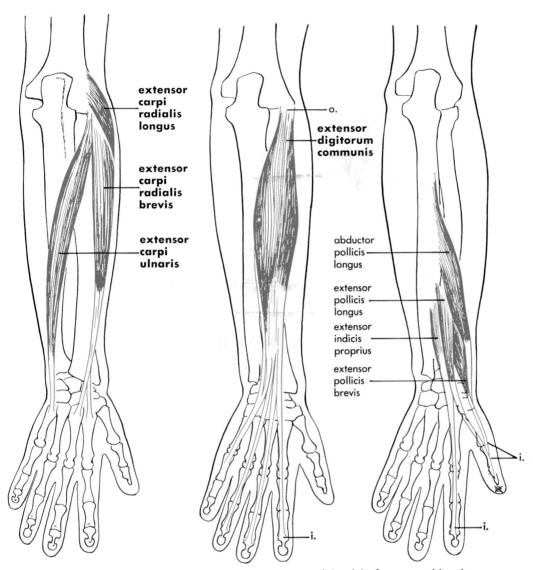

extensor
carpi
radialis
longus

extensor
carpi
radialis
brevis

extensor
carpi
ulnaris

o.

extensor
digitorum
communis

abductor
pollicis
longus

extensor
pollicis
longus

extensor
indicis
proprius

extensor
pollicis
brevis

i.

i.

i.

Fig. 11–46.   Some muscles of the posterior aspect of the right forearm and hand.

tendons of insertion of the flexor and extensor muscles as they pass over this region (Fig. 11–40).

Intrinsic Muscles of the Hand

Reference was made in Chapter 10 to the remarkable capacity and versatility of the human hand and the role it plays in man's control over his environment. This capacity is due largely to the opposable thumb, and to the numerous joints of the wrist and even those of the forearm, arm, and shoulder which enable one to use the hand in many positions. It follows, of course, that the hand must have many muscles within itself to supplement those which enter it from neighboring parts or indirectly influence its action. These intrinsic muscles are the ones which concern us now. Again, they will be presented more by illustration and Table 11–1 than by description of individual muscles. Their names suggest their actions.

They may be organized as follows into

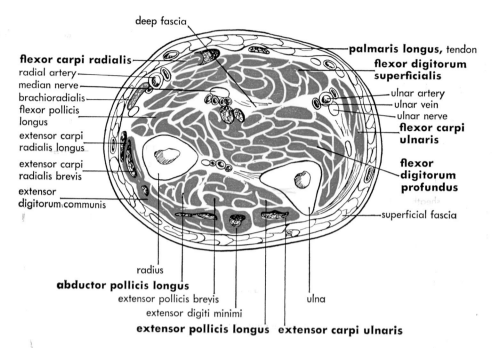

deep fascia

flexor carpi radialis
radial artery
median nerve
brachioradialis
flexor pollicis longus
extensor carpi radialis longus
extensor carpi radialis brevis
extensor digitorum communis

palmaris longus, tendon
flexor digitorum superficialis
ulnar artery
ulnar vein
ulnar nerve
flexor carpi ulnaris
flexor digitorum profundus
superficial fascia

radius
abductor pollicis longus
extensor pollicis brevis
extensor digiti minimi
ulna
extensor pollicis longus   extensor carpi ulnaris

Fig. 11–47.   Cross-section through the middle of the forearm showing muscles and the antebrachial fascia.

three groups: those of the thumb which form the thenar eminence; those of the little finger giving rise to the less prominent hypothenar eminence; and those intermediate in position in the palm and between the metacarpals.

Muscles of the Thenar Eminence (Figs. 11–40, 43, 48, 50)

 Abductor pollicis brevis
 Opponens pollicis
 Flexor pollicis brevis
 Adductor pollicis
  caput obliquum
  caput transversum

Muscles of the Hypothenar Eminence (Figs. 11–40, 43, 48, 50)

 Palmaris brevis
 Abductor digiti minimi
 Flexor digiti minimi brevis
 Opponens digiti minimi

Intermediate Muscles (Figs. 11–40, 43, 51)

Lumbricales
Interossei (dorsales, palmares)

Fascia of the Hand. The fascia of the hand present many modifications in keeping with the mutiplicity and complexity of actions in this structure.

The superficial fascia on the dorsum of the hand is like that of the forearm, movable and delicate. Its two layers are readily identified and it is separated from the deep fascia by a fascial cleft. On the palmar surface, however, the superficial fascia changes abruptly at the wrist and becomes a tough cushion over the palm and palmar surfaces of the digits. It contains a lot of fat and fibrous bands and cannot be separated easily into superficial and deep layers. The superficial fascia adheres also to the deep fascia over the whole palmar surface but is most strongly bound at the creases of the wrist, palm, and digits. The overlying dermis is also very compact and protects the underlying structures.

The **deep fascia** at the wrist, as stated earlier, is thickened to form the annular band or cuff which holds the tendons of the forearm muscles close to the bones. The annular band is divided into palmar and dorsal carpal ligaments (Figs. 11–40, 49). A transverse carpal ligament is also present distal to the palmar carpal ligament, but it is derived more from the ligaments and tendons in the carpus than from the deep fascia. As the tendons pass under

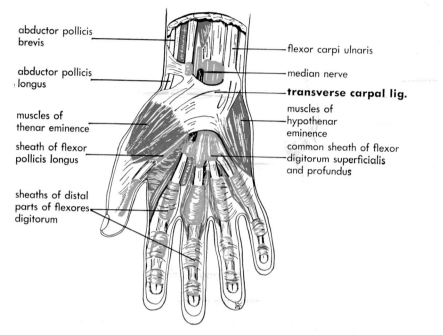

abductor pollicis brevis

abductor pollicis longus

muscles of thenar eminence

sheath of flexor pollicis longus

sheaths of distal parts of flexores digitorum

flexor carpi ulnaris

median nerve

transverse carpal lig.

muscles of hypothenar eminence

common sheath of flexor digitorum superficialis and profundus

FIG. 11–48.   The synovial sheaths of the tendons of the anterior aspect of the right hand.

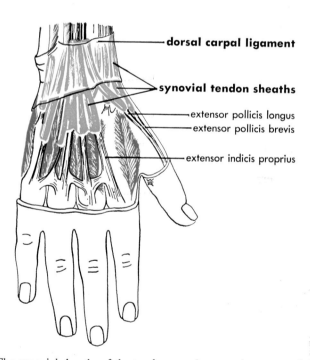

dorsal carpal ligament

synovial tendon sheaths

extensor pollicis longus

extensor pollicis brevis

extensor indicis proprius

FIG. 11–49.   The synovial sheaths of the tendons on the posterior aspect of the right hand.

17

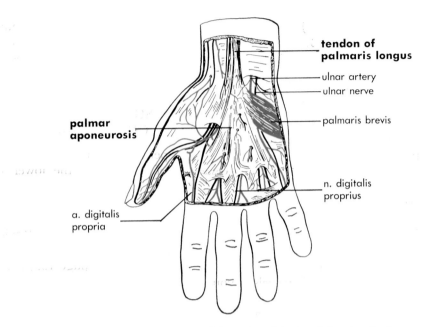

tendon of palmaris longus

ulnar artery

ulnar nerve

palmaris brevis

palmar aponeurosis

n. digitalis proprius

a. digitalis propria

FIG. 11–50.   Dissection of the palm of the hand showing fascia.

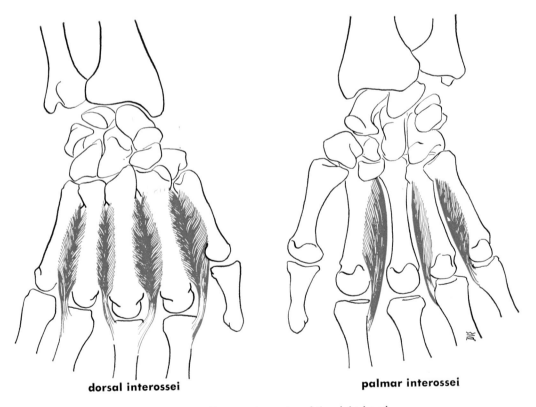

dorsal interossei

palmar interossei

FIG. 11–51.   Interossei muscles of the right hand.

these ligaments they are enclosed in synovial sheaths (Figs. 11–48 to 50).

The deep fascia in the palm is a continuation from the palmar carpal ligament. In the thenar and hypothenar areas it is much like that of the forearm in character, but in the intermediate area of the palm it is fibrous, strong, and complex and is called the palmar aponeurosis (Fig. 11–50).

The hand also contains a number of fascial compartments derived from the investing and penetrating fascial membranes of the fascia. These compartments, of which thenar, hypothenar, and central compartments are examples, may contain muscles, bones, blood vessels, and other structures.

Fascial clefts, which are planes of separation between fascial membranes are also common in the palm of the hand.

Practical Considerations. The structural features of the hand which make it

the hand, therefore, one can understand the likely pathways of spread of infection, and the surgical or other methods needed to treat them.

MUSCLES OF THE LOWER LIMB

The upper limb is noted for its versatility of movement, the lower for stability and strength. The muscles of the lower limb function mainly for locomotion, and for the attainment and maintenance of the erect posture. Review the skeletal and articular relations of this limb in Chapter 9, page 153, before you try to learn and understand the muscles.

MUSCLES OF THE HIP

These muscles may be divided into anterior and posterior or gluteal groups.

Anterior (Fig. 11–19)

Iliacus	Psoas minor
Psoas major	

Posterior (Gluteal) (Figs. 11–52, 53, 54, 56)

Gluteus maximus	Obturator internus
Gluteus medius	Gemellus superior
Gluteus minimus	Gemellus inferior
Tensor fasciae latae	Quadratus femoris
Piriformis	Obturator externus

a tough, versatile, and efficient tool may sometimes be the basis for serious trouble. While the resistant palmar fascia helps to prevent surface infections from reaching the deeper structures of the hand, it also keeps deep infections from easily getting out. Such infections usually travel along the fascial sheaths and clefts and break out to the back of the hand or even through the wrist or fingers. They cannot easily penetrate the palmar fascia. Knowing well the fascial structures of

Anterior Hip Muscles (Fig. 11–19). These muscles lie within the pelvis and are covered by the internal investing layer of deep fascia which follows these muscles as they pass under the inguinal ligament and becomes continuous with the fascia lata which is a part of the external investing fascia of the thigh. The **fascia lata,** as the name suggests, is broad. It is continuous proximally with the thoracolumbar and external abdominal fasciae and attaches to the pelvic

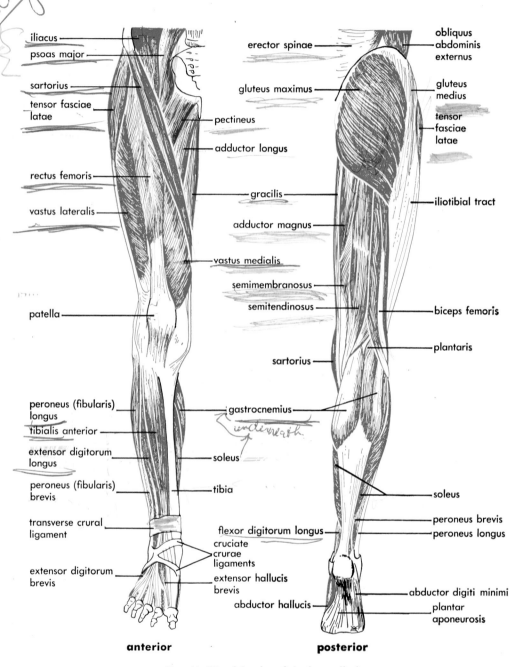

iliacus

psoas major

sartorius

tensor fasciae latae

rectus femoris

vastus lateralis

patella

peroneus (fibularis) longus

tibialis anterior

extensor digitorum longus

peroneus (fibularis) brevis

transverse crural ligament

extensor digitorum brevis

erector spinae

gluteus maximus

pectineus

adductor longus

gracilis

adductor magnus

vastus medialis

semimembranosus

semitendinosus

sartorius

gastrocnemius

soleus

tibia

flexor digitorum longus

cruciate crurae ligaments

extensor hallucis brevis

abductor hallucis

obliquus abdominis externus

gluteus medius

tensor fasciae latae

iliotibial tract

biceps femoris

plantaris

soleus

peroneus brevis

peroneus longus

abductor digiti minimi

plantar aponeurosis

anterior   **posterior**

FIG. 11–52.   Muscles of the lower limb.

bones and inguinal ligament. Distally, it is continuous with the fascia of the leg. Laterally, it is tendinous in character and is known as the **iliotibial band** which attaches below to the lateral condyle of the tibia. Other parts of the fascia lata are similarly strengthened, still others are thin. The **fossa ovalis** is an opening in the fascia lata for the great saphenous vein.

The **Iliacus** is a flat, triangular-shaped muscle which originates in the iliac fossa, from the sacrum, and the anterior sacroiliac and iliolumbar ligaments. Its fibers converge, pass beneath the inguinal ligament, and, in front of the capsule of the hip joint, join the tendon of the Psoas major which inserts into the lesser trochanter of the femur. The **Psoas major,** briefly described on page 208, originates from the bodies and transverse processes and on the intervertebral discs of the last thoracic and all five lumbar vertebrae. Since it shares its tendon of insertion with the Iliacus, these muscles are often called the **Iliopsoas.** Their common function is to flex the thigh and rotate it laterally. The Psoas major may also flex the lumbar spine and bend it laterally.

The Psoas minor is a weak flexor of the pelvis and lumbar spine, but is small and may be lacking in some individuals.

Posterior Hip Muscles (Fig. 11–53). The largest and most superficial of this group of muscles is the **Gluteus maximus.** It is very coarse in structure and quadrilateral in shape. It forms the buttocks, a characteristic feature of man associated with his power of maintaining the trunk in the erect posture. It has its origin from the posterior portion of the outer

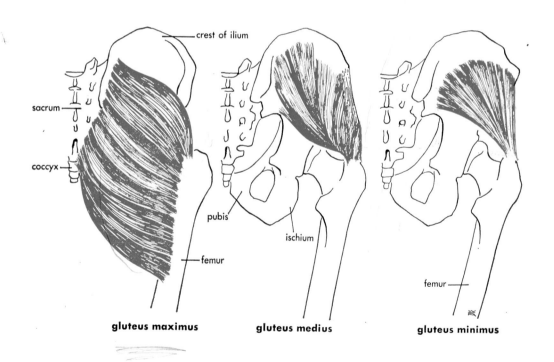

gluteus maximus   **gluteus medius**   **gluteus minimus**

FIG. 11–53. The posterior hip muscles.

lip of the iliac crest, the posterior gluteal line of the ilium, the posterior side of the sacrum and coccyx, and the posterior ligaments of the sacroiliac articulation. It inserts into the gluteal tuberosity of the femur and the iliotibial band of the fascia lata. Its function is to extend and laterally rotate the femur and to brace the knee, through the iliotibial band, when the knee is extended. It is used mostly where great power is needed, as in getting up from the stooped or sitting position, in climbing a hill or stairs, and in running, though not in walking.

The **Gluteus medius** is a thick, triangular-shaped muscle lying partly beneath the Gluteus maximus. It originates from the ilium between the anterior and posterior gluteal lines and is inserted into the lateral surface of the greater trochanter. Its contraction abducts the thigh and rotates it medially. Its ante-

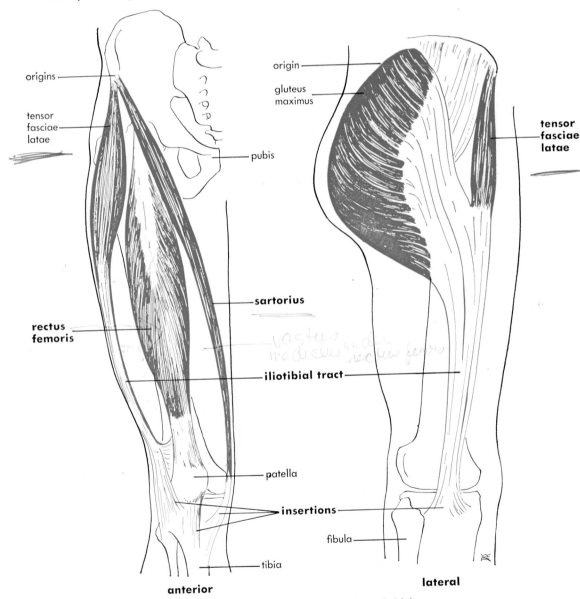

FIG. 11–54. Muscles of the hip and thigh.

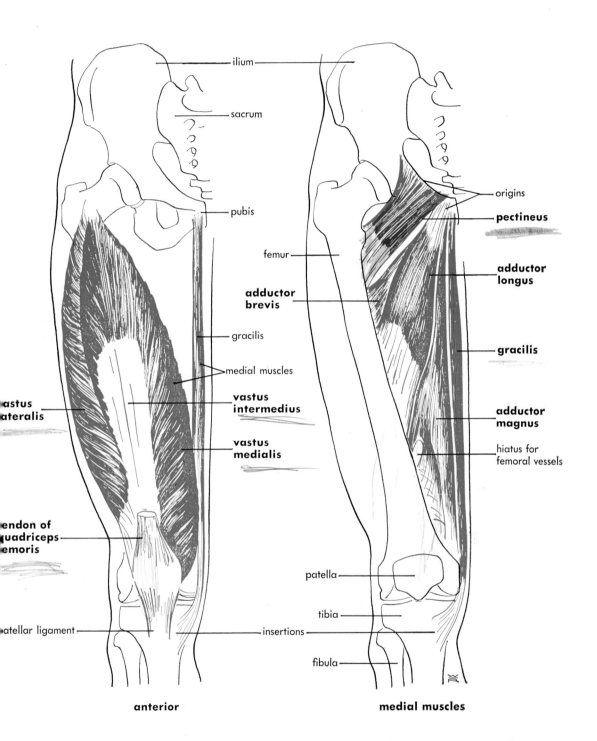

ilium
sacrum
pubis
femur
adductor
brevis
gracilis
medial muscles
vastus
lateralis
vastus
intermedius
vastus
medialis
tendon of
quadriceps
femoris
patellar ligament
insertions

origins
pectineus
adductor
longus
gracilis
adductor
magnus
hiatus for
femoral vessels
patella
tibia
fibula

anterior   **medial muscles**

FIG. 11–55.   Medial and anterior muscles of thigh.

rior fibers alone flex and rotate the thigh medially; its posterior fibers extend and rotate it laterally. When the support of one leg is suddenly removed, it is the Gluteus medius of the opposite side which contracts and keeps the pelvis from falling to the unsupported side. They help to keep the pelvis level when one is walking and prevent a rolling gait or action of the pelvis.

The Gluteus minimus lies beneath the anterior part of the Gluteus medius and is closely related to it in action. It originates on the outer surface of the ilium betweeen the anterior and inferior gluteal lines and inserts on the anterior border of the greater trochanter. It abducts the thigh and rotates it medially.

Between the tendons of insertion of each of the Gluteal muscles and the greater trochanter are bursae which serve to protect the parts from friction. Two

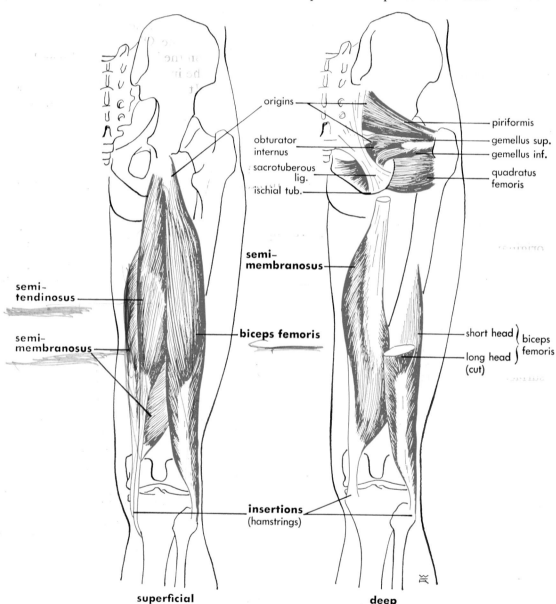

Fig. 11–56.   Posterior thigh muscles (hamstrings).

additional bursae may be present under other parts of the Gluteus maximus.

The **Tensor fascia lata** is a flat, quadrilateral muscle originating from the anterior part of the outer lip of the iliac crest and of the anterior superior spine and inserting between the two layers of the iliotibial band of the fascia lata. It flexes the thigh and rotates it medially. With the Gluteus maximus, which also inserts in part into the iliotibial band, it stabilizes the knee in the weight-bearing position. It also may stabilize the pelvis and trunk on the thighs.

The Piriformis, as suggested by the name, is pear-shaped in form. It originates within the true pelvis by three fleshy digitations on the anterior side of the sacrum and emerges through the greater sciatic foramen to insert into the superior border of the greater trochanter. It rotates the thigh laterally, and also abducts and may extend it.

The Obturator internus, triangular in shape, lies partly within the pelvis and partly at the back of the hip joint. It originates from the area around the obturator foramen, and from the obturator membrane which almost completely closes the foramen. Its fibers converge to the lesser sciatic foramen through which it passes, and its flattened tendon passes across the capsule of the hip joint to insert into the forepart of the medial surface of the greater trochanter above the trochanteric fossa. It rotates the thigh laterally.

The Superior and Inferior gemelli are small muscles which lie above and below the tendon of the Obturator internus, respectively, and join with it for insertion into the medial side of the greater trochanter. They are lateral rotators of the thigh. When the thigh is flexed, they may assist in abducting and extending the thigh.

The Quadratus femoris is a small, rectangular muscle below the Gemelli, and lying over the Obturator externus. It originates on the ischial tuberosity and inserts near the intertrochanteric crest of the femur. It adducts and rotates the thigh laterally.

The Obturator externus, a flat, triangular muscle, is on the outer surface of the anterior wall of the pelvis. It originates around the medial side of the obturator foramen and the outer surface of the obturator membrane. Its fibers converge laterally, extend backward and upward, and its tendon crosses the back of the lower part of the joint capsule to insert into the trochanteric fossa of the femur. It is a lateral rotator of the thigh.

Muscles of the Thigh

These muscles may be divided into three groups—anterior, medial, and posterior.

Anterior (Figs. 11-52, 54, 55)

Sartorius

Quadriceps femoris
Rectus femoris
Vastus lateralis
Vastus medialis
Vastus intermedius
Articularis genu

Medial (Fig. 11–55)

Gracilis   **Adductor longus**   **Adductor magnus**
Pectineus   Adductor brevis

Posterior (Figs. 11–52, 56)

Biceps femoris   Semimembranosus
Semitendinosus

Anterior Thigh Muscles. These muscles are often referred to as the extensor group, for they are primarily concerned with the extension of the leg. However, they also serve other functions, as flexion, abduction, and rotation. The **Quadriceps femoris** makes up the bulk of this mass of muscle, while the **Sartorius** is long and strap-like and runs diagonally across the front of the thigh (Figs. 11–52, 54, 55).

The **Sartorius** is the longest muscle, and has the longest fibers, of any muscle in the body. It extends from its origin on the anterior superior spine of the ilium across to the medial side of the leg where it inserts by a thin, flat, and expansive tendon on the anteromedial surfaces of the head of the tibia. Since it crosses both the hip and the knee joints, it flexes both the thigh and the leg and also abducts and laterally rotates the thigh. It may also rotate the leg medially. The Sartorius is called the "tailor's muscle" because it is used in achieving the tailor's position—seated with the legs crossed. It is a weak muscle and produces the above movements in cooperation with other muscles. It helps to stabilize the thigh and leg in walking.

The **Quadriceps femoris** muscle is the great extensor of the leg. Its Vasti components cover the anterior, lateral, and medial surfaces of the shaft of the femur, and reach posteriorly to the linea aspera. The Rectus femoris lies in front of the Vastus intermedius and between the Vastus medialis and lateralis and unlike them has a double origin—one head to the anterior inferior spine of the ilium and the other to the posterosuperior rim of the acetabulum. These four muscles have a common tendon of insertion into the superior border of the patella which, in turn, is attached to the tuberosity of the tibia, by the ligamentum patellae. Actually, since the patella is a sesamoid bone, which formed in the tendon of the Quadriceps femoris muscle, this muscle can be said to have originally inserted into the tibial tuberosity. The patella, bearing on the condyles of the femur, serves as a lever to improve the angle of pull of the Quadriceps on the tibial tuberosity. The Quadriceps tendon, by sending fibers over the capsule of the knee joint and by the tension of the muscles, helps to strengthen and maintain the integrity of the knee joint. Since the Rectus femoris crosses the hip joint as well as the knee joint, it is a flexor of the thigh as well as an extensor of the knee. It is very important as a kicking muscle for this reason, and the whole group functions in walking, running, climbing, jumping, and kicking.

The Articularis genu is composed of fasciculi from the deep Vastus intermedius, or it may consist of independent fasciculi arising from the anterior surface of the lower end of the femur and inserting into the upper part of the synovial membrane of the knee. It pulls the synovial capsule upward when the knee is extended.

Medial Thigh Muscles. These are the **adductor** muscles of the thigh. With the exception of the Gracilis, which is strap-shaped, these muscles are triangular, with their apices originating on the pubis and/or ischium, and their broad bases inserting on the linea aspera of the femur between the origins of the Vasti muscles. The **Gracilis** originates on the pubis and runs superficially over the medial aspect of the thigh to insert near the Sartorius on the medial side of the upper end of the tibia. It is therefore a flexor of the leg as well as an adductor and flexor of the thigh. The relative sizes and positions of these muscles are shown in Figures 11–52, 55.

Posterior Thigh Muscles (Figs. 11–52, 56). These muscles are often called the hamstrings. They lie behind the Adductor magnus. They are long muscles with their origins on the ischial tuberosity and their insertions in back of the knee joint. The **Biceps femoris** has an extra origin (short head) on the distal portion of the linea aspera of the femur. Its insertion is on the head of the fibula. The **Semimembranosus** inserts on the back of the medial condyle of the tibia, while the long, round tendon of the **Semitendinosus** curves around to the medial side of the knee to insert on the shaft of the tibia close to the Sartorius and Gracilis.

The "hamstrings" are important ex-

Some other functions of thigh muscles might be mentioned at this point. The strap-like Sartorius, Gracilis, and Semitendinosus muscles originate at far-separated points of the pelvic girdle (anterior superior spine, pubis and ischial tuberosity), but insert at a common point on the medial side of the head of the tibia. They form an inverted tripod and with the iliotibial band on the lateral side of the thigh have a stabilizing influence on the pelvis when an individual is standing. The hamstrings also have a ligamentous function, as when one bends over to touch the toes with his hands while the knees are extended, they are a restraining influence.

MUSCLES OF THE LEG

The muscles of the leg may be divided into three groups—the anterior, posterior, and lateral crural muscles.

Anterior Crural Muscles (Figs. 11–52, 57)

Tibialis anterior   **Extensor digitorum longus**
Extensor hallucis longus   Peroneus tertius

Posterior Crural Muscles
Superficial Group (Figs. 11–52, 58)

Gastrocnemius   **Soleus**   Plantaris

Deep Group (Fig. 11–59)

Popliteus   **Flexor digitorum longus**
Flexor hallucis longus   **Tibialis posterior**

Lateral Crural Muscles (Figs. 11–52, 60)

Peroneus longus   Peroneus brevis

tensors of the hip and flexors of the knee and therefore play an important role in walking. The Biceps femoris is also a lateral rotator of the flexed knee, while the other hamstrings, the Sartorius and Gracilis are medial rotators during knee flexion.

Deep Fascia. The muscles of the leg are completely covered by the deep fascia which is continuous above with the fascia lata of the thigh and below with that of the foot. It is fused with the periosteum of the subcutaneous surfaces of the bones such as the anterior border of the tibia. At the knee it is strengthened by slips from the hamstrings and the Sartorius and Gracilis muscles. It is stronger on the front of the leg than on the back, and on the lateral side it gives off from its deep surface two intermuscular septa, the anterior and posterior peroneal septa, which separate the lateral crural muscles from those of the anterior and posterior regions. It also forms a deep transverse fascia between the superficial and deep muscles of the posterior crural region.

Anterior Crural Muscles (Figs. 11–52, 57). The anterior crural muscles serve to dorsiflex the foot and to extend the toes. The **Tibialis anterior** originates on the lateral surface of the head and

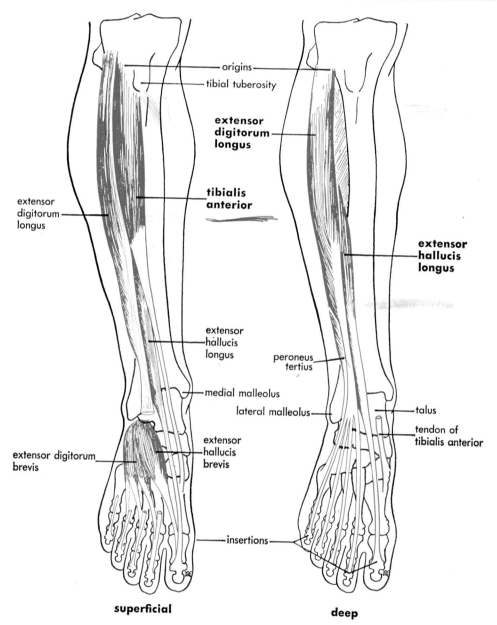

origins
tibial tuberosity
extensor digitorum longus
tibialis anterior
extensor digitorum longus
extensor hallucis longus
extensor hallucis longus
peroneus tertius
medial malleolus
lateral malleolus
talus
tendon of tibialis anterior
extensor digitorum brevis
extensor hallucis brevis
insertions
superficial   **deep**

FIG. 11–57.   Muscles of the anterior aspect of the right leg and foot.

shaft of the tibia and from the interosseous membrane. It inserts into the first cuneiform and the base of the first metatarsal. It is a powerful dorsiflexor of the foot and also inverts (*supinates*) it. In walking, it prevents stubbing the toes as the limb swings forward. It also helps support the median longitudinal arch of the foot.

The **Extensor hallucis longus** lies next to the Tibialis anterior, originating from the middle portion of the anteromedial surface of the fibula and inserting into the base of the distal phalanx of the big toe. Its actions are to extend the big toe and to dorsiflex the foot.

The **Extensor digitorum longus** is a flat, penniform muscle which originates

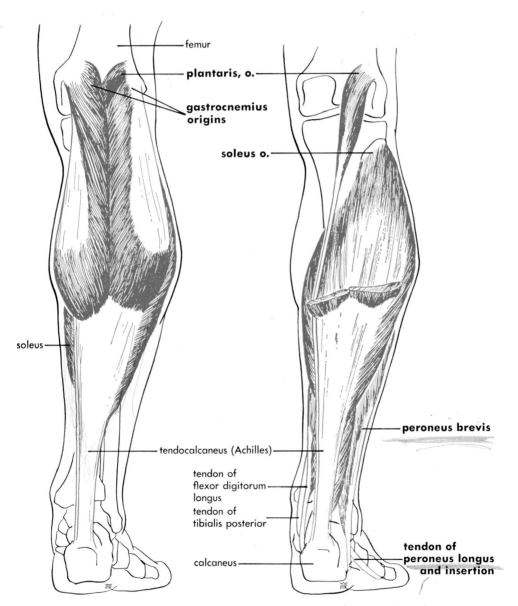

Fig. 11–58. Superficial muscles of the posterior aspect of the right leg.

from the lateral surface of the head of the tibia and from the proximal two-thirds of the anteromedial surface of the fibula and from the interosseous membrane. It lies partly hidden by the edge of the Tibialis anterior. Four tendons of insertion appear at the ankle, cross the dorsum of the foot, and attach to the second and third phalanges of the four lesser toes,

which they extend. This muscle also dorsiflexes and everts (*pronates*) the foot.

The Peroneus tertius is actually the lower part of the Extensor digitorum longus. It is inserted by a tendon into the dorsal surface of the base of the fifth metatarsal. It dorsiflexes and pronates the foot.

The tendons of insertion of these

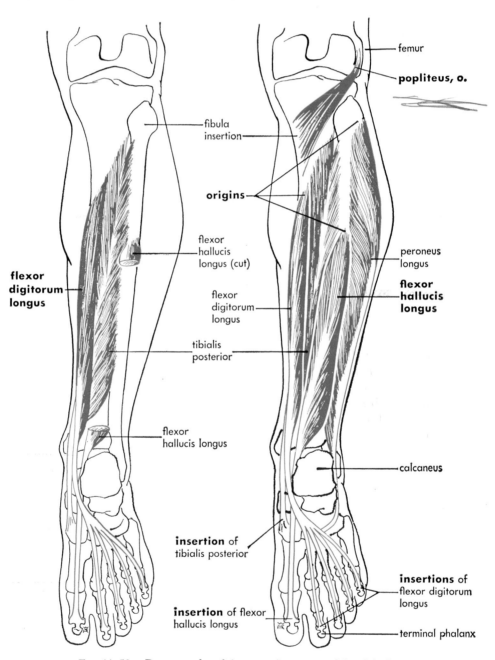

FIG. 11–59.   Deep muscles of the posterior aspect of the right leg.

anterior crural muscles pass under the transverse crural and cruciate ligaments in the ankle region. They are thus held firmly to the underlying bones (Fig. 11–62).

Posterior Crural Muscles—Superficial (Figs. 11–52B, 58). The primary function of these muscles is to extend (*plantar flexion*) the foot at the ankle. The **Gastrocnemius** and **Soleus** together form the "calf of the leg" and they insert by a common tendon, the tendo calcaneus (*tendon of Achilles*) into the tuberosity of the calcaneus or heel bone. These two muscles are sometimes referred to as the Triceps surae. Because the origins of the Gastrocnemius are just above the lateral and medial condyles on the posterior surface of the femur and the muscle crosses the knee joint, it also flexes the knee (Fig. 11–58).

The Plantaris is a small muscle which lies between the proximal ends of the Gastrocnemius and Soleus and inserts into the calcaneus by a long, slender tendon which follows along the medial side of the tendo calcaneus.

Posterior Crural Muscles — Deep (Fig. 11–59). Recall that these muscles are separated from the superficial crurals by the deep transverse fascia. The **Popliteus** muscle arises from the lateral condyle of the femur and inserts into the posterior surface of the shaft of the tibia above the popliteal line. It forms the lower part of the floor of the **popliteal fossa**—the hollow behind the knee. It rotates the tibia medially, thus "unlocking" the knee joint for the initiation of flexion. This is assuming that the knee is in a locked position when it is extended, as in the standing position.

The relative positions of the **Flexor digitorum longus** and **Flexor hallucis longus** are readily seen on Figure 11–59. Their functions are quite evident in their names. The **Tibialis posterior** is the deepest muscle of the posterior group and can be seen on Figure 11–59 between the two preceding muscles. Its distal tendon passes behind the medial malleolus with that of the Flexor digitorum longus to insert on the tuberosity of the navicular and into the plantar surfaces of the cuneiforms and cuboid. The Tibialis posterior extends and inverts the foot. The laciniate ligament, extending between the medial malleolus and calcaneus, holds the tendons of these muscles in position (Fig. 11–63).

Lateral Crural Muscles. The muscles of this group, the **Peroneus longus** and **Peroneus brevis,** extend and evert the foot. The longus is the more superficial and extends along the entire lateral surface of the fibula. The brevis lies under the longus, arising from the middle third of the lateral surface of the shaft of the fibula. The tendons of both muscles pass behind the lateral malleolus to the plantar surface of the foot. The **superior peroneal retinaculum,** stretching between the lateral malleolus and calcaneus, holds them in position (Fig. 11–62). The Peroneus longus, after crossing the foot, inserts into the plantar aspect of the first cuneiform and the base of the first metatarsal. The brevis inserts into the tuberosity of the fifth metatarsal.

Tendon Sheaths of the Ankle. The tendons crossing the ankle joint are all provided with tendon or synovial sheaths which can best be studied in Figures 11–62, 63. They insure the free movement of the tendons over the bones and under the ligaments which hold the tendons in place. Bursae are also present as, for example, under the tendo calcaneus.

INTRINSIC MUSCLES OF THE FOOT

Fascia. The fascia on the dorsal side of the foot is thin and membranous. It is continuous with the transverse and cruciate crural ligaments above; anteriorly, it contributes to a sheath for tendons

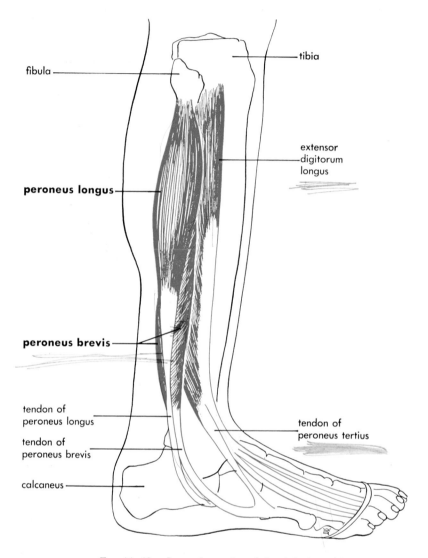

FIG. 11–60.   Lateral muscles of the right leg.

on the dorsum of the foot. Laterally, it blends with the plantar aponeurosis on the sole of the foot (Fig. 11–61). The plantar aponeurosis attaches posteriorly to the tuberosity of the calcaneus and anteriorly it forms fine slips, each of which divides at the metatarsophalangeal joints and attaches to the bases of the first row of phalanges. The flexor tendons pass through the divided slips of the plantar fascia. This structure is a strong supporter of the longitudinal arch of the foot, a function which it shares with the ligaments and the intrinsic and extrinsic muscles of the foot (Fig. 11–61).

In studying the muscles and fasciae of the foot, one should recall that the hand has a basic architecture similar to that of the foot. However, while the hand has remained generalized and capable of great versatility of action, the foot has specialized to the extent that its functions

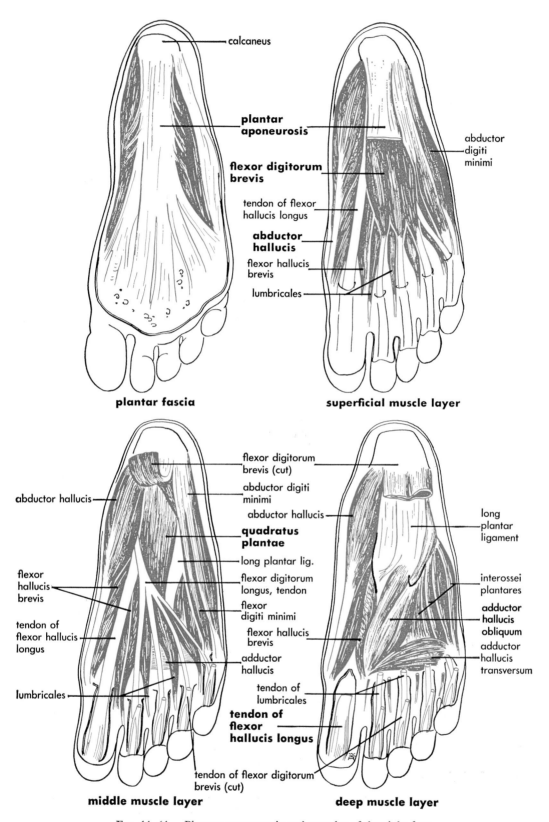

plantar fascia

superficial muscle layer

middle muscle layer

deep muscle layer

Fig. 11-61. Plantar aponeurosis and muscles of the right foot.

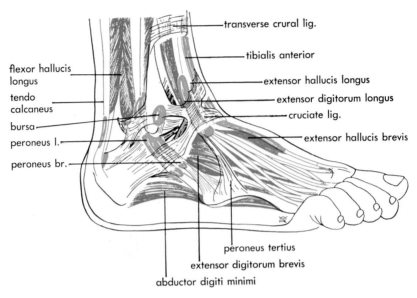

Fig. 11–62.   Lateral aspect of synovial (tendon) sheaths of the right ankle.

are limited—primarily to support and locomotion.

The intrinsic muscles of the foot may be divided into two groups: a dorsal muscle, and the plantar muscles. The latter, in turn are arranged in four layers.

Study carefully the labeled illustrations of these foot muscles, noting particularly their relationship to the tendons of extrinsic muscles of the foot and to the fascia (Fig. 11–61).   Table 11–1 will give you other pertinent data about these muscles.

Dorsal Muscle

Extensor digitorum brevis (Fig. 11–62)

Plantar Muscles

First Layer (Fig. 11–61)

Abductor hallucis   Abductor digiti minimi (quinti)
Flexor digitorum brevis

Second Layer (Fig. 11–61)

Quadratus plantae   Lumbricales

Third Layer (Fig. 11–61)

Flexor hallucis brevis   Flexor digiti minimi brevis
Adductor hallucis

Fourth Layer (Fig. 11–61)

Interossei dorsales   Interossei plantares

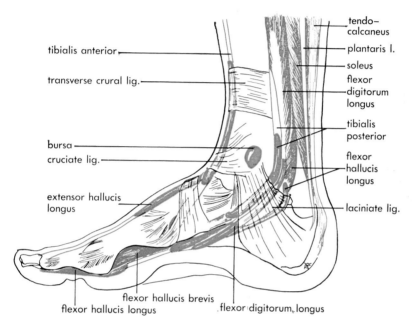

FIG. 11–63.   Medial aspect of synovial (tendon) sheaths of the right ankle.

QUESTIONS

1. What is meant by isotonic and isometric contractions of muscles?
2. Distinguish between a prime mover and a synergist, using an example.
3. What are antagonistic muscles?
4. What is the meaning of the expression that, "the movements, not the muscles, are represented in the central nervous system"?
5. Discuss briefly the skeletal, articular, muscle, and nerve relationships in the movement of the body and its parts.
6. Some muscles cross more than one articulation. What is the functional significance of this? Give an example.
7. What is the most common kind of leverage to be found in the human skeleto-muscular system? Make a diagram to illustrate this type and give examples.
8. Give an example of a lever of the first class in the human body.
9. What are two important advantages gained in the human body by the use of levers?
10. Give examples of each of the following.
    (a) Bipennate muscle   (c) Multipennate muscle
    (b) Radiate muscle   (d) One with parallel, longitudinally oriented fasciculi.
11. What advantages are gained by the various arrangements of muscle fibers or fasciculi listed above?
12. What influence does gravity have on muscle action?
13. What are the three general categories of fascia?
14. What do the fasciae do in the body?
15. Name a few of the facial muscles and state their functions and importance.
16. Name the important muscles of mastication and describe the action of one of them.
17. What do we mean by, "the intrinsic muscles" of the tongue?
18. What are the functions of the Constrictors of the pharynx?
19. What muscles form the ridges which enclose the fossae for the palatine tonsils?
20. Describe the boundaries of the cervical triangles.
21. Give two examples of two-bellied muscles in the cervical region.
22. Explain the muscle arrangement whereby the atlas is rotated on the axis.
23. What is the extent of the Erector spinae muscle?
24. Discuss the position and functions of the diaphragm.
25. Name some of the muscles, other than the diaphragm, which aid in breathing.
26. What is the relationship of the linea alba and the rectus sheath to the abdominal muscles?
27. What is the inguinal ligament? the inguinal canal? the transversalis fascia?
28. Define pelvic diaphragm, perineum, and urogenital diaphragm.
29. Name the muscles which connect the axial skeleton and shoulder girdle.
30. Describe the structure, origin, insertion, and action of the Deltoideus.
31. What is the flexor compartment of the arm? Name the muscles within it.
32. What muscles constitute the "rotator cuff" of the shoulder? What else do they contribute to the shoulder joint?
33. What kind of leverage is illustrated by the Triceps brachii in its relationship to the ulna? What are the advantages in such an arrangement?
34. Describe the extensor and flexor compartments of the forearm.
35. What is significant about the fact that the hand contains so many articulations and so many muscles?
36. Why may deep infections of the hand be so serious?
37. Describe briefly the fascia lata.
38. Describe the Gluteus maximus and discuss its importance in man.
39. Name the components of the Quadriceps femoris muscle and describe its relationship to the patella and knee joint.

40. Describe briefly the deep fascia of the leg, noting particularly the fascial compartments.
41. Name the muscle groups of the leg and state the general functions of each.
42. How do the leg muscles contribute to the support of the arches of the foot?
43. Distinguish between the following: inversion and eversion of the foot; dorsiflexion and plantar flexion; pronation and supination of the foot.
44. Describe the deep fascia of the foot.
45. Compare the hands and feet in terms of their functional anatomy.

TABLE 11-1. MUSCLES OF THE HUMAN BODY. THE MUSCLES ARE ARRANGED ALPHABETICALLY FOR QUICK REFERENCE. Page numbers are shown in italics when they have an illustration of the muscle.

Muscle	Page	Origin	Insertion	Action	Innervation
Abductor digiti minimi	234, 238	Pisiform bone, tendon of Flex. carpi ulnaris.	Base of first phalanx of fifth finger.	Abduction of fifth finger.	Ulnar.
Abductor digiti minimi (quinti)	255, 256	Tuberosity of calcaneus.	Base of first phalanx of fifth toe.	Abduction and flexion of fifth toe.	Lateral plantar.
Abductor hallucis	255, 256	Tuberosity of calcaneus, medial process.	Base of proximal phalanx of first toe, medial side.	Abduction and flexion of first toe.	Medial plantar.
Abductor pollicis brevis	231, 238	Tuberosity of navicular bone; ridge of trapezium bone.	Base of first phalanx of first finger, lateral side.	Abduction of first finger.	Median.
Abductor pollicis longus	235, 237	Lateral and dorsal aspect of radius and ulna, interosseous membrane.	Base of first metacarpal bone, lateral side.	Abduction of first finger and hand.	Deep radial.
Adductor brevis	245, 247	Inferior ramus of pubis.	Upper one-third of linea aspera.	Adduction, flexion and lateral rotation of thigh.	Obturator.
Adductor hallucis	255, 256	(1) Oblique head: Bases of second through fifth metatarsals. (2) Transverse head: Plantar metatarsophalangeal ligaments of third through fifth toes.	Base of proximal phalanx of first toe.	Adduction of first toe.	Lateral plantar.
Adductor longus	245, 247	Crest and symphysis of pubis.	Middle third of linea aspera.	Adduction, flexion, and lateral rotation of thigh.	Obturator.
Adductor magnus	245, 247	Rami of pubis and ischium. Outer inferior ischial tuberosity.	Lower one-third of linea aspera; internal epicondyle of femur and supracondylar line and adductor tubercle.	Powerful adduction of thigh, flexion, lateral rotation, and extension of thigh.	Obturator.
Adductor pollicis	234, 238	(1) Oblique head: capitate, and base of second, third, and fourth metacarpals. (2) Transverse head: body of third metacarpal bone.	Base of first phalanx of first finger, medial side.	Adduction of first finger.	Deep ulnar.
Anconeus	234, 236	Lateral epicondyle of humerus.	Olecranon, upper one-fourth of shaft of ulna.	Extension of forearm.	Radial.

Muscle	Page	Origin	Insertion	Action	Nerve
Articularis genu	248	Lower one-fourth of shaft of femur (anterior aspect).	Synovial membrane of knee joint.	Elevation of articular capsule in leg extension.	Femoral.
Biceps brachii	228, 229	(1) Long head: superior margin of glenoid fossa. (2) Short head: coracoid process.	Tuberosity of radius and deep fascia of forearm.	Flexion of arm and supination of hand.	Musculocutaneous.
Biceps femoris	246, 249	(1) Long head: tuberosity of ischium, sacrotuberous ligament. (2) Short head: linea aspera, lateral side.	Lateral side of head of lateral condyle of tibia.	Flexion and lateral rotation of leg, extension of thigh.	Tibial.
Brachialis (anticus)	229, 230	Lower one-half of anterior surface of shaft of humerus.	Coronoid process of ulna.	Flexion of the forearm.	Musculocutaneous, radial, median.
Brachioradialis	231, 234	Superior two-thirds of the lateral supracondylar ridge of the humerus.	Styloid process of the radius.	Flexion of forearm.	Radial.
Buccinator	196, 200	Alveolar processes of maxilla and mandible and the pterygo-mandibular raphe.	Fibers of this muscle blend with those of the Orbicularis oris.	Compression of cheek. Holds food between teeth.	Facial.
Bulbocavernosus	220	Central point of perineum, and median raphe.	Aponeurosis of the corpus cavernosum penis (clitoris).	Constricts urethral canal in male; constricts vaginal orifice in female.	Perineal br. of pudendal n.
Coccygeus	219, 221	Spine of ischium and sacrospinous ligament.	Lateral margin of sacrum and coccyx.	Draws coccyx forward; support of pelvic diaphragm.	Brs. from pudendal plexus (S4–5).
Constrictor pharyngis inferior	200, 202	Cricoid and thyroid cartilage.	Posterior median raphe of pharynx.	Contraction (constriction) of lower pharynx.	Pharyngeal plexus (IX and X).
Constrictor pharyngis medius	200, 202	Superior border of greater cornu and lesser cornu of hyoid bone.	Posterior median raphe of pharynx.	Contraction (constriction) of middle pharynx.	Pharyngeal plexus (IX and X).
Constrictor pharyngis superior	200, 202	Medial pterygoid plate, pterygo-mandibular raphe, and alveolar process of the mandible.	Posterior median raphe of the pharynx.	Contraction (constriction) of upper pharynx.	Pharyngeal plexus (IX and X).
Coracobrachialis	223, 238	Coracoid process of scapula.	Mid-medial surface of shaft of humerus.	Flexion and adduction of humerus.	Musculocutaneous.
Corrugator	196	Medial end of the superciliary arch of frontal bone.	Integument above mid-region of orbital arch.	Draws eyebrow down to produce frowning expression.	Facial
Cremaster	216, 277	Middle of inguinal ligament, int. oblique m.	Tubercle and crest of pubis.	Elevation of testis.	Genitofemoral.

Muscle	Page	Origin	Insertion	Action	Innervation
Cricothyroideus	200	Cricoid cartilage	Inferior cornu and lower border of thyroid cart.	Tilts cricoid dorsally; tenses vocal cords.	Super. laryngeal.
Deltoideus	225, 226	Clavicle; acromion process and spine of scapula.	Deltoid tuberosity of humerus.	Whole muscle abducts arm; in part, may flex, extend, and rotate arm.	Axillary.
Diaphragm	210, 214	Xiphoid process, cartilages and adjacent parts of last six ribs, and lumbocostal arches.	Central tendon.	Respiration.	Phrenic.
Digastricus	204, 205	(1) Posterior belly: mastoid notch of temporal bone. (2) Anterior belly: lower border of mandible.	Greater cornu of hyoid bone.	Elevation of hyoid; assists in opening of mouth.	Trigeminal.
Dilator naris anterior	196–7	Greater alar cartilage.	Integument of margin of nostril.	Dilation of aperture of nares.	Facial.
Dilator naris posterior	196–7	Margin of nasal notch of maxilla.	Integument of margin of nostril.	Dilation of aperture of nares.	Facial.
Epicranius Frontalis Occipitalis	196, 197	(1) Frontal belly: fibers are continuous with those of Procerus, Corrugator, and Orbicularis oculi. (2) Occipital belly: superior nuchal line and mastoid portion of temporal.	Galea aponeurotica.	Together they provide expression of surprise; frontalis raises eyebrow, wrinkles forehead.	Facial.
Erector spinae	209, 211	See Sacrospinalis.			
Extensor carpi radialis brevis	237	Lateral epicondyle of humerus.	Dorsum of third metacarpal bone.	Extension and abduction of hand (wrist).	Radial.
Extensor carpi radialis longus	234, 237	Lateral supracondylar ridge of humerus.	Dorsum of base of second metacarpal bone.	Extension and abduction of hand (wrist).	Radial.
Extensor carpi ulnaris	234, 237	Lateral epicondyle of humerus, dorsal border of ulna.	Base of fifth metacarpal bone.	Extension and adduction of hand (wrist).	Deep radial.

Muscle		Origin	Insertion	Action	Nerve
Extensor digiti minimi	234, 235	External epicondyle of humerus.	Dorsum of first phalanx of fifth finger.	Extension of fifth finger.	Deep radial.
Extensor digitorum brevis	242, 250	Anterior, superior aspect of calcaneus.	Blends with the tendons of the Extensor digitorum longus.	Extension of proximal phalanges of medial four toes.	Deep peroneal.
Extensor digitorum communis	234, 235	Lateral epicondyle of humerus.	Second and third phalanges of the four lesser fingers.	Extension of phalanges; may extend wrist.	Deep radial.
Extensor digitorum longus	251, 254	Lateral condyle of tibia and from the distal, anterior aspect of the shaft of the fibula, interosseous membrane.	Second and third phalanges of the four lesser toes.	Extension of the proximal phalanges; dorsiflexion and pronation of foot.	Deep peroneal.
Extensor hallucis longus	250, 251	Anterior aspect of fibula and interosseous membrane.	Base of distal phalanx of first toe.	Extension of proximal phalanx of first toe; dorsiflexion of foot, iverts foot.	Deep peroneal.
Extensor indicis	235, 236	Dorsum of shaft of ulna, interosseous membrane.	Blends into tendon of Extensor digitorum communis, index finger.	Extension of index finger.	Deep radial.
Extensor pollicis brevis	235, 237	Dorsum of radius, interosseous membrane.	Base of first phalanx of thumb.	Extension of first phalanx of thumb, abducts hand.	Deep radial.
Extensor pollicis longus	235, 237	Dorsum of shaft of ulna, lateral side.	Base of distal phalanx of thumb.	Extension of thumb, abduction of hand.	Deep radial.
External intercostals (11 pairs)	211, 213	Inferior border of a rib.	Superior border of a rib below.	Draw ribs together—aid in respiration.	Intercostal.
External oblique (abdominis)	215, 276	Anterior, inferior aspect of lower eight ribs.	Linea alba, pubis, and crest of ilium.	Compresses abdominal viscera. Flexion of vertebral column.	Brs. of 8–12 intercostal, iliohypogastric, and ilioinguinal
Flexor carpi radialis	232, 233	Medial epicondyle of humerus.	Base of second metacarpal bone and a slip to the third metacarpal.	Flexion of wrist and forearm, abducts hand.	Median.
Flexor carpi ulnaris	232, 233	(1) Medial epicondyle of humerus. (2) Medial margin of olecranon and distal, dorsal two-thirds of ulna.	Base of fifth metacarpal bone, pisiform and hamate bones.	Flexion and adduction of wrist, flexes forearm.	Ulnar.
Flexor digiti minimi brevis (foot)	255, 256	Base of fifth metatarsal bone, tendon of peroneus longus.	Base of proximal phalanx of fifth toe.	Flexion of proximal phalanx of fifth toe.	Lateral plantar.

Muscle	*Page*	*Origin*	*Insertion*	*Action*	*Innervation*
Flexor digiti minimi brevis (hand)	234, 238	Hamulus of hamate bone.	Base of the proximal phalanx of the fifth finger, medial side.	Flexion of the fifth finger.	Ulnar.
Flexor digitorum brevis	255, 256	Medial process of the tuberosity of the calcaneus.	Second phalanx of the four lesser toes.	Flexion of the four lesser toes.	Medial plantar.
Flexor digitorum longus	249, 252	Posterior surface of shaft of tibia.	Base of the distal phalanges of the four lesser toes.	Flexion of distal phalanges and plantar flexion and supination of the foot.	Tibial.
Flexor digitorum profundus	233, 234	Proximal three-fourths of shaft of ulna, med. coronoid process of ulna.	Bases of distal phalanges of fingers.	Flexion of all phalanges of each finger. Flexes hand.	Volar interosseous of median. Ulnar.
Flexor digitorum superficialis (sublimis)	232, 233	(1) Humeral: medial epicondyle of humerus. (2) Ulnar: coronoid process. (3) Radial: oblique line of the radius.	Second phalanges of each finger.	Flexion of the second phalanx of each finger, flexes forearm and hand.	Median.
Flexor hallucis brevis	255, 256	Plantar surface of cuboid and third cuneiform bones.	Base of proximal phalanx of first toe, medial and lateral sides.	Flexion of proximal phalanx of first toe.	Medial plantar.
Flexor hallucis longus	252, 253	Lower two-thirds of shaft of fibula, interosseous membrane.	Base of distal phalanx of first toe.	Flexion of second phalanx of first toe; flexion and supination of foot.	Tibial.
Flexor pollicis brevis	234, 238	Transverse carpal ligament and the trapezium bone.	Base of first phalanx of first finger, lateral side.	Flexion and adduction of thumb.	Median and deep ulnar.
Flexor pollicis longus	232, 233	Anterior surface of radius and coronoid process of ulna.	Base of distal phalanx of thumb.	Flexion of thumb. Flexes and adducts first metacarpal.	Volar interosseous of median.
Gastrocnemius	251, 253	(1) Lateral head: lateral condyle of femur. (2) Medial head: medial condyle of femur (posterior).	Posterior aspect of the calcaneus.	Plantar flexion of foot; flexion of leg.	Tibial.
Gemellus inferior	246, 247	Tuberosity of ischium.	Greater trochanter of femur.	Lateral rotation of thigh.	Br. of nerve to Quadratus femoris m.

Muscle	Pages	Origin	Insertion	Action	Nerve
Gemellus superior	246, 247	Spine of ischium, outer surface.	Greater trochanter of femur.	Lateral rotation of thigh.	Br. of nerve to Obturator internus m.
Genioglossus	199, 201	Superior mental spine of mandible.	Body of hyoid bone and whole of under surface of tongue.	Retraction, depression, and protrusion of tongue; elevation of hyoid bone.	Hypoglossal.
Geniohyoideus	204, 205	Inferior mental spine of mandible.	Body of hyoid bone.	Forward movement of hyoid bone and tongue.	C1 thru ansa hypoglossi.
Glossopalatinus (Palatoglossus)	201–202	Anterior surface of soft palate.	Lateral border of tongue.	Elevation of back of tongue; constriction of the isthmus of the fauces.	Pharyngeal plexus (XI).
Gluteus maximus	243, 244	Posterior gluteal line of the ilium to crest, sacrotuberous ligament, post. surface of lower sacrum, coccyx.	Fascia lata, and shaft of femur.	Extension and lateral rotation of thigh, braces knee.	Inferior gluteal.
Gluteus medius	243, 244	Outer surface of ilium and crest.	Greater trochanter of femur.	Abduction of thigh, medial rotation.	Superior gluteal.
Gluteus minimus	243, 246	Outer surface of ilium, sciatic notch.	Greater trochanter, anterior border.	Abduction and medial rotation of thigh, weak flexor.	Superior gluteal.
Gracilis	245, 248	Inferior aspect of symphysis pubis.	Proximal medial surface of shaft of tibia.	Adduction of thigh; flexion of leg.	Obturator.
Hyoglossus	199, 201	Body and greater cornu of hyoid bone.	Inferior aspect of the lateral border of the tongue.	Depression of sides of tongue.	Hypoglossal.
Iliacus	208, 243	Iliac fossa, iliac crest, and sacrum; ant. sacroiliac lig.	Lateral side of the tendon of the Psoas major.	Flexion and lateral rotation of the thigh.	Femoral (L2–3),
Iliocostalis cervicis	209, 211	Angles of the third to sixth ribs.	Fourth to sixth cervical vertebra, trans. processes.	Extension of vertebral column.	Brs. of post. rami of cervical nn.
Iliocostalis dorsi	209, 211	See under the following: I. lumborum I. thoracis I. cervicis			
Iliocostalis lumborum	209, 211	Iliac crest, mid-crest of sacrum, spinous processes of lumbar and last two thoracic vertebrae, lat. crest of sacrum.	Inferior borders of angles of lower six or seven ribs.	Extension of vertebral column.	Brs. of post. rami of lumbar nn.

TABLE 11–1. MUSCLES OF THE HUMAN BODY. THE MUSCLES ARE ARRANGED ALPHABETICALLY FOR QUICK REFERENCE. (*Continued*)

Muscle	Page	Origin	Insertion	Action	Innervation
Iliocostalis thoracis	209, 211	Medial portion of upper surface of the angles of the lower six ribs.	Upper borders of the angles of the upper six ribs and transverse process of C7.	Extension of vertebral column	Brs. of post. rami of thoracic nn.
Iliopsoas	208, 243	See under the following: Psoas major Iliacus			
Infraspinatus	225, 227	Infraspinous fossa of scapula.	Greater tubercle of humerus.	Outward rotation of humerus; may assist in both abduction and adduction of humerus.	Suprascapular.
Internal intercostals (11 pairs)	213	Inner surface of a rib.	Superior surface of a rib below.	Draw ribs together. Aid in respiration.	Intercostal.
Internal oblique (abdominis)	216, 277	Lateral portion of inguinal ligament, the iliac crest, and part of the lumbodorsal (thoracolumbar) fascia.	Inferior margins of lower three ribs, linea alba, and xiphoid process.	Compression of abdominal viscera; flexion of vertebral column.	Brs. 8–12th intercostals, iliohypogastric, and ilio-inguinal.
Interossei dorsales (foot)	256	All four arise by two heads, each from adjacent sides of the metatarsal bones 1–4.	Bases of proximal phalanges. 1 and 2 to 2nd toe 3 and 4 to lat. side toes 3 and 4.	Abduction of toes, flexion of proximal and extension of distal phalanges.	Deep br. of lateral plantar.
Interossei dorsales (hand)	238	All four arise by two heads from adjacent sides of the metacarpal bones.	Bases of the first phalanges, fingers 2–4.	Abduction of fingers; flexion of metacarpophalangeal joints; extension of two distal phalanges.	Deep ulnar.
Interossei plantares	238	Proximal end and medial side of third, fourth, and fifth metatarsal bones.	Medial side of base of proximal phalanges of same toes.	Adduction of toes; flexion of proximal and extension of distal phalanges.	Lateral plantar.
Interossei palmares	255, 256	Medial side of metacarpal one, lateral sides of metacarpals four and five.	Base of the first phalanges; medial side phalanx one of index finger, lat. side phalanx one of fingers 4 and 5.	Adduction of fingers; flexion of metacarpophalangeal joint; extension of two distal phalanges.	Deep ulnar.
Interspinales	209, 272	Spinous process of vertebrae above.	Spinous process of vertebrae below.	Extension of vertebral column.	Brs. dorsal rami of spinal nn.

Muscle	Page	Origin	Insertion	Action	Nerve
Intertransversarii	209	Transverse process of vertebrae.	Transverse process of vertebrae.	Lateral bending of vertebral column.	Brs. ventral rami of spinal nn.
Ischiocavernosus	220	Tuberosity of ischium.	Crus of penis (clitoris).	Aids in erection of male or female organ.	Perineal br. of pudendal n.
Latissimus dorsi	222, 224	Spinous processes of lower six thoracic, the lumbar and sacral vertebrae; posterior portion of the crest of the ilium; by muscular digitations from lower three ribs.	Lower portion of the inter-tubercular groove of humerus.	Extension, adduction, and medial rotation of arm. Draws shoulder backward and downward.	Thoracodorsal.
Levator ani Pubococcygeus Iliococcygeus	210, 221	Inner surface of the superior ramus of the pubis, arcus tendineus and ischial spine.	Anococcygeal raphe and the sides of the coccyx, central tendon of perineum.	Support and slight elevation of the pelvic floor.	Pudendal plexus brs. (S4)
Levator scapulae	222, 224	Transverse processes of atlas and axis and posterior tubercles of transverse processes of third and fourth cervical vertebrae.	Upper one-third of medial border of scapula.	Elevation of scapula; slight rotation of scapula. Extends and bends neck laterally.	Dorsal scapular (C3, 4, and 5).
Levator veli palatini	200, 202	Under-surface of apex of petrous part of the temporal bone and the cartilage of the auditory tube.	Aponeurosis of soft palate.	Elevation of soft palate.	Pharyngeal plexus (IX and X).
Levatores costarum (12 pairs)	210, 212	Transverse processes of the seventh cervical and first eleven thoracic vertebrae.	Each slip into the angle of the rib immediately below the processes of origin	Elevation of ribs (respiration); extension of vertebral column; lateral rotation of vertebral column.	Brs. of intercostals.
Longissimus capitis	206, 211	Transverse processes of upper four or five thoracic vertebrae and the articular processes of the lower three or four cervical vertebrae.	Mastoid process of the temporal bone.	Extension of the head; flexion of head to side; slight rotation.	Brs. of dorsal rami of middle and lower cervicals.
Longissimus cervicis	206, 211	Transverse processes of the upper four or five thoracic vertebrae.	Transverse processes of the second to sixth cervical vertebrae.	Extension of vertebral column and flexion to one side.	Brs. of dorsal rami of spinal nn.
Longissimus dorsi	206, 211	See under the following: L. thoracis L. cervicis L. capitis			
Longissimus thoracis	206, 211	Transverse processes of lumbar vertebrae and associated fascia.	Tips of transverse processes of all the thoracic vertebrae and lower nine ribs.	Extension of vertebral column and flexion to one side; draws ribs down.	Brs. of dorsal rami ot spinal nn.

TABLE 11–1. MUSCLES OF THE HUMAN BODY. THE MUSCLES ARE ARRANGED ALPHABETICALLY FOR QUICK REFERENCE. *(Continued)*

Muscle	Page	Origin	Insertion	Action	Innervation
Longus capitis	205, 206	Transverse processes of third to sixth cervical vertebrae.	Basilar part of occipital bone.	Flexion of head.	C1, 2, and 3.
Longus colli	205, 206	(1) Superior oblique portion: transverse processes of third to fifth cervical vertebrae. (2) Inferior oblique portion: bodies of first two or three thoracic vertebrae. (3) Vertical portion: fifth to seventh and first to third thoracic vertebrae.	(1) Anterior arch of atlas. (2) Transverse processes of fifth and sixth cervical vertebrae. (3) Bodies of second, third, and fourth cervical vertebrae.	Flexion of neck; slight rotation of cervical portion of vertebral column.	C2–7.
Lumbricales (foot)	255, 256	Each of four arise from branches of the Flexor digitorum longus.	Tendinous expansions of the Extensor digitorum longus.	Flexion of proximal phalanges and extension of distal phalanges of four lesser toes.	Medial plantar.
Lumbricales (hand)	232, 238	Tendons of the Flexor digitorum profundus, radial side of fingers two and three; adjoining sides of fingers three and four.	Tendinous expansion of the Extensor digitorum communis.	Flexion of metacarpophalangeal joints; extension of two distal phalanges.	Median and deep ulnar.
Masseter	198	(1) Superficial portion: zygomatic process of maxilla. (2) Deep portion: zygomatic arch.	(1) Angle of ramus. (2) Upper portion of ramus and coronoid process of mandible.	Approximates jaws.	Mandibular.
Multifidus	209, 272	Posterior aspect of sacrum. Posterior superior iliac spine, mammillary processes of all lumbar vertebrae, transverse processes of all thoracic and articular processes of last 4 cervical vertebrae.	Spinous processes of the vertebrae, from last lumbar to axis.	Extension and lateral rotation of vertebral column.	Brs. of dorsal rami of spinal nn.
Musculus uvulae	201, 202	Posterior nasal spine of the palatine bones.	Uvula.	Tension of uvula. Raises uvula.	Pharyngeal plexus (IX and X).
Mylohyoideus	205, 206	Mylohyoid line of the mandible.	Body of the hyoid bone and median fibrous raphe.	Elevation of hyoid bone and tongue.	Trigeminal.

Muscle		Origin	Insertion	Action	Nerve
Nasalis	197	(1) Transverse portion: superior and lateral incisive fossa of maxilla. (2) Alar portion: greater alar cartilage.	(1) Aponeurosis on bridge of nose. (2) Integument at apex of nose.	Depression of cartilaginous portion of nose.	Facial.
Obliquus capitis inferior	207	Spinous process of axis.	Transverse process of atlas.	Rotation of atlas.	Br. of suboccipital.
Obliquus capitis superior	207	Transverse process of atlas.	Between superior and inferior nuchal line of occipital bone.	Extension and lateral flexion of head.	Br. of suboccipital.
Obturator externus	247	Medial marginal region of obturator foramen; ischiopubic rami.	Trochanteric fossa of femur.	Lateral rotation of thigh.	Obturator.
Obturator internus	246, 247	Margin of obturator foramen (inner surface), pubis, and ischium.	Greater trochanter, medial surface.	Lateral rotation of thigh, abduction of thigh.	Brs. from L5, S1–2.
Omohyoideus	205, 206	Superior border of scapula (inferior belly). Tendon from clavicle (superior belly).	By tendon to clavicle (inferior belly). Body of hyoid bone (superior belly).	Downward movement of hyoid bone.	C1–3 thru ansa hypoglossi.
Opponens digiti minimi	234, 238	Hamulus of the hamate bone.	Length of metacarpal of fifth finger, medial side.	Abduction, flexion, and rotation of fifth metacarpal bone.	Ulnar.
Opponens pollicis	231, 238	Trapezium bone	Length of metacarpal bone of first finger, radial side.	Abduction, flexion, and rotation of thumb.	Median.
Orbicularis oculi	196, 197	Medial portion of the orbit.	Integument of eyelid.	Closes eyelids; compression of lacrimal sac.	Facial.
Orbicularis oris	196, 197	Complex, involving integument of the lips. A composite of superficial fibers of several facial muscles.	Corners of mouth and, superficially, the integument of the lips without, and the mucous membrane within.	Direct approximation of lips.	Facial.
Palmaris brevis	238, 240	Tendinous fasciculi from transverse carpal ligament.	Integument of palm, medial side.	Draws skin of ulnar side of palm toward midline of palm.	Ulnar.
Palmaris longus	232, 233	Medial epicondyle of humerus.	Transverse carpal ligament and palmar aponeurosis.	Tenses palmar aponeurosis; flexion of wrist and forearm.	Median.
Pectineus	245, 247	Pectineal line and associated fascia of pubis between iliopectineal eminence and tubercle of pubis	Pectineal line of femur.	Flexion, adduction, and lateral rotation of thigh.	Femoral.

TABLE 11-1. MUSCLES OF THE HUMAN BODY. THE MUSCLES ARE ARRANGED ALPHABETICALLY FOR QUICK REFERENCE. *(Continued)*

Muscle	Page	Origin	Insertion	Action	Innervation
Pectoralis major	221, 223	Clavicle, sternum, costal cartilages of true ribs.	Crest of greater tubercle of humerus.	Flexion, adduction, and medial rotation of arm.	Med. and lat. ant. thoracic.
Pectoralis minor	223, 224	Superior margin of third, fourth, and fifth ribs.	Coracoid process of scapula.	Downward rotation of scapula; depression of shoulder.	Med. ant. thoracic.
Peroneus brevis	253, 254	Lower two-thirds of shaft of fibula and intermuscular septum, lateral side.	Base of fifth metatarsal bone.	Pronation and plantar flexion of foot.	Superficial peroneal.
Peroneus longus	253, 254	Head and upper two-thirds of shaft of fibula and intermuscular septum.	Base of first metatarsal bone, and first cuneiform bone.	Pronation and plantar flexion of foot, supports arch.	Superficial peroneal.
Peroneus tertius	253, 254	Distal portion of anterior surface of fibula, interosseous membrane.	Dorsal surface of base of fifth metatarsal bone.	Dorsiflexion and pronation of foot.	Deep peroneal.
Pharyngopalatinus	201, 202	Soft palate.	Posterior border of the thyroid cartilage; lateral post. pharynx.	Elevation of pharynx; helps close nasopharynx.	Pharyngeal plexus (IX and X).
Piriformis	246, 247	Anterior aspect of sacrum, margin of greater sciatic foramen, and sacrotuberous ligament.	Superior border of greater trochanter.	Outward rotation of thigh, abduction.	Br. of S2 or S1 and 2.
Plantaris	251, 253	Lateral portion of linea aspera and popliteal ligament.	Calcaneous bone (via tendo-calcaneus).	Plantar flexion of foot; flexion of leg.	Tibial.
Platysma	197, 198	Fascial investment of Pectoralis major and Deltoid muscles.	Oblique line of mandible and integument of lower regions of face.	Depression of lower lip. Opens jaw.	Facial.
Popliteus	252, 253	Lateral condyle of femur, popliteal ligament.	Above popliteal line of the tibia on its posterior aspect.	Flexion and medial rotation of leg.	Tibial
Procerus	196, 197	Fascia over lower portion of nasal bone.	Integument over the glabella.	Depression of medial angle of eyebrow.	Facial.
Pronator quadratus	233, 234	Distal portion of shaft of ulna.	Lower one-fourth of shaft of radius.	Pronation and rotation of hand.	Volar interosseous of median.
Pronator teres	229, 233	(1) Humeral head: medial epicondyle of humerus. (2) Ulnar head: coronoid process of ulna, medial side.	Middle portion of shaft of radius, lateral side.	Pronation of hand.	Median.

Muscle	Page	Origin	Insertion	Action	Nerve
Psoas major	208, 243	Transverse processes of L1–5, intervertebral fibrocartilages of all lumbar vertebrae and T12.	Lesser trochanter of femur.	Flexion and medial rotation of thigh; flexion of lumbar region of vertebral column.	L2–3.
Psoas minor	208, 243	Twelfth thoracic and first lumbar vertebrae.	Iliopectineal line and eminence and iliac fascia.	Flexion of pelvis on abdomen.	L1.
Pterygoideus lateralis	198, 199	(1) Upper head: great wing of sphenoid bone. (2) Lower head: lateral pterygoid plate.	Condyle of the mandible.	Protrusion of mandible.	Mandibular.
Pterygoideus medialis	198, 199	Lateral pterygoid plate of sphenoid bone, pyramidal process of palatine, and tuberosity of maxilla.	Medial surface of ramus and angle of mandible.	Opposition of jaws.	Mandibular.
Pyramidalis abdominis	219	Pubis, ant. pubic lig.	Linea alba.	Tenses linea alba.	T12.
Quadratus femoris	246, 247	Tuberosity of ischium.	Linea quadrata of femur.	Lateral rotation of thigh.	Special br. from sacral plexus (L4–5, S1).
Quadratus lumborum	208, 219	Iliac crest, iliolumbar ligament.	Lower border of twelfth rib, and transverse processes of upper four lumbar vertebrae.	Draws last rib toward pelvis; flexion of lumbar region of vertebral column.	T12, L1.
Quadratus plantae	255, 256	Anterior aspect of calcaneus.	Tendons of Flexor digitorum longus.	Flexion of four lesser toes.	Lateral plantar.
Quadriceps femoris	242, 248	See under the following: Rectus femoris, Vastus intermedius, Vastus lateralis, Vastus medialis			
Rectus abdominis	217, 278	Crest of pubis and ligaments of symphysis pubis.	Cartilage of fifth through seventh ribs; xiphoid process.	Flexion of vertebral column; compression of abdominal viscera.	Brs. of 7–12th intercostals.
Rectus capitis anterior	205, 206	Lateral mass of the atlas.	Basilar portion of the occipital bone.	Flexion of the head.	C1 and 2.
Rectus capitis lateralis	205, 206	Transverse process of atlas.	Jugular process of the occipital bone.	Lateral flexion of the head.	C1 and 2.
Rectus capitis posterior major	207	Spinous process of the axis.	Inferior nuchal line of the occipital bone.	Extension and rotation of head.	Dorsal ramus–C1.

19

TABLE 11–1. MUSCLES OF THE HUMAN BODY. THE MUSCLES ARE ARRANGED ALPHABETICALLY FOR QUICK REFERENCE. (Continued)

Muscle	Page	Origin	Insertion	Action	Innervation
Rectus capitis posterior minor	207	Posterior arch of the atlas.	Inferior nuchal line of the occipital bone.	Extension of the head.	Dorsal ramus—C1.
Rectus femoris	244, 248	(1) Anterior head: anterior inferior iliac spine. (2) Posterior head: superior margin of acetabulum.	Base of the patella.	Extension of leg and flexion of thigh.	Femoral.
Rhomboideus major	222, 224	Spinous processes of first four or five thoracic vertebrae, supraspinous lig.	Lower one-third of medial border of the scapula.	Adduction of scapula and slight rotation, depresses shoulder.	Dorsal scapular.
Rhomboideus minor	222, 224	Lower part of ligamentum nuchae and spinous process of seventh cervical and first thoracic vertebrae.	Mid-portion of medial border of scapula.	Adduction of scapula and slight rotation. Depresses shoulder.	Dorsal scapular.
Rotatores	209, 212	Transverse processes of all vertebrae from sacrum to axis.	Spinous processes of vertebrae above origin—iong, two above; breves, one above.	Extension and rotation of vertebral column.	Dorsal rami.
Sacrospinalis	209, 211	See under the following: Iliocostalis dorsi Longissimus dorsi Spinalis dorsi			
Salpingopharyngeus	201, 202	Inferior portion of auditory tube.	Blends with the posterior fibers of the Pharyngopalatinus muscle.	Elevation of nasopharynx.	Pharyngeal plexus
Sartorius	244, 248	Anterior, superior iliac spine	Proximal, medial portion of shaft of tibia.	Flexion and lateral rotation of thigh; flexion and medial rotation of leg.	Femoral.
Scalenus anterior	205, 206	Transverse processes of third to sixth cervical vertebrae.	Inner border of first rib.	Elevation of first rib; flexion and slight rotation of neck, inspiration.	C5–8.
Scalenus medius	205, 206	Transverse processes of lower six cervical vertebrae.	Upper surface of first rib.	Elevation of first rib; flexion and slight rotation of neck, inspiration.	C5–8.
Scalenus posterior	205, 206	Transverse processes of fifth to seventh cervical vertebrae.	Outer surface of second rib.	Elevation of second rib; flexion and slight rotation of neck, inspiration	C6–8

Muscle	Pages	Origin	Insertion	Action	Nerve
Semimembranosus	246, 249	Tuberosity of ischium, lateral.	Medial condyle of tibia.	Flexion and medial rotation of leg, extends thigh.	Tibial.
Semispinalis capitis	206, 212	Transverse processes of upper six or seven thoracic and seventh cervical vertebrae.	Between superior and inferior nuchal lines of the occipital bone.	Extension and rotation of head.	Brs. of dorsal rami of spinal nn.
Semispinalis cervicis	206, 212	Transverse processes of upper five or six thoracic vertebrae.	Spinous processes of cervical vertebrae from axis to fifth cervical.	Extension and rotation of vertebral column.	Same as above.
Semispinalis thoracis (dorsi)	206, 212	Transverse processes of sixth to tenth thoracic vertebrae.	Spinous processes of upper four thoracic and lower two cervical vertebrae.	Extension and rotation of vertebral column.	Same as above.
Semitendinosus	246, 249	Tuberosity of ischium, medial side.	Proximal portion of medial aspect of shaft of tibia.	Flexion and medial rotation of leg, extends thigh.	Tibial.
Serratus anterior	223, 224	Muscular digitations from anterior and superior aspect of first eight or nine ribs.	Medial border of scapula.	Abduction of vertebral border of scapula, rotation of scapula.	Long thoracic.
Serratus posterior inferior	213	Spinous processes of last two thoracic and first three lumbar vertebrae; supraspinal lig.	Inferior borders of lower four ribs, lateral to angle.	Draws ribs of attachment, out and down.	Brs. of 9–12th thoracic nn.
Serratus posterior superior	213	Ligamentum nuchae, supraspinal lig., spinous processes of seventh cervical and upper 3 thoracic vertebrae.	Upper borders of ribs, 2–5, lateral to angles.	Raises ribs at attachment increasing thoracic cavity.	Brs. of 1st four thoracic nn.
Soleus	251, 253	Head of fibula and inner border of tibia.	Calcaneus bone (via tendo-calcaneal ligament).	Plantar flexion of foot.	Tibial.
Sphincter ani externus 220 Superficial Deep		Anococcygeal raphe. True sphincter encircling anus. Fibers of two sides decussate, ventral and dorsal to anus.	Central tendon of perineum.	Closes anal orifice. Helps fix central point of perineum.	S4 br. Inferior rectal br of pudendal n.
Sphincter urethrae membranaceae		Inferior ramus of the pubis and ischium.	Median raphe.	Compression of urethra.	Perineal brs. of pudendal n.
Spinalis capitis	209	Associated with that of Semispinalis capitis.	Associated with that of Semispinalis capitis.	Extension of vertebral column.	Brs. of dorsal rami of spinal nn.

Muscle	Page	Origin	Insertion	Action	Innervation
Spinalis cervicis	209, 211	Lower portion of ligamentum nuchae, spinous process of seventh cervical (T1 and 2).	Spinous process of axis.	Extension of vertebral column.	Brs. of dorsal rami of spinal nn.
Spinalis thoracis	209, 211	Spinous processes of first two lumbar and last two thoracic vertebrae.	Spinous processes of upper four to eight thoracic vertebrae.	Extension of vertebral column.	Brs. of dorsal rami of spinal nn.
Splenius capitis	206, 207	Ligamentum nuchae, spinous processes of seventh cervical and upper three thoracic vertebrae.	Below superior nuchal line and mastoid process of temporal bone.	Extension and rotation of neck; flexion of neck to side.	Brs. of dorsal rami of middle and lower cervical nn.
Splenius cervicis	206, 207	Spinous processes of third to sixth thoracic vertebrae.	Transverse processes of upper two or three cervical vertebrae.	Extension and rotation of neck; flexion of neck to side.	Same as above.
Sternocleido-mastoideus	203, 204	(1) Sternal head: Manubrium sterni. (2) Clavicular head: superior medial portion of the clavicle.	Mastoid process of the temporal bone.	Singly, draws head to side; together, flexion of vertebral column, flexion of head on chest, elevation of chin. Rotates head.	Accessory C2 and C3.
Sternohyoideus	205, 206	Medial end of the clavicle and manubrium sterni.	Body of hyoid bone.	Depression of hyoid bone.	C1–3 thru ansa hypoglossi.
Sternothyroideus	205, 206	Manubrium sterni.	Lamina of thyroid cartilage.	Depression of thyroid cartilage.	Descendens cervicis and hypoglossi.
Styloglossus	200, 201	Anterior and lateral aspect of styloid process. Stylomandibular ligament.	Lateral border of tongue, dorsally.	Raises and retracts tongue.	Hypoglossal.
Stylohyoideus	205, 206	Styloid process of temporal bone.	Body of hyoid bone.	Elevation and posterior movement of hyoid bone.	Facial.
Stylopharyngeus	200, 202	Medial side of base of styloid process.	Sides of pharynx; thyroid cartilage.	Elevates and dilates pharynx.	Glossopharyngeal.
Subclavius	223, 224	Medial portion of first rib and costal cartilage.	Subclavian groove of clavicle.	Draws shoulder downward.	Brachiel plexus, lateral trunk (C5–6).
Subcostals	214	Inner surface of lower rib, near angle.	Inner surface of second or third rib below.	Draws ribs together; aids in respiration.	Intercostals.

Muscle	Page	Origin	Insertion	Action	Nerve
Subscapularis	225, 227	Subscapular fossa, medial part.	Lesser tubercle of humerus.	Rotation of head of humerus medially, aids adduction, abduction, flexion, extension.	Subscapular.
Supinator	230, 235	Lateral epicondyle of humerus, ulna below radial notch.	Shaft of radius.	Supination of hand.	Deep radial.
Supraspinatus	226, 227	Supraspinous fossa.	Greater tubercle of humerus.	Abduction of humerus.	Suprascapular.
Temporalis	198	Temporal fossa.	Coronoid process and ramus of mandible.	Closes jaws.	Trigeminal.
Tensor fasciae latae	244, 247	Anterior superior iliac spine and part of notch below it.	Iliotibial band of fascia lata.	Flexion of thigh, medial rotation, abduction.	Superior gluteal.
Tensor veli palatini	200, 202	Scaphoid fossa and spina angularis of sphenoid, lateral auditory tube.	Palatine aponeurosis and palatine bone.	Tension of soft palate.	Trigeminal.
Teres major	225, 228	Dorsal surface of inferior angle of scapula.	Crest of lesser tubercle of humerus.	Adduction, extension, and rotation of arm medially.	Lower subscapular.
Teres minor	222, 227	Axillary border of scapula.	Greater tubercle of humerus.	Rotation of humerus laterally, adduction.	Axillary.
Thyrohyoideus	205, 206	Lamina of the thyroid cartilage.	Greater cornu of the hyoid bone.	Elevation of thyroid cartilage, depression of hyoid.	Descending hypoglossi.
Tibialis anterior	250, 256	Lateral condyle and upper two-thirds of shaft of tibia, interosseous membrane.	First cuneiform and base of first metatarsal bone.	Dorsiflexion and supination of foot.	Deep peroneal.
Tibialis posterior	252, 253	Shaft of tibia and fibula and interosseous membrane, posterior surfaces.	Tuberosity of the navicular bone, the calcaneus, the three cuneiforms, the cuboid bone, and bases of the second to fourth metatarsals.	Plantar flexion and supination of foot.	Tibial.
Transverse thoracis	214	Lower one-third of inner surface of sternum, sternal ends of costal cartilages 4–6.	Inner surface of costal cartilages of second to sixth ribs.	Draws ribs down.	Brs. of intercostals.
Transversus abdominis	217, 218	Inguinal ligament, iliac crest, thoracolumbar fascia, and the cartilages of the lower six ribs.	Xiphoid process, linea alba, inguinal ligament, and the pubis.	Constriction of abdomen.	Brs. 7–12th inter-costals; iliohypogastric and ilioinguinal.
Transversus perinaei profundus	220	Ramus of ischium.	Joins muscle of opposite side.	Support of perineum, compresses memb. urethra.	Perineal branch of pudendal n.

TABLE 11–1. MUSCLES OF THE HUMAN BODY. THE MUSCLES ARE ARRANGED
ALPHABETICALLY FOR QUICK REFERENCE. *(Continued)*

Muscle	Page	Origin	Insertion	Action	Innervation
Transversus perinaei superficialis	220	Tuberosity of ischium, inner part.	Central point of perineum.	Fixation of central tendinous point of perineum.	Perineal branch of pudendal n.
Trapezius	221, 222	Superior nuchal line, occipital protuberance, ligamentum nuchae, seventh cervical spine, spinous processes of all thoracic vertebrae.	Clavicle, spine of scapula, and acromion.	Draws head back, rotates scapula, draws head to side. braces shoulder, adducts scapula.	Spinal accessory, C3 and 4
Triceps brachii	230	(1) Long head: infraglenoid tuberosity of scapula. (2) Lateral head: shaft of humerus. (3) Medial head: posterior shaft of humerus.	Olecranon process of ulna. Fascia of forearm.	Extension of arm and forearm. Long head also adducts arm.	Radial.
Vastus intermedius	245, 248	Front and lateral surfaces of shaft of femur.	Patella.	Extension of leg.	Femoral.
Vastus lateralis	245, 248	Greater trochanter and upper one-half of the lateral margin of the linea aspera.	Lateral border of the patella.	Extension of leg.	Femoral.
Vastus medialis	245, 248	Upper portion of medial aspect of shaft of femur.	Medial border of the patella.	Extension of leg.	Femoral.

"All the vital mechanisms, however varied they may be, have only one object, that of preserving constant the conditions of life in the internal environment."
—Claude Bernard

Chapter 12

The Circulatory System

INTRODUCTION

Cell Environment. The one-celled organism lives in an aqueous environment. In this environment are found the oxygen and food necessary to maintain the life of the cell and into it the cell excretes the waste products of its metabolism. Remove the cell from this fluid environment and it dies.

The cells of multicellular animals including man are no less dependent upon a fluid environment. Their environment is the **tissue fluid,** an intercellular material which provides for their needs. Without the tissue fluid the cells soon die.

The tissue fluid is a part of the **internal environment** of the body. Yet it is far removed from the external environment from which it must receive the materials to sustain the life of the cell. How is this gap bridged?

The **circulatory system** bridges the gap. It is an intricate system of **transportation.** It sends its emissary, the **blood,** to all of those body systems that maintain connections with the external environment and receives from them the materials which they alone in the human body can provide, delivering these in turn to the tissue fluid. The blood delivers to other systems the materials that are detrimental to the life of the cell and they take them for disposal. The digestive system receives and prepares the food for absorption into the blood and lymph through its expansive membranes. The respiratory system provides oxygen and relieves the blood of carbon dioxide. The urinary system, with millions of kidney tubules, effectively removes a variety of waste products and makes other adjustments to maintain the internal environment in a condition favorable to the living cells—and to the total organism.

The effectiveness of a transportation system depends upon its reaching all points needing service. That even the most insignificant cut on the body surface will cause bleeding suggests the efficiency of blood distribution in the body. But it is of equal importance that the circulatory system be so arranged and controlled that it responds to the varying needs of the different tissues or organs of the body. Those with fewer cells and more intercellular material require less blood than those with many more cells such as skeletal muscle or brain. Also, activity varies from one part of the body to another, requiring a differential flow of blood which the circulatory system must satisfy.

Organs of Circulation (Fig. 12–1). The circulatory system is divided into two main parts: (1) the **cardiovascular system** and (2) the **lymphatic system.**

The **cardiovascular system** consists of the **heart,** a pump to maintain the flow of blood; the **arteries,** which carry blood away from the heart; **arterioles,** tiny vessels that lead into the even smaller **capillaries; venules,** tiny vessels into which the capillaries drain; and **veins,** which carry the blood back to the heart. These structures form a complete circuit to distribute the **blood** throughout the body. The arteries and arterioles constitute the high-pressure part of the system; the capillaries, the area of exchange between blood and tissue fluid; the venules and veins, the low-pressure collecting part of the system.

The **lymphatic system** is often treated as a separate system, but is here included as a subsidiary of the circulatory system. It consists of **lymph capillaries,** which begin blindly in the tissues where they collect tissue fluid. They lead into **lymphatic vessels.** The lymphatic vessels empty their **lymph** into large veins above the heart. Along the pathways of lymphatic vessels are the **lymph nodes.** The **spleen, thymus, tonsils** and **solitary** and **aggregated lymph nodules** are also included in the lymphatic system.

Functions. As suggested above, the circulatory system maintains the constancy of the internal environment. In doing so, its primary function is one of **transportation;** of oxygen, nutritive materials, and water to the cells; of carbon dioxide and other waste products to the proper organs of excretion. But also as a part of its transportation function, it plays an important role in the **regulation** and **coordination** of the body because it distributes the **hormones** of the endocrine glands to the cells which they influence. It participates in the **regulation of water** and other substances in the body. It aids in the **regulation of body temperature.** It plays a **protective role** through the phagocytic action of its white blood cells and through the production and transportation of antibodies.

The lymphatic system, while sharing to some degree in the above functions, makes its own contributions. It participates in the **collecting** of material from the **tissue** fluid and returns it to the blood. Its lymph nodes are important "**filters**" which help **prevent** the **spread** of **infection**. They also **produce lymphocytes**. Finally, the lymph capillaries of the intestinal villi **absorb** the **fats** from the digestive system.

The Circulatory Fluids and Tissues. As indicated above, the blood, tissue fluid, and lymph serve as the carriers in the cir-

culatory system. The blood and lymph have been described in Chapter 5, pages 63–66 and should be reviewed at this time. For the descriptive anatomy which follows, the **blood** is defined as that fluid and cells that are confined to the organs of the cardiovascular system; **tissue fluid** is free among the cells and intercellular materials; **lymph** is that fluid tissue confined to the vessels and nodes of the lymphatic system.

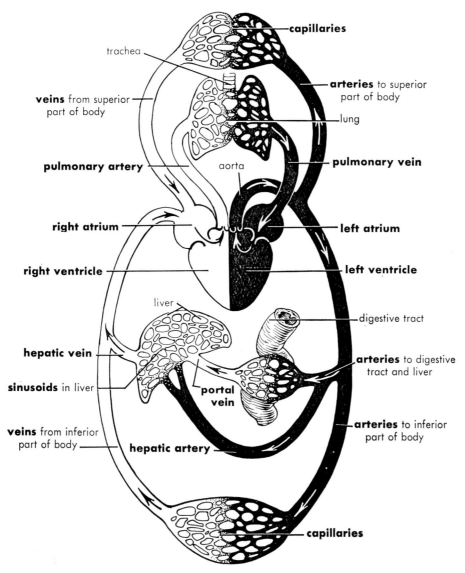

capillaries

trachea

veins from superior part of body

arteries to superior part of body

lung

pulmonary artery

aorta

pulmonary vein

right atrium

left atrium

right ventricle

left ventricle

liver

digestive tract

hepatic vein

arteries to digestive tract and liver

sinusoids in liver

portal vein

veins from inferior part of body

arteries to inferior part of body

hepatic artery

capillaries

Fig. 12–1. Schematic representation of the double system of circulation. Oxygenated blood is shown in black; the nonoxygenated blood in white. The arrows indicate the direction of flow.

GENERAL PLAN OF THE CARDIOVASCULAR SYSTEM
(Fig. 12-1)

Closed System. The cardiovascular system is a **closed system** of circulation. The blood is confined to the heart and associated vessels, never leaving them except when vessels are damaged. This kind of system is more specific in its targets and more critically regulated than the system of arthropods, such as the crabs. Their system is open, and the blood going out into vessels from the heart flows freely into spaces among the tissues.

Double System. The cardiovascular system is also a **double system** of circulation. The heart is a double organ. It serves a pulmonary circuit to the lungs and a systemic circuit to the body. In the lungs the blood receives oxygen and gives off carbon dioxide. The blood then returns to the left side of the heart to be pumped to the body as a whole through a **systemic circuit**. Returning to the heart, the blood—now poor in oxygen and laden with carbon dioxide—enters the right side of the heart to again be sent out through the **pulmonary circuit**. These circuits, of course, operate simultaneously.

Comparative Anatomy. In fish, the cardiovascular system is single rather than double. The blood is pumped to the gills for the exchange of oxygen and carbon dioxide and, before returning to the heart, it circulates through the rest of the body. The heart has only one atrium and one ventricle. It is a relatively low-pressure system, but adequate for the needs of these cold-blooded creatures. In some fish and in the amphibians and reptiles, there are cardiovascular systems which are intermediate between the single and double types. All birds and mammals have double systems.

In the embryological development of the double system, there is a recapitulation of the single and intermediate types, and it is only at birth that the pulmonary and systemic circuits become completely separated. Failure to complete the final stages of separation in man may result in a "blue baby," a condition that will be discussed later (pp. 286, 333). Facts of comparative anatomy and embryology, as we shall discover, will help us understand the normal and abnormal circulatory systems of man.

HEART

The **heart** is a double pump which is of primary importance in the maintenance of the flow and pressure of the blood in the closed and double cardiovascular system. Its contractions start in the early days of embryonic development and must continue for the duration of the individual's life. For the contractions to cease, even for a few minutes, would cause serious damage, or would be fatal. Contracting at an average rate of 72 times per minute as the heart does, with only a fraction of a second pause between contractions, adds up to about 100,000 times a day or about 2,600,000,000 beats in a lifetime of seventy years. In the same time it would pump about 155,000,000 liters or 40,951,000 gallons of blood equaling a weight of about 150,000 tons. If you clench and open your fist alternately at the rate of 72 times per minute, the muscles involved will "feel" tired after about two minutes. This gives some clue to the efficiency level required by the heart to maintain its continuous lifetime action. Indeed, it has been shown that the heart uses about 50 per cent of the available energy for work. An automobile motor, by comparison, has an efficiency of only 25 per cent. The heart, too, is more efficient in its use of oxygen than any other organ of the body. This remarkable organ, and its specialized cardiac muscle tissue (see p. 176) command respect and further attention.

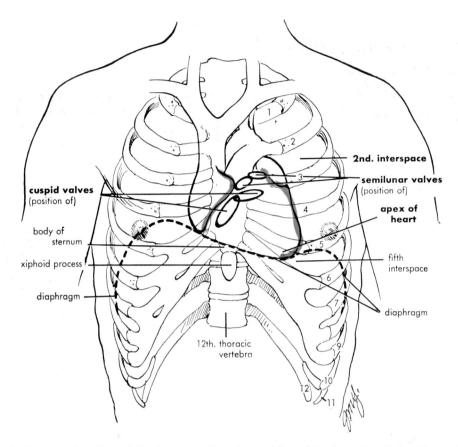

FIG. 12–2.   Anterior view of the thorax to show the position of the heart in relationship to the ribs, sternum, and diaphragm and the position of the heart valves.

Size, Position, and Relationships (Figs. 12–2, 3). The size of the human adult's heart is about that of one's clenched fist; more specifically, 12 cm. long, 9 cm. wide at the broadest point, and 6 cm. thick.

The heart lies obliquely in the middle mediastinum, in the anterior and inferior part of the thorax (Figs. 14–10, 11). It is in front of thoracic vertebrae 5–8. About two-thirds of its substance lies to the left side of the midsternal line, one-third to the right. Its narrow apex is in the fifth intercostal space, 7 to 9 cm. to the left of the midsternal line, its broad base at the level of the third sternochondral attachments. Its sack-like parietal pericardium adheres to the inside of the sternal wall and to the central part of the diaphragm.

The lungs and pleural cavities lie mostly lateral to the heart; the esophagus and descending aorta are posterior.

General Structure and Function. The human heart has four chambers. Two of these, the **atria**, are at the superior side of the heart; the other two, the **ventricles**, are inferior. The boundaries of these chambers are indicated on the external chambers of the heart by the **coronary** (*atrioventricular*) **sulcus**, separating atria from ventricles, and the **anterior** and **posterior longitudinal sulci**, separating right and left ventricles. In these sulci are found the large vessels of the heart, namely, the coronary sinus, veins, and arteries (Figs. 12–3, 8). Internally, the atria are separated by an **interatrial** and the ventricles by an **interventricular**

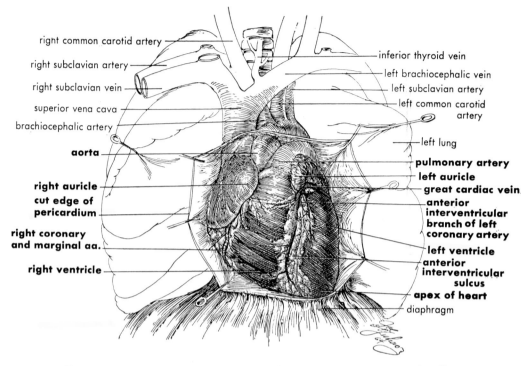

right common carotid artery
right subclavian artery
right subclavian vein
superior vena cava
brachiocephalic artery
aorta
right auricle
cut edge of pericardium
right coronary and marginal aa.
right ventricle

inferior thyroid vein
left brachiocephalic vein
left subclavian artery
left common carotid artery
left lung
pulmonary artery
left auricle
great cardiac vein
anterior interventricular branch of left coronary artery
left ventricle
anterior interventricular sulcus
apex of heart
diaphragm

FIG. 12–3.   Anterior view of heart and main vessels.   The parietal pericardium has been reflected.

septum (Figs. 12–4, 6).   The atria were at one time called **auricles,** a term that is now applied only to the "ear-like" flaps or folds of the atrial walls which are illustrated in Figure 12–3.

The **atria** are the receiving chambers of the heart.   The veins from the systemic circuit, the **venae cavae** and **coronary sinus,** carrying blood deficient in oxygen and laden with carbon dioxide, empty into the **right atrium.**   The **left atrium** receives blood rich in oxygen and poor in carbon dioxide from the pulmonary circuit by way of the **pulmonary veins.**

The **ventricles** are the pumping chambers of the heart.   They take the blood which they receive from the atria and send it out into the systemic and pulmonary circuits, the right into the **pulmonary trunk,** the left into the **aorta.** It is apparent then that the right side of the heart handles non-oxygenated blood, the left heart only blood rich in oxygen.

But this double pump, the heart, would be unable to give direction to the flow of blood indicated above without appropriately placed valves.   There are two pairs of important valves in the heart, the **atrioventricular** (*cuspid*) and the **semilunar valves.**   The cuspid valves lie between atria and ventricles, the semilunar valves between the ventricles and their dispensing vessels, the **aorta** and **pulmonary trunk** (Figs. 12–4, 6).   The details of structure and function of these valves will be discussed along with the more critical descriptions of the heart chambers.

Heart Wall.   The **heart wall** consists of three layers, the inner **endocardium,** the middle **myocardium,** and outer **epicardium** or **visceral pericardium.**

The **endocardium** has a free surface of **endothelium** which lines all of the heart chambers, covers the valve surfaces and continues into the blood vessels

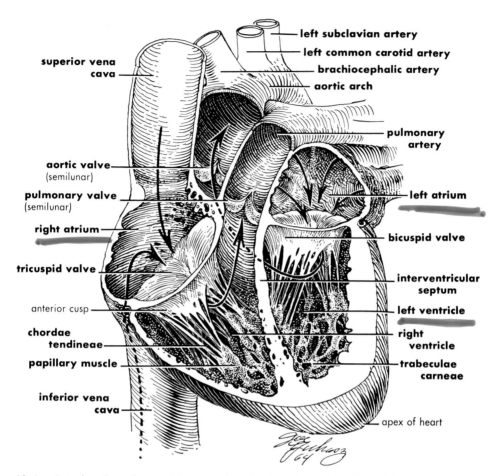

left subclavian artery
left common carotid artery
brachiocephalic artery
aortic arch

superior vena cava

pulmonary artery

aortic valve (semilunar)

pulmonary valve (semilunar)

left atrium

right atrium

bicuspid valve

tricuspid valve

interventricular septum

left ventricle

anterior cusp

chordae tendineae

right ventricle

papillary muscle

trabeculae carneae

inferior vena cava

apex of heart

FIG. 12–4.   Anterior view of opened heart to show its chambers and valves.   The arrows indicate the direction of flow of blood through the heart and to and from its major vessels.

entering and leaving the heart. The endothelium is supported underneath by **fibrous tissue** (*with some elastic and smooth muscle fibers*) which is thickened in the valve cusps. The endocardium supports blood vessels and special impulse-conducting tissue in its deeper layers.

The **myocardium** is the thickest coat in the heart wall and is made up of cardiac muscle (see p. 176). It is thinnest in the atrial walls where the work requirement is minimal. In the ventricles the myocardium is thicker, reaching about an inch in the wall of the left ventricle which must maintain flow in the extensive systemic circuit, a third of that thickness in the right ventricle which

maintains flow in the more limited pulmonary circuit. The ventricular myocardium is also more complicated in fiber arrangement. The muscle fibers are organized into patterns of whorls and spirals, there being no straight fibers running from base to apex of the ventricles. Rather the fibers have both their origins and insertions on the fibrous tissues at the bases of the ventricles. Their contractions therefore result in a wringing action which is very effective in emptying the ventricles.

The myocardium is very irregular on the internal surfaces of the ventricles, being thrown into ridges, folds and bridges called the **trabeculae carneae,** and into conical extensions called **papillary muscles** (Figs. 12–4, 6).

Loose connective tissue supporting lymphatics and blood vessels permeates the myocardium, and the capillary beds are so extensive that more oxygen is freed here by the circulating blood than in any other organ of the body.

Finally, some of the cardiac muscle fibers are modified to form an extensive conductive system, important in the coordination of the contractions of the heart chambers. This mechanism will be described later (pp. 290, 292).

The **epicardium,** often called the **visceral pericardium,** is the thin and transparent outer layer of the heart wall. It is composed of fibrous tissue with an outer covering of **mesothelium.** Blood vessels, lymphatics, and fat are found within its inner portions.

Parietal Pericardium (Fig. 12–5). The **parietal pericardium** consists of two layers, an inner one, the **serous coat,** and an outer **fibrous coat.** It completely encloses the heart and the bases of the vessels entering and leaving the heart.

The **serous coat** is continuous at the base of the heart and around the large vessels with the epicardium or visceral pericardium of the heart surface. These two smooth serous layers are separated

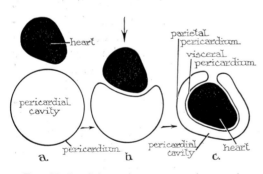

FIG. 12–5. Schematic representation to show how the heart develops its relationship to the pericardium and pericardial cavity. *A*, Heart and pericardium separated; *B*, heart encroaches upon pericardium; *C*, heart and visceral and parietal pericardia in normal relationship (pericardial cavity exaggerated in size).

only by enough watery fluid to lubricate their surfaces and to make for almost frictionless movements of the heart. This potential space between them is the **pericardial cavity.** Disease or injury of the heart may cause fluid to accumulate in this cavity.

The **fibrous coat** of the parietal pericardium reinforces the serous coat and at the base of the heart is attached to the large arteries and veins. As stated on page 281, it also attaches to the central tendon of the diaphragm and to the inside of the sternal wall of the thorax. Laterally, it adheres to the mediastinal parietal pleurae, but is not fused with them. It is a tough, protective membrane for the heart.

Right Atrium. The **right atrium** forms a part of the base, some of the anterior surface, and the upper part of the right margin of the heart. Its cavity is somewhat larger than that of the left atrium and its walls are thinner. Besides its main cavity there is a hollow appendage, the **auricle.** While the lining of its main cavity is mostly smooth, that of the auricle is thrown into parallel folds called the **musculi pectinati** (Fig. 12–6).

Three large openings are found in the wall of the right atrium. The **superior vena cava** enters it on its superior posterior side, its orifice directed downward toward the opening into the ventricle. It brings in blood from the head, thorax, and upper extremities. It has no valve. The **inferior vena cava** enters the right atrium posteriorly and inferiorly, draining the trunk and lower extremities. It is provided with a valve consisting of a crescentic fold of the lining of the atrium. This valve is more prominent in the fetus and is thought to help direct the blood from the inferior vena cava through the foramen ovale of the interatrial septum. In the adult it has no known functional importance.

The **right atrioventricular orifice** leading into the ventricle with its right

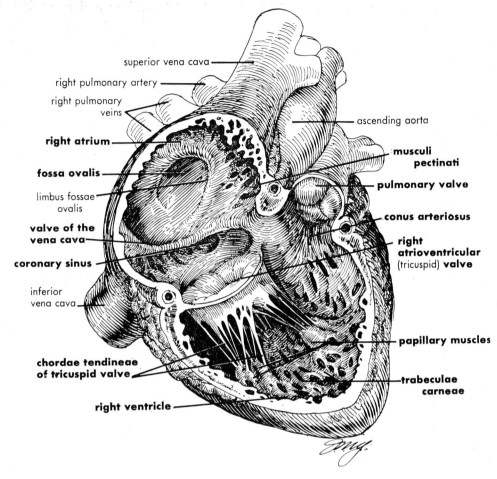

FIG. 12–6. Interior of the right atrium and ventricle.

atrioventricular (*tricuspid*) valve will be described with the right ventricle.

Among the smaller apertures into the right atrium is that of the **coronary sinus** through which the blood from the heart wall is emptied. It is located between the orifice of the inferior vena cava and the right atrioventricular orifice and is provided with a valve consisting of a semicircular fold of the atrial lining attached to the right and inferior lips of the orifice. This valve is capable of no more than partial closing during atrial contraction and is of doubtful functional significance. It is sometimes called the **Thebesian valve.**

A number of smaller openings of veins in the atrial wall, particularly in the septum, are of historical interest. They were the basis for the false conclusion of the ancients that the blood passed from one side of the heart to the other. Harvey's (1628) findings on the circulation of blood refuted such beliefs.

The interatrial septum forming the posterior wall of the right atrium has an oval depression, the **fossa ovalis,** which corresponds to the **foramen ovale** of the fetal heart. In the fetus, the foramen ovale allows the oxygenated and food-laden blood brought to the heart from the placenta by the umbilical veins and the inferior vena cava to be sent directly into the left side of the heart for distribution to the rest of the body. This is necessary since the lungs, prior to birth, are not functioning as respiratory organs. The foramen ovale normally closes at birth. When it does not, a "blue baby" results, a condition in which there is a mixing of oxygenated and non-oxygenated blood. This may cause the death of the infant if the opening is large, or if smaller, it may handicap him. Fortunately many "blue babies" can now be cured of their handicap through surgical procedures. Reference to page 333 and Figure 12–33 will lead to a better understanding of this condition.

Right Ventricle. The triangular-shaped right ventricle occupies the greater part of the anterior or sternocostal surface of the heart and superiorly narrows into a **conus arteriosus** which leads into the **pulmonary trunk.** It is delimited by the coronary sulcus above and the anterior interventricular sulcus to the left. It forms the lower part of the right margin of the heart and extends around to the inferior or diaphragmatic surface (Fig. 12–3).

The right ventricle has two large openings in its walls—one, the **right atrioventricular orifice** connecting it with the right atrium; two, the **orifice** of the **pulmonary trunk.** Each of these has valves (Figs. 12–4, 6).

The atrioventricular orifice is closed by the **right atrioventricular valve** (*tricuspid*) (Fig. 12–4). This valve consists of three triangular leaflets or cusps of unequal size with their bases attached to a fibrous ring around the orifice and their apices hanging down into the ventricle. They are composed of a strong fibrous tissue which is thicker in the center than at the free margins and are covered with endocardium. When the ventricle fills with blood and begins to contract, the valve cusps float into position in the atrioventricular opening and are forced shut by the increasing pressure. To prevent the valve from turning back

into the right atrium and regurgitating blood, fine tendinous cords, the **chordae tendineae,** are attached to the free margins and ventricular surfaces of the cusps. At their other ends, these cords attach to cone-shaped muscles, the **papillary muscles,** projecting from the ventricular wall. These muscles by their contractions maintain the integrity of the valve during ventricular contraction, or **systole,** as heart contraction is usually called.

Closely related to the papillary muscles, in the inner wall of the ventricle, are irregular muscular ridges and bridges, called **trabeculae carneae.** They are abundant, except in the smooth, narrowing, superior part of the ventricle, the **conus arteriosus** (Figs. 12–4, 6).

The second large opening in the wall of the right ventricle is the circular **orifice** of the **pulmonary trunk** near the **base** (*superior surface*) of the heart (Fig. 12–4). Its valve, the **pulmonary valve,** has three **semilunar** pocket-like cusps formed by folds of the endocardial lining of the heart reinforced by fibrous tissue (Figs. 12–4, 6, 9). When the ventricle contracts (*systole*) and places the contained blood under pressure the cusps of the pulmonary valve are pressed against the side walls of the pulmonary artery and blood passes freely. When the ventricle relaxes, called **diastole,** however, the blood from the pulmonary artery tends to return to the ventricle where the pressure is now low, and in doing so fills the pocket-like cusps of the valve, closing it tightly (Figs. 12–7B, 9).

Left Atrium (Figs. 12–3, 4, 7, 8). The **left atrium,** smaller and thicker-walled than the right, forms a large part of the surface of the base and upper posterior part of the heart wall. Its **auricle** curves around the base of the pulmonary trunk enough to be visible on the anterior or sternocostal surface of the heart.

The left atrium receives the four **pulmonary** veins, two from each lung. Their orifices have no valves. The re-

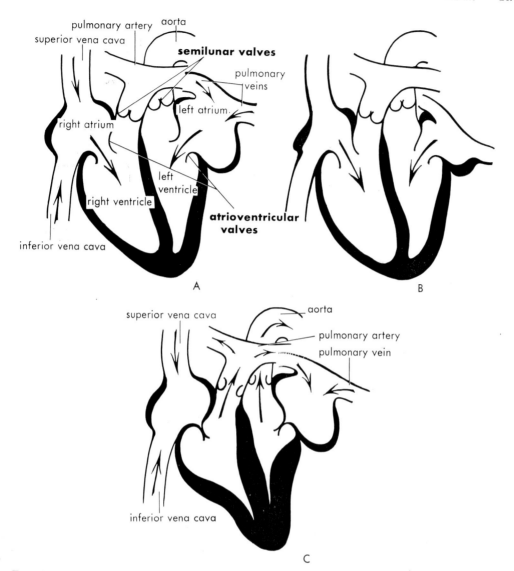

FIG. 12–7.   Stages in the cardiac cycle. *A*, Heart relaxed; semilunar valves are closed; atrioventricular valves are open.   Atria and ventricles are filling with blood.   *B*, Atria are contracting; ventricles are relaxed and filling.   Semilunar valves remain closed; atrioventricular valves are open.   *C*, The atria are relaxed as the ventricles contract.   The atrioventricular valves are closed; the semilunar valves are open and blood is pumped into the pulmonary artery and the aorta.

maining orifice is the **atrioventricular,** guarded by the **left atrioventricular** (*mitral* or *bicuspid*) **valve.**

The lining of the atrium is smooth, except in the auricular portion which has ridges, the **musculi pectinati.** The interatrial septum has a depression bounded below by a crescentic ridge, the **valve** of

the **foramen ovale,** a remnant of the septum which closed the foramen ovale at birth.

Left Ventricle (Figs. 12–3, 4, 7, 8). The wall of the **left ventricle** forms a small part of the sternocostal and about half of the diaphragmatic surfaces of the heart.   It has about the same capacity as

20

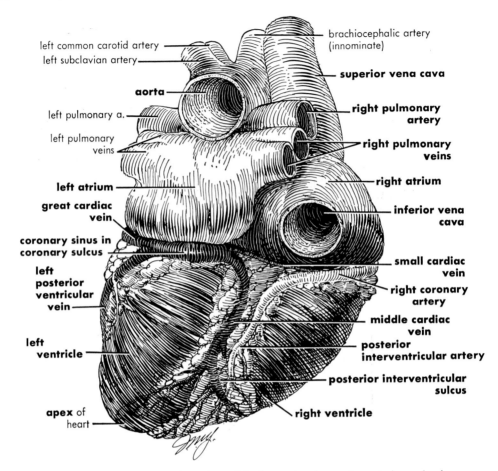

left common carotid artery

left subclavian artery

aorta

left pulmonary a.

left pulmonary veins

left atrium

great cardiac vein

coronary sinus in coronary sulcus

left posterior ventricular vein

left ventricle

apex of heart

brachiocephalic artery (innominate)

superior vena cava

right pulmonary artery

right pulmonary veins

right atrium

inferior vena cava

small cardiac vein

right coronary artery

middle cardiac vein

posterior interventricular artery

posterior interventricular sulcus

right ventricle

FIG. 12–8. Diaphragmatic surface of the heart showing major arteries and veins.

the right ventricle, but its walls are three times as thick. In cross-section its cavity is oval or circular in contrast to the crescent-shaped cross-section of the right ventricle.

There are two openings in the left ventricle, the **left atrioventricular orifice** leading from the left atrium and closed by the **left atrioventricular valve** (*bicuspid-mitral*), and the **aortic** leading into the aorta and closed by the **aortic valve** (*semilunar*). The left atrioventricular valve is constructed and functions essentially like that of the right ventricle, except that it has only two cusps, hence the name **bicuspid**. It is also commonly called **mitral valve** because of the fancied resemblance of its two cusps to a bishop's

mitre. The **chordae tendineae** are thicker, stronger, and less numerous than those in the right ventricle, and the **trabeculae carneae** are more numerous and more closely packed. There are only two **papillary muscles,** but they are of large size.

The **aortic orifice** is circular and closed by the **aortic** (*semilunar*) **valve.** This valve is similar to the pulmonary valve, except that its semilunar cusps are larger, thicker, and stronger. The dilated pockets between the cusps and the aortic wall are called **aortic sinuses** and from two of them the coronary arteries have their origin. Since the pulmonary trunk and the aorta were formed embryologically by an aortic septum dividing **a**

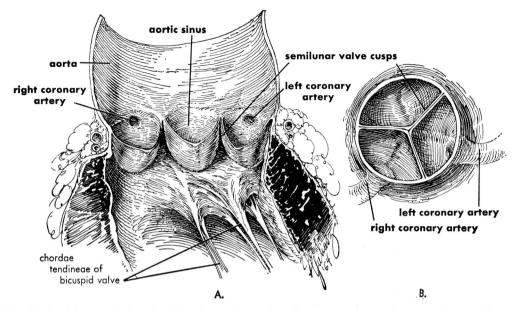

FIG. 12–9.   The aortic (semilunar) valve.   *A*, Aortic orifice cut and spread out to show semilunar and bicuspid (mitral) valves.   *B*, Aortic valve in closed condition as seen from above.

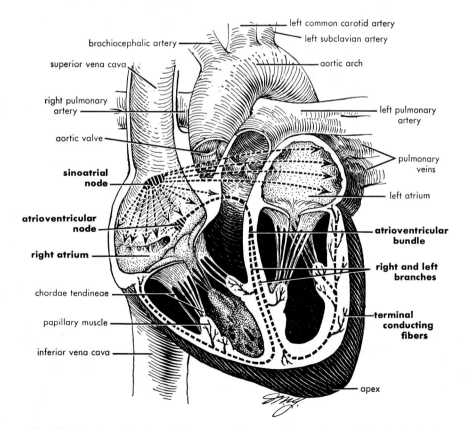

FIG. 12–10.   Anterior view of the opened heart to show schematically the intrinsic conduction mechanism.   Arrows indicate passage of impulses in walls of atria.

common vessel, the **truncus arteriosus,** it is not surprising that the aortic and pulmonary valves lie side by side in the adult heart.

In Summary. The diagram on page 291 may help to summarize and clarify the direction of blood flow through the heart and the relationship of the heart chambers, valves, and vessels (Fig. 12–7).

Interventricular Septum (Fig. 12–4). This septum was mentioned earlier as the structure separating right and left ventricles. It consists primarily of a thick, muscular portion, the **muscular interventricular septum,** and an upper thinner and fibrous portion adjoining the atrial septum, the **membranous interventricular septum.** Embryologically and phylogenetically, the latter is the last part of the septum to close. In some individuals it does not close and the condition is known as **patent interventricular septum,** another common cause of "blue baby."

Skeleton of the Heart (Figs. 12–4, 7). The membranous interventricular septum merges with the fibrous atrioventricular and arterial rings, which support the heart valves, to form a septum separating the muscular walls of the atria and ventricles. In larger animals, such as the ox, cartilage and bone may develop in these structures. In any case, they constitute the skeleton of the heart and as such serve for the attachment of the origins and insertions of the complicated bundles of muscle fibers of both the atria and ventricles. They also serve to separate atrial and ventricular muscle, structurally and functionally, except for some specialized conductive muscle tissue, constituting the **atrioventricular bundle** (*of His*).

Conduction Mechanism of the Heart (Fig. 12–10). Vertebrate cardiac muscle has an "automatic" rhythmic beat which is independent of nervous stimulation. The heart of a vertebrate animal can be severed from its nervous connections or even completely removed from the body and it will continue to contract if cared for properly. The hearts of cold-blooded animals, such as the turtle or frog, are particularly well suited for studies on the origin of this independent rhythmic beat, for they will continue to contract outside the body for long periods of time, even though given relatively little care. For instance, in the turtle heart, which consists of a receiving chamber, the sinus venosus, two atria, and a single ventricle, the heartbeat is normally initiated in the sinus venosus, often called the "pacemaker." It then passes to atria and ventricle, thus giving a coordinated contraction to move the blood through the heart. If, however, the sinus is severed from the atria and the atria from the ventricle, each part of the heart will contract at its own particular rate—the sinus has the fastest rate, the atria next, and the ventricle is slowest.

These facts concerning the behavior of a turtle's heart help us to understand the action of a mammalian heart, such as that of man. Embryological studies show us that the sinus venosus of the mammalian heart is present only during development and is ultimately represented only by a small collection of specialized myocardium in the right atrial wall near the opening of the superior vena cava. We call it the **sinoatrial (S-A) node** because of this relationship. It serves as the point of origin of the heartbeat by initiating an electrical impulse. It is the **"pacemaker."** From the sinoatrial node, this electrical impulse spreads through the myocardium of the atrial walls causing them to contract simultaneously and, of course, to empty their contained blood into the right and left ventricles. The impulse from the atrial myocardium cannot apparently cross the atrioventricular septum. There is, however, in the septal wall of the right atrium near the orifice of the coronary sinus another mass of **nodal tissue,** the

From **systemic** circuit:

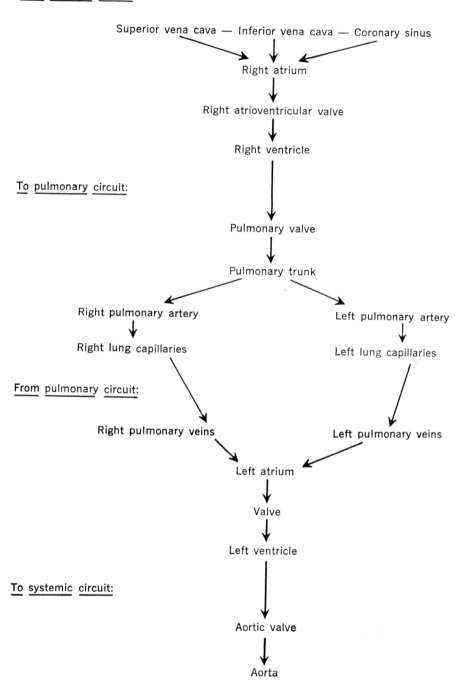

Superior vena cava — Inferior vena cava — Coronary sinus

Right atrium

Right atrioventricular valve

Right ventricle

To pulmonary circuit:

Pulmonary valve

Pulmonary trunk

Right pulmonary artery

Right lung capillaries

Left pulmonary artery

Left lung capillaries

From pulmonary circuit:

Right pulmonary veins

Left pulmonary veins

Left atrium

Valve

Left ventricle

To systemic circuit:

Aortic valve

Aorta

atrioventricular (A-V) node which receives the impulse from the atrial myocardium and after a pause conducts it on into the atrioventricular bundle (*bundle of His*) with which it is continuous. The atrioventricular bundle passes through the atrioventricular septum into the membranous interventricular septum and finally divides to send one branch down each side of the muscular interventricular septum. These branches pass under the endocardium of the right and left ventricles respectively, and each divides into a number of branches, the terminal conducting fibers, or Purkinje fibers, which are continuous with the cardiac muscle fibers at their ends. The electrical impulse is thus conducted from the atria into the ventricles, where it stimulates the ventricular contractions to begin near the apex and move toward the base where the outgoing vessels are located.

The nodal structures, the atrioventricular bundles, and the Purkinje fibers are all composed, not of nervous tissue, but of a modified cardiac muscle with special conducting properties. They bring to the heart a coordinated meaningful contraction which moves the blood efficiently through the heart. If, however, there develops an interruption in this mechanism, a condition called heart block, the atria and ventricles will develop their own rates of contraction, the atria faster than the ventricles, a condition reminding us of the different rates of contraction in the turtle's heart when it was cut apart. The efficiency of the heart is thereby greatly reduced and in severe cases may be fatal.

Extrinsic Nerves of the Heart (Figs. 18–45, 48). Although the nerves of the heart will be considered later, it should be mentioned here that even though the heart possesses the inherent capacities for action just described, to be effective as a body organ its actions must be coordinated with those of the body as a whole. This coordination is achieved by many means—some chemical, some physical, including nerve impulses reaching the heart from branches of the vagus nerve and from the sympathetic system. The vagus nerves are inhibitory to the heart; the sympathetics acceleratory.

Heart Sounds. The heart in its beating produces four audible sounds. The first two sounds, mentioned in almost every textbook of biology, anatomy, and physiology can be heard by placing the ear on the chest wall, or better, by use of a special instrument, the stethoscope. These two sounds are usually described by use of the syllables lubb-dupp. The first sound, lubb, is loud, long and low in pitch and is caused by ventricular systole and the closure of the atrioventricular valves. The second sound, dupp, is shorter and sharper and is caused by the closure of the aortic and pulmonary valves as the blood tends to move back into the ventricles during diastole. The third heart sound is heard in the majority of children and in about 50 per cent of adolescents. This is a soft, low-pitched sound which is best heard at the apex of the heart and becomes louder during inspiration and is loudest after exercise. It is believed to be caused by rapid filling of the ventricles after the opening of the atrioventricular valves. The fourth or "atrial" sound can be heard under certain circumstances at any age. While it is related to atrial contraction, its exact mechanism is not understood.

Practical Considerations. The heart sounds reflect to some degree the condition of the heart and are therefore of clinical importance. It is a misconception, however, that "weak" or "faint" heart sounds, particularly the first sound, indicate weak heart action. It has been shown that the intensity of the first sound depends almost entirely upon the position of the cusps of the atrioventricular valves at the instant of ventricular systole. If the cusps are farther apart, they produce a louder sound than when they are close.

Abnormal sounds, often called "murmurs," may indicate imperfect closure of valves which allows a leaking back or regurgitation of blood, or they may be due to the narrowing (*stenosis*) of the valve ring which interferes with the free movement of blood. It is equally important to know that many "murmurs" are "physiologic" or "innocent"—especially systolic murmurs—and indicate no organic heart disease. This is particularly true in children and adolescents and in conditions of increased cardiac output. When murmurs are detected, it is well

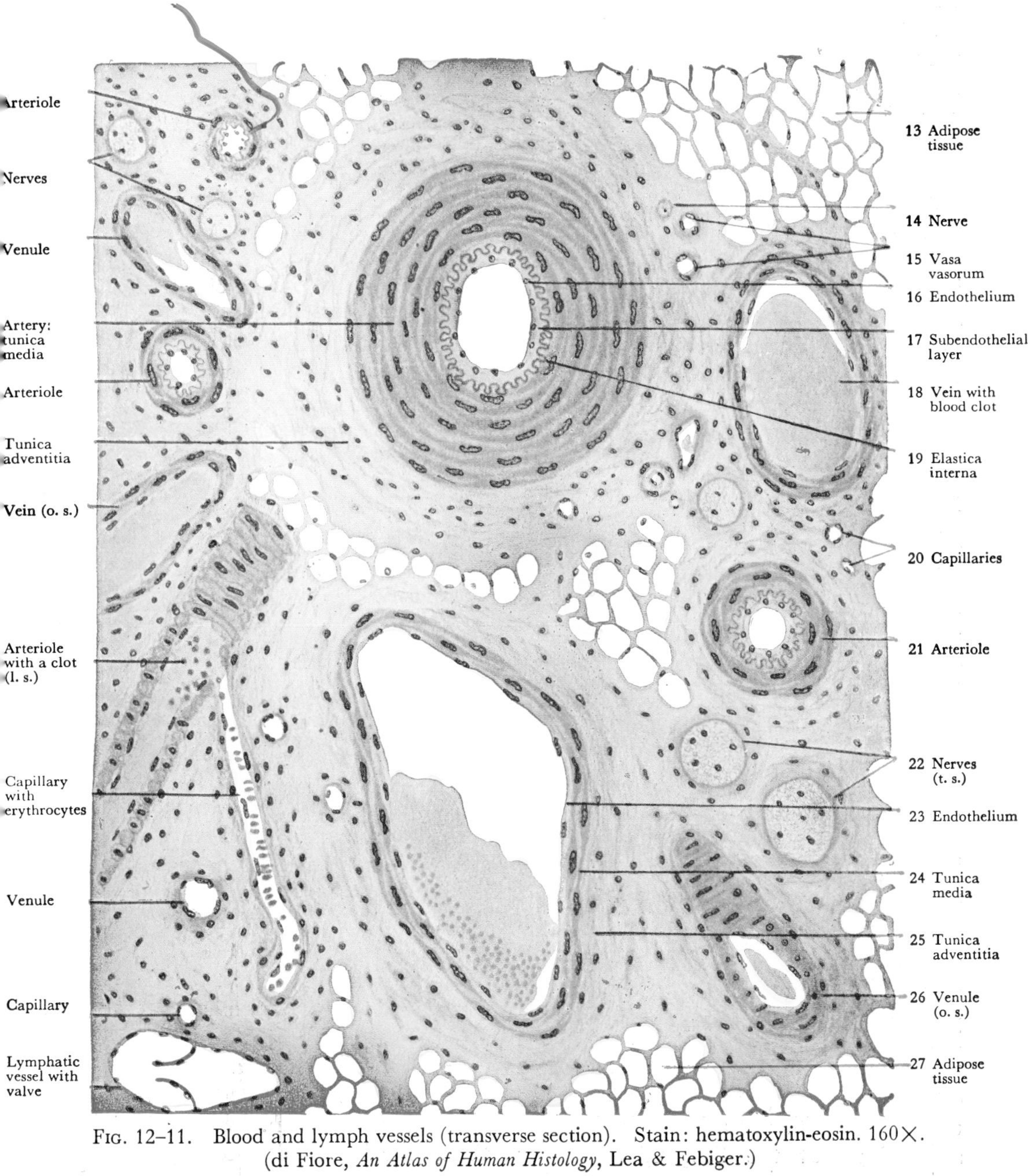
Arteriole
Nerves
Venule
Artery: tunica media
Arteriole
Tunica adventitia
Vein (o. s.)
Arteriole with a clot (l. s.)
Capillary with erythrocytes
Venule
Capillary
Lymphatic vessel with valve
13 Adipose tissue
14 Nerve
15 Vasa vasorum
16 Endothelium
17 Subendothelial layer
18 Vein with blood clot
19 Elastica interna
20 Capillaries
21 Arteriole
22 Nerves (t. s.)
23 Endothelium
24 Tunica media
25 Tunica adventitia
26 Venule (o. s.)
27 Adipose tissue

FIG. 12–11. Blood and lymph vessels (transverse section). Stain: hematoxylin-eosin. 160×. (di Fiore, *An Atlas of Human Histology*, Lea & Febiger.)

to have the patient checked by a specialist and probably re-examined from time to time.

BLOOD VESSELS (Fig. 12–11)

The kinds of blood vessels are named and defined and their relationships described on page 278.

Arteries. Arteries range in size from the aorta which is about 25 mm. (1 inch) in diameter in man to those about 0.5 mm. The structure of their walls is well adapted to the many and varied requirements placed upon them.

As seen microscopically, arteries possess three coats in their walls (Fig. 12–11). The innermost is the **tunica intima** which is thin and consists of a surface layer of smooth endothelium with a basement membrane and some connective tissue. The thicker, intermediate layer, the **tunica media,** is composed of smooth muscle and connective tissue, largely of the elastic type. These components are usually circularly arranged. The outer layer, the **tunica externa** (*adventitia*), somewhat thinner than the media, contains white fibrous connective tissue and in some cases a few smooth muscle fibers, all arranged longitudinally. The three coats of the arteries may be separated by **internal** and **external elastic membranes.**

As would be expected, the coats in the walls of arteries vary widely. In general, the larger arteries, such as the aorta, pulmonary, innominate, common carotids, subclavians and common iliacs, have a thicker tunica intima with an **internal elastic membrane,** and the tunica media is very thick, containing smooth muscle fibers, but so much elastic tissue that the muscle fibers are obscured. The externa is thin but strong to limit the stretch of the arterial walls. Such arteries are often called **elastic arteries** and are ideally suited to act as shock-absorbers, smoothing out the flow of blood by *stretching passively* with each systole of the heart and recoiling during diastole. Without this elasticity the flow of the blood would be intermittent rather than continuous through the vascular channels. Also, the blood pressure would be elevated, as indeed happens when these vessels, through aging or due to pathological conditions become hardened, a condition known as **arteriosclerosis.**

Other arteries may be called **muscular arteries,** because they have more smooth muscle than elastic tissue in the tunica media, as many as 40 layers in some and even more in arteries of the lower extremities. They are active rather than passive in their influence upon blood flow, pressure, and distribution. There is no hard and fast distinction to be drawn between "elastic" and "muscular" arteries. They grade one into another and generally as the vessel diameter decreases. Distally, the smooth muscle almost completely replaces the elastic tissues. Such is the case in the **arterioles,** arbitrarily defined as arteries of under 0.5 mm. in diameter. These are the arteries closest to the capillaries and are of prime importance in regulating capillary flow. Muscular arteries also have the capacity to grow in size as conditions demand. If the main arteries to an area occlude, smaller **anastomosing arteries,** of which there are many examples in the human body, may increase in size to take care of the demands of the tissues involved. Thus there is established a **collateral circulation.** Also, muscular arteries contract spastically when injured, which is a protection against hemorrhaging.

Vessels of the Microcirculation— Capillaries (Fig. 12–12). In 1628 William Harvey demonstrated that blood flows continuously from arteries to veins. In 1661 Marcello Malpighi first discovered the capillaries which Harvey's work suggested must be there, but which he had never seen. Since Malpighi's time, our knowledge of capillaries has

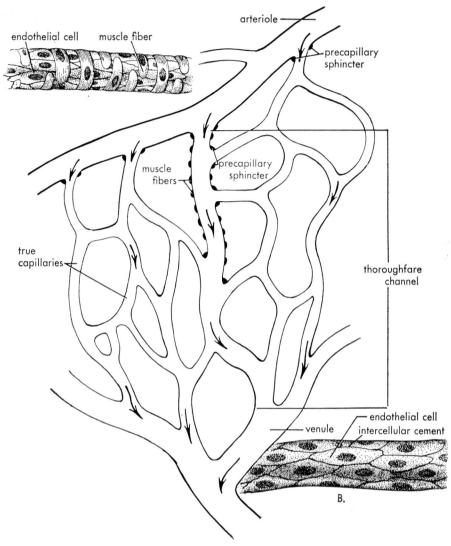

FIG. 12–12.   Schematic representation of a capillary bed.   Insert *A* shows some of the muscle fibers of the proximal part of a thoroughfare channel.   Insert *B* shows a part of a true capillary.

increased, helped by relatively recent techniques in microsurgery and microchemistry.

A capillary is not much more than a millimeter long and only about 8 to 10 microns in diameter, about the same diameter as a red corpuscle. The capillary wall consists of endothelial cells, forming a layer one cell thick. Their inner surface is in contact with the circulating blood. Their outer surface may rest on a basement membrane of amorphous ground substance lying between the cells and the surrounding loose connective tissue containing tissue fluid.

Though each capillary is small, they form vast networks, giving in the total body a surface area of about one and one-half acres, or 6,000 sq. meters. Placed end-to-end the capillaries would make a tube 60,000 miles long. If they should all be open at once, they would hold all the blood in the body, or if they could be collected into one mass, they would be the largest organ in the body, twice the size of the liver. These conditions, plus the slow rate of blood flow in the capillaries, allow for an efficient interchange between blood and cells.

The capillaries penetrate *almost* every tissue in the body—some tissues, the most active such as skeletal muscle, having many more than others.

Many students have reported independent movement within the capillaries. Special cells have been seen—the cells of Rouget, placed among the endothelial cells, and it was thought that these were responsible for reported movements. Microsurgical techniques have shown that in mammals the cells of Rouget do not produce independent capillary movement, but that among the capillaries are minute vessels called **thoroughfare channels** which do have a few scattered muscle fibers, especially on the side closest to the arterioles. These channels run between arterioles and venules. From the thoroughfare channels the capillaries take their origin and at each such origin of a capillary there is a ring of smooth muscle, the **precapillary sphincter,** which regulates the flow of blood into the capillaries. The activity of these sphincter muscles and the fibers in thoroughfare vessels may account for the capillary movements described.

This relatively new concept that the vessels between arterioles and venules are not all alike is justification for thinking of them in terms of a system of interrelated parts, which we might call a **capillary bed.**

Capillary beds constitute a most vital part of the circulatory system, for it is through the walls of these vessels that the essential exchanges of materials take place to maintain the constancy of the internal environment. It has been said that all the other organs of the circulatory system exist only to serve the capillary beds; a statement of considerable truth.

Venules.   Venules closest to the capillary beds are much like large capillaries except that their walls have some fibrous tissue outside the endothelial lining. The larger venules possess a few circularly arranged smooth muscle fibers and the still larger ones an outer tunica externa, containing some longitudinally arranged smooth muscle cells and connective tissue.

Veins (Fig. 12–11). The same coats as were described for the arteries are present in some veins but are seldom as well defined. In some, the tunica media is very thin or entirely absent. The tunica externa is usually the thickest tunic in veins and in some contains large amounts of smooth muscle. Elastic tissue is at a minimum. In general, veins are of greater diameter, but thinner-walled than the arteries with which they travel. Very often there will be two veins accompanying one artery. Veins without accompanying arteries are found beneath the skin, especially in the extremities—the **cutaneous veins.**

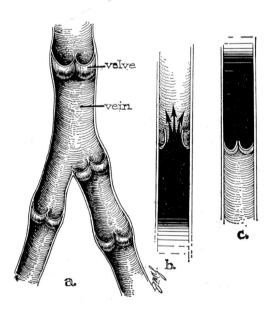

Fig. 12–13. *A,* Veins opened to show structure and position of valves. *B,* Valve opened as blood flows through. *C,* Valve closed as blood fills cusps and backflow of blood is prevented.

Valves. **Valves** are found in those veins which carry blood against the force of gravity, especially in the veins of the lower extremities. The arteries have no valves. The valves are folds of the tunica intima which allow the blood to flow freely toward the heart, but fill and come together to occlude the vessel when the blood tends to reverse its direction of flow (Fig. 12–13). Valves are absent in the venae cavae and the veins of the pulmonary and portal systems. Because veins depend largely upon the contractions of surrounding muscles for the movement of blood in them, the valves are necessary to insure flow toward the heart.

Vasa Vasorum (Fig. 12–11). The blood vessels require oxygen and nourishment as do other tissues and therefore have blood vessels in their own walls. These are called the **vasa vasorum.** Lymphatics are also present in vessel walls.

PULMONARY CIRCUIT

The pulmonary circuit takes the blood from the right ventricle of the heart to the lungs for exchange of respiratory gases and returns it to the left atrium of the heart (Figs. 12–1, 6).

The **pulmonary trunk** arises from the conus arteriosus of the right ventricle. It lies in front of the ascending aorta and then passes to its left and under the aortic arch where it divides into the **right** and **left pulmonary arteries.**

The **right pulmonary artery** is larger and longer than the left and passes horizontally in front of the right bronchus to the root of the right lung, where it divides into two branches, the lower and larger one branching into the lower and middle lobes of the lung, the other to the upper lobe.

The **left pulmonary artery** passes horizontally in front of the descending aorta and left bronchus to the root of the left lung, where it divides into two branches, one for each lobe of the lung.

The left pulmonary artery is attached to the concavity of the aortic arch by the **ligamentum arteriosum** which is a fibrous remnant of the **ductus arteriosus** of the fetus. The ductus arteriosus is the distal part of the left pulmonary arch (*6th aortic arch*); the pulmonary trunk is the proximal part. If the ductus arteriosus does not close and become fibrous soon after birth, there is a continued mixing of oxygenated and nonoxygenated blood—again the so-called "blue baby" condition. This condition can be treated surgically.

The arterioles of the pulmonary arteries open into the capillary beds of the lungs, which course closely over the alveoli. They then join the venules, which in turn lead to the pulmonary veins. Thus is provided the proper relationship of air and blood passageways to permit adequate external respiration to provide for the oxygen requirements of the individual.

The **pulmonary veins,** two from each lung, enter the posterior upper side of the left atrium. These veins are without valves.

SYSTEMIC CIRCUIT

SYSTEMIC ARTERIES

Aorta and Its Branches (Fig. 12–14). The **aorta** is the largest artery in the body and supplies oxygenated blood to all the arteries of the systemic circuit. At its beginning, at the left ventricle of the heart, it is about 3 cm. in diameter. It extends upward from the aortic valves, arches to the left and dorsalward, passing in front of the root of the left lung. It then descends, lying to the left side of the thoracic vertebrae, passes through the hiatus of the diaphragm and at the inferior border of the fourth lumbar vertebrae bifurcates to the right and left common iliac arteries. It has gradually diminished in size until its inferior portion is about 1.75 cm. in diameter.

Review Diagram — Pulmonary Circuit

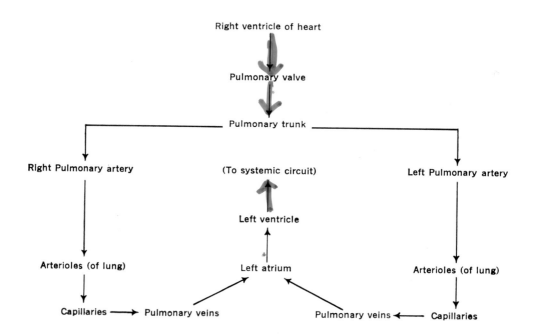

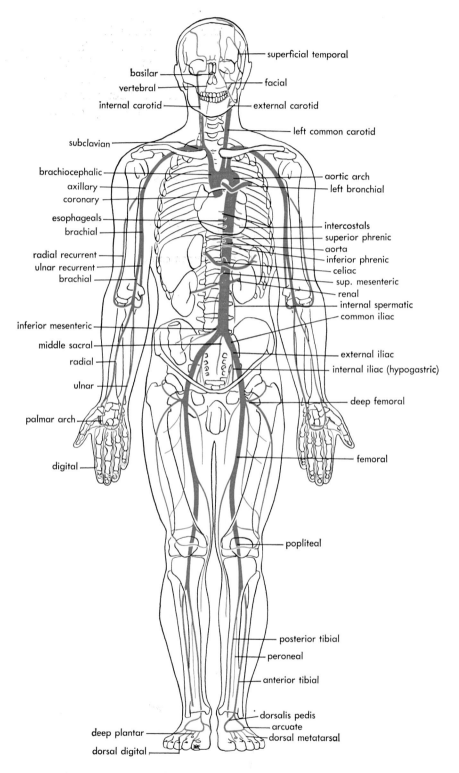

FIG. 12-14.   Diagram of arterial system.

For purposes of description, the aorta is divided into the **ascending aorta,** the **aortic arch,** and the **descending aorta.** The descending aorta is further divided into the **thoracic** and **abdominal aortae.**

The main branches of the aorta are the following:

Ascending Aorta (Figs. 12–3, 4). The ascending aorta is about 5 cm. (2 in.) long and 3 cm. in diameter. Its outer surface is covered with visceral pericardium, hence it is within the pericardial sac. It lies between the pulmonary artery which partly covers it and the

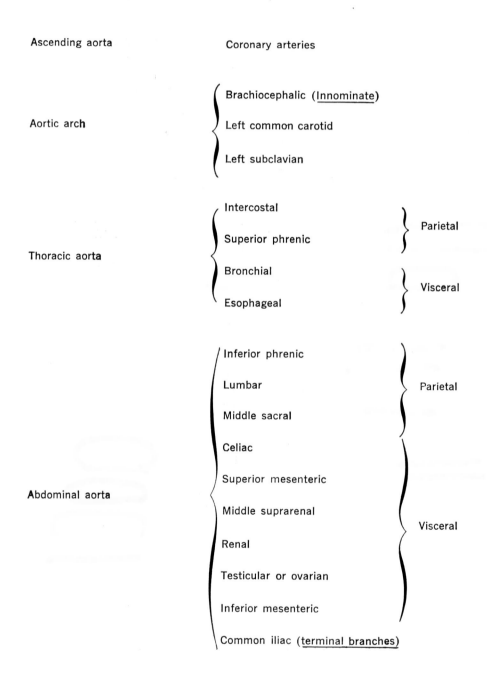

Ascending aorta — Coronary arteries

Aortic arch
- Brachiocephalic (Innominate)
- Left common carotid
- Left subclavian

Thoracic aorta
- Intercostal — Parietal
- Superior phrenic — Parietal
- Bronchial — Visceral
- Esophageal — Visceral

Abdominal aorta
- Inferior phrenic — Parietal
- Lumbar — Parietal
- Middle sacral — Parietal
- Celiac — Visceral
- Superior mesenteric — Visceral
- Middle suprarenal — Visceral
- Renal — Visceral
- Testicular or ovarian — Visceral
- Inferior mesenteric — Visceral
- Common iliac (terminal branches)

superior vena cava. It curves upward and to the right to about the level of the sternal angle.

The **right** and **left coronary arteries** are the only branches of the ascending aorta (Figs. 12–3, 8, 9). They arise in bulbous swellings at the beginning of the ascending aorta, the **aortic sinuses.** Since their openings are behind the aortic valve cusps, blood can enter them only when

the left ventricle is relaxed. The right coronary artery arises in the right anterior aortic sinus, passes between the conus arteriosus and the right auricle and into the coronary sulcus. It follows the sulcus around to the diaphragmatic surface of the heart and ends in branches beyond the interventricular sulcus. It thus forms an almost complete circle or crown around the heart, as its name

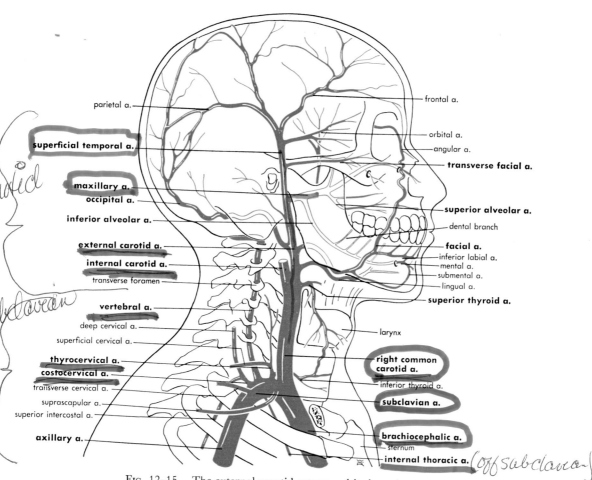

Fig. 12–15.   The external carotid artery and its branches.

suggests. The left coronary artery arises in the left posterior aortic sinus. It has two branches, one passing along the anterior interventricular sulcus to the apex and diaphragmatic surface, the other passing in the left coronary sulcus.

The coronary vessels with their many branches supply all parts of the heart. Fortunately, their branches anastomose freely, for stoppage of blood flow to any appreciable portion of the heart muscle would be fatal. Even so, coronary thrombosis (*clotting*) remains one of the common causes of sudden death.

Aortic Arch and Its Branches (Fig. 12–4). Leaving the pericardial sac, the aorta bends to the left and arches backward to the root of the left lung and ends at the left side of the body of the fourth thoracic vertebra. This is the **aortic arch.**

The three branches of the aortic arch from right to left are the **brachiocephalic** (*innominate*), the **left common carotid artery,** and the **left subclavian artery.** They arise from the top of the arch. Wide variations are found in the numbers and positions of these vessels, often reminding one of phylogenetic or embryonic conditions.

The **brachiocephalic** (*innominate*) **artery** is the largest branch of the aortic arch but is only about 5 cm. in length (Fig. 12–4). It arises from the right side of the arch and extends upward, backward, and slightly to the right where, posterior to the right sternoclavicular joint, it divides into the **right common carotid** and **right subclavian arteries.**

The **left common carotid artery** is the next branch of the aortic arch, followed closely by the **left subclavian artery.** These two vessels are in a sense the counterpart of the brachiocephalic and like its branches, they supply the head and neck and upper limb. They ascend vertically from the aortic arch on the left side of the trachea and esophagus to the area behind the left sternoclavicular

joint. Since from this point on the carotids and subclavians of the right and left sides follow essentially the same course, one description will do for both.

The **common carotids** course upward along the trachea, enclosed in a common sheath with the internal jugular veins and vagus nerves. Inside the common sheath, derived from the cervical fascia, each component has its own fibrous covering. The common carotid arteries have no side branches but at the level of the superior border of the thyroid cartilage they bifurcate into the **external** and **internal carotid arteries.** At this point of bifurcation there is a small oval body with abundant nerve supply called the **carotid body.** It is a chemo-receptor, probably sensitive to the changing levels of oxygen in the blood, which reflexly accelerates respiration. There is also at this point, either in the terminal part of the common carotid and the beginning of the internal carotid, or in the internal carotid alone, a small dilation called the **carotid sinus** which is an important receptor for the regulation of blood pressure.

The **external carotid artery** (Fig. 12–15). This artery supplies structures in the head and neck regions with the exception of the contents of the orbital and cranial cavities and structures of the lower neck. The orbital and cranial cavity structures are supplied by the internal carotid, the lower neck by branches of the subclavian.

The external carotid artery passes upward and enters the substance of the parotid salivary gland, and in this gland, at the level of the neck of the mandible, divides into its terminal branches, the **maxillary** and the **superficial temporal arteries.**

The **maxillary** passes deep, medial to the neck of the mandible, to the infratemporal fossa. It sends branches to the muscles of mastication, the mucous membranes of the pharynx, palate, and nose, and to the external auditory canal and external surface of the tympanic membrane. **Inferior** and **superior alveolar branches** supply the lower and upper teeth respectively and a prominent **middle meningeal branch** passes upward through the foramen spinosum of the skull floor to travel in grooves on the inner surface of the cranial wall to supply the meninges (*membranes around brain*).

The **superficial temporal artery** enters the scalp in front of the ear and supplies many structures in this area. Its pulse can be felt just in front of the ear above the base of the zygomatic process. It is considered the continuation of the external carotid artery.

Between its origin and its terminal branches the external carotid artery gives rise to:

Superior thyroid artery
Lingual artery
Facial artery
Occipital artery
Posterior auricular artery
Ascending pharyngeal arteries

The areas supplied by these vessels are suggested by their names. The pulse in the facial artery can be felt on the lower border of the mandible about 2 cm. in front of the mandibular angle.

Internal Carotid Artery (Fig. 12–16). The internal carotid passes upward in front of the transverse processes of the upper three cervical vertebrae and enters the cranial cavity through the carotid canal in the petrous portion of the temporal bone (Fig. 8–3). It forms an S curve around the body of the sphenoid bone and below the brain gives off **posterior communicating** and **ophthalmic arteries,** the latter going to the contents of the orbital cavity, the eye, and to nearby structures. At its terminus the internal carotid divides into **anterior** and **middle cerebral arteries** which contribute to the **circle of Willis** which will be more completely described below

21

(Fig. 12–17). These vessels help to supply the brain with blood.

Subclavian Arteries (Fig. 12-16). The **right** and **left subclavian arteries,** as you remember, arise differently but have the same distribution on the two sides of the body. From behind the sternoclavicular articulation the subclavian arches above the clavicle, in front of the apex of the lung and behind the scalenus anterior muscle. It then runs laterally and downward to the outer border of the first rib where it becomes the axillary artery.

The most important branches of the subclavian artery are:

Vertebral   **Internal thoracic**
Thyrocervical   **Costocervical**

The **vertebral artery** is the first branch of the subclavian. It passes upward through the transverse foramina of the upper six cervical vertebrae and turning behind the superior articular process of the atlas enters the foramen magnum. The vertebral artery gives off spinal branches which pass downward in the vertebral canal to supply the spinal cord and some also to the cerebellum of the brain. The vertebral then joins its counterpart from the opposite side, at the posterior rim of the pons of the brain, to form the **basilar artery.** The **basilar artery** continues forward to the anterior rim of the pons and divides into the **posterior cerebral arteries.** The posterior cerebral arteries join the posterior communicating arteries derived from the internal carotids, and with the anterior cerebrals and the **anterior communicating arteries** they complete an arterial ring beneath the brain, the **circle of Willis** mentioned above. Branches from this "circle" are distributed to various parts of the brain (Fig. 12–18).

The **internal thoracic** (*internal mammary*) **artery** (Fig. 12–16) arises from the concavity of the arch of the subclavian artery and descends on the

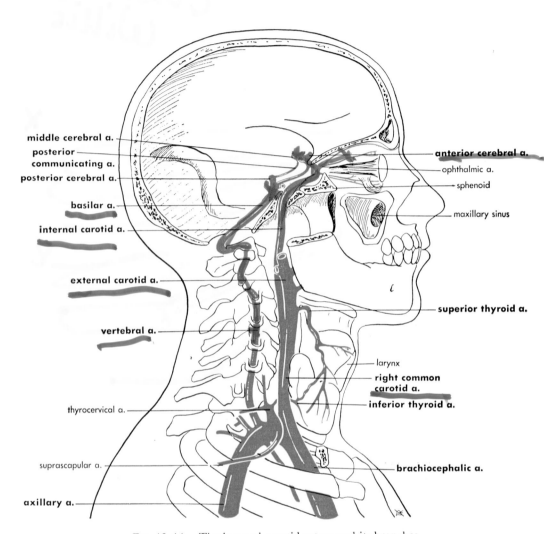

middle cerebral a.
posterior communicating a.
posterior cerebral a.
basilar a.
internal carotid a.
external carotid a.
vertebral a.
thyrocervical a.
suprascapular a.
axillary a.

anterior cerebral a.
ophthalmic a.
sphenoid
maxillary sinus
superior thyroid a.
larynx
right common carotid a.
inferior thyroid a.
brachiocephalic a.

FIG. 12–16.   The internal carotid artery and its branches.

posterior sides of the cartilages of the upper six ribs a little over a centimeter from the margin of the sternum. At the level of the sixth intercostal space it divides into the **musculophrenic** and **superior epigastric arteries.** As the internal thoracic crosses the ends of the first six intercostal spaces it gives off two branches to each space, the **anterior intercostal arteries.** These anastomose with posterior intercostal arteries from the descending thoracic aorta. Other branches of the internal thoracic supply the diaphragm, pericardium, mediastinum and muscles of thoracic and upper abdominal wall. The **superior epigastric artery,** as it descends from the internal thoracic, pierces the sheath of the Rectus

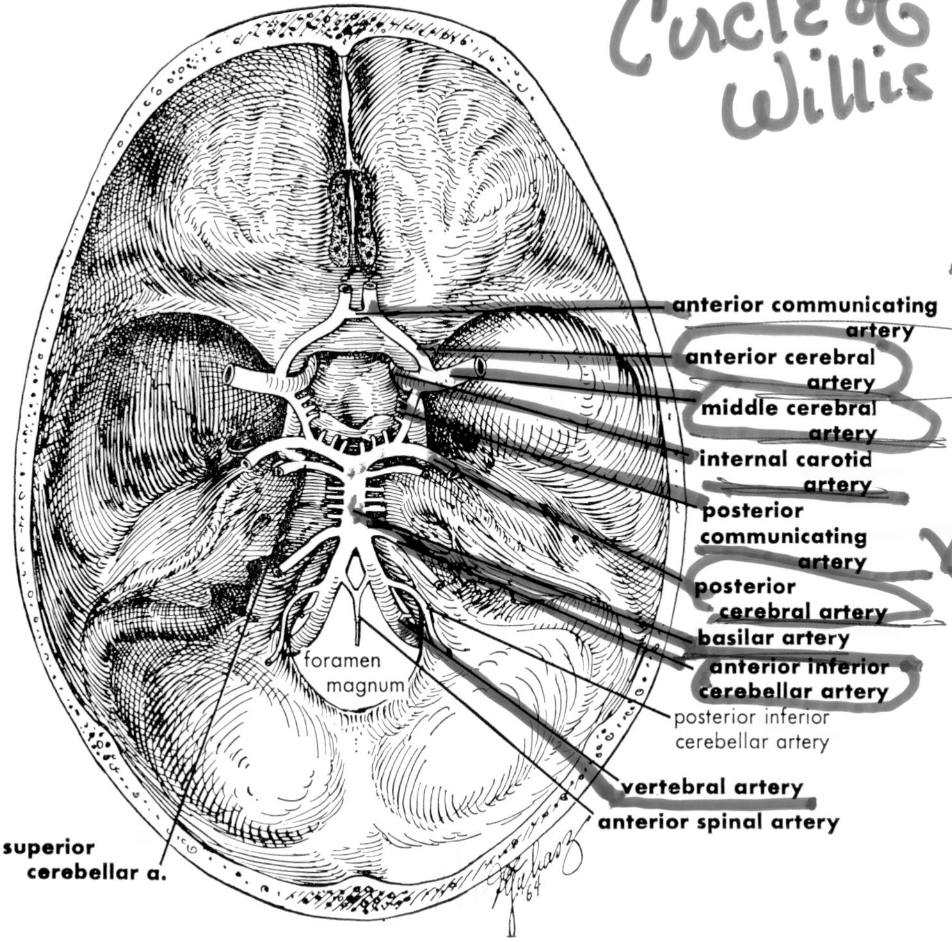

Circle of Willis

anterior communicating artery

anterior cerebral artery

middle cerebral artery

internal carotid artery

posterior communicating artery

posterior cerebral artery

basilar artery

anterior inferior cerebellar artery

posterior inferior cerebellar artery

foramen magnum

vertebral artery

anterior spinal artery

superior cerebellar a.

FIG. 12–17.   Arterial circulation of the base of the brain shown in relationship to the floor of the cranial cavity.   The internal carotid, cerebral, and communicating arteries form the circle of Willis.

abdominis muscle and anastomoses with the inferior epigastric artery to be described later.

The **thyrocervical artery** sends a branch to the thyroid gland, the inferior thyroid artery. It also sends branches to the muscles of the scapula.

The **costocervical artery** sends branches to the upper intercostal muscles, to muscles in the back of the neck, and to the spinal cord and its membranes.

Arteries of the Upper Limb (Fig. 12–19).   The **axillary artery** is a continuation of the subclavian artery beyond the first rib in the axilla. It is closely involved with the complex of nerves called the brachial plexus. It continues to the lower border of the tendon of the Teres major muscle, at which point it becomes the **brachial artery** extending on out into the arm.

The **brachial artery** ends about 1 cm. beyond the bend of the elbow where it divides into **radial** and **ulnar arteries**. The blood pressure is usually taken from this artery.

The **radial artery** continues down the radial side of the anterior surface of the forearm to the wrist where it is often used for taking the pulse. It terminates in

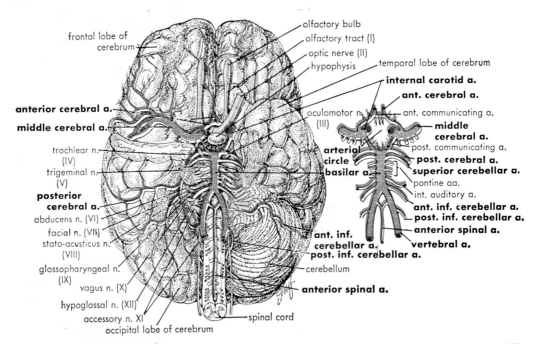

Fig. 12–18.   The base of the brain showing the arteries and their relationship to the cranial nerves.   The right temporal lobe of the cerebrum and the right cerebellar hemisphere have been removed.

the deep palmar arch of the hand which lies beneath the long flexor tendons of the digits.

The **ulnar artery** passes down the ulnar side of the forearm to the wrist where it then continues into the hand, ending in a superficial palmar arch. These palmar arches, deep and superficial, each give off palmar branches which go distally and unite at the clefts between the fingers.   From these common vessels digital arteries run down the adjacent sides of the fingers to end in networks of small vessels at the fingertips (Fig. 12–19).

Descending Thoracic Aorta and Its Branches (Figs. 12–14, 20).   The thoracic aorta lies to the left of the bodies of thoracic vertebrae five through twelve, turning mediad near the diaphragm.   It is about 20 cm. (8″) long.   It is provided with **parietal** branches to the body wall, the posterior intercostal arteries and the superior phrenic arteries; bronchial and esophageal branches, which are **visceral,** supply organs within the cavity.

The **posterior intercostal arteries** are paired, originating along the posterior border of the thoracic aorta and running into the intercostal muscles.   In the muscles they anastomose with the anterior intercostal arteries from the internal thoracic arteries.   The **superior phrenic arteries** supply the posterior part of the superior surface of the diaphragm.

The **right** and **left bronchial arteries** are distributed chiefly among the bronchi of the lungs and are nutrient vessels.   The four or five **esophageal arteries** arise from the anterior surface of the thoracic aorta and form networks of vessels about the esophagus.

Descending Abdominal Aorta and Its Branches (Figs. 12–14, 20).   **The abdominal aorta** is continuous with the thoracic aorta at the hiatus of the diaphragm and descends in front of the vertebral column.   It terminates in front of the fourth lumbar vertebra by dividing into the common iliac arteries.   It is convex anteriorly, since it lies upon the

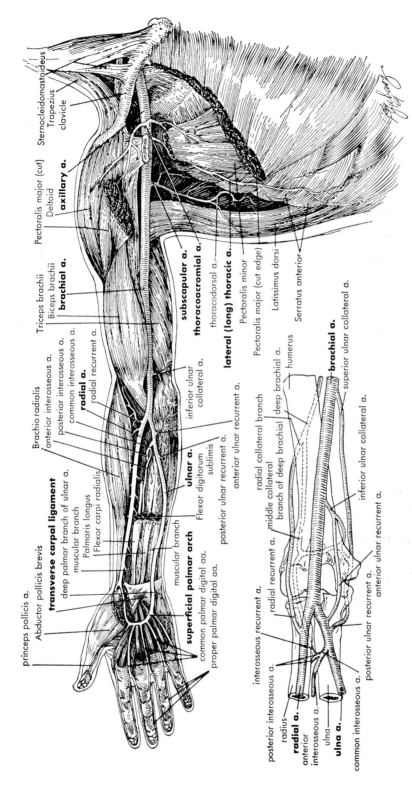

FIG. 12–19. Arteries of the right shoulder and upper extremity. The insert shows the relationships of the anastomosing arteries around the elbow joint.

lumbar curve of the vertebral column. Because its branches are large, it diminishes rapidly in size.

The branches of the abdominal aorta may be divided into three groups— parietal, visceral, and terminal.

The **parietal branches** are the inferior phrenic, lumbar, and middle sacral arteries.

The **inferior phrenics** vary a great deal in their origin. They may arise separately from the aorta or from one of the renal arteries. Sometimes they have a common trunk from the aorta or from the celiac artery. They supply the inferior surface of the diaphragm and some of the neighboring structures, such as the inferior vena cava, esophagus, and

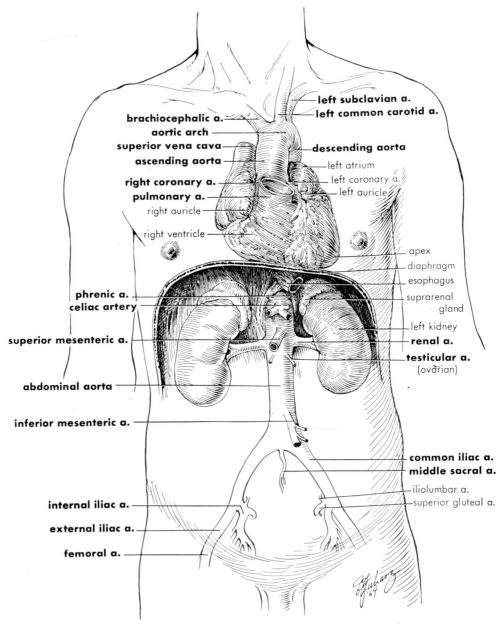

Fig. 12–20.   The aorta and its main branches and relationships.

suprarenal glands. Their branches anastomose with each other and with other vessels in the area.

The **lumbar arteries,** four pairs in number, arise from the posterior side of the abdominal aorta opposite the bodies of the upper four lumbar vertebrae. They lie in series with the intercostal arteries above. They supply the posterior and lateral abdominal walls.

The **middle sacral artery** is a small vessel which arises from the posterior side of the abdominal aorta just before it divides into the common iliacs. It descends on the midline of the anterior surface of the fourth and fifth lumbars, the sacrum, and the coccyx. It anastomoses freely with other vessels in the area and supplies blood to the pelvic wall and coccygeal gland.

The **visceral branches** of the abdominal aorta, named from above downward, are the celiac, superior mesenteric, middle suprarenal, renal, testicular or ovarian, and inferior mesenteric arteries. The celiac, superior mesenteric and inferior mesenteric are *single* arteries coming from the anterior surface of the aorta and supplying organs of the digestive tract, its accessory organs, and the spleen.

The **celiac artery** is a short, stout vessel which arises from the abdominal aorta just below the diaphragm (Figs. 12–20, 21). It divides into three conspicuous branches, the left gastric, the lienal (*splenic*) and the common hepatic arteries. The **left gastric** is the smallest and passes upward to the cardiac end of the stomach and descends along its lesser curvature, supplying the lower end of

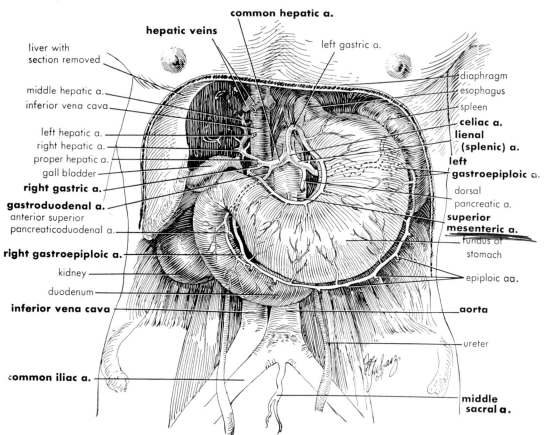

FIG. 12–21. The celiac artery and its branches showing organs served.

the esophagus and part of the stomach. The **lien al artery** is the largest branch of the celiac. It extends to the left along the pancreas to which it sends branches. Branches are also sent to the fundic stomach, and its terminal branches reach the spleen. A very large branch, the **left gastroepiploic artery,** reaches the greater curvature of the stomach near the spleen and travels along the curvature to the right, supplying also some branches to the greater omentum. The **common hepatic artery** extends to the right to the pyloric stomach and duodenum. There it turns upward and travels in the free edge of the lesser omentum to reach the liver. It sends a branch to the gall-

bladder, then divides into right and left hepatic arteries to nourish the liver (see pp. 376–383). It also sends branches to the stomach, duodenum and pancreas with some interesting interconnecting of its branches with those of the left gastric and splenic arteries (Fig. 12–21).

The **superior mesenteric artery** arises from the abdominal aorta a little over a centimeter behind the celiac artery. Occasionally, the superior mesenteric and celiac have a common origin. The superior mesenteric artery is long, extending obliquely through the mesentery toward the right iliac fossa. Its branches are shown in Figure 12–22. Notice the anastomoses, one, with one of its own

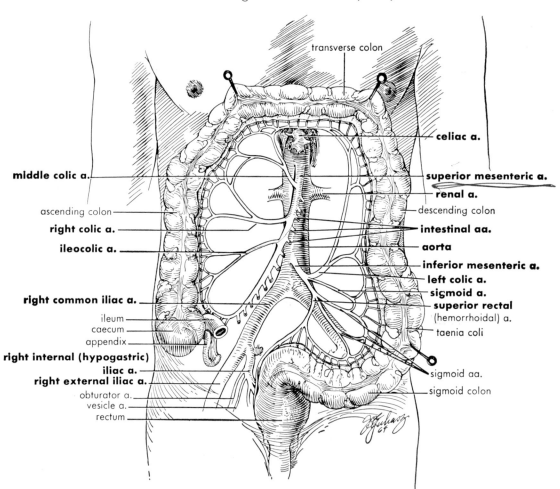

Fig. 12–22. Superior and inferior mesenteric arteries and their branches.

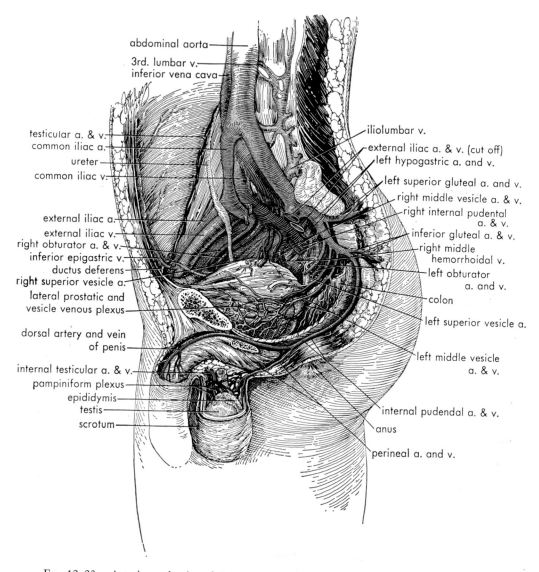

abdominal aorta

3rd. lumbar v.

inferior vena cava

testicular a. & v.

common iliac a.

ureter

common iliac v.

external iliac a.

external iliac v.

right obturator a. & v.

inferior epigastric v.

ductus deferens

right superior vesicle a.

lateral prostatic and
vesicle venous plexus

dorsal artery and vein
of penis

internal testicular a. & v.

pampiniform plexus

epididymis

testis

scrotum

iliolumbar v.

external iliac a. & v. (cut off)

left hypogastric a. and v.

left superior gluteal a. and v.

right middle vesicle a. & v.

right internal pudental
a. & v.

inferior gluteal a. & v.

right middle
hemorrhoidal v.

left obturator
a. and v.

colon

left superior vesicle a.

left middle vesicle
a. & v.

internal pudendal a. & v.

anus

perineal a. and v.

FIG. 12–23. Arteries and veins of the male pelvis, perineum, and external genital organs.
Viewed from the left side.

branches, another with a branch of the inferior mesenteric artery. It supplies blood to all of the small intestine except the proximal part of the duodenum. It also supplies the cecum and appendix, the ascending, and part of the transverse colon.

The **inferior mesenteric artery** arises from the abdominal aorta 3 to 4 cm. above its bifurcation. It is a smaller artery than the superior mesenteric. It supplies the left half of the transverse colon, the descending colon, sigmoid colon, and most of the rectum. Its left colic branch anastomoses with the middle colic branch of the superior mesenteric artery (Fig. 12–22).

The middle suprarenal, renal, and ovarian or testicular arteries are *paired* vessels.

The **middle suprarenal** (*adrenal*) **arteries** arise from the abdominal aorta at about the level of the superior mesenteric artery. They pass laterally and slightly upward to the medial sides of the suprarenal (*adrenal*) glands. They anastomose fully with the suprarenal branches from the inferior phrenic and renal arteries.

The **renal arteries** (Fig. 12–20) are short but wide vessels in keeping with the high vascular requirements of the kidneys. They arise from the abdominal aorta about 1 cm. below the superior mesenteric artery and pass transversely to the hilum of the kidney. The left renal artery is placed somewhat higher and is shorter than the right. Near the

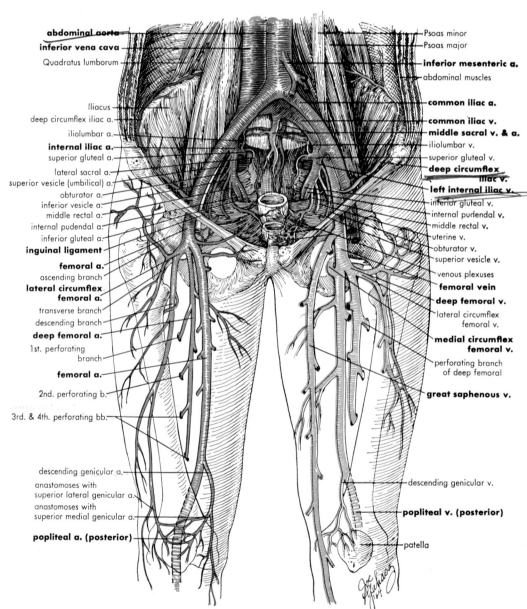

FIG. 12–24.   Arteries and veins of the pelvis and thigh.

hilum of the kidneys the renal arteries each break up into four or five branches which enter the sinus of the kidney and distribute through its substance in a definite pattern (Fig. 15–3). Accessory renal arteries are common, especially on the left side. Instead of entering at the hilum, they usually penetrate the superior or inferior part of the kidney.

The **testicular** (*internal spermatic*) **arteries** (Fig. 12–20) are two long, slender vessels which arise from the front of the abdominal aorta below the renal arteries. They pass downward and lateralward beneath the peritoneum to the abdominal inguinal ring, pass through the inguinal canal with other components of the spermatic cord, and supply the testes and other nearby structures. Branches to the ureter and cremaster muscle are also provided by the testicular artery.

The **ovarian arteries** (Fig. 12–20) are the female counterpart of the testicular arteries of the male. They have the same origin as the testicular arteries but are shorter and pass to the ovaries in the pelvis. They give off branches to the ureters and uterine tubes and terminally join the uterine arteries.

The **terminal** branches of the abdominal aorta are the **common iliac arteries** (Figs. 12–20, 21, 22). They arise at the level of the middle of the body of the fourth lumbar vertebra, a little to the left of the median plane. They end by dividing into the external and internal iliac (*hypogastric*) arteries at the level of the lumbo-sacral junction, about 5 cm. from their origin. The common iliac arteries give off small branches to the peritoneum, ureter, areolar tissue, and psoas muscles but their chief contribution is the formation of the external and internal iliac arteries.

The **internal iliac** (*hypogastric*) **arteries** (Figs. 12–23, 24) are 3 to 4 cm. long. They descend into the pelvis and at the upper margin of the greater sciatic foramen divide into two divisions, ante-rior and posterior. The internal iliac and its branches are highly variable. In general, the **posterior division** passes posteriorly and gives rise only to parietal branches. These are the iliolumbars and lateral sacrals to the pelvic wall and the superior gluteals to the gluteal region. The **anterior division,** a direct continuation of the internal iliac, gives rise to both parietal and visceral branches. The three parietal branches, the obturator, internal pudendal, and inferior gluteal, go to the thigh and perineum. The visceral branches are the umbilical, inferior vesicle, and middle rectal, supplying the pelvic viscera. In the female, **vaginal** and **uterine arteries** are present. The uterine artery, as we have noted above, anastomoses with the ovarian artery.

Special consideration of the **internal iliac arteries** of the **fetus** is appropriate at this point. They are twice as large as the external iliac artery in the fetus and are in direct line with the common iliac, essentially a continuation of it. They pass up either side of the urinary bladder to the inside of the anterior abdominal wall and then on to the umbilicus where they enter the umbilical cord. We now call these vessels the umbilical arteries. They coil around the umbilical vein and finally reach the placenta. They carry fetal blood to the placenta for release of waste products of metabolism and to obtain food and oxygen, which is brought back to the fetus by the umbilical vein.

When the placental circulation ceases at birth, the pelvic portion of these arteries remain active as the internal iliacs and a part of the superior vesical (*umbilical*) arteries of the adult; the remaining portions are converted into fibrous cords, the **lateral umbilical ligaments,** extending from the bladder to the umbilicus beneath the peritoneum.

The **external iliac artery** on either side, essentially a continuation of the corresponding common iliac, is larger in the adult than is the internal iliac artery.

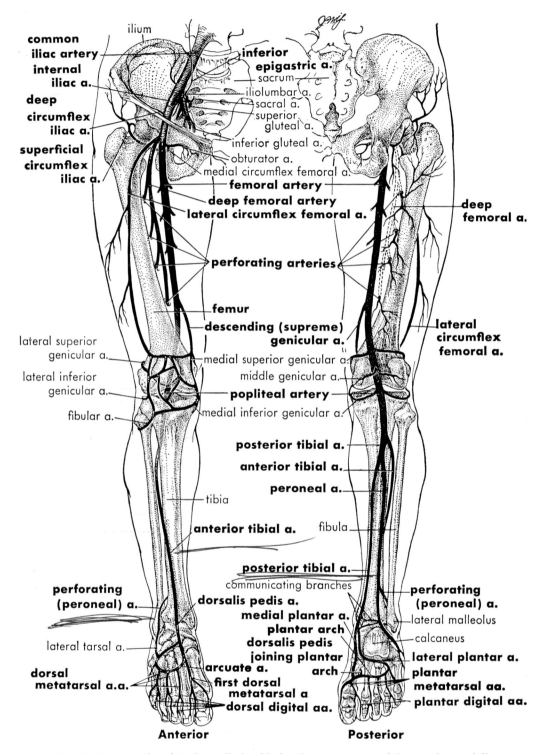

ilium

common
iliac artery
internal
iliac a.
deep
circumflex
iliac a.
superficial
circumflex
iliac a.

inferior
epigastric a.
sacrum
iliolumbar a.
sacral a.
superior
gluteal a.
inferior gluteal a.
obturator a.
medial circumflex femoral a.
femoral artery
deep femoral artery
lateral circumflex femoral a.

deep
femoral a.

perforating arteries

femur
**descending (supreme)
genical a.**
medial superior genicular a.
middle genicular a.
popliteal artery
medial inferior genicular a.

lateral
circumflex
femoral a.

lateral superior
genicular a.
lateral inferior
genicular a.
fibular a.

posterior tibial a.
anterior tibial a.
peroneal a.

tibia

anterior tibial a.

fibula

posterior tibial a.
communicating branches

perforating
(peroneal) a.

dorsalis pedis a.
medial plantar a.
plantar arch
**dorsalis pedis
joining plantar
arcuate a.   arch
first dorsal
metatarsal a
dorsal digital aa.**

perforating
(peroneal) a.
lateral malleolus
calcaneus

lateral tarsal a.

dorsal
metatarsal a.a.

lateral plantar a.
**plantar
metatarsal aa.
plantar digital aa.**

Anterior   **Posterior**

FIG. 12–25. Arteries of the lower limb. Notice the anastomoses of the vessels especially
at the knee and in the foot.

It travels in a lateral and inferior direction along the medial side of the Psoas major muscle to a point midway between the symphysis pubis and the anterior superior iliac spine, where it passes underneath the inguinal ligament to become the femoral artery. The external iliac artery gives off small branches to the Psoas major, and two major branches, the inferior epigastric and deep iliac circumflex.

The **inferior epigastric artery** (Fig. 12–25) arising from the external iliac just above the inguinal ligament ascends obliquely toward the umbilicus in the extraperitoneal tissue of the anterior abdominal wall. It pierces the transversalis fascia and enters the sheath of the Rectus abdominis. It supplies vessels to the abdominal muscles and skin and superiorly anastomoses with the superior epigastric and lower posterior intercostal arteries.

The **deep circumflex iliac artery** (Figs. 12–24, 25) arises from the lateral aspect of the external iliac near the inguinal ligament. It travels obliquely behind the inguinal ligament toward the anterior superior spine of the ilium and turns upward into the lateral abdominal wall to supply the abdominal muscles. It anastomoses with the superficial circumflex iliac, the iliolumbar, and other body wall vessels.

Arteries of the Lower Limb (Figs. 12–24, 25). The arterial blood supply to the lower limb has much that reminds us of that of the upper limb. There the subclavian, axillary, and brachial arteries formed one continuous vessel from trunk to forearm with terminal branches, the radial and ulnar arteries. In the lower limb the external iliac, femoral, and popliteal are continuous from trunk to leg, and the anterior and posterior tibials are the terminal branches.

The **femoral artery** (Figs. 12–24, 25) starts at the inguinal ligament where it enters the **femoral triangle** along with the femoral vein and nerve. From the apex of the femoral triangle, it passes through a fascial compartment on the medial side of the thigh under the Sartorius muscle and between the Quadriceps and Adductor muscles, the **adductor canal**. At the lower end of the adductor canal, the femoral artery passes through an opening in the tendon of the Adductor magnus muscle, the **adductor hiatus,** and enters the popliteal fossa behind the knee. It is now called the **popliteal artery**.

The femoral artery gives off a number of branches, the largest of which is the **deep femoral** arising about 5 cm. below the inguinal ligament on the lateral side. Other branches are shown in Figure 12–25 and should be studied to get an impression of the extent of the thigh circulation. The femoral artery and its branches supply the skin and muscles of the thigh, the lower part of the abdominal wall, the hip joint, and external genital organs.

The **popliteal artery** (Fig. 12–25), the continuation of the femoral artery, traverses the popliteal fossa obliquely, giving off a number of genicular branches to supply the muscles, the knee joint, and skin. At the interval between the tibia and fibula at the lower border of the Popliteus muscle it divides into **anterior** and **posterior tibial arteries.**

The **posterior tibial artery** (Fig. 12–25) extends obliquely down the posterior and tibial side of the leg. In the foot, below the medial malleolus, it divides into **medial** and **lateral plantar arteries** to supply the sole of the foot. Other branches of the posterior tibial artery supply the muscles and skin of the leg. Its largest branch, the **peroneal artery,** arises at its superior end and passes toward the fibula and travels down the medial side of that bone to supply structures of the lateral side of the leg and in the calcaneal region of the foot.

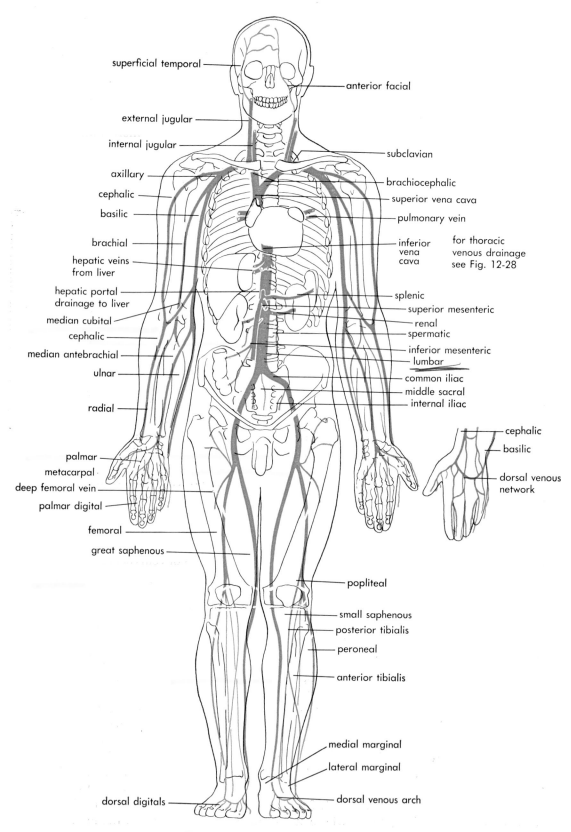

superficial temporal

anterior facial

external jugular

internal jugular

subclavian

axillary

cephalic

basilic

brachial

hepatic veins
from liver

hepatic portal
drainage to liver

median cubital

cephalic

median antebrachial

ulnar

radial

palmar

metacarpal

deep femoral vein

palmar digital

femoral

great saphenous

dorsal digitals

brachiocephalic

superior vena cava

pulmonary vein

inferior
vena
cava

for thoracic
venous drainage
see Fig. 12-28

splenic

superior mesenteric

renal

spermatic

inferior mesenteric

lumbar

common iliac

middle sacral

internal iliac

cephalic

basilic

dorsal venous
network

popliteal

small saphenous

posterior tibialis

peroneal

anterior tibialis

medial marginal

lateral marginal

dorsal venous arch

Fig. 12–26. Diagram of venous system.

The **anterior tibial artery** (Fig. 12–25) after arising from the popliteal artery passes forward above the interosseous membrane and runs deep in the front of the leg along the anterior surface of the interosseous membrane. At the lower end it comes to lie on the front of the tibia. As it passes in front of the ankle joint, it is more superficial and becomes the **dorsalis pedis artery.**

The **dorsalis pedis artery** forms two branches on the dorsum of the foot, the first dorsal metatarsal and the deep plantar arteries. The **deep plantar artery** penetrates into the sole of the foot and unites with the lateral plantar artery to complete the **plantar arch** of the foot (Fig. 12–25).

Systemic Veins (Fig. 12–26)

Recall that veins are best defined as blood vessels which carry blood toward the heart, regardless of the quality of the blood within them. The pulmonary veins of the pulmonary circuit carry fully oxygenated blood; the inferior vena cava, blood laden with nutrients from the viscera, though low in oxygen; the superior vena cava, blood deficient in both food and oxygen.

Since the systemic veins constitute a drainage system with "headwaters" in the capillary beds and venules, it is appropriate to speak of the vessels that a vein receives as it goes toward the heart as its *tributaries*, rather than its branches. Each vein, though larger in cross-sectional area than any *one* of its tributaries, is smaller in that dimension than that of *all* of its tributaries combined. Therefore, the rate of flow of blood in the systemic veins tends to increase as it gets closer to the heart—much like in a stream draining a lake, where the flow is faster than the water in the lake. The opposite, of course, is the situation in the systemic arteries where as they divide and subdivide the rate of flow decreases until they

reach the "lake," the capillary beds, where flow is very slow indeed—a nice adaptation to facilitate the efficient exchange of materials between blood and tissue fluid.

Veins are larger and more numerous than arteries, giving the venous system a greater capacity than the arterial. Veins also anastomose more freely with one another—connecting vessels being present between the larger trunks as well as between the smaller vessels. This is especially true among the venous sinuses of the cranium, the veins of the neck, vertebral canal, and the large venous plexuses of the abdomen and pelvis.

There are three kinds of systemic venous channels, the **superficial** and **deep veins** and the **venous sinuses.**

The **superficial veins** (*cutaneous veins*) lie between the layers of the superficial fascia beneath the skin. They receive the blood from these structures and communicate with the deep veins by penetrating the deep fascia.

The **deep veins** accompany the arteries, except in certain areas such as those in the skull and vertebral column, the hepatic veins of the liver, and the larger veins from bones. Arteries and veins traveling together usually are enclosed in a common sheath. Some of the smaller arteries have pairs of veins accompanying them, one on each side, called the **venae comitantes.** Such is the case with the radial, ulnar, brachial, tibial, and peroneal. Larger arteries, such as the popliteal, femoral, axillary and subclavian, have only one accompanying vein.

Venous sinuses, such as are found in the skull, are channels between the two layers of the dura mater, the outer membrane around the brain. They are lined with endothelium continuous with that of the veins and have an outer fibrous coat. They receive the cerebral veins and are drained by the internal jugular vein (Figs. 12–27, 8–4). There are also emissary veins which pass through the cranial wall connecting venous sinuses with veins outside the cranium.

The **systemic veins** fall logically into three groups: (1) those of the **heart,** ending in the **coronary sinus;** (2) those of the **head, neck, thorax,** and **upper extremity,** ending in the **superior vena cava;** (3) those of the **abdomen, pelvis,** and **lower extremity,** ending in the **inferior vena cava** (Fig. 12–26).

Since the veins travel with the arteries to such a large extent they will not be considered in as much detail. Emphasis will be given to the larger collecting vessels, to those with no companion arteries, and to that special system of vessels carrying blood from certain of the abdominal viscera, the **portal system.**

Veins of the Heart (Fig. 12–8). The **coronary sinus** is the largest venous channel of the heart. It lies in the coronary sulcus on the posterior side of the heart and empties into the right atrium between the orifice of the inferior vena cava and the atrioventricular aperture. Its orifice is guarded by an incompetent semilunar valve, the **valve of the coronary sinus** (*valve of Thebesius*). The coronary sinus receives a number of tributaries draining nearly all parts of the heart wall, the **cardiac veins,** most of which have valves at their orifices. A few smaller veins empty independently into the right atrium, still fewer may empty directly into the ventricles.

Veins of the Head, Neck, Thorax and Upper Limb. The **superior vena cava,** the drainage point for this area, is a large vessel about 7 cm. in length which is formed just below the first right costal cartilage by the **right** and **left brachiocephalic** (*innominate*) **veins** (Fig. 12–27). It descends to the right atrium of the heart into which it opens. The lower half of the vessel lies within the pericardium. Just before piercing the pericardium the superior vena cava receives the large **azygos vein** (Fig. 12–28).

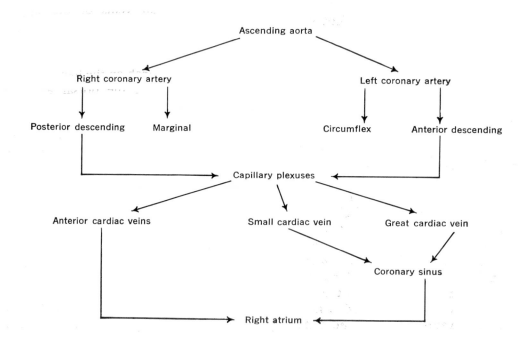

Review Diagram — Coronary circuit

The **azygos vein** begins in the abdominal region opposite the first or second lumbar vertebra as a continuation of the ascending lumbar vein and sometimes by a branch from the right renal. It passes through the diaphragm with the aorta and thoracic duct (*lymphatic*) in the aortic hiatus, and enters the thorax. It passes along the right side of the vertebral column, receiving tributaries from the intercostal areas, the esophagus, bronchi, mediastinum, and pericardium. At the level of the fourth thoracic vertebra it arches anteriorly to join the superior vena cava.

The **brachiocephalic veins** (*innominates*), right and left, are formed by the union of the **internal jugular** and **subclavian veins.** The right brachiocephalic is only about 2.5 cm. long; the left is 6 cm. The left brachiocephalic runs obliquely downward from its origin behind the sternal end of the clavicle to where it joins the right brachiocephalic to form the superior vena cava.

The brachiocephalic veins each receive ~~vertebral~~, **internal thoracic** (*internal mammary*), and **inferior thyroid veins,** and the left receives, in addition, the highest intercostal vein.

The left brachiocephalic vein sometimes joins the right atrium directly, instead of uniting with its right counterpart to form the superior vena cava. This condition is normal in the early fetus as it is in birds and some adult mammals.

The head and neck regions are drained largely by two prominent veins, the **external** and **internal jugular veins.**

The **external jugular vein** (Fig. 12–27) receives its tributaries from the scalp, face, and neck, the areas supplied by the external carotid artery. It is smaller than the internal jugular vein and lies superficially in the lateral aspect of the neck. It joins the subclavian vein inferiorly just before that vein joins the internal jugular to form the brachiocephalic vein.

22

The **internal jugular** is a larger and deeper vein of the neck than the external jugular. It has its origin at the base of the skull as the continuation of the transverse sinus through the jugular foramen. The transverse sinus receives blood from the brain and is one of a complex system of sinuses around the brain. The internal jugular is enclosed in a common sheath with the common carotid artery and vagus nerve. It receives a number of tributaries from the face and neck before joining inferiorly with the subclavian vein (Fig. 12–27).

The **veins of the upper limb** drain into the **subclavian vein** which accompanies the subclavian artery. In the arm and forearm the veins may be divided into two categories, the deep and superficial veins. The deep veins follow the arteries and have the same names, **axillary, brachial, radial,** and **ulnar.** They differ in that they are paired, one member of each pair lying on each side of the corresponding artery. This arrangement is referred to as **venae comitantes.** The deep veins have more valves than the superficial ones.

The superficial veins of the upper limb, lying between the two layers of the superficial fascia, begin with a **dorsal venous arch** or network, or a **palmar venous arch** on the hand. From the lateral side of the dorsal arch, a **cephalic vein** passes up the front of the lateral side of the limb until it reaches the groove between the Deltoid and Pectoralis major muscles where it turns mediad along the groove and then plunges deep to join the axillary vein. From the medial side of the dorsal venous arch, a **basilic vein** passes up the medial posterior side of the forearm. A little below the elbow it reaches the front of the arm, where it shows several conspicuous tributaries, and a little farther proximad on the arm it goes deep to run with the brachial artery, and finally joins the brachial vein to form the axillary vein. The median

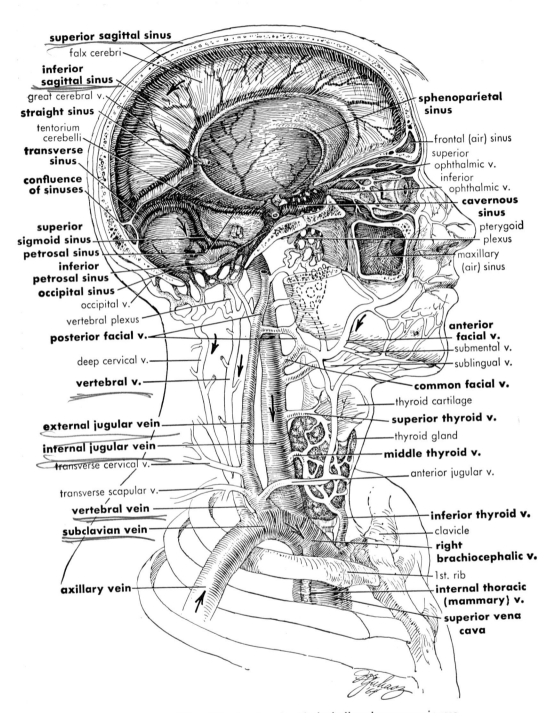

superior sagittal sinus
falx cerebri
inferior sagittal sinus
great cerebral v.
straight sinus
tentorium cerebelli
transverse sinus
confluence of sinuses
superior sigmoid sinus
petrosal sinus
inferior petrosal sinus
occipital sinus
occipital v.
vertebral plexus
posterior facial v.
deep cervical v.
vertebral v.
external jugular vein
internal jugular vein
transverse cervical v.
transverse scapular v.
vertebral vein
subclavian vein
axillary vein

sphenoparietal sinus
frontal (air) sinus
superior ophthalmic v.
inferior ophthalmic v.
cavernous sinus
pterygoid plexus
maxillary (air) sinus
anterior facial v.
submental v.
sublingual v.
common facial v.
thyroid cartilage
superior thyroid v.
thyroid gland
middle thyroid v.
anterior jugular v.
inferior thyroid v.
clavicle
right brachiocephalic v.
1st. rib
internal thoracic (mammary) v.
superior vena cava

Fig. 12–27. Veins of the head and neck, including the venous sinuses.

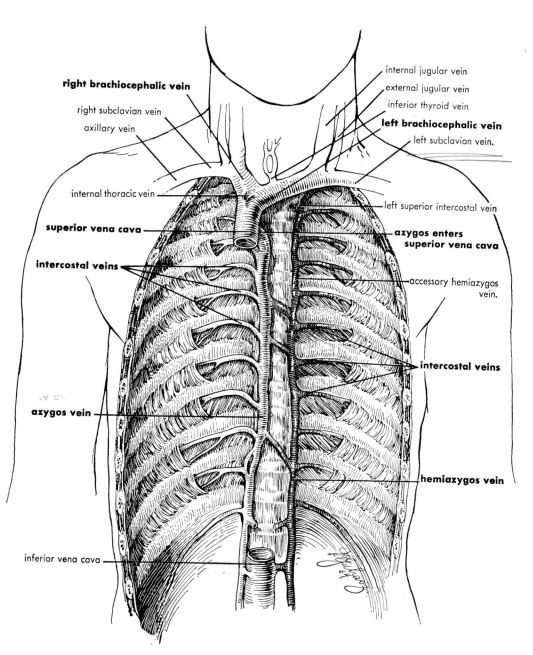

right brachiocephalic vein

right subclavian vein
axillary vein

internal thoracic vein

superior vena cava

intercostal veins

azygos vein

inferior vena cava

internal jugular vein
external jugular vein
inferior thyroid vein
left brachiocephalic vein
left subclavian vein.

left superior intercostal vein

azygos enters
superior vena cava

accessory hemiazygos
vein.

intercostal veins

hemiazygos vein

FIG. 12–28.   Venous drainage of posterior wall—the azygos and hemiazygos veins.

Review Diagram — Circulation for Brain

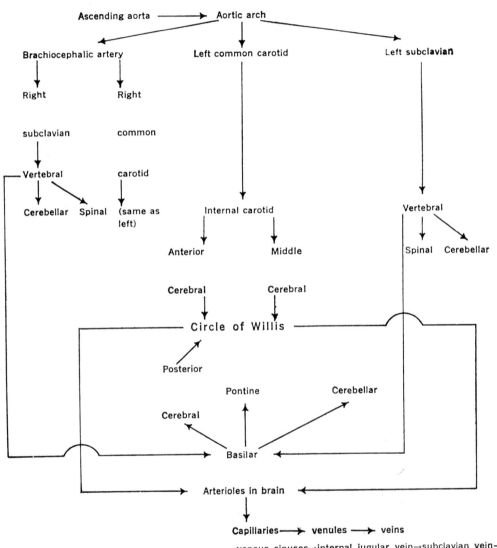

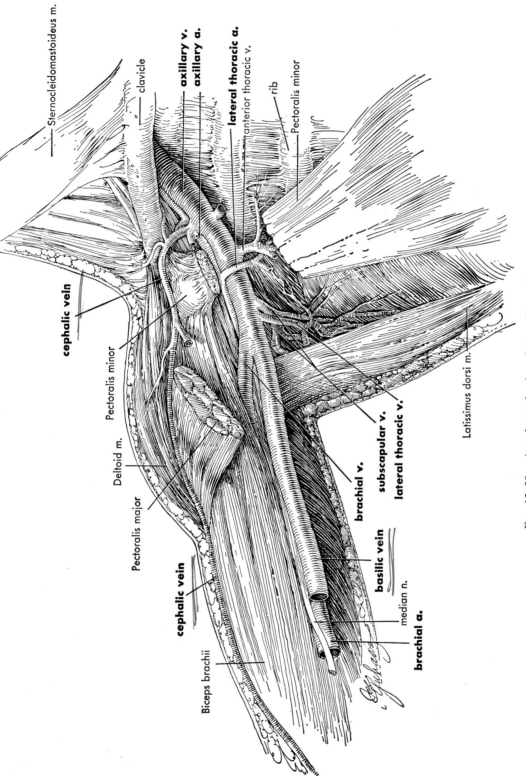

Sternocleidomastoideus m.

clavicle

axillary v.
axillary a.

lateral thoracic a.

anterior thoracic v.

rib

Pectoralis minor

cephalic vein

Pectoralis minor

Deltoid m.

Pectoralis major

Latissimus dorsi m.

brachial v.

subscapular v.
lateral thoracic v.

cephalic vein

basilic vein

median n.

brachial a.

Biceps brachii

Fig. 12–29. Arteries and veins of the axilla and arm.

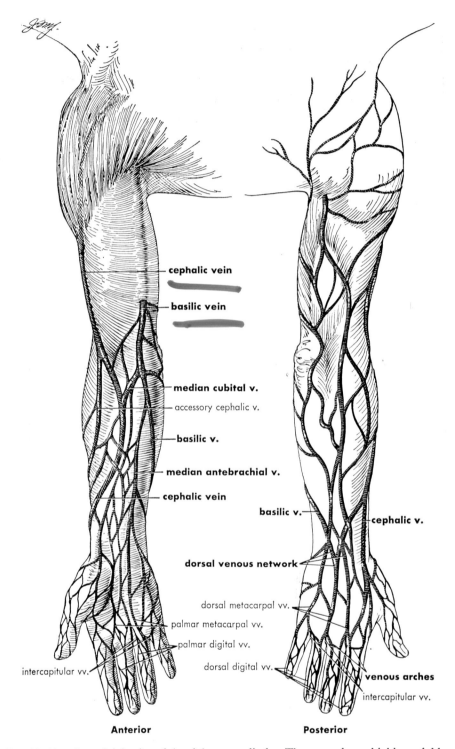

cephalic vein

basilic vein

median cubital v.

accessory cephalic v.

basilic v.

median antebrachial v.

cephalic vein

basilic v.

cephalic v.

dorsal venous network

dorsal metacarpal vv.

palmar metacarpal vv.

palmar digital vv.

dorsal digital vv.

venous arches

intercapitular vv.

intercapitular vv.

Anterior

Posterior

Fig. 12–30. Superficial veins of the right upper limb. These vessels are highly variable.

Review Diagram — Circulation in Left Upper Limb

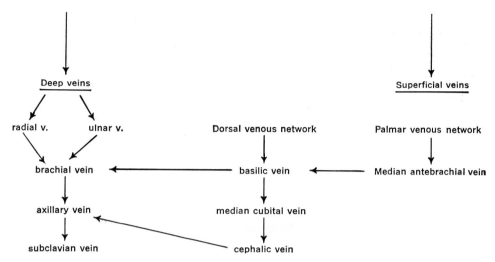

Ascending aorta→aortic arch→left subclavian artery→axillary artery→brachial artery⟨radial a. / ulnar a.⟩→superficial **and**

deep palmar arches→digital arteries→arterioles→capillaries→venules→superficial and deep palmar venous **arches**

cubital vein, lying in the antecubital fossa, runs obliquely between the cephalic and basilic veins. The median antebrachial vein begins in the palmar venous network of the hand and passes up the anterior surface of the forearm on the ulnar side and joins the basilic vein (Figs. 12–29, 30)

Veins of the Abdomen, Pelvis, and Lower Limb (Figs. 12–23, 24, 31). The large vein which directly or indirectly receives the blood from this area to return it to the right atrium is the inferior vena cava.

The **inferior vena cava** is the largest blood vessel in the body. It is formed by the confluence of the two common iliac veins on the right side of the fifth lumbar vertebra. It ascends on the right side of the lumbar vertebral bodies, and to the right of the aorta lying behind the parietal peritoneum. When it reaches the liver it passes in a groove along its posterior surface and then penetrates the diaphragm and the parietal pericardium to enter the inferior and posterior part of the right atrium.

In its course through the abdomen, the inferior vena cava receives tributaries which correspond to many of the arteries which branch from the aorta; **lumbars, renal, suprarenal,** and **testicular** or **ovarian.** Note, however, that it does **not** receive veins direct from the digestive tract, the pancreas, and spleen It does receive **hepatic veins** from the liver.

The paired vessels entering the inferior vena cava also show some interesting asymmetries. Because the inferior vena cava lies to the right of the midline the left renal vein is longer than the right and passes over the anterior surface of the aorta. The left testicular or ovarian and the suprarenal veins enter the left renal vein rather than the inferior vena cava directly. Their right counterparts enter the inferior vena cava.

The **common iliac veins** which by their joining form the inferior vena cava are themselves formed by the **external** and **internal iliac** (*hypogastric*) **veins.** The tributaries of these veins in the pelvis and lower limb follow closely the pattern of the arteries and have the

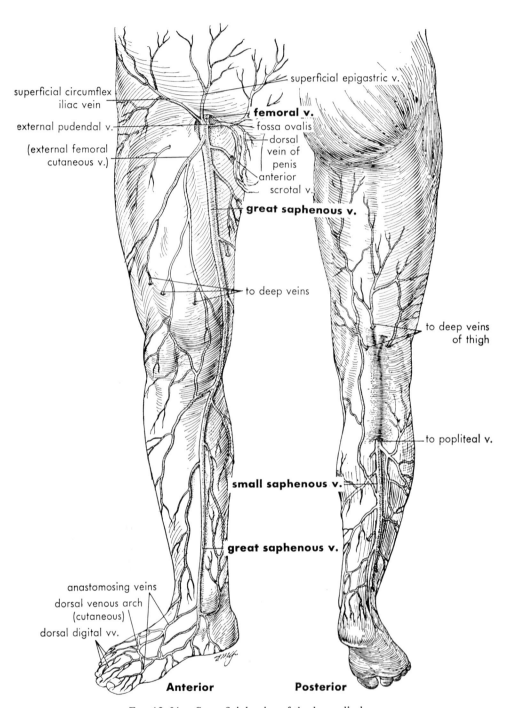

superficial epigastric v.

superficial circumflex
iliac vein

external pudendal v.

(external femoral
cutaneous v.)

femoral v.

fossa ovalis

dorsal
vein of
penis

anterior
scrotal v.

great saphenous v.

to deep veins

to deep veins
of thigh

to popliteal v.

small saphenous v.

great saphenous v.

anastomosing veins
dorsal venous arch
(cutaneous)
dorsal digital vv.

Anterior   **Posterior**

FIG. 12–31.   Superficial veins of the lower limb.

Review Diagram — Circulation in Lower Limb

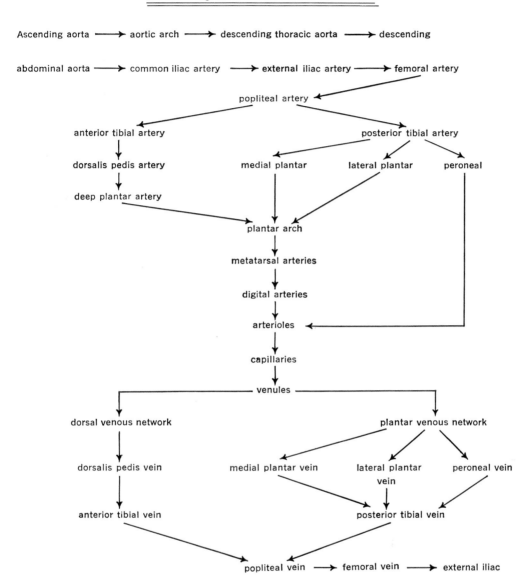

Ascending aorta ⟶ aortic arch ⟶ descending thoracic aorta ⟶ descending

abdominal aorta ⟶ common iliac artery ⟶ external iliac artery ⟶ femoral artery

popliteal artery

anterior tibial artery   posterior tibial artery

dorsalis pedis artery   medial plantar   lateral plantar   peroneal

deep plantar artery

plantar arch

metatarsal arteries

digital arteries

arterioles

capillaries

venules

dorsal venous network   plantar venous network

dorsalis pedis vein   medial plantar vein   lateral plantar   peroneal vein
                                                           vein

anterior tibial vein   posterior tibial vein

popliteal vein ⟶ femoral vein ⟶ external iliac

vein ⟶ common iliac vein ⟶ inferior vena cava ⟶ right atrium

Superficial Veins

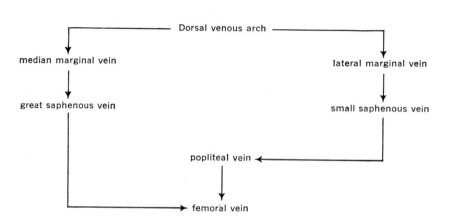

same or similar names (see review diagram, p. 327).

Two **superficial** or **cutaneous veins** of the lower limb require special attention, since they have no companion arteries. They are the **great** and the **small saphenous veins.** Like corresponding veins in the upper limb, they arise at the medial and lateral extremities of a **dorsal venous arch,** which lies a little proximal to the toes on the dorsal side of the foot. The dorsal venous arch connects with a **plantar cutaneous venous arch** on the sole of the foot. Both venous arches connect laterally and medially into **lateral** and **medial marginal veins.**

The **great saphenous vein,** the longest vein in the body, arises from the medial marginal vein on the dorsal side of the foot, ascends along the medial side of the leg and thigh to enter the femoral vein about 3 cm. below the inguinal ligament. It has numerous connections with superficial and deep veins of the lower extremity (Fig. 12–31).

The **small saphenous vein** is a continuation of the lateral marginal vein behind the lateral malleolus. It ascends along the lateral side of the calcaneal tendon, then crosses it to the middle of the posterior side of the leg at the popliteal

fossa and penetrates the deep fascia to join the **popliteal vein** (Fig. 12–31). It communicates with deep and superficial veins of the foot and leg. Both the small and large saphenous veins have many valves, as do the deep veins of the lower limb. See diagram above.

Portal System (Fig. 12–32). The portal system consists of those veins which drain the blood from the abdominal part of the digestive tract, the pancreas, spleen, and gallbladder. From these veins the blood is carried into the **portal vein** which conveys the blood to the capillary-like **sinusoids** of the liver. It is then collected by the **hepatic veins** which empty it into the inferior vena cava. In the portal system, unlike other parts of the systemic circuit, the blood passes through two sets of minute vessels instead of one from the time it leaves arterioles until it is returned to the heart. One set of minute vessels, the capillaries, are those in the walls of the organs of the digestive tract, pancreas, spleen and gallbladder; the other set, the sinusoids of the liver.

The **portal vein** is a large vessel formed at the level of the second lumbar vertebra by the joining of the superior mesenteric and lienal veins. It passes upward and travels in the free border of

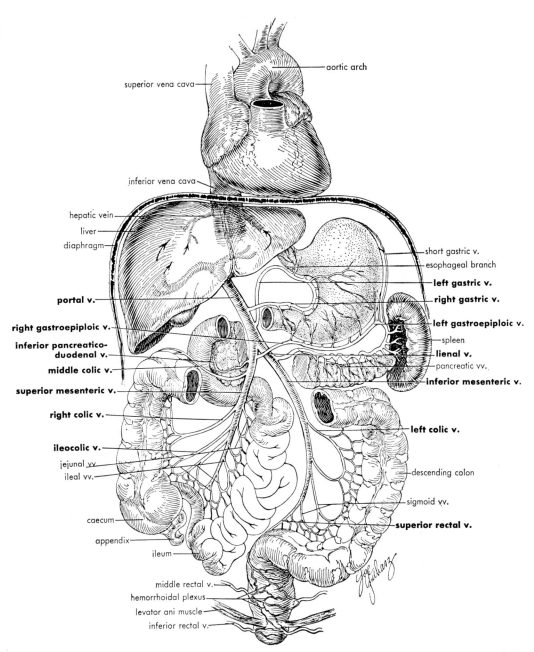

FIG. 12–32.   The portal vein and its tributaries.   Arrows suggest the sinusoids of the liver and the collection of "portal" blood by the hepatic veins of the systemic circulation.

Review Diagram — Portal Circulation

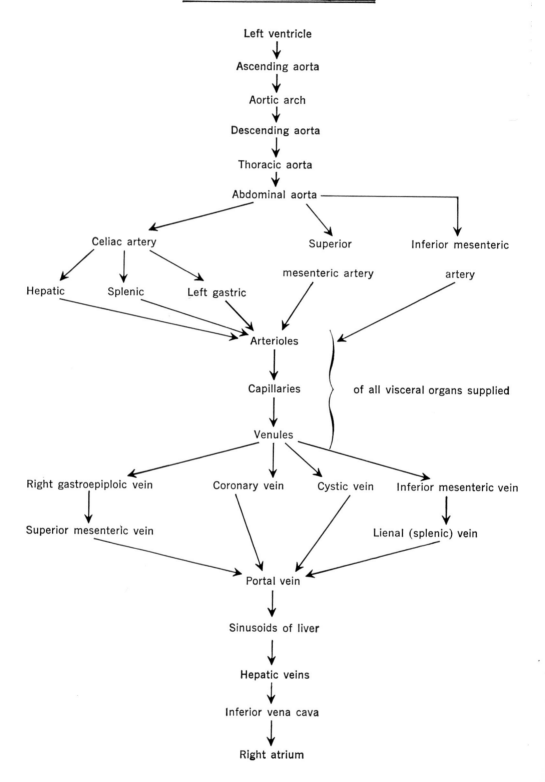

the lesser omentum to the porta of the liver where it divides into right and left branches which accompany the branches of the hepatic arteries into the liver.

The principal tributaries of the portal vein are the **superior mesenteric** and the **lienal** (*splenic*), which, in turn, have large tributaries. The general arrangement of the vessels of the portal drainage is shown in Figure 12–32. Its relationship to the systemic arteries is shown in the review diagram on page 330.

Fetal Circulation

In the foregoing account of the cardiovascular system every opportunity has been seized upon to interpret adult structure in terms of embryological development (*ontogeny*). It seems appropriate at this point to collate these references by a very general description of fetal circulation.

A first basic fact to remember is that the fetus is dependent entirely upon an organ outside of its own body for oxygen and nutrients, and for waste disposal, namely, the **placenta.** This arrangement places somewhat different demands upon the transportation system than does that of the individual who has had its fetal ties with the maternal parent severed. Perhaps the general plan of fetal circulation can best be shown by tracing the course of blood from the placenta through the fetus and back again to the placenta (Fig. 12–33).

The blood leaves the placenta through the **umbilical vein** which passes through the umbilical cord to the umbilicus of the fetus where it enters the body. The umbilical vein ascends along the free margin of the falciform ligament to the inferior surface of the liver where it gives off two or three branches to the left, caudate, and quadrate lobes of the liver. At the porta of the liver it divides, one branch joining the portal vein, the other, called the **ductus venosus,** enters the inferior vena cava directly.

The blood in the inferior vena cava, coming from the lower extremities and abdominal wall of the fetus is mixed with that coming from the hepatic veins and ductus venosus. The inferior vena cava enters the right atrium where the blood is diverted by the valve of the vena cava through the **foramen ovale** into the left atrium, where it mixes with a small quantity of blood returned by the pulmonary veins from the lungs. It then passes into the left ventricle from which it is pumped to the aorta which distributes it largely to the head, neck, and upper extremities. From the head, neck, and upper extremities the blood is returned by way of the superior vena cava to the right atrium where it may mix with a small quantity of the blood from the inferior vena cava. It passes next into the right ventricle and is pumped into the pulmonary trunk. Since the lungs of the fetus are inactive, only a small amount of the blood is distributed to them by the pulmonary arteries; the greater part passes through the ductus arteriosus into the aorta where it mixes with blood from the left ventricle. The aorta now distributes a part of this blood to the abdominal and pelvic viscera and to the lower extremities, the greater amount, however, goes to the placenta by way of the **umbilical arteries.**

From the preceding account of fetal circulation, these facts are evident:

(1) Only the umbilical vein and its branches to liver, portal vein, and ductus venosus contain "pure" blood, the placenta serving as respiratory, nutritive, and excretory organ.

(2) Nearly all the blood from the placenta goes through the liver before entering the inferior vena cava, accounting for the large size of the liver.

(3) The valve of the inferior vena cava in the right atrium functions effectively to direct the blood from the placenta and lower part of the body into the left atrium allowing that blood from the

superior vena cava to descend freely into the right ventricle.

(4) The blood from the umbilical vein and lower part of the body goes from left atrium to left ventricle where it is pumped to the aorta and most of it distributed to the head, neck, and upper extremities.

It is almost always true that the head end of an animal receives the best quality of blood available.

(5) The descending aorta contains mostly blood which has already been circulated to the upper part of the body.

Changes in the Cardiovascular Sys-

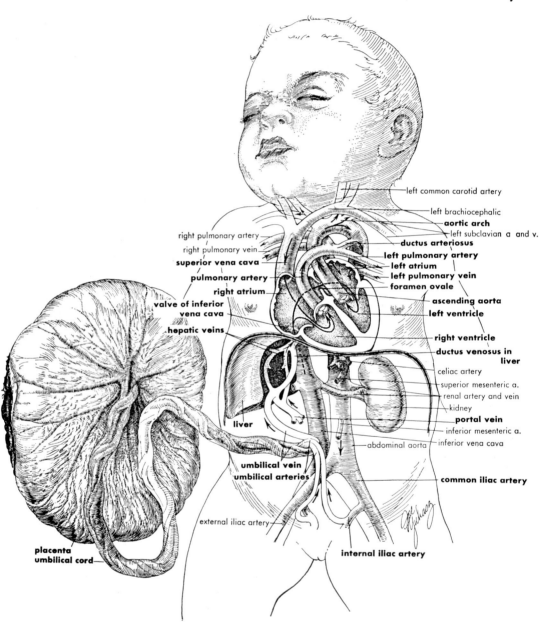

FIG. 12–33. General plan of the cardiovascular system in the fetus. The placenta and umbilical cord are shown at the left. Arrows indicate the direction of blood flow. Notice especially the route which the blood takes through the heart.

tem at Birth. At birth, when pulmonary respiration is initiated, the ductus arteriosus contracts and the blood previously shunted to the aorta now goes to the lungs. More blood, therefore, also returns to the left atrium and the flow through the foramen ovale decreases and finally ceases as the atrial intakes are equalized. The ductus arteriosus remains as a fibrous **ligamentum arteriosum.** The foramen ovale gradually closes and may be fully closed before the end of the first year, or a small opening may persist indefinitely. Its position is marked in the adult by the **fossa ovalis.** With the ligation and severing of the umbilical cord, flow in the umbilical vein and arteries ceases and the vessels are gradually converted to fibrous cords. The umbilical vein between the body wall and liver becomes the **ligamentum teres** (*round ligament*), the ductus venosus becomes the **ligamentum venosum** of the liver. The proximal portions of the umbilical arteries remain as the functioning internal iliac arteries; the distal portions extending between the urinary bladder and umbilicus under the peritoneum become the **lateral umbilical ligaments** of the bladder.

By these changes the cardiovascular system has become a complete double system. Failure of the ductus arteriosus or the foramen ovale to close, results, as we have stated before, in a "blue baby" and incomplete double circulation, incompatible with the health, or in severe cases, even with the life of the individual.

LYMPHATIC SYSTEM

Brief mention was made of the lymphatic system as a part of the circulatory system on pp. 278–279. Its component structures and functions were outlined. We return to it now for a closer view. After completing our study of the cardiovascular organs, we are perhaps better able to understand its relationships (Figs. 12–34, 35, 36).

The lymphatic system offers an alternative route for the return of **tissue fluid** to the blood stream. It may be said to have its beginnings in the networks of **lymph capillaries** which collect from the tissue fluids. From them the **lymphatic vessels** arise to conduct the lymph centrally, through **lymph nodes,** to the **right lymphatic** or the **thoracic ducts.** These ducts, emptying into large veins above the heart, represent the terminal part of the system. The lymphatic system is sometimes called a one-way system because it does not form a complete circuit like the cardiovascular system. It has no structures corresponding to the arteries.

Lymphatic Capillaries (Fig. 12–34). Like the blood capillaries, the lymphatic capillaries are the specific functional units of the system. They are simple endothelial tubes which begin blindly and form complex anastomoses in the tissues. They vary greatly in size from a few micra to almost a millimeter in diameter. They ordinarily contain no valves. Their distribution in the body is very uneven. They are most abundant in the inner and outer surface layers of the body such as the dermis of the skin, and the mucosal and submucosal layers of the digestive and respiratory systems, where they form continuous networks from nares and lips to the anal opening. They are also numerous under the serous membranes, both parietal and visceral. Muscles, bones, and fascia have a limited supply of lymphatic capillaries, while none are found in the central nervous system, meninges, epidermis, subcutaneous tissue, eyeball, cornea, internal ear, cartilage, and spleen.

Special lymphatic capillaries extend as blind ends into the intestinal villi and are known as **lacteals** because, during fat absorption from the intestine, the lymph they contain takes on a milky

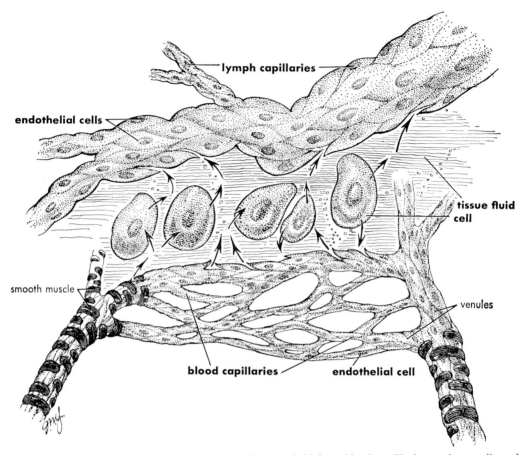

FIG. 12–34. Diagram to show the circulation of tissue fluid from blood capillaries to tissue cells and back into blood capillaries or into lymph capillaries.

appearance and is called **chyle.** They have a small amount of smooth muscle and are contractile.

Lymphatic Vessels (Fig. 12–35). The lymph flows from the lymphatic capillaries of the various parts of the body into lymphatic vessels, most of which are small and contain **valves.** The lymphatic vessels are so small and thin-walled that they are seldom seen in laboratory dissections. They are quite like veins in the tissue make-up of their walls, but are even thinner and the tunics are less well defined. They have more valves and frequent anastomoses. They travel in the loose connective tissue between organs, the subcutaneous and subserous tissues, and in the submucosa of the digestive,

respiratory, and urogenital tracts. They have a drainage pattern similar to that of veins.

The **thoracic duct** is the largest lymphatic vessel of the body. All of the lymphatic vessels except those of the right half of the head, neck, and thorax and the upper right extremity drain into it. It has its origin in a triangular sac, the **chyle cistern** (*cisterna chyli*), which lies in front of the second lumbar vertebra. The thoracic duct passes through the aortic hiatus of the diaphragm and enters the thorax. Opposite the fifth thoracic vertebra it turns toward the left side and ascends to the point of junction of the left internal jugular and left subclavian veins where it empties. The **chyle cistern**

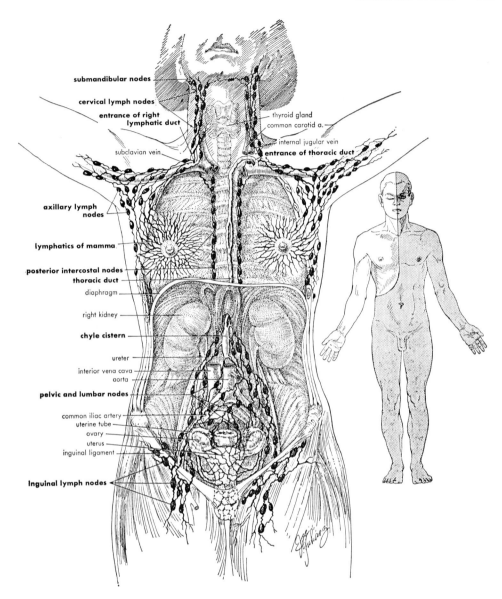

submandibular nodes

cervical lymph nodes

entrance of right
lymphatic duct

thyroid gland
common carotid a.

internal jugular vein

subclavian vein

entrance of thoracic duct

axillary lymph
nodes

lymphatics of mamma

posterior intercostal nodes
thoracic duct

diaphragm

right kidney

chyle cistern

ureter

interior vena cava
aorta

pelvic and lumbar nodes

common iliac artery
uterine tube
ovary
uterus
inguinal ligament

Inguinal lymph nodes

FIG. 12–35.   Scheme of lymphatic system.   Insert shows area drained by the thoracic duct in stipple, that drained by the right lymphatic duct in white.

(Fig. 12–35) receives lymph from the right and left lumbar lymphatic trunks and from the intestinal trunk. These together drain most of the lower parts of the body. The thoracic duct receives other tributaries as it ascends.

The **right lymphatic duct** drains the right upper part of the body and it enters the venous system at the junction of the
23

right internal jugular and right sub-clavian veins.

Movement of Lymph. The lymph flows from the tissues toward the large lymphatic ducts. The flow is maintained by the milking action of the muscle tissues of the body on the adjacent or contained lymphatic capillaries and vessels. The valves insure its moving in the

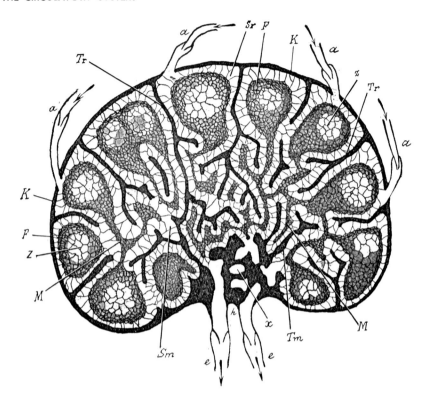

FIG. 12–36. Diagram of lymph node. *a*, afferent and, *e*, efferent lymphatic vessels with **valves;** the arrows indicate the direction of lymph flow; *F*, cortical tissue; *h*, hilus; *K*, capsule; *M*, medullary cords; *Sm*, medullary and, *Sr*, cortical sinuses; *Tm*, medullary trabeculae, continuous with those of the cortex; *Tr*, trabeculae originating in the capsule and dividing the cortex into ampullae; *x*, lymphatic vessels in the dense connective tissue of the hilus; *Z*, nodules. (Bloom and Fawcett, *A Textbook of Histology*, courtesy of W. B. Saunders Co.)

right direction. The difference in pressure at the two ends of the system is very influential, that in the thoracic region being zero or negative. The breathing movements aspirate lymph into the upper thorax. The lymph may also be moved by contractions of the vessel walls.

Lymphatic Nodes (Figs. 12–36, 37). Lymphatic nodes or glands are small oval or bean-shaped collections of lymphatic tissue interposed in the course of lymphatic and lacteal vessels. The tissue of the node is enclosed in a strong fibroelastic capsule from which septa or **trabeculae** push into the substance of the node to partially subdivide it into a number of compartments. A network of reticular fibers with reticulo-endothelial

cells extends from the trabeculae to all parts of the node. The outer part of the node, the **cortex,** contains closely packed masses of lymphocytes, the **lymph follicles,** and cords of cells occupy the center or **medulla.** Around the lymph follicles and permeating the whole gland are **lymphatic sinuses.** Several **afferent lymphatic vessels** enter the node on its convex surface and release lymph into the lymphatic sinuses of the node. It moves slowly through the node, the reticulo-endothelial cells, by phagocytosis, "filtering" out foreign particles, including bacteria, thus preventing their entrance into the blood stream. Lymphocytes produced in **germinal centers** of the lymph follicles enter the lymph

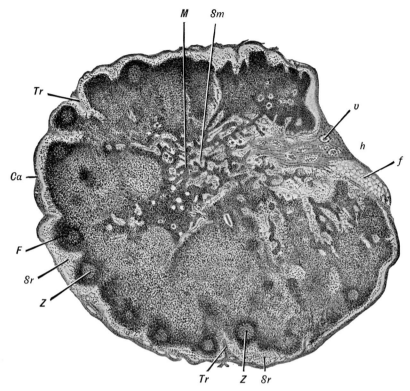

FIG. 12–37.   Section through small jugular lymph node of man. *Ca*, capsule; *F*, nodules with their centers (*Z*); *f*, fat tissue; *h*, hilus; *M*, medullary cord; *Sm*, medullary and, *Sr*, subcapsular sinus; *Tr*, trabeculae; *v*, blood vessel. × 18. (Redrawn and slightly modified from Sobotta.) (Bloom and Fawcett, *A Textbook of Histology*, courtesy of W. B. Saunders Co.)

stream.   At the hilus, fewer **efferent lymphatic vessels** leave the node to continue on toward the venous system. Valves present at the orifices of both afferent and efferent lymphatic vessels at the node insure the proper direction of flow.   Blood vessels enter and leave the nodes at the hilus.

In summary, the lymphatic system performs five functions.

(1) Produces lymphocytes.

(2) "Filters" the lymph and thus keeps foreign materials out of the blood.

(3) Receives the fatty food in absorption from the small intestine.

(4) Aids in the return of tissue fluid.

(5) Origin of immunological activity of body.

A general idea of the locations of lymphatic nodes and hence of lymphatic drainage can be obtained by studying Figure 12–35.   Fortunately, some of the nodes are palpable at the body surface and often offer clues as to the presence and location of infections.   Most of them are deep.   Among the nodes which are palpable are those in the neck region, in the axilla, and in the inguinal area.

Other Lymphatic Organs (Fig. 12–38).   We have noted other lymphatic structures in the course of our study of human anatomy.   The **tonsils** are located and described in the study of the digestive and respiratory systems.   Solitary and aggregated lymph nodules are discussed in connection with the intestinal tract.   The **thymus** is located above the heart in the upper thorax and lower neck regions. Its functions are not fully worked out but recent studies suggest that it produces

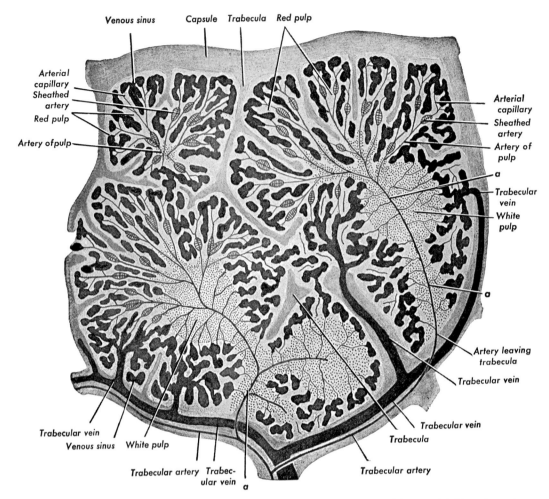

FIG. 12–38.   Diagram of lobules in the spleen.   Two complete lobules (right and left) and portions of two others (above and below) are shown.   *a*, arteries surrounded by lymphatic tissue (white pulp). (A.A.M.)   (Blood and Fawcett, *A Textbook of Histology*, courtesy of W. B. Saunders Co.)

lymphocytes which give rise to cells which have marked antibody potentialities.   All of these structures differ from the lymph nodes by not being in the lymphatic stream, but they do have lymphatic connections.   The spleen will be discussed at this time (Fig. 12–38).

The **spleen** is an organ about 12 cm. long and ovoid in shape which lies between the fundus of the stomach and the diaphragm and is protected anteriorly by the ninth, tenth, and eleventh left ribs.   Its diaphragmatic surface is convex and smooth; its medial surface is marked by three concavities named for the organs they contact, the gastric, renal, and colic.   It is almost completely invested in visceral peritoneum.   Its **hilus** is medial and slit-like and here it is connected to the greater curvature of the stomach by the **gastro-splenic** and to the left kidney by the **lieno-renal** **"ligaments."**   Its blood vessels enter at the hilus.

Inside the serous or peritoneal covering of the spleen is a fibroelastic coat which at the hilus is reflected inward upon the blood vessels as a sheath.

From it are given off in all directions small trabeculae which unite with each other to form the framework of the spleen. The small spaces enclosed by the trabeculae are the **areolae** and contain the **splenic pulp.** Because the framework of the spleen is so elastic the spleen is pliable and capable of great variation in size, depending upon demands placed upon it. There is also a small amount of smooth muscle in the fibroelastic framework. The splenic tissue is of two kinds; a **white pulp** of quite typical lymphatic tissue with reticular fibers, histiocytes, and lymphocytes arranged in nodules; a **red pulp** of strands and cords of cells with wide sinuses filled with blood in between the cords. It, too, contains histiocytes and many erythrocytes which give it its color.

The spleen is a large blood filter and reservoir placed in the pathway of a wide blood stream, the lienal (*splenic*) artery. It removes dead and worn-out erythrocytes, other debris, and bacteria. It manufactures lymphocytes and monocytes, and, during fetal life all types of blood cells. Red cells are stored in large quantities and discharged as needed. By contracting, it adds blood to the general circulation. It may also produce antibodies.

QUESTIONS

1. Why does man require a circulatory system?
2. What are the functions of the circulatory system as we find it in man?
3. Define closed and double circulation.
4. Distinguish between blood, tissue fluid, and lymph.
5. Describe the position and relationships of the heart.
6. Describe the valves of the heart, correlating their structure with function.
7. Define atrium, auricle, trabeculae carneae, fossa ovalis, and interatrial septum.
8. Name and briefly describe the layers in the heart wall.
9. What is the relationship of the parietal pericardium to the heart?
10. Diagram and label the conduction mechanism of the heart and explain how it functions.
11. Why is it necessary for the heart to have an extrinsic nerve supply?
12. What causes the "heart sounds"?
13. Describe the coats in the walls of large arteries and arterioles and relate the differences in them to the functions they perform.
14. Discuss the adaptations of capillaries for blood-tissue exchange.
15. List some of the important differences between veins and arteries.
16. A good way to test your knowledge of the relationships of blood vessels and the structure of the heart is to trace blood from one part of the body to another, naming all vessels, heart chambers, and valves in proper sequence. Practice on the following and then work out some more for yourself.
    (*a*) From the inferior mesenteric vein to the left coronary artery.
    (*b*) From the dorsal venous arch of the foot to the right pulmonary artery.
    (*c*) From the ascending aorta to the coronary sinus.
    (*d*) From the transverse sinus of the cranial cavity to the right atrium.
    (*e*) From the left ventricle to the circle of Willis—give alternate routes, if any.
    (*f*) From the arch of the aorta to the right ulnar artery.
    (*g*) From the dorsal venous arch of the hand to the right atrium.
17. Describe the portal circulation emphasizing how it differs from circulation elsewhere in the body.
18. How does the fetal heart differ from that of the adult heart?

19. What happens to the umbilical arteries, umbilical veins, ductus venosus and ductus arteriosus after birth?
20. Trace the course of a drop of lymph from a lymph capillary to the left brachiocephalic vein.
21. Describe the structure and function of a lymph node.
22. Compare the structure and function of the spleen with that of a lymph node.

For it's rare that a man thinks of anything so seriously as his dinner!

—Ben Jonson

Chapter 13

The Digestive System

INTRODUCTION

THE digestive system is the oldest of the body. In embryological development, it is initiated at the time of gastrulation. Phylogenetically, the gastrovascular cavity and its endodermal lining of the "gastrula-like" coelenterates, such as Hydra, mark its beginnings.

Basically, the digestive system of man, as of other vertebrates and most invertebrates, is a tube within a tube—an inner tube (Fig. 13–1). It is about 30 feet long in man and extends from mouth to anus with no membranous barriers to be crossed in traveling the whole route. The lumina (*insides*) of its organs are in direct communication with the outside world and are themselves, in a very real sense, a part of the external environment, though with their own peculiar ecologies. To gain access to the internal environment by whatever route—digestive, respiratory, integumentary, or excretory—material must pass through living membranous barriers—the linings of organs or the integument itself. Therefore, food, before

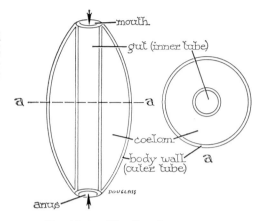

FIG. 13–1. The digestive system—
a tube within a tube.

it is available to the body, must be **digested** to enable it to pass through the walls of the stomach or intestine (*absorption*) to reach the blood or lymph components of the internal environment of the body.

Man is selective about the food which he introduces into the mouth, but nevertheless he is biologically considered to be omnivorous. He eats all. Actually, because he is an animal of culture, he is influenced in his selection and processing of food by the culture he, by accident, is born into. He may thus become the victim of bad cultural habits and accept a diet which is not the best for him. To some degree the digestive system protects him by refusing items which he introduces or too large quantities of them. Sometimes the refusal is direct and dramatic as indicated by regurgitation, or indigestible materials are simply egested (*eliminated*) at the anal aperture. Unfortunately, it does not give full protection, for man does at times "eat himself to death" or poison himself by careless introductions at the mouth.

Another factor, a very realistic one in man's world, is economic. In our so-called civilized world we are so indoctrinated or educated, and so numerous that to eat it is necessary to have the wherewithal to pay for our food, or become a ward of society. Uncivilized man, as we choose to call him, may still be sufficiently subject to the laws of nature and its controls that, if left alone, he feeds directly from the fruits of the earth and may even be happy in doing so.

These observations point up the functions of the digestive system and something of its basic structural arrangements. Some of the economic problems relative to the feeding of man are suggested, for they, too, must be taken more seriously in the future if man is to survive and enjoy the experience.

Organs of Digestion (Fig. 13–2). The organs of digestion are those that carry out the functions suggested above—the functions of **ingestion, digestion, absorption,** and **egestion.** Ingestion is merely the taking in of food at the mouth. Digestion involves a series of changes, some mechanical, as chewing; others chemical, through the agency of enzymes, whereby the large molecules of food are reduced to smaller, simpler, or soluble molecules capable of being absorbed. The passing of the food through the walls of digestive organs to reach the blood and lymph is absorption. For adequate absorption, the digestive system must provide extensive surfaces and, as we shall see, some of the organs have special means for providing such surface areas. Egestion is elimination at the inferior end of the digestive tube, through the anus. Another problem which the digestive system must meet is that of moving the food mass through the tubes at a pace commensurate with efficient digestion and absorption.

The organs which together carry out these functions are the digestive organs, or the organs of the alimentary canal and accessory glands. Listed from the superior to the inferior end in proper sequence they are the mouth, pharynx, esophagus, stomach, small intestine, large intestine, the rectum, anal canal, and anus. The

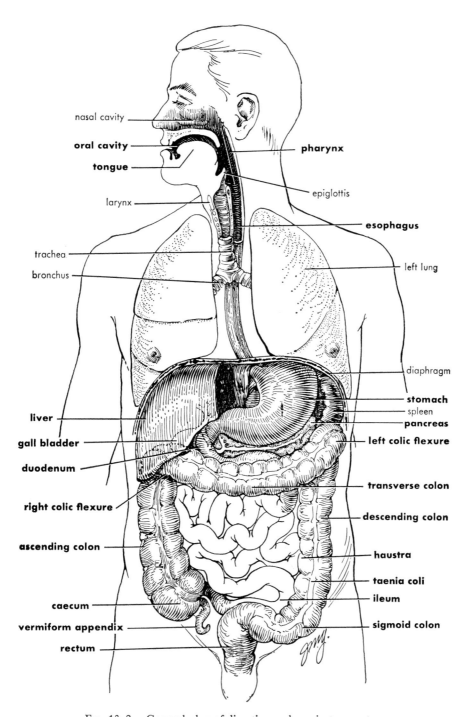

Fig. 13–2. General plan of digestive and respiratory systems.

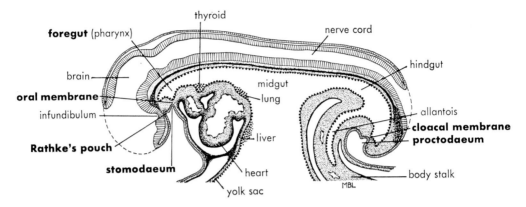

Fig. 13–3. Sketch of a sagittal section of a human embryo of about 22 days to show stomodaeum and proctodaeum and related structures. (Modified from Patten, *Human Embryology*.)

macroscopic accessory glands are the salivary glands emptying into the mouth, and the liver and pancreas which empty into the small intestine. Many types of microscopic glands are, of course, found within the walls of the digestive tract and will be considered with the appropriate organs.

DEVELOPMENT OF THE DIGESTIVE SYSTEM (Fig. 13–3)

In Chapter 4, page 31, the initial stages in the development of the digestive system were described. The embryo, as it was left at that point, was a cylindrical structure with an outer ectoderm, and inside the beginnings of a digestive tube lined with **endoderm** (Fig. 4–7). The digestive tube extended well forward into the head process as a blind **foregut** and back into the tail region as a blind **hindgut.** In between foregut and hindgut the ventral wall opened out through the intestinal portal to the yolk sac. This central area is sometimes called the midgut, but is short-lived in man for by the end of the fifth week the intestinal portals have come together and closed and the yolk sac has become detached from the gut. The foregut and hindgut now grow in length and breadth to keep up with the growth of the embryo.

Each end of the primitive endodermal gut comes into contact ventrally with the ectoderm to form the **oral** and **cloacal membranes.** The oral membrane is at the bottom of a deep depression called the **stomodaeum** which is bounded by maxillary, mandibular, and frontonasal processes (Fig. 4–15). During the fourth week of development, the oral membrane breaks through and the stomodaeum and foregut become continuous. After the oral membrane breaks, it is impossible to tell the exact boundaries of the ectoderm and endoderm. It is likely that the epithelium of the vestibule, oral cavity, nasal passageways, palate, and front of tongue all come from the ectoderm. Also, the ectoderm gives rise to the primary tissue of the salivary glands, the enamel of the teeth, and through an invagination at the back of the stomodaeum called **Rathke's pouch,** to the anterior lobe of the hypophysis (*pituitary*).

At the caudal end of the hindgut the **cloaca** develops. It is a chamber common to the digestive, urinary, and reproductive systems in many lower vertebrates and a transitory embryonic structure in man. The allantois communicates with the cloaca, although it developed much earlier. Before the urinary and genital connections with the cloaca are established, it begins to divide into a dorsal rectum and a ventral bladder and urogenital sinus. The cloacal membrane also separates into anal and urogenital portions, which soon break through forming two openings, the urogenital sinus and anal canal. The cloacal membrane, having pushed in toward the rectum before rupture (the **proctodaeum**), contributes some ectodermal lining to the future anal canal (Fig. 13–3).

The primitive gut, being made up mainly of endoderm, with minor ectodermal contributions at each end, is the **primary alimentary canal.** It differentiates into three main segments, the **mouth, pharynx** and **digestive tube,** the latter including the esophagus, stomach, small intestine, large intestine, rectum, and anal canal. It should be clearly understood that the primary alimentary canal gives rise only to the **epithelial linings** of the above organs, and to the primary **epithelial tissues** of such structures or organs that develop from it by pushing outward into surrounding areas. Examples of such structures are the salivary glands, liver, and pancreas—large glands whose development will be elaborated upon later; tonsils, thyroid, parathyroids and thymus, developed from the pharyngeal region; the respiratory tract, discussed in the next chapter; and finally, a large variety of microscopic glands confined to the walls of the alimentary canal, from mouth to anus.

The outer or accessory layers of the alimentary canal, the connective tissues and muscle, are **secondary investments** derived from the splanchnic layer of the mesoderm (Fig. 4–10). While studying this diagram, notice the relationship of the alimentary canal to the coelom and the dorsal and ventral mesenteries. This applies, as we shall see, only to that part of the alimentary canal inferior to the diaphragm.

It is beyond the scope of this book to pursue in detail the further development of the digestive system. In a few instances where it does seem important, further discussion is introduced. In general, from this point on in development, the digestive tube elongates; it enlarges in some areas, constricts in others, develops a variety of glands and special structures which increase its efficiency in digestion and absorption. In accordance with general development of the body, differentiation and specialization start sooner and progress faster at the superior or cranial end of the body.

MOUTH (Figs. 13–4, 5)

The mouth is divided into two parts, the first appropriately called the **vestibule** lies just inside the **orifice** and is the narrow interval between the lips and cheeks externally, and the gums and teeth internally. The second and major part of the mouth, the **oral cavity,** lies central to the gums and teeth and is limited superiorly by the hard and soft palates, inferiorly by the tongue, the lower jaw and intermediate mucous membranes, and posteriorly by the glosso-palatine arch. The mouth opens posteriorly through the **isthmus of the fauces** into the **oropharynx.**

When the mouth is closed, communication between the vestibule and oral cavities is through **interdental spaces** and clefts behind the molar teeth, the **diastemata.** This is important in cases where liquid feeding is necessary, as when spasms of the muscles of mastication lock the jaws or when a jaw is fractured and wired together.

Lips and Cheeks. The **lips** are two highly movable folds that surround the mouth orifice. Their musculature has been described on page 197. They are covered externally by skin, internally by moist mucous membrane. The intermediate or **red margin** is covered with dry, nonglandular, translucent mucous membrane, the red color being due to

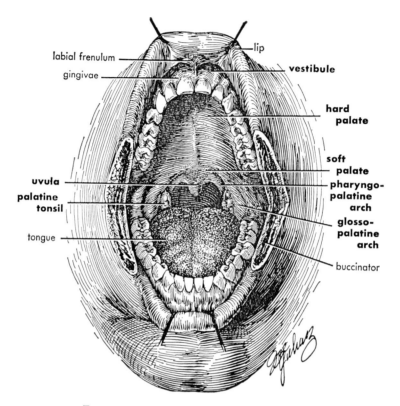

labial frenulum

gingivae

lip

vestibule

hard palate

soft palate

pharyngo-palatine arch

glosso-palatine arch

buccinator

uvula

palatine tonsil

tongue

FIG. 13–4.   Anterior view of the mouth cavity.

blood showing through from the underlying capillaries. The lips are highly vascular and extremely sensitive. Internally, the submucosa contains closely packed **labial glands** which empty mucus into the vestibule. Both the upper and lower lips have medial folds of mucous membrane which run from the gums to the inner surface of the lips, the superior and inferior **labial frenula**. They can easily be felt by using the tip of the tongue. Some individuals have smaller folds between the cheeks and gums, the buccal frenula. These can be felt by tongue in cheek.

The **cheeks** are similar in structure to the lips and have the **buccinator** as their principal muscle (page 197). The **buccal glands,** smaller than the labial glands but otherwise like them, lie in the submucosa. The subcutaneous tissue of the

cheek is quite loose and may contain considerable fat. The duct of the parotid gland opens through the cheek by means of a palpable pore at the level of the second upper molar tooth.

The lips and the cheeks play important roles in moving the food between vestibule and oral cavity during mastication and in the articulation of speech.

Gingivae and Teeth (Figs. 13–4 to 7). The **gingivae** are the gums and consist of a dense, fibrous connective tissue that attaches to the underlying bone—the alveolar margins of maxillae and mandibles. It also attaches to the necks of the teeth. It is covered with a smooth mucous membrane which is continuous with the mucous membrane of the lips and cheek and is reflected into the tooth sockets where it is continuous with the periosteum (*periodontal membrane*).

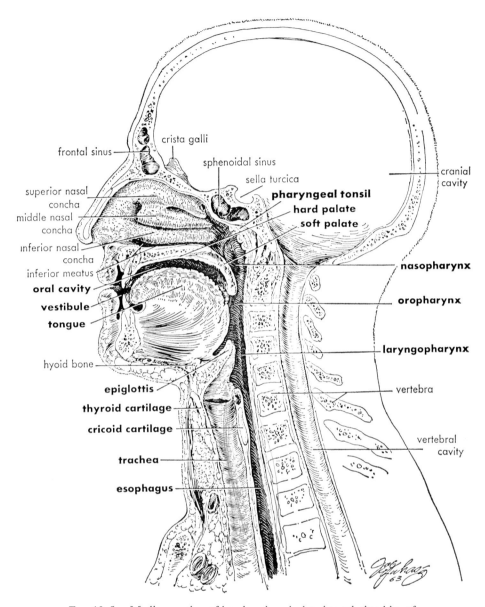

crista galli
frontal sinus
sphenoidal sinus
sella turcica
superior nasal concha
middle nasal concha
inferior nasal concha
inferior meatus
oral cavity
vestibule
tongue
hyoid bone
epiglottis
thyroid cartilage
cricoid cartilage
trachea
esophagus

pharyngeal tonsil
hard palate
soft palate

cranial cavity

nasopharynx

oropharynx

laryngopharynx

vertebra

vertebral cavity

FIG. 13–5.   Median section of head and neck showing relationships of
upper digestive and respiratory systems.

The **teeth** are the organs of mastication and are also influential in articulate speech. They rest in **alveoli** (*sockets*) along the alveolar margins of the maxillae and mandible and form the superior and inferior **dental arches.** Each tooth is made up of an exposed portion, the **crown,** a **root** which embeds in the bony alveolus, and between these, at the gum

level, a slightly constricted **neck.** The root is provided at its end with an **apical foramen** which is the entrance to a **root canal** leading to the **pulp cavity.** Blood vessels and nerves enter the apical foramen and pass to the pulp cavity, where a rich vascular connective tissue makes up the **pulp.** The root is covered with a **cementum** which is connected to the

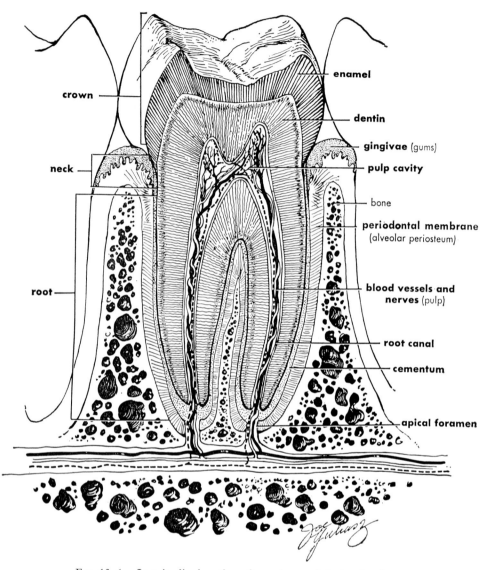

FIG. 13–6.   Longitudinal section of a molar tooth in its alveolus.

alveolar bone by the **periodontal membrane.** **Dentin** makes up the bulk of the tooth, and the crown is covered with **enamel** (Fig. 13–6).

Teeth of man are varied in form. The front teeth or **incisors** are chisel-shaped for cutting; to either side of these are the more pointed **canines** for tearing, followed by the **premolars** or bicuspids and finally by the **molars.** Premolars and molars have broad crowns with tubercles or cusps for grinding purposes (Fig. 13–7).

Two sets of teeth appear in the life of the individual, a **deciduous** (*milk, temporary*) set, followed by the **permanent** teeth. There are twenty deciduous teeth and thirty-two permanent ones, as indicated by the dental formulae, on page 349, taken from *Gray's Anatomy.*

There is considerable variation in the time of eruption of deciduous and permanent teeth. The listing on page 350 gives approximate times.

The time of tooth eruption is dependent

Deciduous Teeth

	Mol.	Can.	In.	In.	Can.	Mol.	
Upper Jaw	2	1	2	2	1	2	} Total
Lower Jaw	2	1	2	2	1	2	20

Permanent Teeth

	Mol.	Pr. Mol.	Can.	In.	In.	Can.	Pr. Mol.	Mol.	
Upper Jaw	3	2	1	2	2	1	2	3	} Total
Lower Jaw	3	2	1	2	2	1	2	3	32

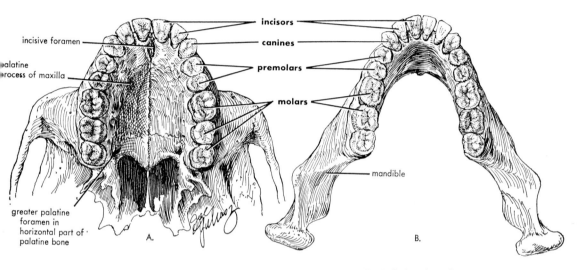

Fig. 13–7.   Permanent teeth.   A, Superior dental arch.   B, Inferior dental arch.

in part upon endocrine and nutritional conditions. Hypothyroidism or nutritional deficiency, as lack of vitamin D, may retard growth. The third molars are often delayed by impaction and may not erupt until the twenty-fifth year or sometimes not at all. Tooth eruption may be earlier than expected in some individuals.

Practical Implications. Both deciduous and permanent teeth initiate their development during embryonic or fetal life, the former at about six weeks, the latter at four months. Calcification of deciduous teeth takes place partly during intrauterine life and partly after birth. Permanent teeth calcify during infancy and childhood. The calcium for deciduous teeth may be derived in part by withdrawal from the bones of the mother, but no evidence is available that it is withdrawn from her teeth. After birth, the child's diet should be rich in calcium, phosphorus, and vitamin D to insure sound tooth development. It has also been shown that after tooth development is complete, the level of calcium in the diet does not affect tooth decay. While tooth decay (*dental caries*) is the most common ailment of teeth, the consensus seems to be that a high-carbohydrate diet more than anything else favors bacterial growth and may cause caries.

Frequent examination of the teeth to insure early recognition of dental caries, as well as other tooth problems, is important.

Tongue (Figs. 13–4, 5, 8, 9). The tongue is a highly muscular organ and

Tooth Eruption	Deciduous		Permanent	
	Mandibular (*Mo.*)	*Maxillary* (*Mo.*)	*Mandibular* (*Yr.*)	*Maxillary* (*Yr.*)
Central incisors	6	8	6–7	7–8
Lateral incisors	7	9	8	8–9
Canines	16	19	10	11
First premolars			9	10
Second premolars			10	11
First molars	12–14	13–14	6	6–7
Second molars	20–24	20–24	11–13	12–13
Third molars			17–21	17–21

therefore very versatile in its movements. These movements are put to a variety of uses as **mastication, deglutition** (*swallowing*), **articulate speech** and even **whistling,** the latter not an unfamiliar means of communication among the younger set and a device for giving expression to certain emotions. Another important function of the tongue, with receptors in the mucous membrane, particularly in some of the papillae, is the **sense of taste.** The diversity of activities of the tongue testify to its importance.

The muscles of the tongue fall nicely into two groups—the extrinsic and intrinsic muscles (Fig. 11–12). The extrinsic muscles originate on three widely separated bones, the mandible, styloid processes of the temporal, and the hyoid, thus pulling the tongue in different directions. The intrinsic muscles, i.e. those entirely within the tongue, are divided into **superior** and **inferior longitudinal, transverse,** and **vertical** and by combining their actions one can modify, in a multitude of ways, the form of the tongue.

Besides the attachments of the tongue through its extrinsic muscles, it is attached by a fold of mucous membrane to the epiglottis and to the soft palate by the glossopalatine arches (*pillars of the fauces*). Ventrally, the tongue is connected to the floor of the mouth by a fold of mucous membrane, the **lingual frenulum** (Fig. 13–10). If this frenulum is so short as to hamper the movements of the tongue in articulate speech, the condition is called **tongue-tie.**

The mucous membrane of the dorsum of the tongue is as diverse as its muscles. A V-shaped groove, with the apex of the V directed backward, separates the anterior two-thirds from the posterior one-third of the tongue. This groove is the **terminal sulcus.** Over its anterior part the mucous membrane is thin and closely attached to the underlying muscular tissue and contains numerous tiny papillae. On the posterior one-third, the mucous membrane is thick and freely movable and its submucosa contains an aggregate of lymph nodules, forming the **lingual tonsil.** The surface of the anterior two-thirds is marked by a variety of papillae which give it a characteristic texture. They are the vallate (*circumvallate*), fungiform, filiform, and even a few rudimentary foliate papillae. There are only about twelve of the large **vallate papillae** which are so placed as to form a V-shape, just in front of the V-shaped **sulcus terminalis** (Fig. 13–8). The vallate papillae are flattened structures with a moat-like trough around each one. They are richly supplied with taste buds. **Fungiform papillae** are most numerous on the sides and tip of the tongue and more irregularly spaced over its surface. They are globular and highly vascular, hence have a pinkish color. **Filiform papillae** are small and slender and have

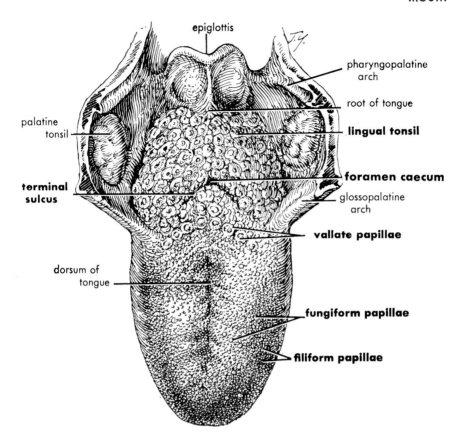

FIG. 13–8.   The dorsum of the tongue and related structures.

a scaly surface. They are quite evenly distributed over the anterior two-thirds of the tongue (Fig. 19–5).

Back of the terminal sulcus is a small pit on the center of the tongue—the **foramen caecum,** which marks the junction between the embryonic primary (*anterior two-thirds*) and the secondary (*posterior one-third*) portions of the tongue and is itself the point of attachment of the embryonic thyroglossal duct which gives rise to the thyroid gland.

The tongue also has a number of glands opening on its surface—**serous** (von Ebner's) **glands** in the region of the vallate papillae, **mucous glands** at the back of the tongue, and **mixed glands** on the under-surface near the tip.

Palate (Figs. 13–4, 8). The palate is divided into the hard palate anteriorly,

24

which is underpinned by the horizontal processes of the maxillae and palatine bones, and the soft palate posteriorly. They form the roof of the mouth. The **hard palate** also supports the floor of the nasal cavities. The hard palate is covered with a dense mucous membrane fused to the underlying periosteum, which forms a midline ridge, the **palatine raphe.** Anteriorly, this raphe ends in the **incisive papilla** over the region of the incisive foramen. Transverse ridges also cross the anterior part of the hard palate, the **palatine rugae. Palatine glands** are found in the submucosa at the posterior part of the hard palate.

The **soft palate** is a flexible musculo-membranous curtain which projects backward and downward from the hard palate with its superior and posterior sur-

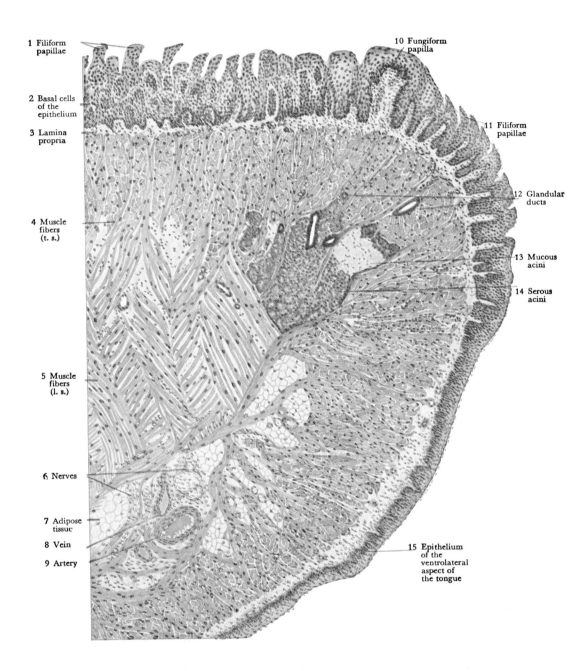

1 Filiform papillae

2 Basal cells of the epithelium

3 Lamina propria

4 Muscle fibers (t. s.)

5 Muscle fibers (l. s.)

6 Nerves

7 Adipose tissue

8 Vein

9 Artery

10 Fungiform papilla

11 Filiform papillae

12 Glandular ducts

13 Mucous acini

14 Serous acini

15 Epithelium of the ventrolateral aspect of the tongue

FIG. 13–9. Tongue (panoramic view, transverse section). Stain: hematoxylin-eosin. 38×.
(di Fiore, *An Atlas of Human Histology*, Lea & Febiger.)

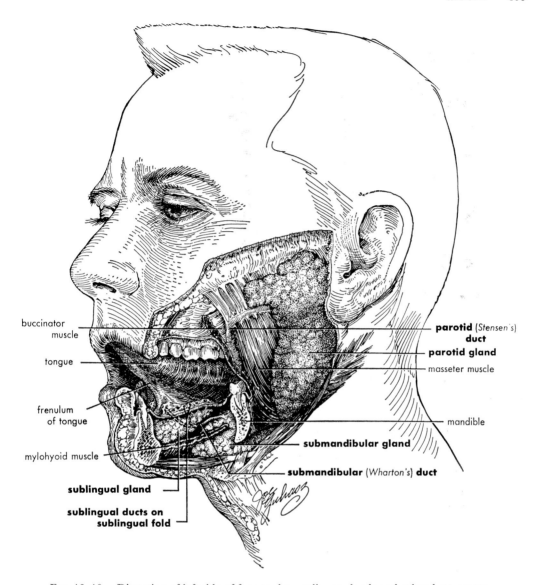

buccinator muscle

tongue

frenulum of tongue

mylohyoid muscle

sublingual gland

sublingual ducts on sublingual fold

parotid (*Stensen's*) **duct**

parotid gland

masseter muscle

mandible

submandibular gland

submandibular (*Wharton's*) **duct**

FIG. 13–10. Dissection of left side of face to show salivary glands and related structures.

face toward the pharynx and its inferior and anterior surface toward the mouth. It attaches to the tongue laterally by the **glossopalatine arches** and to the lateral wall of the oropharynx by the **pharyngopalatine arches**. From the free margin of the soft palate medially is a small, conical projection, the **uvula** (Fig. 13–4). Lateral to the opening into the pharynx and between the glossopalatine and pharyngopalatine arches is a **fossa** housing the **palatine tonsils**. The muscles of the soft palate should be reviewed on page 200.

Salivary Glands (Figs. 13–10, 11). There are three pairs of salivary glands, the parotid, submaxillary, and sublingual.

The **parotid gland** lies below and in front of the external ear and within the cervical fascia. Its duct, the **parotid** (*Stensen's*) **duct** crosses the Masseter muscle and, at its anterior border, bends

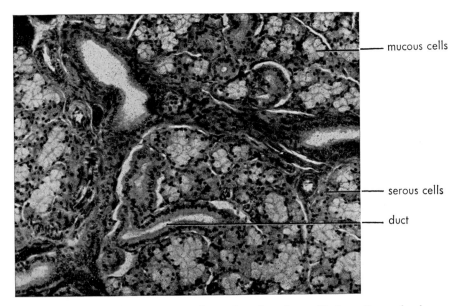

mucous cells

serous cells

duct

Fig. 13–11.   Photomicrograph of a section of a submandibular salivary gland.

inward to penetrate the cheek and open at the level of the second upper molar tooth.

The **submandibular gland** lies below the body of the mandible and the mylohyoid muscle. Its duct (*Wharton's*) arises from a lobe of the gland under the floor of the mouth from which it pushes upward and forward to open on the underside of the tongue in the floor of the mouth. The **sublingual gland** lies in a fold of mucous membrane—the **sublingual fold** in the floor of the mouth. It empties its secretions by a number of small ducts on the sublingual fold, often by a major duct emptying alongside the submandibular duct. The other glands of the mouth mentioned above also contribute to the saliva.

The salivary glands are of the compound tubuloalveolar type, consisting of the glandular tissue or parenchyma and supporting connective tissue elements. The alveoli are composed of cuboidal glandular cells, the excretory ducts of the glands of pseudostratified columnar epithelium, and the various intermediate ducts of simple columnar or simple

squamous epithelium. The secretory cells are all of the serous type in the parotid gland, whereas both **serous** and **mucous** cells are found in the submandibular and sublingual glands.

The salivary glands are activated only by stimuli associated with food, whereas the other mouth glands, which contribute to saliva, secrete continuously to keep the mucous membranes moist. About 1500 cc. of saliva are secreted daily by all of the above glands. Saliva aids in mastication, in the dissolving of the food which makes taste possible, and in lubricating the bolus (*food mass*) for swallowing. It also, through the enzyme **ptyalin,** initiates carbohydrate digestion. Through buffering agents it tends to keep the contents of the mouth about neutral.

PHARYNX (Figs. 13–5, 12)

The pharynx is an organ which is common to the digestive and respiratory systems. It lies behind the nasal cavities, the mouth, and the larynx, hence the names of its three parts: **nasal pharynx,**

oral pharynx, and **laryngeal pharynx.**
It extends from the underside of the
skull to the level of the sixth cervical
vertebra behind, and the cricoid cartilage of the larynx in front.   Its posterior wall is continuous and is connected by loose areolar tissue to the
cervival vertebral column and the prevertebral fascia.   Its anterior wall is
discontinuous, having openings into the
nasal cavities by the **choanae;** the mouth,
by the **isthmus of the fauces;** and the
larynx by the **laryngeal orifice.**   It is
continuous inferiorly into the esophagus.
It communicates with seven cavities:
two nasal, two tympanic, the mouth, the
larynx, and the esophagus.

Nasal Pharynx.   The **nasal pharynx**
is primarily respiratory, communicating
anteriorly with the nasal cavities through
the choanae.   On its lateral walls it
receives the openings of the **auditory**
(*Eustachian*) **tubes** which connect with
the tympanic cavities of the middle ears.
Behind each auditory tube opening is a
raised area, the **torus,** from which prominent folds of the mucous membrane
extend downward to the soft palate and
pharyngeal wall.   Behind and above the
torus is the deep **pharyngeal recess.**   On
the posterior wall is a mass of lymphoid
tissue, the **pharyngeal tonsil.**   Below,
the nasal pharynx communicates with
the oral pharynx through the somewhat
narrowed pharyngeal isthmus.

Oral Pharynx.   The **oral pharynx**
extends from the soft palate to the level
of the hyoid bone.   It receives food from
the mouth through the isthmus of the
fauces and air from the nasal pharynx.
In its side walls between the two palatine
arches is the **palatine tonsil** mentioned
earlier.

Laryngeal Pharynx.   The **laryngeal
pharynx** extends from the hyoid bone to
the esophagus.   It communicates superiorly with the oral pharynx and anteriorly with the laryngeal aperture.   It is
here related to the epiglottis and the
aryepiglottic folds, and the arytenoid
and cricoid cartilages (Fig. 13–12).

There are three coats in the walls of the
pharynx—an inner mucosa, a middle
fibrous, and outer muscular.   The **mucosa**
has a surface of **stratified squamous epithelium** in the oral and laryngeal parts
of the pharynx, as in the mouth.   The
nasal pharynx has a **ciliated pseudostratified epithelium.**   The lamina propria of the mucosa consists of reticular
and loose fibrous connective tissue with
an abundance of lymphoid tissue, diffuse
in some areas and aggregated in others,
as in the tonsils.   Dense fibrous connective
tissue with many elastic fibers is present
between the mucosa and the muscular
layer, the **pharyngobasilar fascia.**   The
muscles have been described in Chapter
11, page 202.

Tonsils (Figs. 13–4, 8, 12).   Tonsils
are formed of aggregates of lymphatic
follicles.   Three tonsils—the **palatine,
lingual,** and **pharyngeal,** have been
located in the pharynx.   The palatine
tonsils are compound aggregates of lymphatic follicles encapsulated in dense
fibrous connective tissue which is continuous with that of the submucosa.   The
free surface is covered with stratified squamous epithelium which is invaginated
into the underlying lymphatic tissue to
form **tonsilar crypts** of varying depths.
The crypts may divide and subdivide
through the tonsils.   The lamina propria
is made up mostly of lymphocytes which
obscure the fibrous tissue.   Lymphatic
follicles develop many germinal centers
beneath the tonsilar epithelium, and
lymphocytes escape into the epithelium
and crypts.   The crypts also accumulate
desquamated epithelial cells, detritus and
bacteria.

The pharyngeal tonsils have much the
same structure as the palatine except that
their covering epithelium is pseudostratified.   Enlargement of the pharyngeal tonsil is called "adenoids" and may,
as it pushes forward into the nasal

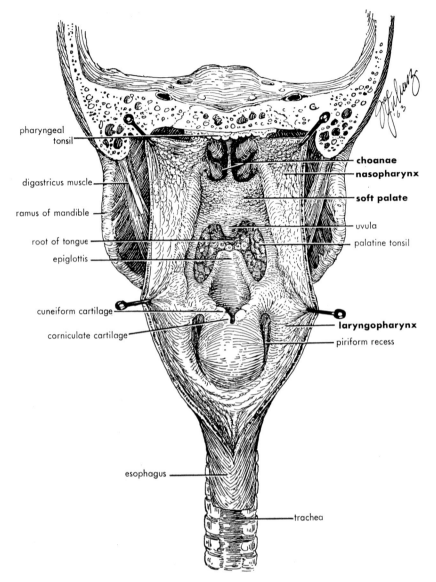

pharyngeal tonsil

digastricus muscle

ramus of mandible

root of tongue

epiglottis

cuneiform cartilage

corniculate cartilage

esophagus

choanae

nasopharynx

soft palate

uvula

palatine tonsil

laryngopharynx

piriform recess

trachea

FIG. 13–12.   Pharynx.   Posterior wall cut and reflected laterally to reveal the orifices and structures of the anterior wall.

pharynx, interfere with breathing; consequently, it is a cause of mouth breathing.

The tonsils form a ring of lymphoid tissue around the upper digestive and respiratory systems and are considered to be of a protective nature against infection. In the process, however, they may become infected and it often becomes necessary to remove them surgically. Unlike other aggregates of lymphatic follicles, such as the lymph nodes, the tonsils lack afferent lymphatic vessels. Their efferent vessels convey great numbers of lymphocytes to the blood stream and this is perhaps the chief function of tonsils—lymphocyte formation.

Mechanisms of Chewing and Swallowing (Fig. 13–5).   The chewing or **mastication** of food is peculiar to mammals, among the vertebrates. The cheeks

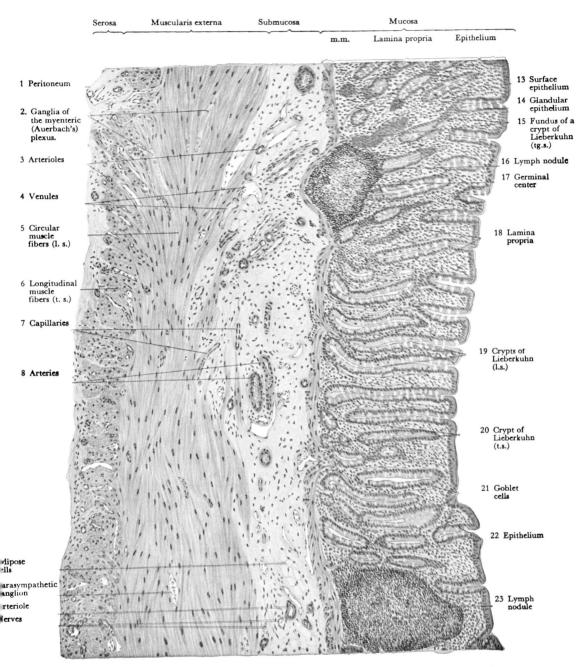

Serosa   Muscularis externa   Submucosa   Mucosa

m.m.   Lamina propria   Epithelium

1 Peritoneum

2. Ganglia of
the myenteric
(Auerbach's)
plexus.

3 Arterioles

4 Venules

5 Circular
muscle
fibers (l. s.)

6 Longitudinal
muscle
fibers (t. s.)

7 Capillaries

8 Arteries

Adipose
cells

Parasympathetic
ganglion

Arteriole

Nerves

13 Surface
epithelium

14 Glandular
epithelium

15 Fundus of a
crypt of
Lieberkuhn
(tg.s.)

16 Lymph nodule

17 Germinal
center

18 Lamina
propria

19 Crypts of
Lieberkuhn
(l.s.)

20 Crypt of
Lieberkuhn
(t.s.)

21 Goblet
cells

22 Epithelium

23 Lymph
nodule

FIG. 13–13.   Large intestine: wall (transverse section).   Stain: hematoxylin-eosin.
53×.   (di Fiore, *An Atlas of Human Histology*, Lea & Febiger.)

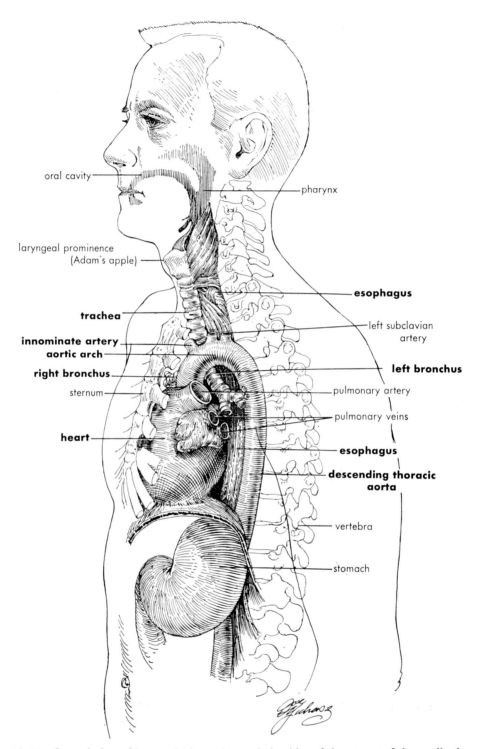

oral cavity

pharynx

laryngeal prominence
(Adam's apple)

esophagus

trachea

left subclavian
artery

innominate artery
aortic arch

left bronchus

right bronchus

pulmonary artery

sternum

pulmonary veins

heart

esophagus

descending thoracic
aorta

vertebra

stomach

FIG. 13–14.   Lateral view of human body to show relationships of the organs of the mediastinum.

(*and lips*) which help to keep the food in the mouth when it is open are important aspects of the mechanism. Besides this, there are molars for grinding food and an arrangement of muscles and articulations which enables the jaws not only to open and close, but to move sideways, forward, and back. When one experiences the paralysis of some of the muscles of mastication, those of the cheeks, for example, and cannot even keep the food in the mouth for chewing, he realizes the importance of the mechanism and its close coordination.

Swallowing or **deglutition** is even more involved than chewing. To analyze it serves as a review of the anatomy which has been previously described in this and other chapters (Figs. 11–13, 14). The mechanism once set into action functions as a reflex. It is usually initiated voluntarily. The first step is the forcing of the bolus through the isthmus of the fauces by the pressure of the tongue against the palate, the elevation of the mandible, hyoid, and thyroid cartilage, and the compression of the cheeks by the buccinator muscle. Secondly, the bolus moves from the oral pharynx down into the esophagus. In this procedure the pharynx is first expanded and elevated by the Stylopharyngeal muscles; the isthmus of the fauces is closed by the Glossopalatine and Pharyngopalatine muscles; the soft palate is elevated to shut off passage into the nasal pharynx by action of the Levator and Tensor palatine and the muscle of the uvula; and the superior laryngeal aperture is closed by bringing the arytenoid cartilages of the larynx closer to the epiglottis by a complex of muscles not described in this book. It was thought at one time that the epiglottis was brought down over the laryngeal aperture like a cover on a box. By placing a finger on the "Adam's apple" (*thyroid cartilage*) and swallowing, one will find that the cartilage, and therefore the larynx, does move upward toward

the epiglottis. The Mylohyoid, Thyrohyoid, and Geniohyoid muscles aid in this movement. Now the pharyngeal constrictors contract, and since the only passage left open is the esophagus, the food passes into it. In the last step, rings of constriction of the esophagus preceded by rings of relaxation, a process known as **peristalsis,** moves the bolus down the esophagus and beyond the control of voluntary muscles. That this complex chain of events can be interrupted is indicated by the not uncommon incident, perhaps when not giving enough attention to one's eating, of food getting into the larynx, which sets off another reflex response causing one to cough and forcefully eject the irritating substance, sometimes with embarrassing consequences.

DEVELOPMENT OF THE PALATE AND PHARYNX

Palate. The development of the mouth was outlined in an earlier section of this chapter. To clarify some of the major relationships in this area relative to both the digestive and respiratory systems, a brief consideration is here given of the development of the **palate.**

The palate forms the roof and part of the posterior wall of the mouth. It is a device for separating nasal passageways from the mouth, enabling young mammals to suck and breathe and, later, to chew at the same time. Two olfactory pits appear above the mouth orifice and maxillary processes and enlarge to form blind sacs which remind the zoologist of the nasal sacs of sharks (Fig. 4–15). As the sacs deepen, they come to lie over the roof of the primitive mouth and are separated from it only by a thin membrane. These membranes rupture at about the seventh week, forming internal nares, the **primitive choanae,** which open into the oral cavity—a permanent condition in amphibians. However, in

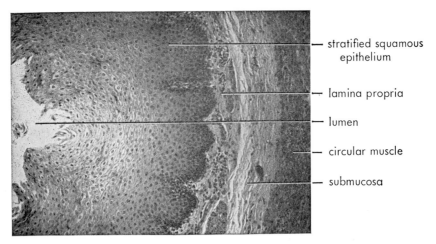

— stratified squamous epithelium

— lamina propria

— lumen

— circular muscle

— submucosa

FIG. 13–15.   Cross-section of a part of the wall of the esophagus.   100×.

mammals, the nasal passageways become separated from the mouth by means of a horizontal partition that subdivides the primitive mouth cavity, and at the same time extends the nasal passageways back to communicate with the pharynx by **secondary** or **permanent choanae.** This horizontal partition is the palate. Its forepart is formed from **palatine processes** from the region of the maxillary process. These unite in the midline from in front backward and superiorly unite with the nasal septum. They ossify in their forward part to form the **hard palate,** but caudad remain membranous to form the **soft palate.** Failure of these lateral palatine processes to unite on the midline results in **cleft palate.** The extent of cleft palate varies greatly and often is accompanied by a **cleft lip** or **harelip** (Fig. 4–22).

Pharynx. One gets little appreciation and understanding of the pharynx and its contributions to adult anatomy, unless he follows its embryology or its evolutionary history. It is, as we have seen, a relatively simple structure in adult man, one which serves primarily for the passage of air and food. In lower chordates, it serves for the concentration of microscopic food particles and directs them into the digestive tube and also is an essential organ of respiration. In lower vertebrates, such as fish, it is also the chief organ of respiration, and is provided with **gills.** But when we describe or illustrate, even briefly, the embryology of this organ we are able to see it in a "transitional period" from fish to man— a recapitulation, perhaps.

In Chapters 1 and 8, brief consideration was given to the **branchial arches** and the **pharyngeal pouches** which characterize the developing pharynx (Figs. 1–1, 3; 8–7). That the pharynx is developed from the foregut is suggested earlier in the present chapter.

The epithelial derivatives of the pharyngeal pouches are shown in Figure 17–5. Some of them are described in greater detail in Chapter 17, as is also the thyroid from the epithelium of the midpharyngeal floor.

TUBULAR DIGESTIVE ORGANS

Structural Plan (Fig. 13–13). The esophagus and the remaining tubular organs of the digestive system have the same basic arrangement of layers in their walls, though each organ varies as to details. The layers are arranged as follows, starting with the innermost and working outward.

Tunica mucosa (mucous membrane) ⎡Epithelium
⎢Lamina propria (connective tissue)
⎣Muscularis mucosae (smooth muscle)

Tela submucosa (connective tissue)

Tunica muscularis (externus) (smooth muscle) ⎡Circular muscle
⎣Longitudinal muscle

Tunica serosa ⎡Supporting connective tissue
⎣Mesothelium (simple squamous epithelium)

Glands. The epithelium of the hollow organs is the source of their glands. Except for the deep glands of the esophagus and the duodenal (*Brunner's*) glands which push down into the submucosa, and the liver and pancreas which push beyond the walls of the hollow organs, the glands are limited to the tunica mucosa.

ESOPHAGUS (Figs. 13–5, 12, 14, 15)

The esophagus is a muscular tube about 25 cm. (10 inches) long connecting the laryngeal pharynx with the stomach. It starts at the level of the sixth cervical vertebra. Lying anterior to the vertebrae and the prevertebral muscles and posterior to the trachea it passes into the mediastinum (*central cleft*) of the thorax. At the level of the superior end of the heart, it swings anteriorly against the pericardium to allow the aorta to assume the prevertebral position. As it nears the diaphragm, it swings slightly to the left before penetrating that organ to join the stomach. The esophagus is the only digestive organ to be represented in three body regions—the neck, thorax, and abdomen.

The mucosa of the esophagus has a surface of stratified squamous epithelium (Fig. 13–15). Its lamina propria contains superficial esophageal glands, while deep esophageal glands extend also into the submucosa. The muscularis mucosa is thick, especially toward the lower end of the esophagus.

In the muscular tunic of the esophagus,

we see the gradual changeover from voluntary to involuntary muscle. That of the upper third of the esophagus is largely voluntary (*skeletal*), especially the longitudinal layer which is continuous with the inferior constrictor of the pharynx. The middle third is mixed, the lower third is smooth muscle.

The outer tunic of the esophagus, except for the lowermost 1 to 2 cm. in the abdomen, has no serosa and is called **tunica adventitia.**

The esophagus makes no contribution to the chemical digestion of food, its glands secreting a mucous material for the lubrication of the mucosa which makes for easier passage of the bolus.

ABDOMEN AND PERITONEAL CAVITY (Figs. 13–16, 17)

Review at this time the section on the development of the coelom (pp. 33 and 34). It will help to understand the relationships of the remainder of the digestive organs, the peritoneal cavity, and the supporting membranes. Refer also to Figures 4–10 and 13–16, 17.

The **diaphragm** separates the thoracic and peritoneal cavities. The **peritoneal cavity** extends from the diaphragm to the pelvic floor. It is arbitrarily divided into **abdominal** and **pelvic portions** at the brim of the true or lesser pelvis (Fig. 9–14). It is typically a closed cavity, except in the female where it opens to the outside through the reproductive organs. It is lined with peritoneum and has only potential space within it because the visceral organs have everywhere pushed in, though not broken through, its walls. It thus appears crowded with organs, though actually it contains only a bit of serum-like fluid which moistens and lubricates its inner surface.

Under the dome of the diaphragm on the right side is a large gland, the **liver;** on the left is the **stomach.** Hanging from the greater curvature of the stomach is a large apron-like fold of mesentery laden with fat, the **greater omentum.** It contains between its folds a potential space, the **lesser peritoneal cavity,** or omental bursa which can be reached from the greater peritoneal cavity only by a small opening just above the first part of the duodenum, the **epiploic fora-**

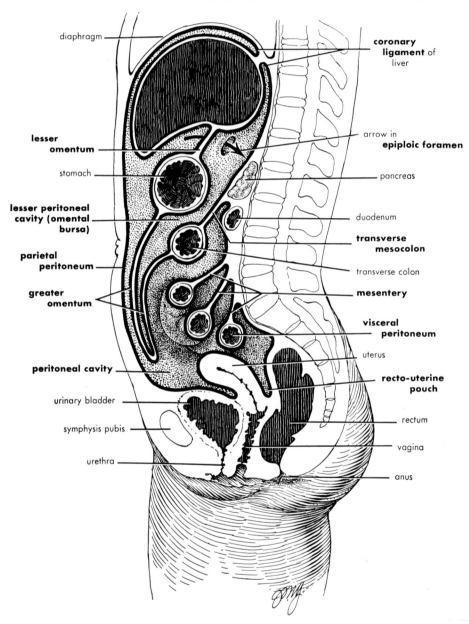

FIG. 13–16.   Schematic presentation of the peritoneum and peritoneal (coelomic) cavities.

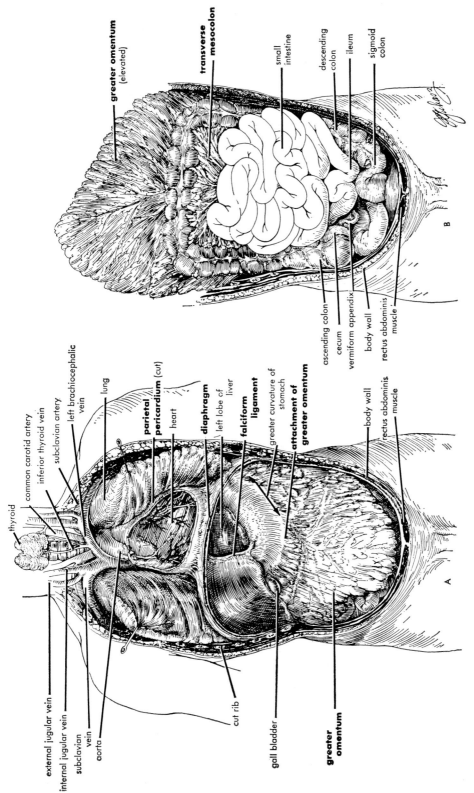

Fig. 13–17. Body cavities and related viscera. *A,* Greater omentum in position over intestines. *B,* Greater omentum elevated with transverse colon to show underlying intestines.

men (*foramen of Winslow*) (Fig. 13–16). The greater omentum obscures the small and most of the large intestines (Fig. 13–17*A*). When this is lifted up or removed, the remainder of the abdominal and pelvic viscera can be seen (Fig. 13–17*B*). They are held loosely or, in some cases, closely to each other or to the abdominal and pelvic walls by double layers of serous membrane, the **mesenteries, omenta,** and **ligaments.** At the surface of the organs, these double-layered serous membranes separate to cover the organs as their **visceral peritoneum** (*serosa*), whereas at the body wall they separate to form the **parietal peritoneum** (Fig. 13–16).

STOMACH (Figs. 13–17*A*, 18)

The stomach is often described as a J-shaped organ, and the most dilated of the digestive tube. Actually, it varies greatly in shape, depending in part upon the amount and condition of its contents and the condition of the neighboring intestines. Its position also varies, except at its relatively fixed beginning and end. Its intermediate portion is but loosely held by the broad membranous greater and lesser omenta and its position varies with stature, position of the body, contents of the stomach and neighboring organs, and respiratory movements.

The stomach has two orifices—an upper cardiac and a lower pyloric. It has a right concave border, the **lesser curvature,** and a left convex border, the **greater curvature.** The left and larger part of the stomach is the body, the right and narrowing portion, the **pyloric region.** The part of the **body** which bulges above the cardiac orifice to the left is the **fundus.** Gas sometimes collects in the fundus and causes pressure on the heart. The area immediately around the cardiac orifice is sometimes called the **cardia.**

Microscopic Anatomy (Figs. 13–18, 19, 20). The walls of the stomach, while conforming to the usual four layers, have some special features. The mucosa and submucosa are thrown into prominent folds, the **rugae,** which flatten to some extent when the stomach is expanded with food. In addition to the circular and longitudinal muscles, there is an incomplete layer of oblique fibers internal to the circular muscle. At the juncture between pylorus and small intestine, the circular muscle is thickened to form a **pyloric sphincter** or **valve** which regulates passage of the stomach contents.

Microscopic studies reveal that at the cardiac orifice the stratified squamous epithelium of the esophagus abruptly gives way to the **simple columnar epithelium** of the stomach mucosa. Simple columnar epithelium lines the remainder of the digestive tube to the anal canal where stratified squamous epithelium is again found, which is continuous with that of the skin.

Mucus-secreting cells of the stomach epithelium dip down into the lamina propria to form the **gastric pits,** into the bottoms of which the gastric glands empty. Three kinds of gastric glands are present, the cardiac, pyloric and fundic (*principal*). The **cardiac glands** are found in the immediate area of the cardiac orifice, resemble superficial esophageal glands and are relatively unimportant. **Pyloric glands** are limited to the pyloric end of the stomach and are of the short, branched tubular type which may also coil. They are lined by clear, columnar cells and their chief contribution is mucus.

The **fundic glands** are the principal glands of the stomach and occur throughout the body and fundus. They are long, straight tubes which may bend and branch at their deep ends. Their necks, which open into the gastric pits, tend to be constricted. They are lined at their neck ends by mucus-secreting columnar cells, and deeper by two kinds of cells— the **chief cells** which produce the

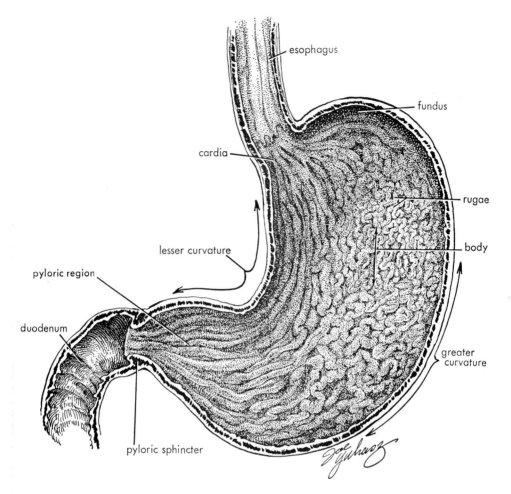

esophagus

fundus

cardia

rugae

lesser curvature

body

pyloric region

duodenum

greater curvature

pyloric sphincter

Fig. 13–18. Frontal section of the stomach.

enzymes **pepsin, rennin** and **lipase,** and the large **parietal cells** "producing" hydrochloric acid. The **gastric juice** is primarily the product of the fundic glands.

Functions. The stomach is a temporary storage chamber along the digestive route. While liquids move through the stomach quickly, a regular meal may be one to four hours in passing. Little absorption takes place through the stomach wall, alcohol being an exception. Most of the foods are not yet digested, nor is the stomach well adapted for absorbing. Its principal function is digestion.

Enzymes. Necessary agents in digestion are **enzymes** which are **catalysts.** Catalysts are substances which accelerate chemical reactions without themselves being permanently changed in the reactions. Enzymes are very specific, acting only on certain substances. They require equally specific environmental conditions under which to work. Temperature is important, also the acid-base conditions. Table 13–1 gives basic information relative to the important enzymes of digestion.

Movements of the Stomach. The movements of the stomach during digestion are mostly in the form of peristaltic waves which, when the stomach is full, start about midway between the cardiac

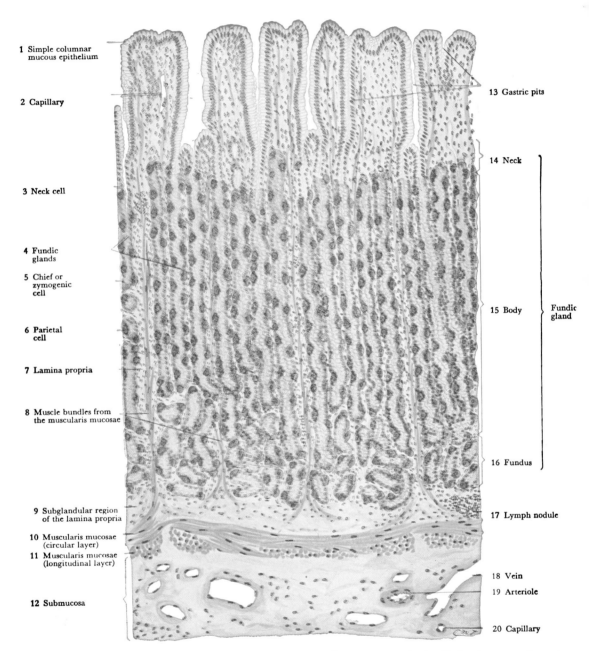

1 Simple columnar mucous epithelium

2 Capillary

3 Neck cell

4 Fundic glands

5 Chief or zymogenic cell

6 Parietal cell

7 Lamina propria

8 Muscle bundles from the muscularis mucosae

9 Subglandular region of the lamina propria

10 Muscularis mucosae (circular layer)

11 Muscularis mucosae (longitudinal layer)

12 Submucosa

13 Gastric pits

14 Neck

15 Body

16 Fundus

Fundic gland

17 Lymph nodule

18 Vein

19 Arteriole

20 Capillary

FIG. 13–19.   Stomach: gastric mucosa of the fundus (transverse section).   Stain: hematoxylin-eosin. 180×.   (di Fiore, *An Atlas of Human Histology*, Lea & Febiger.)

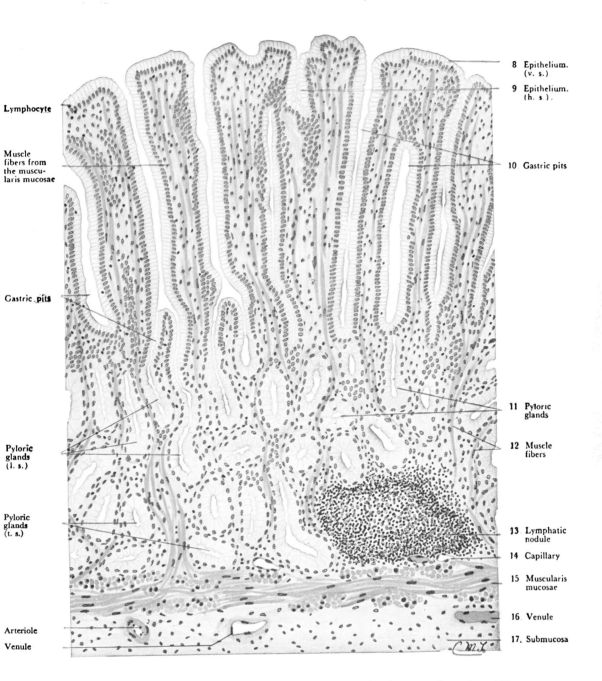

Lymphocyte

Muscle
fibers from
the muscu-
laris mucosae

Gastric pits

Pyloric
glands
(l. s.)

Pyloric
glands
(t. s.)

Arteriole

Venule

8 Epithelium.
(v. s.)

9 Epithelium.
(h. s.).

10 Gastric pits

11 Pyloric
glands

12 Muscle
fibers

13 Lymphatic
nodule

14 Capillary

15 Muscularis
mucosae

16 Venule

17. Submucosa

FIG. 13–20.   Stomach: gastric mucosa of the pyloric region.   Stain: hematoxylin-eosin.   100×.
(di Fiore, *An Atlas of Human Histology*, Lea & Febiger.)

TABLE 13–1.   DIGESTIVE ENZYMES AND THEIR ACTIONS

Environment	Enzyme	Where Found	Substance Changed	Products Formed
Neutral	Ptyalin (salivary amylase)	Saliva—mouth	Boiled starch—dextrins	Dextrin and maltose
Acid (HCl)	Pepsin Rennin	Gastric juice—stomach " " " "	Protein Casein (milk)	Proteoses and Peptones Insoluble paracasein (curd)
Alkaline	Trypsin Amylopsin Steapsin	Pancreatic juice in small intestine " " " " " " " "	Protein, Proteoses Peptones Starch and dextrin Fats	Peptones Peptids Maltose Fatty acid; glycerol
Alkaline	Erepsin Enterokinase Maltase Sucrase Lactase	Intestinal juice in small intestine " " " " " " " " " " " " " " " "	Peptones Peptids Trypsinogen Maltose Sucrose Lactose	Amino acids Trypsin Glucose Fructose and glucose Galactose and glucose

and pyloric orifices and move toward the latter. As digestion progresses, the waves start closer and closer to the left side of the stomach until the process is completed. As the food mass become digested, it takes on the appearance of a creamy mass called **chyme** which, bit by bit, is forced through the pyloric sphincter into the small intestine.

Stomach Disorders. Besides the usual digestive upsets to which the whole system is subject, the stomach is a frequent site of ulcers or cancer. **Ulcers** may be mild and yield to dietary and other treatment, or they may work their way through the tunics of the stomach wall and ultimately break into the peritoneal cavity and set up a peritonitis. Ulcers require careful treatment to avoid perforation and often have to receive surgical care. They are common, too, in the duodenum. **Cancer** is an insidious condition in the stomach, for it may go for some time without causing symptoms. The only effective treatment known is surgery, and this should be done as early as possible.

One hears so much advertising relative to alkalizing the stomach that one forgets that the normal gastric juice is acid, and needs to be, for proper stomach function. It is true that under certain circumstances, ulcers for example, it is important to reduce stomach acidity; but this should be done under a doctor's directions.

There is evidence that emotional states such as fear, tension, and worry may cause ulcers. They are common in our high-pressure world.

SMALL INTESTINE
(Figs. 13–2, 17B)

The small intestine extends from the pyloric valve of the stomach to the ileocecal (colic) valve where it joins the large intestine (Fig. 13–2). It is the longest and most convoluted part of the intestine, probably averaging about 21 feet in length (7 meters), although extremes of from 11 to 25 feet are of record. It is divided into a short upper portion, the **duodenum,** about 10 inches (25 cm.) in length, a central **jejunum** of

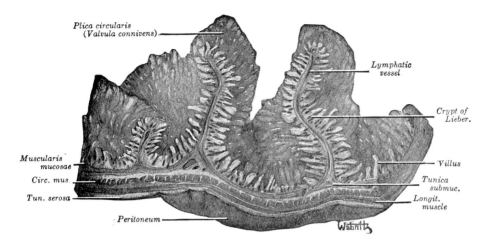

Fig. 13–21.   Section through wall of small intestine (jejunum) of adult cadaver. ×5. (*Gray's Anatomy.*)

about 8 feet and the terminal two-thirds, the **ileum,** of about 12 feet. Except for the duodenum which is in back of the peritoneum (*retroperitoneal*), the tube is covered by peritoneum and loosely supported by the **mesentery** from the dorsal wall.

There are the typical four tunics in the walls of the small intestine, but the mucosa and submucosa have features different from the other digestive organs. These features also enable one to distinguish, in a general way, between the duodenum, jejunum and ileum.

Microscopic Anatomy (Figs. 13–21, 22, 23). Circular folds (*plicae circulares*), villi and intestinal glands (*crypts of Lieberkühn*) are diagnostic characteristics of the small intestine. **Circular folds** involve both the mucosa and submucosa and are valve-like folds which project from 3 to 10 mm. into the intestinal lumen. They are transversely placed, some of them extending only one-half to two-thirds around the tube, others going all the way around, a few spiraling for two or three turns. High and low folds tend to alternate. These are permanent folds and do not flatten out, like the rugae of the stomach, when the organ is distended.

Villi are finger-like projections of the mucosa into the lumen of the small intestine. They are so closely packed as to give the surface a velvet-like appearance. They are barely within the visual capacity of the naked eye. They vary in size and to some extent in shape, and tend to flatten when the intestine is distended. Like all of the mucosal surface the villi are covered with simple columnar epithelium in which the free surfaces of the cells have **striated borders.** There are numerous **goblet cells.** Inside a villus is a cone of areolar and reticular connective tissue giving support to blood capillaries which form a network, and usually to a single lymphatic capillary which begins blindly in the villus (Fig. 13–22, 23). The lymphatic capillaries in the villi are called **lacteals** because they contain, during the absorption of fat material, a milk-white substance called **chyle.**

Intestinal glands are simple, straight, tubular glands which open into the depressions between the villi and terminate at the muscularis mucosa (Fig. 13–22). The simple columnar epithelium and goblet cells turn into the gland, but at its terminus there is a group of special secreting cells (*cells of Paneth*)

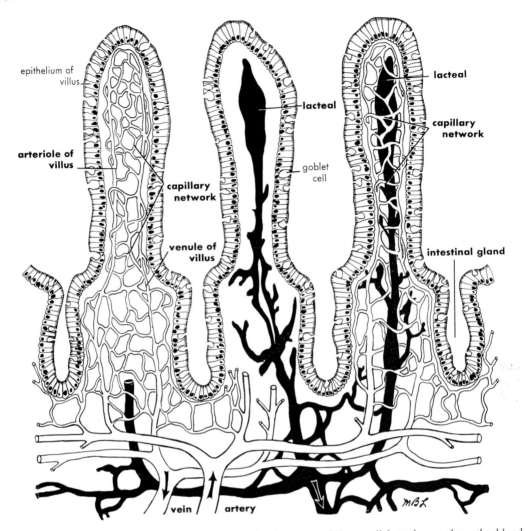

epithelium of
villus

arteriole of
villus

capillary
network

venule of
villus

lacteal

goblet
cell

lacteal

capillary
network

intestinal gland

vein   artery

FIG. 13–22. Schematic section of mucosa and submucosa of the small intestine to show the blood and lymph circulation in the villi. The villus to the left shows only the blood vessels; the villus in the center only the lymph vessel; the one on the right has both, which is the normal condition for all villi.

which probably produce the digestive enzymes of the intestinal juice (Table 13–1).

Lymphocytes and solitary lymph nodules are found in the submucosa of the small intestine, as they are throughout the digestive tube. Lymphocytes also invade the mucosa, sometimes heavily.

Duodenum (Figs. 13–2, 24, 34). The duodenum differs from other parts of the small intestine by having **duodenal** (*Brunner's*) **glands** in the submucosa (Fig. 13–24). These are compound tubulo-

alveolar glands and their ducts, after penetrating the muscularis mucosa, enter the intestinal glands. These glands are most numerous near the stomach end of the duodenum and gradually diminish until they are lacking near the jejunum. The pyloric end of the duodenum is lacking in circular folds, but they become very numerous and high at its other end. Villi are especially large and numerous. The duodenum also receives the pancreatic secretion and the bile from the liver through a common duct enter-

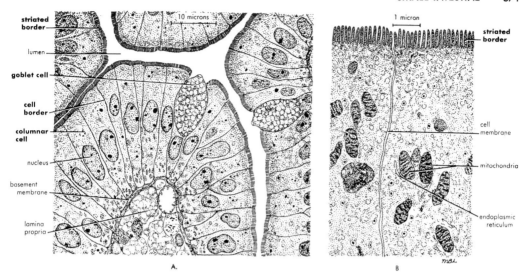

FIG. 13–23.   Electronmicrograph of the surface epithelium of the small intestine.
A. Jejunum (about 1250×).   *B.* Duodenum (about 7500×).

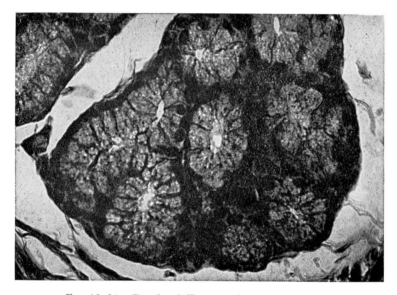

FIG. 13–24.   Duodenal (Brunner's) glands. (×450.)

ing on an elevation of the mucosa—the **duodenal papilla**—about 4 inches below the pylorus.   The opening on the duodenal papilla is the ampulla of Vater (Fig. 13–34).

Jejunum.   The jejunum is not clearly differentiated histologically from either the duodenum or ileum.   The circular folds and villi, large and numerous at its beginning, gradually decrease in number and size toward the ileum.

Ileum (Figs. 13–2, 25, 26).   The circular folds and villi are less numerous in the ileum than in the jejunum, and circular folds are usually entirely lacking at its lower end.   The most characteristic features of the ileum are the **Peyer's patches,** aggregates of lymph nodules in

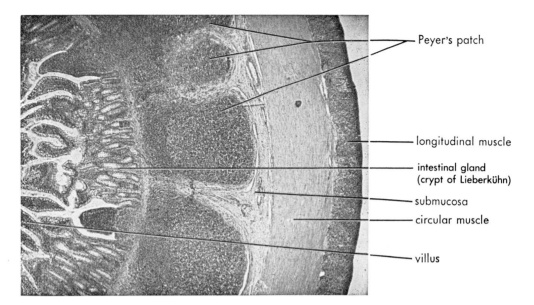

Peyer's patch

longitudinal muscle

intestinal gland
(crypt of Lieberkühn)

submucosa

circular muscle

villus

FIG. 13–25.   Photomicrograph of transverse section of ileum. (×48.)

the mucosa. These patches may be 1 cm. wide and 5 cm. long. They are largest and most frequent at the distal end and are found only occasionally in the jejunum. They are particularly affected by typhoid. The ileum opens into the ascending colon through the **ileocaecal** (*colic*) **valve.**

Function of the Small Intestine. The **digestive process** is brought to completion in the small intestine through the agency of the various enzymes from intestinal glands and pancreas (Table 13–1). The liver has contributed bile which helps to create a proper alkaline environment for the enzymes, and aids in the emulsification and saponification of fats for greater ease of digestion and absorption. The other primary function of the small intestine is **absorption,** and the tremendous increase in surface area created by the circular folds, villi, and great length of the tube makes this a very efficient process—the greater the surface area, the greater the volume of intake in a given unit of time.

The movements of the small intestine, which make their contribution to the digestive and absorptive functions, are **peristalsis,** already described for stomach and esophagus, and segmentation. **Segmentation** refers to local contractions of the circular muscle which divide the food column into masses and these in turn are subdivided by other constrictions so that digestive juices are brought into contact with all parts of the food, and the digested food given full **exposure** to the absorbing tissues.

LARGE INTESTINE
(Figs. 13–2, 17, 26, 27)

The large intestine is about 5 feet (*1.5 m.*) in length, extending from the ileocaecal valve to the anus. It is composed of the cecum, colon, rectum, and anal canal. It is greater in diameter than the small intestine and is readily recognized by its sacculations, the **haustra.** It also has little tag-like, fat-filled peritoneal bodies hanging from its outer walls, especially on the descending colon. These are the **epiploic appendages.**

Cecum. The **cecum,** at the beginning of the large intestine, is a large blind pouch, into which the ileum empties

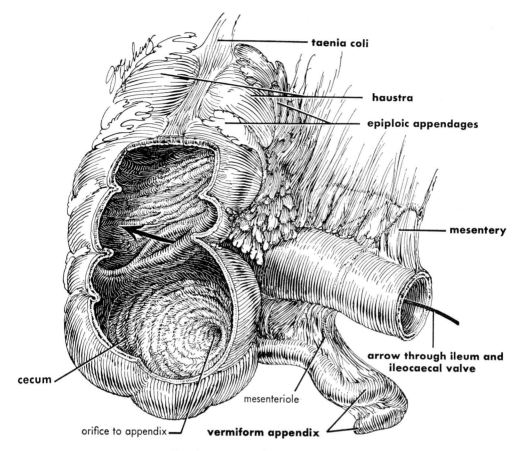

taenia coli

haustra

epiploic appendages

mesentery

arrow through ileum and
ileocaecal valve

cecum

mesenteriole

orifice to appendix

vermiform appendix

FIG. 13–26.   The ileocaecal region.

(Fig. 13–26). To the cecum is attached a narrow, worm-like tube, the **vermiform process** or **appendix.**

Colon (Figs. 13–2, 26, 27). The **colon** consists of four parts, the ascending, transverse, descending, and sigmoid which form a kind of enclosure around the coils of the small intestine. The **ascending colon** starts at the cecum in the right iliac region, passes upward through the lumbar to the right hypochondriac region. It lies against the posterior wall and therefore has no mesentery. In the hypochondriac region, just below the liver, it turns abruptly forward and to the left, forming the **right colic** (*hepatic*) **flexure.** It now becomes the **transverse colon** which is longer than the distance it must traverse and so hangs down into the umbilical region when the individual is standing. It has a mesentery, the mesocolon, and is attached to the posterior surface of the apron-like greater omentum. It lies below the stomach and in front of the upper coils of the small intestine and when reaching the left hypochondriac region makes an acute turn, the **left colic** (*splenic*) **flexure** to become the descending colon. The **descending colon,** the narrowest portion of the large intestine, pushes back to the posterior wall to become retroperitoneal and descends through the left lumbar and iliac regions to the pelvic brim where it is continuous with the sigmoid colon. The **sigmoid colon** swings in an S-shaped form to the midline, supported in part by a mesentery, the **sigmoid mesocolon.**

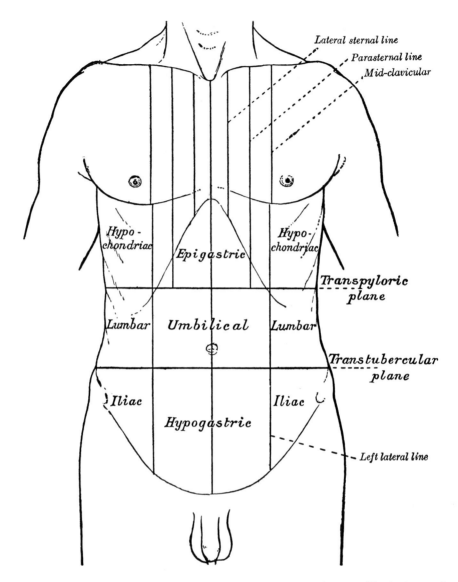

Fig. 13–27.   Surface lines of the front of the thorax and abdomen.   (*Gray's Anatomy.*)

Rectum and Anal Canal (Fig. 13–28). The sigmoid colon becomes continuous with the rectum at the level of the third sacral vertebra. The **rectum** passes in the curve of the lower sacrum and coccyx, dilates at its lower end, then bends abruptly backward to become the **anal canal** which opens at the anus. The anus is provided with an **internal sphincter** of smooth muscle and an **external sphincter** of skeletal or voluntary muscle.

Refer to Figures 11–29, 30 for a review of some of the relations in this important area.

Microscopic Anatomy (Figs. 13–13, 26, 28). The diagnostic features of the walls of the large intestine are the lack of villi and the closely placed intestinal glands with many **goblet cells.** Also, the longitudinal muscle layer is discontinuous, being in the form of three bands running from the cecum to, but not

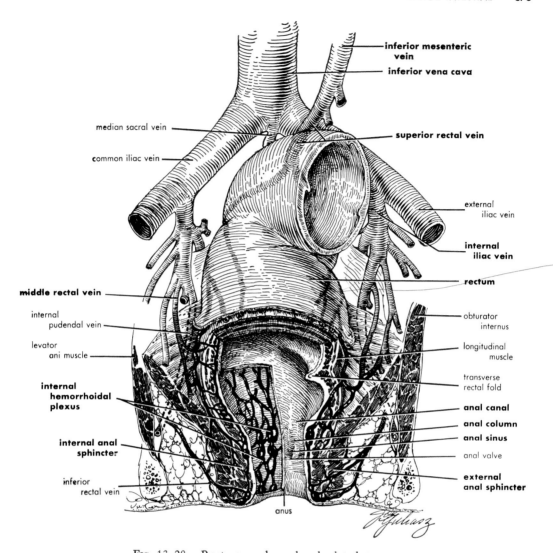

Fig. 13–28.   Rectum, anal canal and related structures.

including, the rectum. These bands, called the **taeniae coli,** are shorter than the intestinal wall, hence cause it to form sacculations, the **haustra.** The walls between the haustra are thrown into folds which show on the inside of the colon as the **semilunar folds.** Lymphocytes and solitary lymphatic nodules are of frequent occurrence in the mucosa.

The **appendix** has a complete longitudinal layer and its mucosa and submucosa are quite thick, with abundant lymphocytes and lymphatic nodules. It is some-

times called the tonsil of the abdomen and may, like other tonsils, become infected and require removal.

The **mucosa** of the **rectum** and **anal canal** is thick and highly vascular. In the anal canal at its upper end the mucosa and some of the muscle tissue are thrown into folds, the **rectal columns.** They are separated from each other by furrows, the **anal sinuses,** which end below in valve-like folds, called **anal valves.** Beneath the rectal columns are dilated veins which become tortuous.

They are derived from the superior and inferior rectal (*hemorrhoidal*) veins and may cause difficulty when over-enlarged and blood becomes stagnated in them, a condition called **hemorrhoids** or **piles.** Finally, just above the anal opening the epithelium changes abruptly from simple columnar to stratified squamous, and at this line numerous, large modified sweat glands are found, the **anal glands.**

Functions. The large intestine absorbs water from the fluid material which comes to it from the ileum, and thereby concentrates the mass into a solid or semisolid consistency, the **feces.** The feces contain food residues, epithelial cells, leukocytes, bacteria, and secreted material from the stomach and intestinal glands, the liver, and the pancreas. Calcium, magnesium, phosphates and iron (iron sulfide) are included in the feces.

The glands of the large intestine secrete mucus, but no enzymes, which, when mixed with the feces, helps to hold them together and aids in their elimination.

The movements of the large intestine involve **segmentation,** or **churning,** and **peristalsis,** much as in the small intestine In addition, there are mass movements or mass peristalsis occurring two or three times in twenty-four hours which carry the intestinal contents along for some distance. These often follow the introduction of food into the stomach.

The sigmoid colon, the main storehouse for the feces, fills from its lower end upward. The rectum usually remains empty until shortly before defecation. The feces move into the rectum when the colon becomes overloaded or usually following mass peristalsis. **Defecation,** the emptying of the rectum, involves a strong contraction of the rectal muscles and a relaxation of the anal sphincters, the outer one of which is controlled voluntarily. The process is aided by the contractions of the diaphragm and the abdominal muscles.

LIVER (Figs. 13–2, 27, 29)

The liver is the largest gland in the body and is located in the upper right hypochondriac and the epigastric regions, fitting closely against the inferior surface of the diaphragm. It is made up of two principal lobes, the large **right** and the small **left,** the right having, besides the main lobe, two small lobes associated with it. These are the inferiorly placed **quadrate lobe,** and the posterior **caudate lobe** (Fig. 13–29B). The right and left lobes are separated by the **falciform ligament** anteriorly and superiorly, which attaches the liver to the anterior abdominal wall and the diaphragm. In the free border of the falciform ligament is the **ligamentum teres hepatis** (*round ligament*), a fibrous cord homologous to the umbilical vein of the fetus. It extends from the umbilicus to the liver where it may be traced to become continuous with the **ligamentum venosum,** homologous to the fetal ductus venosus. On the superior and posterior surface of the liver are the two widely separated folds of the **coronary ligament** and to either side the **left** and **right triangular ligaments,** which hold the liver to the diaphragm. Between the folds of the coronary ligament is a **bare area,** the only appreciable part of the liver surface which is not covered with **visceral peritoneum.** It rests against the diaphragm.

On the inferior surface of the liver, between the quadrate and caudate lobes, is a transverse fissure, the **porta** (*door*), which transmits the hepatic portal vein, the hepatic artery and nerves to the liver, the hepatic duct and lymphatics from the liver. These same structures pass from the porta in a fold of peritoneum, the **lesser omentum,** to or from the lesser curvature of the stomach and the duodenum.

Gallbladder and Bile Ducts (Figs. 13–28B, 34). The gross biliary ducts or passageways consist of two **hepatic ducts,**

right and left, coming from the major lobes of the liver. These join at the porta to form the **common hepatic duct.** It then enters the lesser omentum where it is joined by the cystic duct, and continues, as the **common bile duct,** to enter the duodenum at the duodenal papilla. The **cystic duct** is the duct of the gallbladder, the two representing a diverticulum from the biliary tract. The **gallbladder** is a pear-shaped structure lying on the inferior surface of the liver with its **fundus** slanting downward and to the right. The peritoneum of the liver passes over its free surface. It is about 7 to 10 cm. long and has a capacity of 30 to 35 cc. The **body** and **neck** of the gallbladder are directed to the left and upward toward the porta. The neck makes an S-shaped curve and becomes continuous with the cystic duct. The walls of the gallbladder have three coats, the **serous,** the **fibromuscular** and **mucous.** The **serous coat** covers all surfaces of the fundus but only the inferior surfaces of the body and neck.

The **fibromuscular coat** consists of dense fibrous tissue mixed with smooth muscle fibers, some longitudinal, others transverse. It forms a thin, but strong, framework for the sac.

The **mucous coat** is loosely connected with the fibromuscular layer and is elevated into minute rugae. The neck of the gallbladder leads into the cystic duct where the folds of the mucous coat become especially prominent and form a **spiral fold.** The epithelium lining the bladder is simple columnar with striated border and continues all the way to the duodenum and into the liver ducts.

Microscopic Anatomy of the Liver (Figs. 13–30 to 32). Underneath the peritoneal coat is a thin **fibrous capsule** encasing the liver and sending fibrous partitions into its substance, dividing it into units known as liver **lobules.** At the porta the capsule is continuous with loose areolar tissue which supports the portal vein, hepatic arteries, and hepatic ducts and their subdivisions. The lobules range from 1 to 2.5 mm. in diameter so it can be seen with the naked eye in a section of liver. Each lobule consists of sheets of cells radiating out from a common center occupied by a single tributary of the **hepatic vein,** and called an **intralobular** (*central*) **vein.** Between adjacent lobules are found tiny tributaries from the **hepatic portal vein, hepatic artery,** and **bile duct** which may be termed **interlobular veins, arteries,** and **bile ducts** respectively. Between the interlobular veins and arteries and the central intralobular vein, and running between the sheets of liver cells are **sinusoids** lined with modified endothelium and containing many macrophages (*Kupffer cells*) (Fig. 13–32). Blood from the interlobular veins and arteries circulates through the sinusoids of the lobule and makes its contributions of food and oxygen to the liver cells, picking up other materials. It then passes into the intralobular vein which carries it to a **sublobular vessel.** This vessel takes the blood finally into a hepatic vein which leads to the inferior vena cava on the posterior side of the liver. Thus the liver receives blood from two sources, the hepatic portal vein and hepatic artery; but it is all returned to the hepatic veins.

The interlobular bile vessels received from the lobule many tiny bile ducts which come from **biliary canaliculi** (*bile capillaries*) and carry the bile secretions from the liver cells. The interlobular bile vessels return the bile ultimately to the hepatic ducts.

Development of the Liver (Fig. 13–33). The liver develops as a hollow outgrowth from the ventral surface of the gut, the part which eventually becomes the descending part of the duodenum. The endoderm of the gut extends into this diverticulum which grows upward into the transverse septum, a mass of mesoderm between the pericardium and the

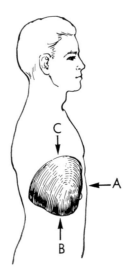

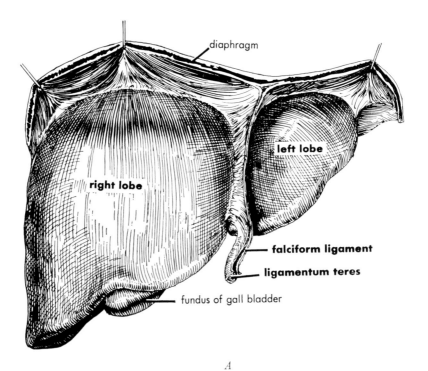

diaphragm

left lobe

right lobe

falciform ligament

ligamentum teres

fundus of gall bladder

A

Fig. 13–29.   (*Continued on opposite page.*)

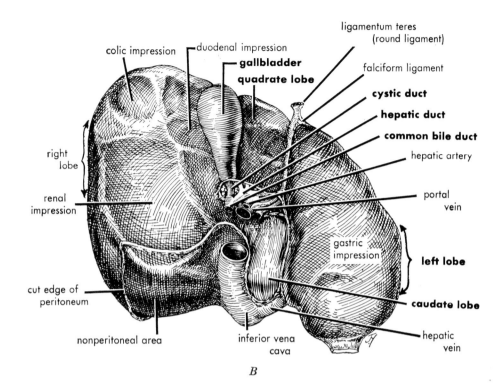

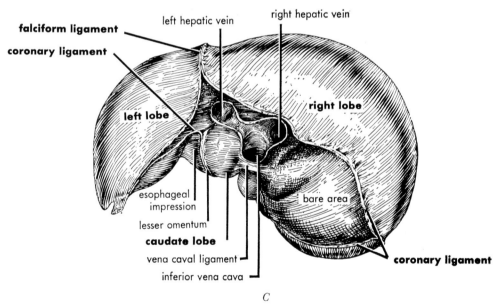

FIG. 13–29. (*1st figure*) The sketch of the human figure indicates the directions from which the following drawings of the liver were made. *A*, Anterior view of the liver. *B*, Inferior surface of the liver. *C*, Superior or diaphragmatic surface of the liver.

1. Hepatic lobule.

2 Bile duct.

3. Kiernan's
   fissure.

4. Branch of
   the hepatic
   portal vein
5. Branch of the
   hepatic artery
6 Bile duct.

7. Portal area,
   interlobular
   space, or
   Kiernan's
   space

8 Bile duct

9 Branch of
  the hepatic
  portal vein

10 Central
   vein

11 Liver cord

12 Branch of
   the hepatic
   artery

13 Branch of
   the hepatic
   portal vein

14 Portal area

15 Bile duct

16 Liver cords.

17 Sinusoids.

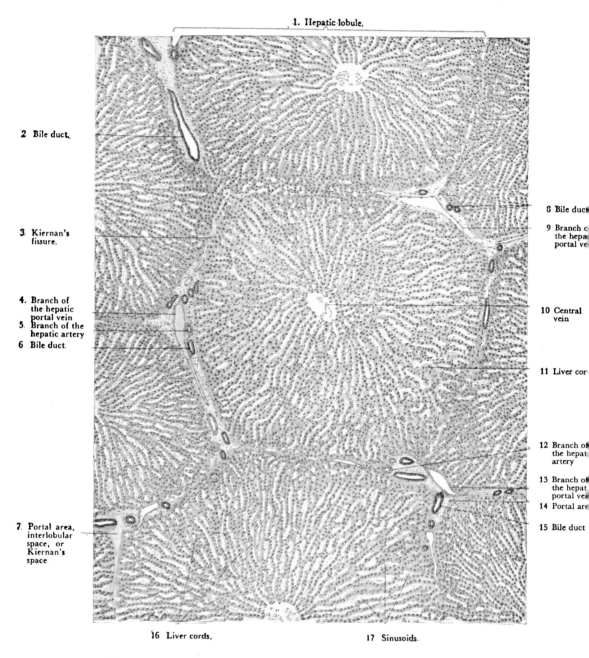

FIG. 13–30.   Liver lobule (panoramic view).   Stain: hematoxylin-eosin.   × 45.
(di Fiore, *An Atlas of Human Histology*, Lea & Febiger.)

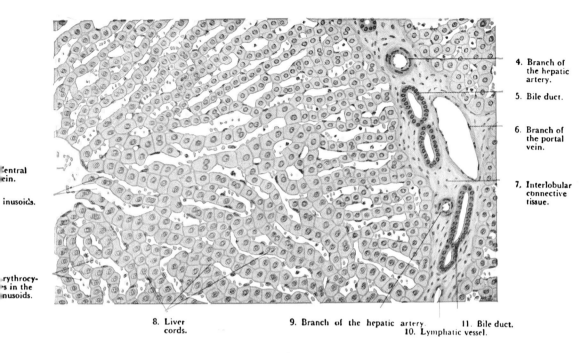

4. Branch of
the hepatic
artery.

5. Bile duct.

6. Branch of
the portal
vein.

7. Interlobular
connective
tissue.

Central
vein.

Sinusoids.

Erythrocy-
tes in the
sinusoids.

8. Liver
cords.

9. Branch of the hepatic artery.
10. Lymphatic vessel.

11. Bile duct.

FIG. 13–31. Liver lobule (sectional view). Stain: hematoxylin-eosin. × 285.
(di Fiore, *An Atlas of Human Histology*, Lea & Febiger.)

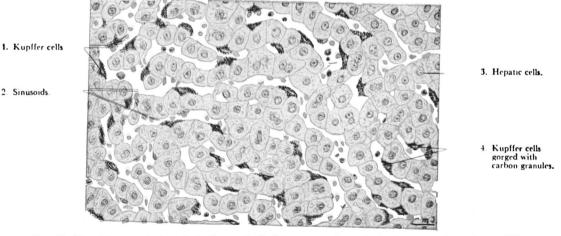

1. Kupffer cells

2. Sinusoids.

3. Hepatic cells.

4. Kupffer cells
gorged with
carbon granules.

FIG. 13–32.   Liver: reticuloendothelium. India ink preparation. Stain: hematoxylin-eosin. ×350.
(di Fiore, *An Atlas of Human Histology*, Lea & Febiger.)

vitelline duct, which finally contributes to the diaphragm. At the end of the liver diverticulum, two solid cords of endodermal cells form, from which develop the complex cellular masses which become the right and left liver lobes. In their development they involve the um-

bilical and vitelline vessels, causing them to form capillary-like vessels, the **sinusoids.** In this way the intimate association of liver cells and blood vessels described above is established. The vitelline vessels in part become the hepatic portal and hepatic veins.

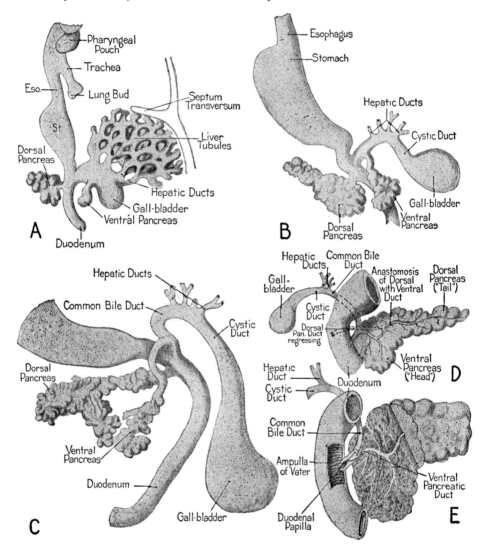

Fig. 13–33. Development of hepatic and pancreatic primordia. *A,* Semi-schematic diagram based, in part, on Thyng's reconstructions of a 5.5.-mm. pig embryo. Stage of development is comparable to that in a human embryo early in fifth week. *B,* Reconstruction from a 9.4-mm. pig embryo. Stage of development is equivalent to that of a human embryo early in sixth week. *C,* Schematized from Thyng's reconstruction of a 20-mm. pig embryo—developmentally equivalent to a human embryo of seven weeks. *D,* Schematic sketch showing dropping out of proximal part of dorsal pancreatic duct after its anastomosis with ventral duct. *E,* Manner in which common bile duct and pancreatic duct beome confluent in ampulla of Vater and discharge through duodenal papilla. (From *Human Embryology* by Bradley M. Patten, 2nd Ed., Copyright, 1953. McGraw-Hill Book Co. Used by permission.)

The original diverticulum from the gut forms the common bile duct. From it the cystic duct and gallbladder arise as a solid outgrowth which later acquires a lumen. Because the gut rotates during development, the common bile duct, having come from the ventral wall originally, now enters the dorsal wall of the duodenum.

As the liver enlarges, it outgrows the transverse septum and pushes its posterior wall backward. At the same time the gut is lengthening and pulling back from the septum to form a definite ventral mesentery. The liver in this way pushing from the septum into the ventral mesentery develops relations with each; namely, the lesser omentum (*ventral mesentery*), falciform ligament (*ventral mesentery*) and coronary ligament (*transverse septum*).

Functions of the Liver. The functions of the liver involve it in the activities of many of the body systems. The function most obvious from our discussion so far is the **formation of the bile.** Bile is stored and concentrated in the gallbladder and released when food appears in the duodenum. The role of bile in the digestion and absorption of fats has already been discussed. The liver converts glucose to glycogen and stores it for later conversion, as needed, to glucose, to again be released into the blood. The liver makes and stores vitamin A. It elaborates heparin, an agent in the prevention of blood-clotting, as well as fibrinogen and prothrombin, which are involved in clot formation. It is concerned in fat and protein metabolism, producing urea, for example. It contains phagocytic cells (*Kupffer*). It is a detoxifying agent, a blood reservoir, and produces blood cells in the embryo. It stores iron and copper and breaks down the hemoglobin from dead red blood cells. This partial list of functions indicates the importance of this organ.

Disorders of the Liver. Some disorders or anomalies of development occur.

26

Only rarely are the liver lobes increased or reduced in number. The gallbladder and main ducts are sometimes duplicated. The gallbladder is quite often absent, a normal condition in some mammals, like the horse and elephant. There may be embryonic occlusions of the gallbladder or ducts.

The gallbladder is often the site of the formation of gallstones. Bile is stored and concentrated there and one of its constituents, cholesterol, only slightly soluble, may precipitate out and form the center for a gallstone, which is added to as further precipitation takes place. Small stones may pass from the bladder to the duodenum without causing any difficulty; large ones may remain in the gallbladder and cause no symptoms. However, when stones do block the neck of the bladder or ducts, painful symptoms do appear. Stones, inflammatory conditions, or tumors may block the common duct or the hepatic ducts resulting in a backing-up of bile, its absorption into the blood and a yellow jaundice color of the skin. Surgical removal of gallstones is the only known treatment.

Infections of the liver (*hepatitis*) or of the gallbladder (*cholecystitis*) are not uncommon. Tumors and cancer of the liver also occur.

PANCREAS (Fig. 13–34)

The **pancreas** is a gland about 12.5 to 15 cm. (*6 to 7 inches*) long which may be divided into a **head, neck, body** and **tail.** It lies back of the peritoneum, behind the stomach and in front of the inferior vena cava, aorta and left kidney. Its broad head better than fills the loop formed by the duodenum; its tail reaches to the spleen.

The **pancreatic duct** (*duct of Wirsung*) runs the full length of the gland from tail to head, receiving branches from the various lobules. It joins the common bile duct, and together they run obliquely through the duodenal wall to open upon the summit of the **duodenal papilla.** The pancreatic and common bile ducts do sometimes have separate openings. An **accessory pancreatic duct** (*duct of Santorini*) sometimes branches from the pancreatic duct in the neck region and empties into the duodenum about 2.5 cm. above the duodenal papilla.

Microscopic Anatomy (Fig. 13–35). The pancreas is a **compound tubulo-alveolar gland** resembling the parotid salivary gland. The capsule of the pancreas is thin and indefinite and composed of loose fibrous connective tissue. It sends septa into the gland to divide it

into lobules and to support blood and lymphatic vessels and nerves.

The alveoli (*acini*) are separated from one another only by a little reticular tissue. They are purely serous cells. The beginnings of the pancreatic duct system are in the alveoli where flattened cells are seen which constitute the intercalated ducts. These lead into the interlobular ducts, and into the main ducts of the pancreas. All of the excretory ducts of the pancreas have simple columnar epithelium with some goblet cells. The pancreatic and accessory ducts have a thick coat of dense fibrous connective tissue.

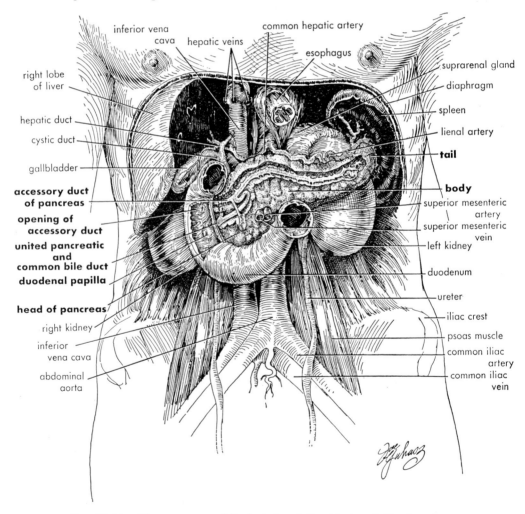

FIG. 13–34. The pancreas and its ducts in relationship to neighboring viscera.

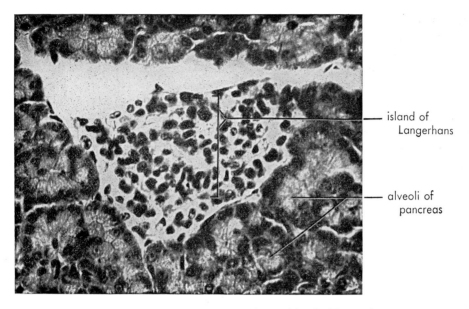

island of
Langerhans

alveoli of
pancreas

Fig. 13-35.   Section of pancreas with an island of Langerhans.

Lighter-staining groups of cells, the **islands of Langerhans,** are seen among the alveoli. They are endocrine glands and as such empty their secretions into the blood rather than into the duct system. About a million of these islands occur in the pancreas and they are richly supplied with blood.

Functions. The pancreas is a double gland, partly exocrine, partly endocrine. The exocrine portion produces the pancreatic juice which is emptied into the duodenum. It contains digestive enzymes and is alkaline in reaction (Table 13-1). The alkaline juice helps to establish the right environment for the intestinal enzymes.

Secretion of the exocrine gland is initiated by a hormone **secretin** which is liberated by the intestinal epithelium, stimulated by the entrance of the acid stomach contents.

The endocrine island cells produce the hormone **insulin** which regulates carbohydrate metabolism and blood-sugar levels. When the glands fail, the body cannot use sugar, which increases in the blood and is excreted by the kidneys.

This is called **diabetes mellitus** or sugar diabetes and can be readily controlled, but not cured, by artificial use of insulin.

Development (Fig. 13-33). The pancreas develops from dorsal and ventral diverticula of that part of the gut which later becomes the duodenum. They appear about the fourth week of embryonic life. The dorsal diverticulum grows upward and backward into the dorsal mesogastrium (*mesentery*) and forms part of the head and neck and all of the body and tail of the pancreas. Its duct becomes the **accessory pancreatic duct.** The ventral diverticulum becomes attached to the lengthening liver diverticulum and they form a common duct. Due to differential growth of the wall of the duodenum, the bile duct and ventral pancreas are brought to the dorsal side. The ventral and dorsal buds now join and their ducts intermingle, the ventral bud forming the remaining parts of the head and neck of the pancreas. The part of the duct of the dorsal pancreas between the point where it joins the duct of the ventral pancreas and the duodenum remains small during further development, that of the ventral pancreas increases in size and importance.

Secretory alveoli begin to appear as terminal and side outgrowths from the endodermal ducts at about the third month. At the same time, other groups of cells develop, also from the duct system, but these ultimately lose their connections with the ducts and are the **island cells.** Again we see cells with very different functions coming from essentially the same embryonic material, only to become unable to convert from one type to another. We are reminded of parietal and chief cells of the fundic glands of the stomach and the mucous and serous cells of the submaxillary salivary glands.

As with the liver, the connective tissue components of the gland derive from the mesoderm (*mesenchyme*) which surrounds the growing endodermal buds. The blood and lymphatic supply has a similar mesodermal origin.

Although the pancreas, growing out into the dorsal mesogastrium, was at first completely covered with peritoneum, later events placed it in contact with the posterior wall, and its peritoneum on that side was resorbed. Hence, in the adult it lies behind the parietal peritoneum (*retroperitoneal*).

QUESTIONS

1. Explain the idea of the digestive system as a tube within a tube.
2. Now that you have studied the entire digestive system, explain the concept that the inside of the digestive organs is a part of the outside environment.
3. What is indicated by the phrase that in its lumen each digestive organ has its "own peculiar ecology"?
4. What are some of the relationships between the digestive system, culture, economics, and survival?
5. Define digestion and absorption.
6. What are some of the "built-in" features that a digestive system must have in order to carry out digestion and absorption efficiently?
7. What constitutes the internal environment of an animal?
8. How do the lips and cheeks serve in mastication?
9. Name the types of teeth in the human mouth and describe the anatomy of a tooth.
10. What is the relationship between calcium and tooth development and decay?
11. Name the important functions of the tongue. Account for its versatility of movement.
12. What is the relationship of the lingual frenulum to speech?
13. Describe the hard and soft palates and their relationships to neighboring organs.
14. Where do the ducts of the salivary glands enter the mouth?
15. Name the cavities with which the pharynx communicates.
16. Describe the nasal pharynx.
17. Describe the palatine tonsils. What are adenoids?
18. Describe chewing and deglutition.
19. Name the tunics in the walls of a typical tubular organ of the digestive system and their tissue components.
20. Where do the primary tissues of the digestive glands come from?
21. What is different about the muscle tissue in the esophagus relative to that in the small intestine?
22. Describe the fundic glands of the stomach.
23. Describe the gross features of the stomach.
24. What are the boundaries of the abdominal and pelvic cavities? With what are they lined?
25. What is a mesentery? Define visceral and parietal peritoneum.

26. What is the lesser peritoneal cavity and where does it open?
27. List some of the ways by which the absorptive surface of the small intestine has been increased.
28. Describe and diagram the microscopic structure of a villus.
29. What are some of the movements involved in the small intestine during digestion and absorption?
30. Describe the special glandular components which identify the duodenum.
31. What is the function of bile in the digestive system?
32. What are solitary lymph nodules and Peyer's patches?
33. Elaborate the functions of the small intestine.
34. What are some of the special structural features of the large intestine?
35. Describe the functions of the large intestine.
36. What are the parts of the large intestine?
37. What is the relationship of the colon to mesenteries?
38. What are some of the muscular actions involved in defecation?
39. Name and locate the lobes of the liver.
40. Name and locate the ligaments of the liver. What do they come from or what are their homologues?
41. What is the porta of the liver? Indicate some of the important structures which enter and leave this door.
42. Describe the relationships of the gallbladder and the various bile-carrying ducts.
43. Describe a liver lobule and its relationships to bile and blood channels.
44. Explain the statement: The liver has a double blood supply.
45. What do the developments of the liver and pancreas have in common?
46. List some of the functions of the liver.
47. What are gallstones and how do they form? What is the best treatment?
48. The pancreas is a double gland with a twofold function. Explain this statement.
49. Where does the connective tissue of the pancreas come from—embryologically speaking?
50. What systems are most closely related to the digestive system—anatomically and physiologically?

Each person is born to one possession which outvalues all his others—his last breath.

—Mark Twain

Chapter 14

The Respiratory System

INTRODUCTION

Functions. The respiratory system performs two vital functions for the body. It makes essential **oxygen** constantly available and carries away the **carbon dioxide** which results from respiration processes within the body cells. It serves a number of accessory functions, an important one being **vocalization,** in which advantage is taken of the expired air as it passes over the vocal cords in the larynx. This may be in the form of talking, singing, laughing or other sounds in some of which a great deal of training in muscular control is involved, as the great singer or the orator. It should be remembered that organs other than the larynx are involved in these vocalizations. **Straining** is another function, a device used in defecation, in urination, and in childbirth. The abdominal muscles and the muscles of the evacuating organ are used, but in the case of the abdominal muscles, they are effective largely because of the deep breath followed by the closing of the glottis and the fixing of the diaphragm.

Coughing and sneezing are reflex responses to irritants along the respiratory mucosa in the pharynx, larynx, or in the nose, respectively. These are essentially protective responses.

The flow of venous blood and lymph are favored by breathing, for inspiration results in lowered intrathoracic pressures and blood and lymph are aspirated into that area. The contraction of the diaphragm in breathing also compresses the abdominal viscera, including the blood and lymph vessels, thus pushing their contents toward the heart and large thoracic duct. These processes are effective because both venous and lymphatic vessels have valves which allow flow only toward the heart area.

Small quantities of water and heat are lost through the lungs.

But let us return to the prime function of exchange of the respiratory gases, oxygen and carbon dioxide, which we may call **external respiration.** This takes place between the air in the lungs and the blood circulating through thin-walled capillaries. Actually, all of the other organs of the respiratory system, the **accessory organs,** serve the lungs, the **essential organs** in external respiration. Their function is to maintain a constant flow of air over the extensive respiratory membranes of the lungs, a process we may call pulmonary ventilation or **breathing.** And here again we are dealing with a problem which should remind us of those just considered under the digestive system, namely, the need of the organism for something available in the external environment, in this case, oxygen. The so-called insides of the respiratory organs, like those of the digestive system, are a circumscribed part of the outside world in which conditions are favorable for the necessary transfers. There is the same problem of surface-volume relationships. The lungs provide the extensive surface-area needed to allow the diffusion of adequate quantities of oxygen. The fact is that the lung membranes are more than adequate in providing a factor of safety. One can live on a single healthy lung. The lungs, of course, must be moist to allow this transfer of gases, and this is provided for in the construction of the system—tucked away as they are in the thoracic cavity, away from the drying influence of the air. The reliance of the respiratory system upon circulation is obvious. The blood takes the oxygen in combination with the hemoglobin of its red cells to the tissues and cells, involving again transfers across living membranes. This is called **internal respiration.** Carbon dioxide takes the trip in the opposite direction from the tissues to the lungs. The term **respiration,** without modifying adjectives, refers to the union of the oxygen with the food in the cells (*oxidation*) with the subsequent release of energy for doing work, heat, and ultimately carbon dioxide and water.

Respiratory Organs (Fig. 13–2). The organs which directly or indirectly serve the above functions are, named in logical sequence, the nasal cavities (*nose*), pharynx, larynx, trachea, bronchi, and lungs. A number of muscles are essential in the operations of this system, important ones being the **diaphragm** and the **intercostal muscles;** but they belong to the system of skeletal muscles and have been described on pages 210 to 214.

NASAL CAVITIES
(Figs. 13–4, 11; 14-1, 2, 3)

These consist of complex passageways connecting the **nares** externally to the nasopharynx into which they open by the **choanae.** The form of the nasal cavities is established by the bony framework of the nose which was studied in Chapter 8. The principal bones involved are the vomer and perpendicular plate of the ethmoid which divide the nose into right and left sides. Cartilage serves to further extend this partition anteriorly. The lateral masses of the ethmoid form shelf-life structures on each side, the **superior** and **middle nasal conchae,** and below them are the **inferior nasal conchae.** Beneath these "shelves" are recesses—the **superior, middle,** and **inferior meatuses.** Above the superior cocha is the **spheno-ethmoid recess,** which lies just below the **cribriform plate** which forms the roof of the nasal cavity. As you will recall, the nasal, maxillaries, palatines, lacrimals, frontal, and sphenoid all contribute to the nose structure.

Inside the naris is a dilated area, the **vestibule,** which is lined by skin containing hairs and sebaceous glands. Above the vestibule, the nasal fossa is divided into an **olfactory region** that consists of the superior nasal concha and the adjacent part of the septum, and a **respiratory region** made up of the rest of the fossa or cavity (Fig. 19-4).

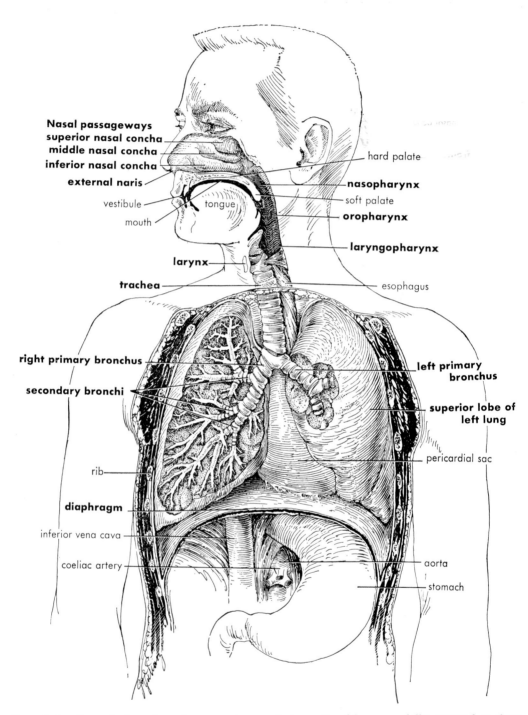

Nasal passageways
superior nasal concha
middle nasal concha
inferior nasal concha
external naris
vestibule
mouth
tongue
larynx
trachea

hard palate
nasopharynx
soft palate
oropharynx
laryngopharynx
esophagus

right primary bronchus
secondary bronchi
rib
diaphragm
inferior vena cava
coeliac artery

left primary bronchus
superior lobe of left lung
pericardial sac
aorta
stomach

FIG. 14–1. General view of respiratory system—anterior walls of lungs partially removed to show branching of bronchi and bronchioles.

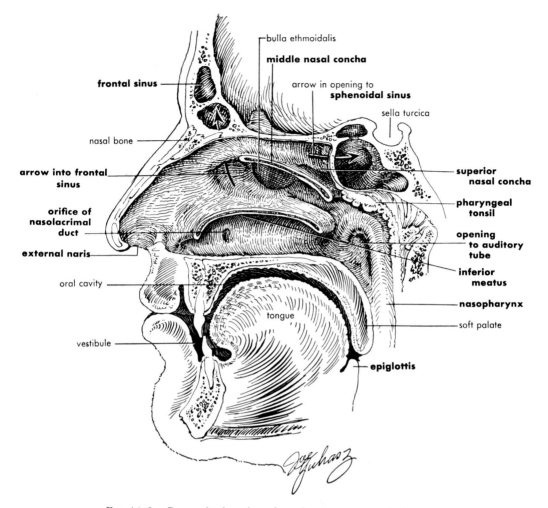

bulla ethmoidalis

middle nasal concha

arrow in opening to
sphenoidal sinus

sella turcica

frontal sinus

nasal bone

**arrow into frontal
sinus**

**orifice of
nasolacrimal
duct**

external naris

oral cavity

tongue

vestibule

**superior
nasal concha**

**pharyngeal
tonsil**

**opening
to auditory
tube**

**inferior
meatus**

nasopharynx

soft palate

epiglottis

Fig. 14–2. Parasagittal section of nasal passageways and pharynx.

Into the nasal cavities open the passageways from the **paranasal sinuses,** which are spaces in the bones, from which the sinuses take their names—**maxillary, frontal, sphenoidal** and **ethmoidal sinuses.** The **nasolacrimal canal** also empties into the nasal cavity (Figs. 14–2; 19–13).

The nasal cavities are lined with a mucous membrane that is very closely related to the underlying periosteum and perichondrium. It is continuous with the mucous membrane of the nasal pharynx through the choanae and with the skin at the vestibule. It is continuous also through the various openings into the paranasal sinuses and the nasolacrimal duct. It varies considerably in thickness, being quite thin in the meatuses, on the floor of the cavities, and in the sinuses. It is thick on the septum and conchae. It has a surface covering of **pseudostratified ciliated columnar epithelium** with many **goblet cells** and rests on a basement membrane. The underlying areolar tissue is heavily infiltrated with lymphocytes. The lamina propria contains glands near the surface, and a layer of blood vessels next to the periosteum. The vessels and blood spaces over the conchae remind one of the erectile tissue of the penis and they become dilated

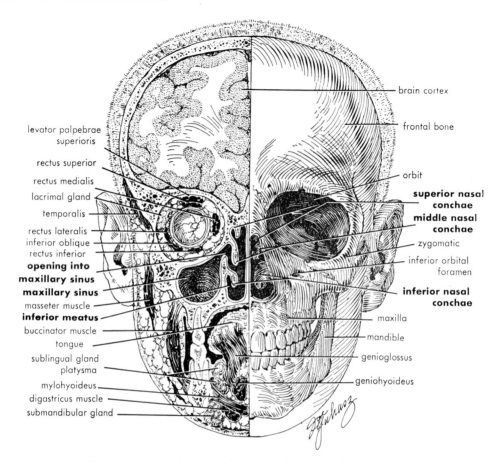

FIG. 14–3. Anterior view of head; the left side in frontal section.

very readily as a result of irritation. In the olfactory region, the mucous membrane lacks cilia and goblet cells and there are **olfactory cells** among the columnar cells. These will be described in more detail under the organs of sense (Fig. 19–4).

Special Considerations. The nasal fossae are so constructed as to modify the air which passes through them to the pharnyx. The hairs at the nares serve to screen out gross objects which might otherwise enter the nasal fossae. The mucous membrane, by its expansive surface, glandular, and highly vascular condition serves to warm and moisten the air, while the cilia and mucus strain out dust and other foreign matter.

The mucous membrane is subject to infection, to inflammation due to the "common cold" or rhinitis, for example. It becomes extremely swollen and interferes with breathing. The infection may spread into the paranasal sinuses or even into the nasal pharynx and through the auditory tubes to the middle ears. From here it could move into the mastoid cells of the temporal bone. A cold therefore should be given proper treatment to avoid these possible complications.

The nasal fossae and sinuses also influence the voice and when one has a cold and these chambers are closed, or partially so, one's voice sounds quite different.

PHARYNX
(Figs. 13–2, 4, 11; 14–1, 2)

The pharynx has been described in the previous chapter. Since both the nose and mouth lead into the pharynx, the mouth is used for breathing when the nasal passageways or nasopharynx are closed by inflammation, adenoids, or other causes.

LARYNX (Figs. 14–5 to 8)

The larynx is a part of the air-conducting system to and from the lungs. It is

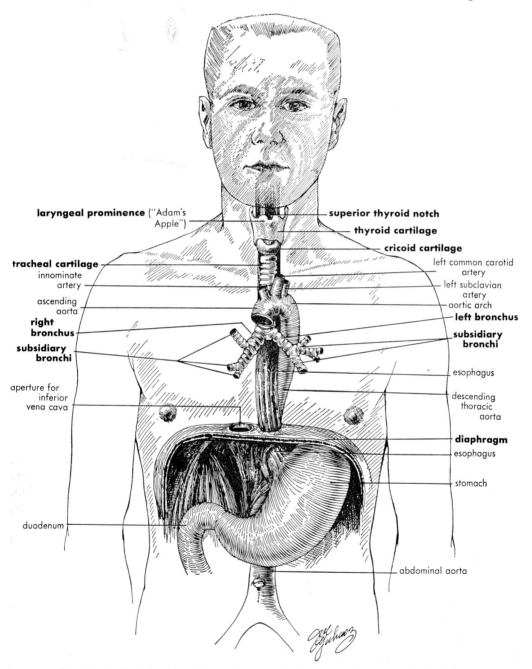

laryngeal prominence ("Adam's Apple")

superior thyroid notch

thyroid cartilage

cricoid cartilage

tracheal cartilage

left common carotid artery

innominate artery

left subclavian artery

ascending aorta

aortic arch

right bronchus

left bronchus

subsidiary bronchi

subsidiary bronchi

aperture for inferior vena cava

esophagus

descending thoracic aorta

diaphragm

esophagus

stomach

duodenum

abdominal aorta

Fig. 14–4.   Relationships of central organs of thorax superimposed on body form.

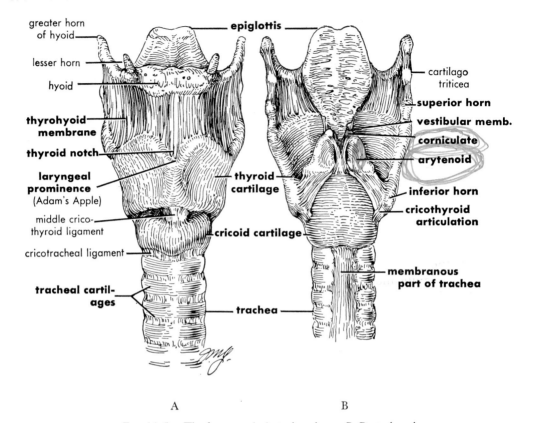

greater horn
of hyoid

epiglottis

lesser horn

hyoid

cartilago
triticea

superior horn

vestibular memb.

**thyrohyoid
membrane**

corniculate

arytenoid

thyroid notch

**laryngeal
prominence**
(Adam's Apple)

thyroid
cartilage

inferior horn

**cricothyroid
articulation**

middle crico-
thyroid ligament

cricotracheal ligament

cricoid cartilage

**membranous
part of trachea**

**tracheal cartil-
ages**

trachea

A   B

Fig. 14–5.   The larynx: *A*, Anterior view. *B*, Posterior view.

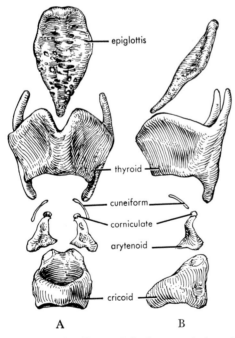

epiglottis

thyroid

cuneiform

corniculate

arytenoid

cricoid

A   **B**

Fig. 14–6.   Cartilages of the larynx. *A*, Anterior
view; *B*, Left lateral view.

also far more than this, because it acts
as a valve which prevents the passage of
solids and liquids into the air passages
below and regulates the flow of air from
the lungs in sound production and for
other purposes. By means of its lining
membranes, it acts as a kind of air-
conditioning system for the parts below.
It is provided with a series of **cartilages**
which give it form, and muscles which,
acting upon its cartilages and ligaments,
enable it to modify its apertures.

The larynx lies in the neck region in
front of the fourth, fifth and sixth cervical vertebrae, and its position is made
obvious anteriorly by the **laryngeal
prominence** or "Adam's apple" (Fig.
13–2). The larynx is large and triangular
in cross-section at its upper end, but is
smaller and circular below. It lies
between the carotid sheaths, and the

lateral lobes of the thyroid gland cover it anterolaterally. Posteriorly, it is covered by the mucosa of the laryngeal pharynx which lies behind it. Above, it is closely related to the hyoid and base of the tongue; below, it continues into the trachea.

There are three relatively large and important single cartilages in the larynx and three pairs of small ones. The former are the **thyroid,** the largest; the **epiglottis,** leaf-like in shape; and the **cricoid,** shaped like a signet ring with the signet posteriorly. Of the smaller paired ones the **arytenoids** are most important, the **corniculates** and **cuneiforms** of lesser significance (Figs. 14–4,5,6).

The **thyroid cartilage** is the largest in the larynx. It is composed of two laminae which join together on the anterior midline to form the **laryngeal prominence** which is larger in the male than the female. Just superior to the prominence the laminae diverge to form the **superior thyroid notch** which can be easily palpated. The posterior borders of the laminae are thick and rounded and have conspicuous **superior** and **inferior horns,** of which the superior is longer. The superior horns are joined to the greater horns (*cornua*) of the hyoid bone by the **thyrohyoid ligaments.** The inferior horns of the thyroid articulate by synovial joints to elevated facets on the sides of the cricoid cartilage. The superior border of the thyroid cartilage forms a wide curve from which the **thyrohyoid membrane** extends upward to the hyoid bone. The lower border near the middle is connected to the cricoid cartilage by the **middle cricothyroid ligament** (Fig. 14–5).

The **cricoid cartilage** is thicker and smaller than the thyroid and forms a complete ring. It consists of a deep, broad posterior portion and a narrowing band around the sides to the anterior side. On the midline posteriorly, there is a vertical ridge to which longitudinal fibers of the esophagus attach. Lateral to the ridge are fossae for muscle attachment. On its side it has articular surfaces for the inferior horns of the thyroid cartilage. Its lower border articulates with the first tracheal cartilage by the **cricotracheal ligament.** Its superior border posteriorly has articular surfaces for the arytenoid cartilages; anteriorly it attaches to the cricothyroid ligament, and laterally to the **conus elasticus,** at the upper margin of which is the **true vocal fold.**

The **arytenoid cartilages** are two small cartilages, pyramidal in form, which articulate to the superior border of the cricoid cartilage at the back of the larynx. On the apex of each is a corniculate cartilage. The sides of the arytenoids are triangular. They have anterior angles which attach to the vocal folds and a lateral angle for muscular attachment. Through muscular action the arytenoids can be turned and the vocal cords approximated or moved apart.

The **epiglottis** is a leaf-shaped **elastic cartilage** which attaches by its narrow stalk to the inside of the thyroid cartilage, just below the superior thyroid notch. Its broad, free upper part projects upward behind the root of the tongue and in front of the entrance to the larynx. Between the lateral borders of the lower epiglottis and the arytenoid cartilages are the paired **vestibular** (*quadrangular*) **membranes.** Their upper free borders circumscribe the laryngeal orifice and are known as the **aryepiglottic folds.** The lower free borders of the vestibular membranes form the **ventricular folds** (*false vocal cords*) which have an attachment on the inside of the angle of the thyroid cartilage (Figs. 14–5, 7, 8).

The **corniculate cartilages** are small and conical and each rests upon the apex of an arytenoid cartilage. They are composed of yellow elastic cartilage.

The **cuneiform cartilages** are two small and elongated pieces of elastic

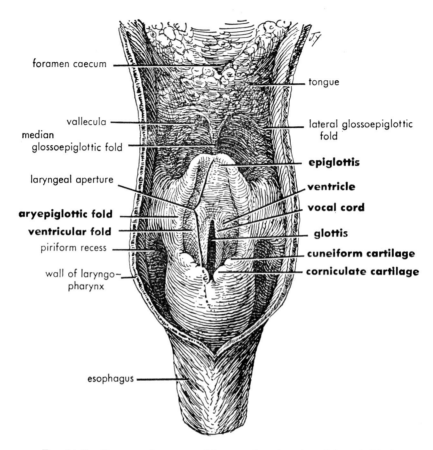

foramen caecum

tongue

vallecula

lateral glossoepiglottic fold

median glossoepiglottic fold

epiglottis

laryngeal aperture

ventricle

vocal cord

aryepiglottic fold

ventricular fold

glottis

piriform recess

cuneiform cartilage

wall of laryngo-pharynx

corniculate cartilage

esophagus

FIG. 14–7.   Laryngopharynx and laryngeal cavity viewed from behind.

cartilage which lie in the aryepiglottic folds in front of the arytenoids.

Laryngeal Cavity (Fig. 14–7).   One gains entrance to the **laryngeal cavity** through the triangular-shaped **laryngeal aperture** or **aditus laryngis.**   It is bounded anteriorly by the epiglottis, laterally by the aryepiglottic folds, and posteriorly by the arytenoid cartilage and the corniculate and cuneiform tubercles which are produced by the cartilages of the same name.   The first part of the cavity is the vestibule which is marked inferiorly by the **vocal folds** (*true vocal cords*) and the slit between them, the **glottis** (*rima glottidis*).   Between the vocal folds and the **ventricular folds** (*false vocal cords*) are the **ventricles of the larynx.**   This is a restricted cavity which

has an **appendix** anteriorly between it and the thyroid cartilage.   The appendix has a highly glandular epithelium which secretes a material that lubricates the surfaces of the vocal folds.

Below the vocal folds the remainder of the laryngeal cavity is at first elliptical in form, as seen in cross-section, but becomes circular before becoming continuous with the trachea below.

Muscles of the Larynx.   The muscles of the larynx may be divided into extrinsic and intrinsic.   Those called extrinsic connect the larynx with some neighboring part and some of them have been mentioned in Chapter 11.   The intrinsic muscles, totally within the larynx, are listed below, but it is beyond the scope of this account to describe them in

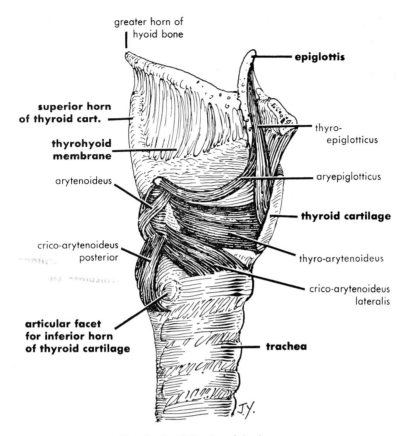

greater horn of
hyoid bone

epiglottis

superior horn
of thyroid cart.

thyro-
epiglotticus

thyrohyoid
membrane

arytenoideus

aryepiglotticus

thyroid cartilage

crico-arytenoideus
posterior

thyro-arytenoideus

crico-arytenoideus
lateralis

articular facet
for inferior horn
of thyroid cartilage

trachea

J.Y.

FIG. 14–8.   Muscles of the larynx.

detail.   However, their names suggest the cartilages which they connect and give some clue to their functions as does Figure 14–8.

Cricothyroideus
Arytenoideus
Thyro-arytenoideus
Crico-arytenoideus lateralis
Crico-arytenoideus posterior
Aryepiglotticus
Thyro-epiglotticus

Their actions are concerned with (1) opening and closing the glottis in keeping with its use at any moment and (2) increasing or decreasing tension on the vocal cords for high or low pitch of the voice.

Microscopic Anatomy.   All parts of the laryngeal cavity are provided with a **mucous membrane.**   The epithelium forming its superficial layer is stratified squamous on the posterior surface of the epiglottis, the upper part of the aryepiglottic folds, and over the vocal folds. The remainder is covered with pseudo-stratified ciliated columnar epithelium. The mucous membrane is closely adherent to the epiglottis and is thin and adherent to the vocal ligaments.   It actually forms much of the substance of the ventricular and aryepiglottic folds.

The mucous membrane is well supplied with mucus-secreting glands almost throughout, the free edges of the vocal folds being an exception.   They are particularly numerous on the epiglottis and small pits can be seen where they lodge on the cartilage itself.

The laryngeal mucosa is also extremely

sensitive and hence is readily stimulated by foreign particles, such as gas, smoke, or food substances, resulting in the protective cough reflex which forcefully expels such materials.

The thyroid, cricoid, and most of the arytenoid cartilages are of hyaline cartilage and usually begin to calcify at about twenty years of age and may in late life become bony. The remainder of the arytenoids, the epiglottis, corniculates, and cuneiforms are of elastic cartilage and do not ordinarily calcify or ossify.

TRACHEA AND BRONCHI
(Fig. 14–4)

The **trachea** or **windpipe** is a fibroelastic tube about 11 cm. ($4\frac{1}{2}$ inches) in length extending from the larynx, at the

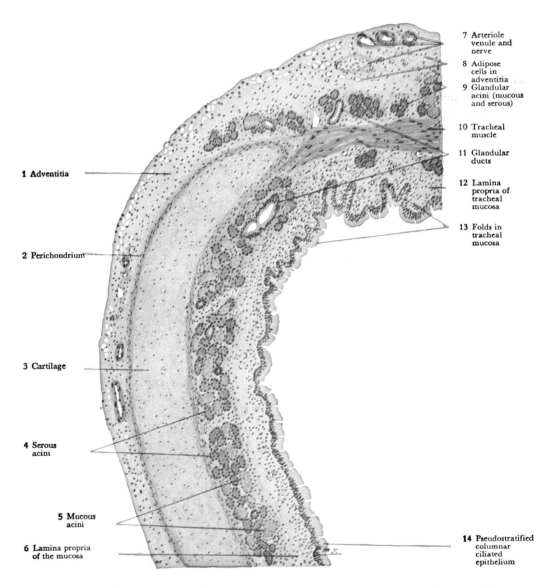

1 Adventitia

2 Perichondrium

3 Cartilage

4 Serous acini

5 Mucous acini

6 Lamina propria of the mucosa

7 Arteriole venule and nerve

8 Adipose cells in adventitia

9 Glandular acini (mucous and serous)

10 Tracheal muscle

11 Glandular ducts

12 Lamina propria of tracheal mucosa

13 Folds in tracheal mucosa

14 Pseudostratified columnar ciliated epithelium

FIG. 14–9.   Trachea (panoramic view, transverse section).   Stain: hematoxylin-eosin. 50×.
(di Fiore, *An Atlas of Human Histology*, Lea & Febiger.)

level of the sixth cervical vertebra, to the fifth thoracic vertebra where it divides into right and left bronchi. It is supported, except posteriorly, by **C-shaped hyaline cartilages** which prevent its collapse. The C-shaped cartilages are closed posteriorly by the **fibroelastic membrane** which also contains a few internal transverse and external longitudinal smooth muscle fibers.

The trachea lies against the anterior surface of the esophagus and is flanked by the large arteries, veins, and nerves of the neck. The isthmus of the thyroid gland crosses its anterior superior surface and in the mediastinum of the thorax it lies behind the thymus gland and the manubrium of the sternum. It is in close relationship to the large vessels entering and leaving the heart (Figs. 13–13; 14–4).

The **bronchi** are similar in construction to the trachea. The **right bronchus** is shorter and wider than the left and leaves the trachea at less of an angle. It divides into three subsidiary bronchi. The **left bronchus,** about twice as long as the right, divides into two subsidiary bronchi (Fig. 14–4).

Microscopic Anatomy (Fig. 14–9). The **mucous membrane** of the trachea and bronchi is similar to that of the larynx, nasopharynx, and nasal passageways. It is lined with **pseudostratified ciliated columnar epithelium** the cilia of which beat upward to carry mucus and contained materials into the pharynx. There are numerous goblet cells. There is a well-defined basement membrane under which there is a lamina propria of reticular tissue and elastic fibers containing many lymphocytes, small blood and lymphatic vessels and nerves.

The **submucosa** is a fibrous connective tissue with larger blood vessels and tubuloacinous glands, the **tracheal glands,** whose ducts open through the mucosa to the free surface of the epithelium.

MEDIASTINUM (Figs. 14–10, 11, 12)

The **mediastinum** is located in the middle of the thorax between the lungs and their pleurae. It is in the form of a broad septum, the **mediastinal septum,** comprising the expanded subserous fascia and thoracic viscera, exclusive of the

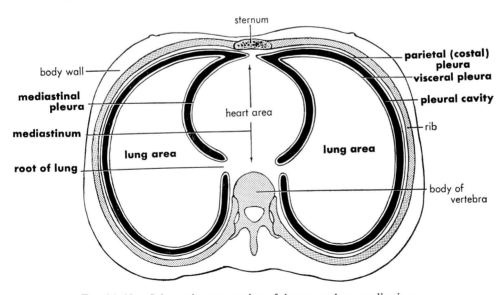

FIG. 14–10.   Schematic cross-section of thorax to show mediastinum, pleural membranes and pleural cavities.

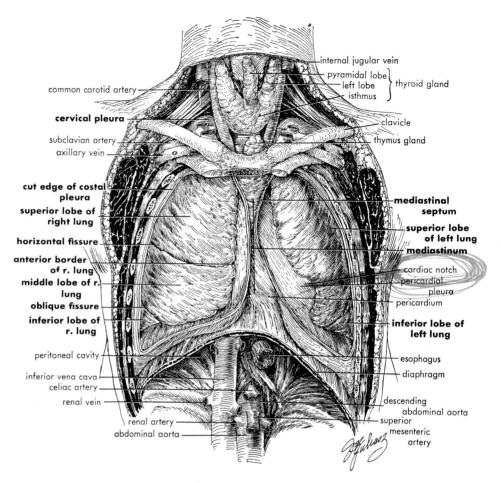

common carotid artery

internal jugular vein
pyramidal lobe
left lobe } thyroid gland
isthmus

cervical pleura

clavicle

subclavian artery
axillary vein

thymus gland

**cut edge of costal
pleura**
**superior lobe of
right lung**

**mediastinal
septum**
**superior lobe
of left lung**
mediastinum

horizontal fissure

cardiac notch
pericardial
pleura
pericardium

**anterior border
of r. lung**
**middle lobe of r.
lung**
oblique fissure
**inferior lobe of
r. lung**

**inferior lobe of
left lung**

peritoneal cavity

esophagus
diaphragm

inferior vena cava
celiac artery
renal vein

descending
abdominal aorta
superior
mesenteric
artery

renal artery
abdominal aorta

FIG. 14–11.   Anterior view of the opened thorax.   The parietal pleurae
have been removed anteriorly to reveal the lungs.

lungs.   It extends from the superior aperture of the thorax to the diaphragm and from the sternum anteriorly to the vertebral column posteriorly.   It is divided arbitrarily by a transverse plane extending from the sternal angle to the lower border of the fourth thoracic vertebra into a **superior mediastinum** above this plane and an "inferior" mediastinum below it.   The **"inferior" mediastinum** is further subdivided into anterior, middle, and posterior mediastina.

The **superior mediastinum** contains the thymus, large arteries and veins above the heart, parts of the trachea and

esophagus, the thoracic duct, important nerves such as the vagus, phrenic, and cardiac, and a few lymph nodes.

The **anterior mediastinum** is a narrow region in front of the heart containing a few lymph nodes and vessels and a thin layer of subserous fascia.   There is a firm attachment in its inferior part, the pericardio-sternal ligament.

The **middle mediastinum** is the broadest part of the mediastinum, since it contains the heart and its pericardial cavity.   It also contains the ascending aorta, the inferior half of the superior vena cava, the superior end of the azygos vein, the pulmonary artery, and the

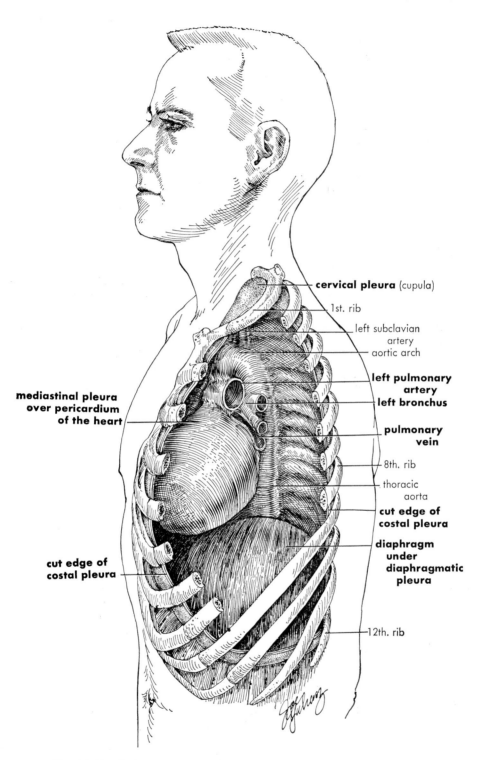

cervical pleura (cupula)

1st. rib

left subclavian
artery

aortic arch

**left pulmonary
artery**

left bronchus

**pulmonary
vein**

8th. rib

thoracic
aorta

**cut edge of
costal pleura**

**diaphragm
under
diaphragmatic
pleura**

12th. rib

**mediastinal pleura
over pericardium
of the heart**

**cut edge of
costal pleura**

FIG. 14–12.   Left thoracic cavity and mediastinal septum.   Left lung and
most of costal pleura removed.

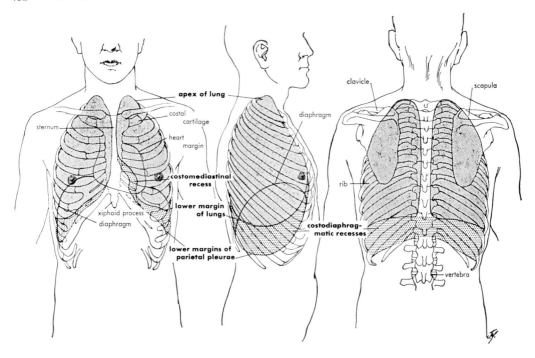

Fig. 14–13.   Diagrams of anterior, lateral, and posterior views of the thorax showing relationship of the pleurae and lungs to the chest wall.   The heavily stippled areas represent the full extent of the pleural cavities beyond the resting lungs.

pulmonary and phrenic nerves.   The end of the trachea, two bronchi and the roots of the lungs are considered by some anatomists to lie in this area.

The **posterior mediastinum** contains the thoracic aorta, azygos and hemiazygos veins, esophagus, thoracic duct, lymph nodes, vagus, and splanchnic nerves.

PLEURAE (Figs. 14–10, 11, 12, 13)

A double-walled sac of delicate serous membrane encloses each lung.   The innermost wall, the **pulmonary** or **visceral pleura,** is adherent to the lung and dips down into the fissures between its lobes.   The outer wall, the **parietal pleura,** lines the wall of the chest, covers the diaphragm, and is reflected over the viscera of the mediastinum. Between the parietal and visceral pleura

is a potential space, the **pleural cavity** (Figs. 14–10, 11).

Parts of the parietal pleura, although a continuous membrane, receive special names based upon their relationships to different parts of the thoracic cage. The **costal pleura** lines the inner surface of the ribs and intercostal muscles; the **diaphragmatic pleura** covers the convex surface of the diaphragm; the **cervical pleura** (*cupula, apical*) rises into the lower neck over the apex of the lung; the **mediastinal pleura** separates the other thoracic viscera from the lungs.

The parietal and visceral pleurae are continuous at the **root of the lung** where the blood vessels and bronchi connect with the lung.   At the lower border of the root of the lung the investing layers of pleura come together to form a mesenteric fold which is called the **pulmonary ligament.**   It extends from the lower part of the mediastinal surface of the lung to the pericardium and ends in a free border above the diaphragm. It holds the lower part of the lung in position. The parietal pleurae of the two sides make their closest contact just behind the sternum between the second and fourth costal cartilages.

The lung with its covering visceral pleura does not push into all the recesses formed by the parietal pleura. One such recess is where the periphery of the diaphragm is close to the costal wall, the **costodiaphragmatic recess,** right and left. The other, unpaired and on the left side behind the sternum and rib cartilages, the **costomediastinal recess** (Fig. 14–13).

The free surfaces of the pleurae are mesothelial and are smooth and moistened with a serous fluid, making a relatively friction-free arrangement for the movements involved in breathing.

LUNGS (Figs. 14–10, 11, 12)

The lungs are the essential organs of respiration. They provide the extensive epithelial surface, some 70 square meters, and the capillary networks which make possible the exchange of oxygen and carbon dioxide—**external respiration.**

The lungs normally lie free in their pleural cavities attached only at their roots and pulmonary ligaments. Adhesions are frequently found, however, between parietal and visceral pleurae due to pleurisy, an inflammation of the pleurae. Because the liver pushes the diaphragm higher on that side, the right lung is shorter and broader than the left, though it is also larger in volume. The left lung is longer, but the heart occupies a greater proportion of the left thorax, resulting in a smaller volume in the left lung.

The lungs are light and spongy and highly elastic. They show on their smooth surfaces many polygonal areas marking the lobules of the organ. The color of the lung varies from the almost red color in the fetus, to pink in childhood, to the various shades of gray or almost black in adults, depending upon the amount of impregnation with dust or carbon particles.

Borders and Surfaces. The form of the lung reflects the outline of the thoracic cage in which it is found. Each lung is somewhat conical in shape with a narrow apex and a broad base, costal and mediastinal surfaces, and anterior, posterior, and inferior borders.

The **apex** is narrow and rounded and extends into the root of the neck, from 1.5 to 2.5 cm. above the sternal end of the clavicle (Figs. 14–11, 12). The **base** is broad and concave to fit upon the convex surface of the diaphragm. The concavity is greater on the right lung than on the left. The base is bounded laterally by the thin, sharp **inferior border** which extends into the costodiaphragmatic recess of the parietal pleura during inspiration. The medial portion of the inferior border is blunt and rounded. The **costal surface** is large and convex to fit the curvature of the ribs and intercostal muscles. The **mediastinal surface** lies against the mediastinal pleura and contains a depression, the **hilum** through which the structures of the **root of the lung** enter and leave. The root of the lung is invested with pleura which, below the hilum, forms the **pulmonary ligament.** A concavity below the hilum is the **cardiac impression** for the heart, the impression being larger and deeper on the left lung. The mediastinal surface of the left lung also shows a groove for the aortic arch and the thoracic aorta. On the right lung surface the esophagus leaves a groove as do the superior vena cava and azygos vein. Near the apex on both lungs the subclavian arteries and innominate veins leave impressions. Other surface markings may be studied on Figures 14–11 to 13.

The **posterior border** of the lung is broad and rounded and rests in the deep groove on either side of the vertebral column. Below, it projects into the costo-diaphragmatic recess. The **anterior border** is thin and sharp and overlaps the pericardium, that of the right lung being almost vertical. A deep indentation on the left lung is the **cardiac notch** in which the pericardium is exposed. This brings the anterior margin at this point some distance lateral to the line of reflection of the corresponding pleura.

Fissures and Lobes. The fissures and lobes of the lungs are best understood by reference to Figure 14–14. The left lung is divided into two lobes, the right lung into three. Each lung is divided by an **oblique** (*interlobar*) **fissure** which starts about 6 cm. below the apex on the posterior border and runs downward and forward over the costal surface, reaching the inferior border near its anterior extremity. This divides the lungs into **inferior** and **superior lobes.** The right lung is further divided by a **horizontal fissure** which begins in the oblique fissure near the posterior border of the lung and runs forward to cut the anterior border on a level with the sternal end of the fourth costal cartilage. This gives the right lung **superior, middle** and **inferior lobes,** of which the middle is the smallest.

Bronchopulmonary Segments. The concept of smaller units or **segments** has been emphasized by thoracic surgeons and others in the healing arts. It is not within the scope of this book to present in detail the nature of "segments" but only to outline the anatomical basis for them. The right lung has 3 lobes and to provide for these the right bronchus has 3 **secondary** or **lobar bronchi.** From these secondary bronchi there are formed 10 **tertiary** or **segmental bronchi,** 2 for the upper, 3 for the middle and 5 for the lower lobe. A similar situation is seen in the left lung. It has 2 lobes and 2 secondary bronchi to service them. Each secondary bronchus furnishes 4 tertiary bronchi to take care of the 8 segments in the left lung. It is possible, knowing of this segmental identity within the lobes of the lungs, to remove diseased segments intact, with minimal influence upon the others (Fig. 14–14).

Minute Structure of the Lung (Figs. 14–15, 16). As the bronchi divide and subdivide to form the "respiratory tree" they become progressively smaller until they reach about 1 mm. in diameter when we call them **bronchioles.** As these divisions have taken place, the nature of the walls has remained similar, even to the presence of cartilaginous supports, although these cartilages have become more irregular in form, and, of course, less in amount. As the amount of cartilage has decreased, the relative amount of smooth muscle has increased. No cartilage is found in bronchioles of 1 mm. or less in size.

Ultimately, the subdivisions of the "respiratory tree" lead to **terminal bronchioles** which, with their dependent structures, make up the **respiratory units** of the lung. There may be fifty to one hundred of these units in each lobule and many hundreds of lobules in a lung.

A respiratory unit consists of the terminal bronchiole which gives rise to two or more **respiratory bronchioles** that have a few **alveoli** along their sides. The respiratory bronchiole gives rise to several **alveolar ducts** with a great number of alveoli and these are connected in turn with a variable number of **atria** which are irregular spherical spaces that also have alveoli about them. Finally, the atria have a number of **alveolar sacs** connected with them, with alveoli all around their circumferences. The alveolus is the ultimate subdivision of the respiratory tree and only its thin, plate-like epithelium and the endothelium of the blood capillaries, which cover them liberally, lie between the air and the blood. According to some histologists

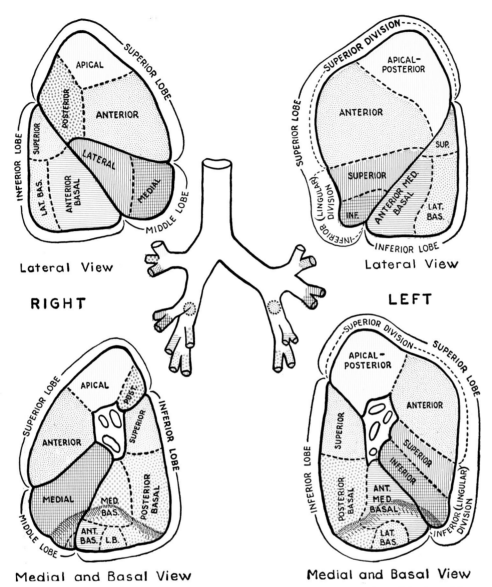

Lateral View

RIGHT

Lateral View

LEFT

Medial and Basal View

Medial and Basal View

FIG. 14–14.   The bronchopulmonary segments.   The segmental branches of the bronchi are shown in corresponding colors.   (After J. F. Huber by W. B. Stewart.)

the epithelium of the alveolus may be discontinuous, bringing air and blood even closer for diffusion.   **To recapitulate,** the respiratory units consists of these parts—terminal bronchiole, respiratory bronchioles, alveolar ducts, atria, alveolar sacs, and alveoli.   Alveoli are, of course, found all along the way from the respiratory bronchioles (Fig. 14–16).   The alveoli and their extensive capillary

supply constitute the units for the vital exchange of oxygen and carbon dioxide (Figs. 14–15, 16).

Breathing and Lung Function.   The lungs are passive in the process of breathing.   The pleural cavities around them are closed, whereas the insides of the lungs are in free communication with the outside atmosphere and subject to its pressure.   The closed thorax and pleural

LUNG (PANORAMIC VIEW)

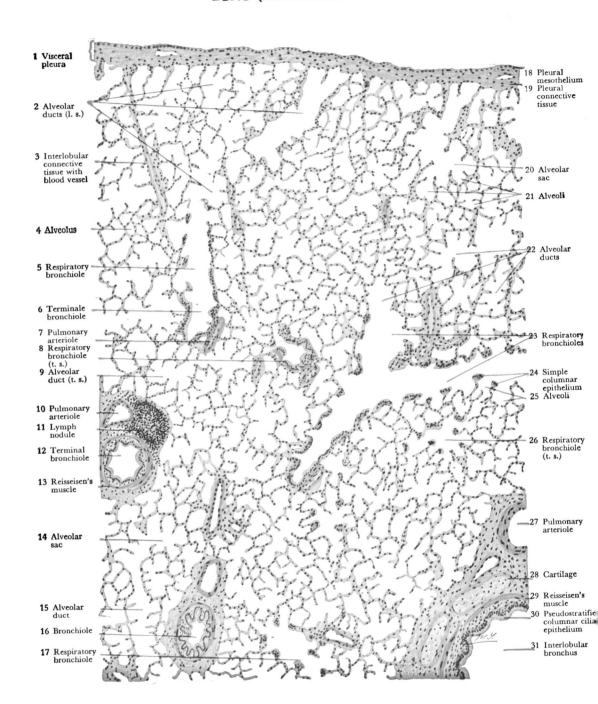

1 Visceral pleura

2 Alveolar ducts (l. s.)

3 Interlobular connective tissue with blood vessel

4 Alveolus

5 Respiratory bronchiole

6 Terminale bronchiole

7 Pulmonary arteriole
8 Respiratory bronchiole (t. s.)
9 Alveolar duct (t. s.)

10 Pulmonary arteriole
11 Lymph nodule

12 Terminal bronchiole

13 Reisseisen's muscle

14 Alveolar sac

15 Alveolar duct

16 Bronchiole

17 Respiratory bronchiole

18 Pleural mesothelium
19 Pleural connective tissue

20 Alveolar sac

21 Alveoli

22 Alveolar ducts

23 Respiratory bronchioles

24 Simple columnar epithelium
25 Alveoli

26 Respiratory bronchiole (t. s.)

27 Pulmonary arteriole

28 Cartilage

29 Reisseisen's muscle

30 Pseudostratified columnar cilia epithelium

31 Interlobular bronchus

Fig. 14–15.   Lung (panoramic view).   Stain: hematoxylin-eosin.   30×.
(di Fiore, *An Atlas of Human Histology*, Lea & Febiger.)

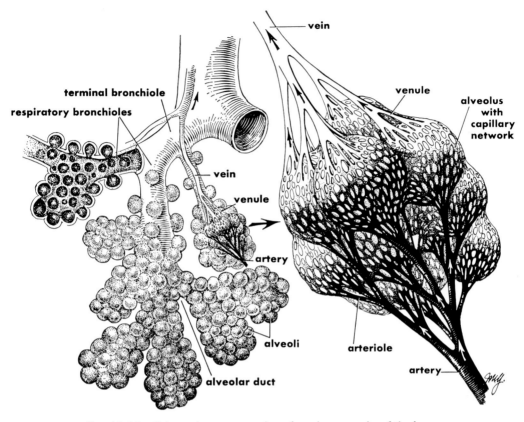

Fig. 14–16.   Schematic representation of respiratory units of the lungs.

cavities are enlarged by contraction of the diaphragm, intercostals, and other muscles, causing the internal pressure to fall and the lungs to enlarge. As the lungs enlarge the pressure drops within them, causing the outside air to rush in to fill the partial vacuum. This is **inspiration.** When the breathing muscles relax, the cavities in the thorax become smaller, the elastic tissue of the lung recoils, and air is forced out—**expiration.** Expiration in normal easy breathing is quite passive and the amount of air exchanged is about 500 cc. and is called **tidal air.** One can take a "deep breath" and follow this by a forced expiration, in which case the exchange of air may be 3700 cc. or more. This is called **vital capacity.** After forced expiration about 1200 cc. of air still remains in the lungs. This is **residual air.** Even when a lung is

collapsed, some air will remain which is trapped in the alveoli by the recoil of the elastic tissue of the lung. This is **minimal air.**

Since breathing is dependent upon the enlargement of closed cavities and the pressure relationships resulting, it is possible, by introducing air or otherwise destroying the vacuum, to prevent breathing. This is sometimes done to rest a lung which is infected with tuberculosis. It is called **pneumothorax.**

Breathing is accomplished by the action of voluntary muscles. We can control breathing at will. Yet, as anyone knows who has considered it at all, breathing goes on without the conscious attention of the individual. It is, in fact, critically regulated through a special center in the medulla oblongata of the brain, the **respiratory center.** This center is influenced by the amount of carbon dioxide in the blood. As the carbon dioxide level in the blood rises, it stimulates the center and impulses go to the respiratory muscles, and breathing, and therefore external respiration, are accelerated. In adults during rest, the normal rate of breathing is between 12 to 20 per minute. It is from 20 to 25 in children.

DISORDERS OF THE RESPIRATORY SYSTEM

Only a few of many disorders of this system are mentioned here as a kind of review of some of the facts of anatomy. A system as open as this one, from mouth or nose to alveoli of the lungs, is vulnerable to infections. Microorganisms are present in the mouth and nose from birth until death and may attack any part of the system. Inflammation of the nose, **rhinitis;** of the pharynx, **pharyngitis;** of the larynx, **laryngitis;** of the paranasal sinuses, **sinusitis;** and going deeper, **tracheitis,** and **bronchitis,** are all examples and perhaps manifestations of the common cold caused by virus. The ciliated and mucus-producing epithelia of many of the respiratory organs aid in keeping these passageways free, clear, and uninfected.

Special organisms cause special diseases such as pneumonia and tuberculosis involving the lungs. The latter may cause secondary infection in some other part of the body. Diphtheria and scarlet fever are dangerous not only because of local damage in the throat but because of metabolic products which do damage —sometimes fatal—in other parts of the body.

There are other disorders of the respiratory system which are noninfectious but none the less disturbing and sometimes fatal. So-called "hay fever" and asthma are among these. They are manifesta-tions of high sensitivity, in some individuals, to foreign proteins. Pollens, food, and dusts are common causes. In **hay fever** the mucous membranes of the nose are usually involved and they become enlarged with blood and secrete large quantities of mucus, causing one to sneeze and blow one's nose frequently. In **asthma** the smooth muscle tissue, so abundant in the small passageways of the lungs, is caused to contract, resulting in difficulty with breathing, particularly with expiration, which is normally passive. One feels as if he is suffocating because of the necessity of actively forcing out the air.

DEVELOPMENT OF THE RESPIRATORY SYSTEM

The development of the nose and pharynx has been considered in Chapter 13, page 359.

The remainder of the system starts development as a median longitudinal groove in the **endoderm** of the gut just caudal to the pharyngeal pouches. This groove deepens and closes from below upward and splits away from the gut to become the **laryngo-tracheal tube.** The anterior end of the tube, which maintains a slit-like opening (*glottis*) into the pharynx, is now at the level of the fourth branchial arch. The anterior end of the laryngo-tracheal tube will develop into the larynx, the middle portion into the trachea, and the lower rounded end, the **lung bud,** into the bronchi and lungs. The lung bud very soon bifurcates into right and left lung primordia (Fig. 14–17).

The cartilages, muscles and connective tissues of the **larynx** and **trachea** develop from the **branchial arches** of the pharynx, primarily arches four and five. The branchial arch material forms a dense mesenchyme under the epithelial lining of the larynx and in it the cartilages develop as follows:

Epiglottis	{	Third branchial arch (in part)		
		Fourth " " (?)		
Arytenoid	{	Fifth " "		
		Fourth " " (?)		
Thyroid	{	Fourth " "		
		Fifth " " (?)		
Cricoid		Fifth " "		
Cuneiform		Fourth " "		
Corniculate		Fifth " "		

The **tracheal tube,** with the **lung buds,** now elongates rapidly, reaching down about eight body segments beyond its original position. It is interesting that at this stage the right bronchus or lung bud extends quite directly caudad, whereas the left one makes a sharper angle with the trachea, a condition which prevails into adult life and accounts for the fact that when one swallows something "the wrong way," if it gets beyond the larynx, it usually ends up in the right bronchus (Fig. 14–17).

The **tracheal rings** develop in the condensed mesenchyme underlying the pseudostratified ciliated columnar epithelium. Glands also form, as invaginations from the epithelium. This occurs at about the fourth month.

The ends of the lung buds soon become lobulated, three **lobules** developing on the right and only two on the left. These subdivisions are the foundation for the lobes of the lungs which are seen in the adult. As growth continues the lobules undergo repeated subdivisions and become increasingly bush-like in appearance, ultimately ending in **alveolar sacs,** at the sixth month. The growing lung has become increasingly involved with the blood capillaries and units for external respiration are established. After birth and until middle childhood new alveolar sacs are formed and the parent ones are converted into **respiratory bronchioles.**

Our description so far has dealt mainly with the endodermal respiratory structures. These were located early in a

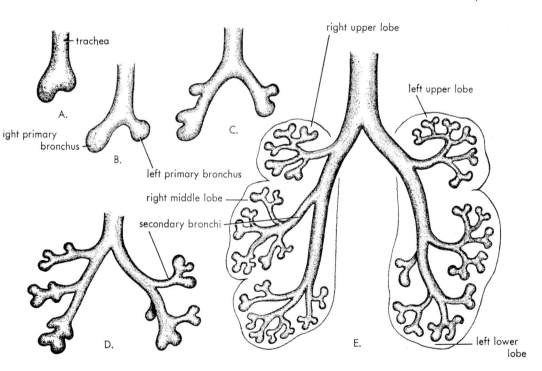

FIG. 14–17. Development of the lobes and bronchi of the human lungs.

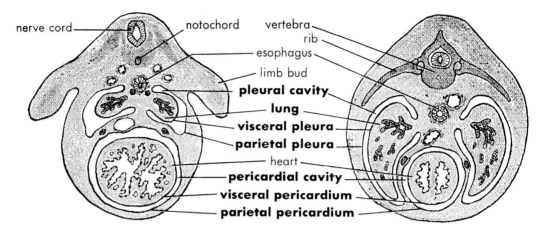

FIG. 14–18.   Scheme showing growth and development of lungs and pleural cavities in relationship to the heart and pericardial cavity.   (Modified from Patten.)

median mass of mesenchyme, dorsal and cranial to the main peritoneal cavity, in a broad mesentery-like structure later to become the **mediastinum.** As development progressed, the lung buds pushed out laterally and carried with them the mesenchyme and the surface **mesothelium.** As branching took place, the mesenchyme became intimately associated with the endodermal structures, forming their coats of muscle, connective tissue, and cartilage plates.

With the continued expansion of the lungs, they push into the spongy tissue of the body wall and migrate in a caudal direction, coming to lie on either side of the heart. When completed, the connective tissue and mesothelial covering of the lung become the permanent **visceral pleura,** that lining the thoracic wall, the **parietal pleura.** The visceral pleura came from the splanchnic mesoderm, the parietal pleura from the somatic mesoderm of the embryo (Fig. 14–18).

QUESTIONS

1. What is the primary function of the respiratory system and where is it carried out?
2. What is the role of the respiratory system in human communication?
3. In what way might the respiratory system aid the digestive system in some of its activities?
4. Of what significance are coughing and sneezing?
5. How does breathing aid the flow of blood and lymph?
6. Name the accessory organs of the respiratory system.
7. What is internal respiration? Respiration?
8. What are the requirements that must be met by an organ of external respiration?
9. Describe some of the special features of the nasal passageways which make them effective in warming, moistening, and cleansing the incoming air.
10. What function, other than those concerned with breathing, is served by the nose?
11. What separates the oral and nasal cavities?
12. What and where are the paranasal sinuses and how are they related to the nose?
13. How would you describe the location of the larynx?
14. Name the cartilages of the larynx.
15. Describe the vocal and ventricular folds.
16. How are the vocal cords manipulated to produce sounds of high and low intensity?

17. Define the aperture of the larynx.
18. What are some of the structural connections between the thyroid cartilage and the hyoid bone?
19. How are the vocal cords operated to regulate air flow through the larynx?
20. What are the ventricles of the larynx?
21. What kind of epithelium lines most of the accessory organs of respiration?
22. What is the importance of tracheal cartilages?
23. What contribution is made by the cilia and the mucous glands of the trachea?
24. Compare the right and left bronchi.
25. Describe the mediastinum.
26. Does the trachea have a serous membrane?
27. Describe the relationship of the lungs to the pleural membranes and pleural cavity.
28. What is contained in the root of the lung?
29. What is the significance of bronchopulmonary segments?
30. Describe a respiratory unit.
31. Describe breathing.
32. What are some of the possible consequences of a common cold as made possible by the anatomy of the respiratory system and related structures?
33. What is asthma and what feature of the anatomy of the respiratory system does it involve to produce its symptoms?
34. What contributions do the branchial arches make to the anatomy of the respiratory system?
35. From what germ layer is the epithelium of the lung derived?
36. Where does the laryngo-tracheal tube come from?

Chapter 15

The Urinary System

INTRODUCTION

Pathways of Excretion. As the cells of the body use food and oxygen, so also do they produce metabolic wastes. Of these wastes, carbon dioxide leaves the internal environment through the lungs along with small quantities of water and heat. The alimentary canal accounts for the loss of some carbon dioxide, water and heat, and, in addition, salts and the secretions of certain glands. The skin plays only a minor role in excretion.

Organs of the Urinary System. Of primary importance for the excretion and elimination of waste products are the organs of the urinary system. Of these organs, the kidneys perform the complex task of removing from the blood not only the toxic waste products of protein metabolism as urea and uric acid, but non-toxic material as water and inorganic salts. In doing this, the kidneys are serving the function of excretion and also that of regulating the composition of the internal environment, in terms of water content, osmotic relations, and acid-base balances. The other organs of this system, the **ureters, urinary bladder** and **urethra,** are involved only in the relatively simple processes of transportation, storage, and elimination of the urine (Fig. 15–1).

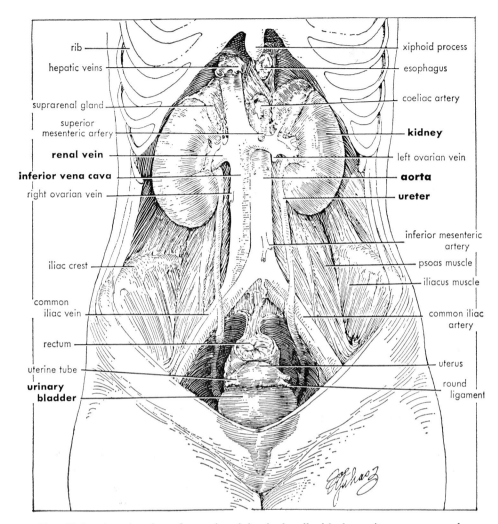

Fig. 15–1.   Anterior view of posterior abdominal wall with the peritoneum removed showing kidneys, ureters and related organs.

Special Considerations. A very close relationship exists between the excretory and reproductive systems. This relationship is closer among the lower classes of vertebrate animals than it is in man. Also, it is more apparent in embryological development than in the adult. Therefore, in many courses these systems are studied together as the urogenital system (Fig. 15–2). There is evidence that in the evolution of vertebrate animals there was a tendency for these systems to separate. In the adult females of primate animals, including man, the separation is quite complete and the reproductive and urinary systems have separate openings to the outside. In the males, however, the urethra still serves a double function, conducting sperms as well as urine. The **ductus deferens** (*sperm duct*) itself serves to carry urine in lower animals where it is called the mesonephric duct. This would appear to be an inefficient arrangement, for urine is not conducive to the well-being of sperm cells. Hence, special secretions from seminal vesicles and prostate gland are produced to give the sperms a more favorable

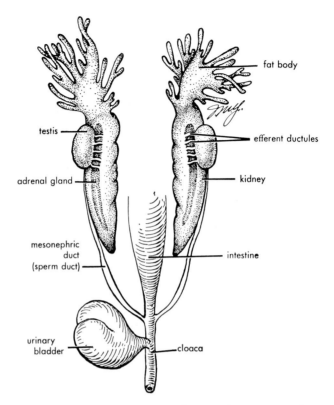

FIG. 15–2. The urogenital system of a male frog—ventral view.

environment for passage through the **urethra.**

KIDNEYS

Size, Location, and Attachments. The kidneys in man are in life reddish-brown organs which lie anterior and lateral to the twelfth thoracic and the first three lumbar vertebrae and behind the abdominal coelom. They are **retroperitoneal** which means that they lie behind the peritoneum. Although subject to variation, the left kidney is usually placed a little higher than the right (Fig. 15–1). They are approximately 11.25 cm. ($4\frac{1}{2}$ inches) in length, 5 to 7.5 cm. (2 inches) wide, and 2.5 cm. ($1\frac{1}{4}$ inches) thick. They are convex laterally and concave medially. The concavity on the medial side is called the hilus. The renal artery, vein, nerves, and the ureter join the kidney at the hilus, passing into the **renal sinus** (Fig. 15–3).

The kidneys, lying as they do behind the parietal peritoneum, are held loosely in place by a mass of fat called the **perirenal fat** or **adipose capsule** and by double layers of the **subserous fascia** between which the kidneys are placed. The latter is also known as the **renal fascia** (Fig. 15–4).

Each kidney is invested by a firm, strong, **fibrous capsule** which can be easily removed. In removing the capsule, small blood vessels and connective tissue processes are torn. An infant's kidney thus exposed is found to be lobed, a characteristic which is lost in the adult. Such lobing is found in the kidneys of many lower vertebrates.

Macroscopic Section. A frontal section of the kidney shows the three general regions of which it is composed, the

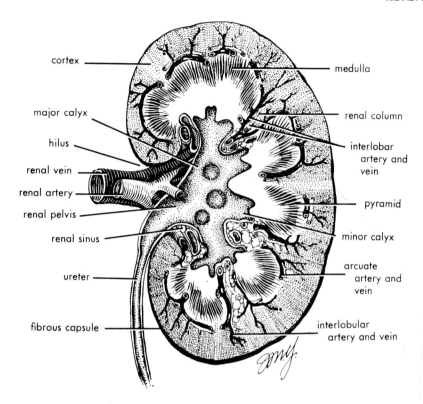

cortex

medulla

major calyx

renal column

hilus

interlobar
artery and
vein

renal vein

renal artery

renal pelvis

pyramid

renal sinus

minor calyx

arcuate
artery and
vein

ureter

fibrous capsule

interlobular
artery and vein

FIG. 15–3.   Posterior view of a frontal section of the kidney of man.

pelvis, the **medulla,** and **cortex.** These regions can be easily distinguished without the use of lenses (Fig. 15–3).

The **pelvis** is the expanded upper end of the ureter and, within the renal sinus of the kidney, it subdivides to form the **major** and **minor calyces.** The fibrous capsule of the kidney lines the renal sinus and is reflected back over the calyces and pelvis as their outer covering. The renal sinus also contains blood vessels and nerves, and the intervening spaces are filled with loose connective and fat tissue. There are from 4 to 13 minor calyces which lead into 2 or 3 major calyces.

The **medulla** is made up of the **renal pyramids.** These vary in number from 8 to 18 and have a striated appearance. Their bases are directed toward the periphery of the kidney, their apices (*papillae*) project into the minor calyces.

The **cortex** has a granular appearance and forms the peripheral layer of the kidney just under the fibrous capsule. The cortex arches over the bases of the pyramids of the medulla as the **cortical arches** and pushes down between the pyramids as the **renal columns.**

Microscopic Anatomy. The kidney is composed of closely packed tubular units, the **nephrons** or **renal tubules, collecting tubules,** and blood vessels, lymphatics, and nerves. A small amount of connective tissue extending inward from the fibrous capsule binds these structures together. The renal tubules and blood vessels are in close association and are very large in total surface area. This allows an efficient exchange of materials between them.

The Renal Tubules (*nephrons*). By microdissection methods, a renal tubule may be teased out from its close neigh-

28

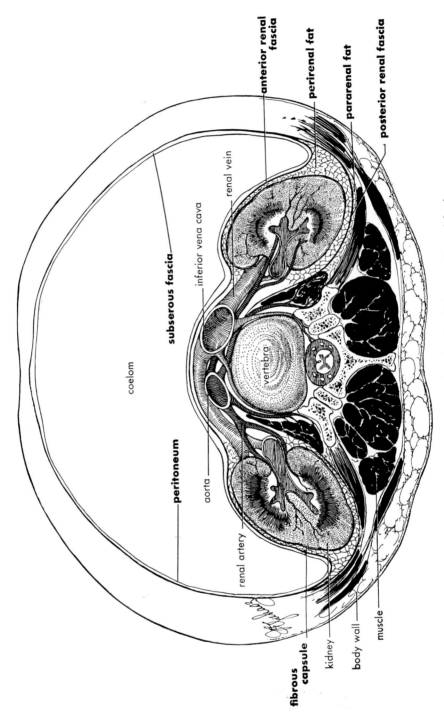

anterior renal fascia

perirenal fat

pararenal fat

posterior renal fascia

renal vein

subserous fascia

inferior vena cava

coelom

vertebra

peritoneum

aorta

renal artery

fibrous capsule

kidney

body wall

muscle

Fig. 15–4.   Transverse section showing kidneys and renal fascia.

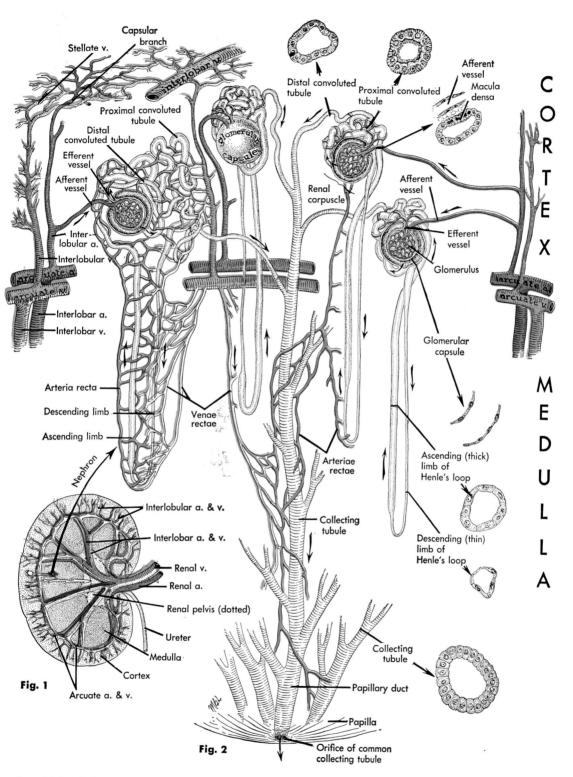

Fig. 1

Fig. 2

Fig. 15–5.   Schematic representation of kidney tubules and their relationship to the blood vessels.

bors and studied. Its beginning or proximal end is found in the cortex. It pursues a very irregular route in the cortex and medulla. Finally, being joined by other nephrons, it opens by a common collecting tubule at the apex of a pyramid into one of the minor calyces of the pelvis. There are one and one-fourth million or more of these nephrons in each kidney. They are the functional units of the kidney (Fig. 15–5).

The proximal end of the renal tubule is pushed in on itself to form a **glomerular** (*Bowman's*) **capsule.** This part of the tubule is composed of a single layer of flat or squamous cells, a **simple squamous epithelium,** which allows an easy filtration of materials from the blood (Figs. 15–5, 6). Leading from the glomerular capsule is a **proximal convoluted tubule** in which the cells are cuboidal, have central nuclei and a **brush border** on the lumen side (Fig. 15–5). The proximal convoluted tubule changes abruptly

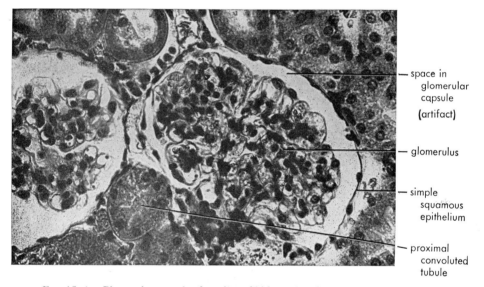

space in glomerular capsule (artifact)

glomerulus

simple squamous epithelium

proximal convoluted tubule

FIG. 15–6.   Photomicrograph of section of kidney showing two renal corpuscles.

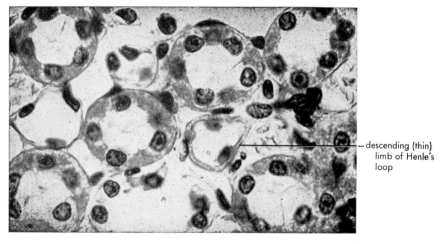

descending (thin) limb of Henle's loop

FIG. 15–7.   Photomicrograph of section of kidney showing thin loops of Henle and distal convoluted tubules.

as it nears the medullary region. The cells become flat, the tube narrows and dips into a pyramid as the **descending loop of Henle** (Fig. 15–7). Then the tube bends back on itself, enlarges, its cells become cuboidal, and as the **ascending loop of Henle** it returns to the cortical region (Fig. 15–5). Back in the cortex, the tube again becomes convoluted, the **distal convoluted tubule** (Fig. 15–5). Its cells are cuboidal with quite distinct cell boundaries and no brush borders (Fig. 15–7). This part joins a collecting tubule (Fig. 15–5).

As a distal convoluted tubule passes the root of the glomerulus, the area where afferent and efferent arterioles enter and leave its glomerular capsule, the tubule shows on one side, adjacent to the afferent arteriole, a heightening of its cells and a concentration of nuclei; a structure called the **macula densa.** One is shown in the upper right corner of Figure 15–5.

In the afferent arteriole adjacent to the macula densa the smooth muscle cells of the tunica media become modified. Their nuclei instead of being elongate are rounded and their cytoplasm, instead of containing myofibrils, contains granules. They are the **juxtaglomerular cells.** (See Fig. 15–5 adjacent to macula densa—not labeled.) The afferent arteriole in this area also loses its internal elastic lamina bringing its juxtaglomerular cells even closer to the lumen and its contained blood. The basement membrane of the distal convoluted tubule is absent in the macula densa thus bringing its cells into more intimate relationship to the juxtaglomerular cells. These two structures are sometimes referred to as the **juxtaglomerular mechanism.**

Although it is still in part speculative the following is the way in which the juxtaglomerular mechanism may work. The juxtaglomerular cells produce a substance **renin.** Its production appears to be increased when the blood pressure is low and when the juxtaglomerular cells are not stretched. Renin initiates a complex series of biochemical reactions which result in the elevation of blood pressure (*hypertension*). These reactions are briefly summarized as follows: renin + angiotensinogen (*from liver*) → angiotensin I

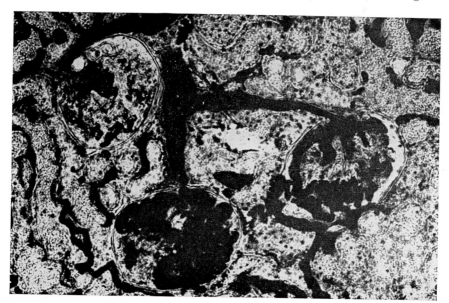

Fig. 15–8.   Photomicrograph of section of injected kidney showing afferent and efferent arterioles and glomeruli.

+ enzyme → angiotensin II which causing arteriolar constriction (*peripheral resistance*) → elevation of blood pressure. Normal kidneys also produce an enzyme, angiotensinase, which inactivates angiotensin II. In diseased kidneys the influence of renin may cause a permanent hypertension.

Collecting Tubules. The collecting tubules receive the distal terminations of many nephrons or renal tubules. They open into the minor calyces of the pelvis. Their lining cells are cuboidal proximally, and gradually, as the tubules become large and approach the calyces, the cells become columnar (Fig. 15–5).

Looking at the nephrons and collecting tubules in relationship to the gross regions of the kidney, one finds that the glomerular capsules and convoluted tubules lie within the cortex, while the loops of Henle and collecting tubules are in the medulla. Though each nephron is only about 3 cm. in length, it is estimated that the over two million found in human kidneys, if placed end to end, would extend about 50 miles, and would have an extensive surface area.

Blood Vessels. A casual glance at the size of the **renal arteries** and **veins** would make one realize the importance of the kidneys in the economy of the body. About 20 per cent of the blood volume of the body is pumped to the kidneys, an estimated 1700 liters per day.

The **renal arteries** branch from the abdominal aorta. Before reaching the kidneys they often divide into two vessels —an anterior and posterior—although there may be considerable variation. There are other subdivisions of the arteries in the sinus region of the kidney which then enter the kidney substance between the pyramids. These are the **interlobar arteries.** The interlobar arteries next form incomplete arches around the bases of the pyramids, the **arcuate arteries.** These, in turn, give off vertical branches into the

cortex, the **interlobular arteries.** The interlobular arteries give off nutrient branches to the cortical substance and to the capsule of the kidney (Fig. 15–3). Also, and most important, branches go to the glomerular capsules where, as **afferent arterioles,** they form knots of capillaries within the glomerular capsules called **glomeruli.** Here the relationship between glomerular capsules and blood capillaries is very close (Fig. 15–8). From the glomeruli, **efferent arterioles** lead out only to again form a network of capillaries, this time around the convoluted tubules and Henle's loop. These capillaries converge into **venules** which carry the blood into a system of veins (Fig. 15–5). The veins follow in much the same pattern as the arteries and have corresponding names. They lead into the **renal veins** which carry the blood into the **inferior vena cava** (Fig. 15–3).

Innervation. The kidney receives its **sympathetic** (*thoracolumbar*) nerve supply from the tenth to twelfth **thoracic nerves** by way of the **splanchnic nerves** and the **celiac plexus.** The **parasympathetic** (*craniosacral*) **nerves** are branches from the **vagus** which also run through the celiac plexus. The nerves enter the kidney with the blood vessels through the hilus. They terminate on the walls of the blood vessels rather than in the tubules of the kidney. Their function is believed to be vasomotor. Sympathetic stimulation results in reduced flow of urine. Vagus stimulation may influence the amount of solids excreted. A completely denervated kidney, however, continues to produce urine. The **hypothalamus** also participates in control of the flow of urine (Figs. 18–24, 45).

Functional Considerations. The relationships of the kidney tubules to the blood supply of the kidney and their histological structure give good clues as to how they function in excretion. The function may be divided into three parts. 1. **Filtration.** This takes place under the influence of *blood pressure* through the renal corpuscles. Examination of the filtrate from glomerular capsules demonstrates that it is similar to blood plasma, lacking only the plasma proteins. As much as 180 liters of the filtrate may be produced in a day. Obviously, much of this must be returned to the body, since it does not appear as urine. 2. **Reabsorption.** This complex process takes place through the walls of the convoluted tubules and Henle's loops. It results in reclaiming about 178 liters of the 180 given up by filtration. It is a highly selective process, requiring an energy output by the cells involved, in contrast to the passive nature of filtration. Water, glucose, salts and other products are thus reclaimed by the blood. 3. **Secretion by tubular epithelium.** The list of products excreted in this way is growing. Among them are hippuric acid and other derivatives of benzoic acid.

Kidney Disorders. Kidney problems may be in the form of tumors, stones, abscesses, and tuberculosis or degenerative diseases such as nephritis. In the first group, stones are most common, usually resulting from the deposit of salts on some foreign body. They may be found not only in the kidney pelvis but in the ureters or bladder as well.

Any agent which modifies the epithelium of the nephrons or the blood vessels of the kidney will modify kidney function. These may be infectious agents or their products or hardening of the arteries (*arteriosclerosis*), a common accompaniment of old age.

The examination of the urine is one of the oldest means of detecting and determining the nature of kidney disorders. Various substances, easily detectable in the urine, may be given to patients to test the efficiency of their kidneys. Blood tests also give valuable information about the kidneys.

Kidney operations were, in general, more successful than others in the early days before the development of aseptic surgery because of the position of the kidneys behind the peritoneum. The operation could be carried out without cutting the sensitive peritoneum, which is highly subject to infection.

There is a large "factor of safety" built

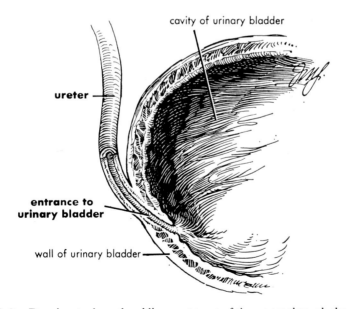

Fig. 15–9.   Drawing to show the oblique entrance of the ureter through the wall of the urinary bladder.

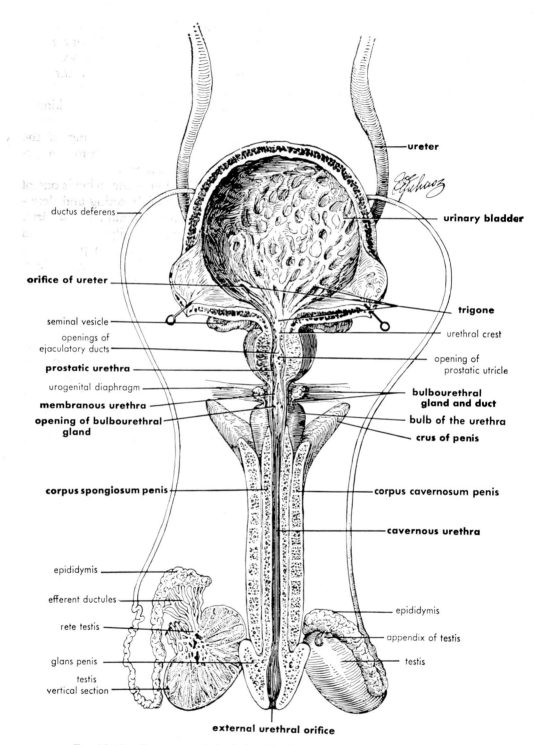

FIG. 15–10. Some urogenital relationships in the human male—anterior view.

into our kidneys, too. A person can live quite normally on one kidney.

URETERS

General. The urine comes from the collecting tubules into the minor calyces of the kidney pelvis. It then passes into the major calyces and into the funnel-shaped pelvis proper which tapers into the **ureter.**

The ureters are 28 to 35 cm. (10 to 12 inches) long. They run from the medial side of the kidney to the urinary bladder (Fig. 15–1). They lie behind the peritoneum and descend vertically in the subserous fascia anterior to the medial part of the Psoas major muscle. Entering the region of the true pelvis, they turn mediad, usually crossing over the common iliac arteries near the beginnings of the external iliac arteries. Dipping deeper into the pelvic cavity, they then swing anteriorly to join the urinary bladder at its posterolateral angles. They run for about 2 cm. obliquely through the bladder wall before opening into its lumen at its base. This arrangement serves as a valve to prevent the backflow of urine (Fig. 15–9). Just before the ureters reach the urinary bladder in the male, a **ductus deferens** loops from the lateral to the medial side around each one. The ductus deferens carries **sperm cells** (Fig. 15–10).

Microscopic Anatomy (Fig. 15–11). The ureters are not of uniform caliber, varying from 1 to 10 mm. as they course

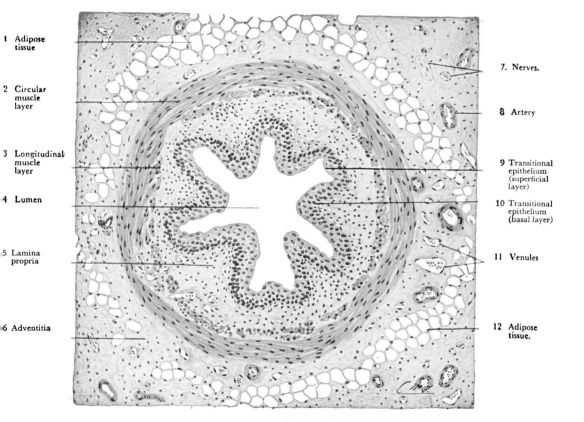

1 Adipose tissue
2 Circular muscle layer
3 Longitudinal muscle layer
4 Lumen
5 Lamina propria
6 Adventitia

7. Nerves.
8 Artery
9 Transitional epithelium (superficial layer)
10 Transitional epithelium (basal layer)
11 Venules
12 Adipose tissue.

FIG. 15–11. Ureter. Transverse section. Stain: hematoxylin-eosin. 50×.
(di Fiore, *An Atlas of Human Histology*, Lea & Febiger.)

from the kidneys to the urinary bladder. They have three coats in their walls; an inner **mucosa** lining the lumen, a middle **muscular,** and an outer **fibrous.**

The mucosa has a surface **transitional epithelium** of four to five layers of cells. There are no glands, nor is there a **basement membrane.** The epithelium lies directly on a **lamina propria** which is of dense connective tissue in its outer layers and contains networks of elastic fibers. There is no submucosa.

The **muscular coat** consists of inner longitudinal and outer circular layers. In the lower third of the ureter, an outer longitudinal layer is added. The muscle layers are not as distinct as in the walls of the intestine.

The **fibrous coat** (*tunica adventitia*) continues into the subserous fascia. It is related to the fibrous capsule of the kidney at the upper end and is continuous with the fibrous layer of the bladder below.

Function. The ureters actively propel the urine toward the urinary bladder by peristaltic contractions. This also gives enough pressure to overcome the resistance at the points where the ureters pass obliquely and for some distance through the bladder wall. Also, within the wall of the bladder where the ureters lose their circular muscle fibers, the contractions of longitudinal fibers in the lamina propria serve to open the lumina of the ureters.

URINARY BLADDER

General. The bladder is a hollow, muscular organ which stores the urine. In the empty or contracted condition, it lies in the true pelvis, posterior to the symphysis pubis, from which it is separated by a prevesical space. This space is filled with loose connective tissue that allows the filling bladder to move craniad. In the male, the bladder lies anterior to the rectum, seminal vesicles, and ductus

deferentes. In the female, it is anterior to the uterus and upper vagina (Fig. 15–1). When the bladder is filled, its base becomes a little depressed as the superior surface pushes into the abdominal region.

Internally, there are three openings in the urinary bladder wall, the two ureters described above and a urethra in the base. These three openings mark the corners of a triangle and delimit an area called the **trigone** (Fig. 15–10).

The urinary bladder is held loosely in position by true ligaments at its base and vortex and also by folds of the peritoneum which are reflected from the bladder to the wall of the abdomen. The bladder is also enclosed by the loose subserous fascia.

Microscopic Anatomy. The histology of the urinary bladder wall, though similar to that of the ureter, differs in the following ways:

1. There is a submucosa layer of loose (*areolar*) connective tissue.

2. The transitional epithelium of the mucosa has six to eight layers of cells. There are also some mucus-secreting glands. The transitional epithelium flattens as the bladder is filled (Figs. 15–12, 13).

3. There is a prominent external longitudinal layer of smooth muscle.

4. A dense mass of **smooth** (*involuntary*) **muscle fibers** forms a circle around the internal opening to the urethra—the **internal sphincter of the bladder.**

5. There is an outer serous layer over the upper part of the bladder. It is formed by the peritoneum.

URETHRA

The urethra extends from the urinary bladder to the body surface. It differs considerably in females and males.

The **female urethra** is about 4 cm. (1½ inches) long (Fig. 16–4). It is closely applied to the front wall of the

vagina and opens just anterior to the **vaginal orifice.** Its lining is thrown into longitudinal folds. Its mucosa is lined with transitional epithelium near the bladder, which gradually becomes stratified squamous epithelium at its external orifice. There are mucus-secreting cells present. Its **lamina propria** is of loose connective tissue with many elastic fibers. The lamina propria also has an extensive venous plexus, and hence a **cavernous** or **erectile** character. The **muscular coat**

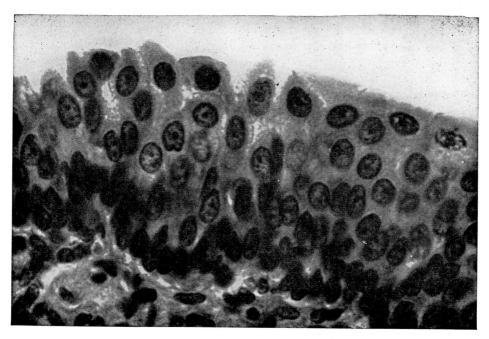

FIG. 15–12. Photomicrograph of section of urinary bladder showing contracted transitional epithelium such as is seen in the empty bladder.

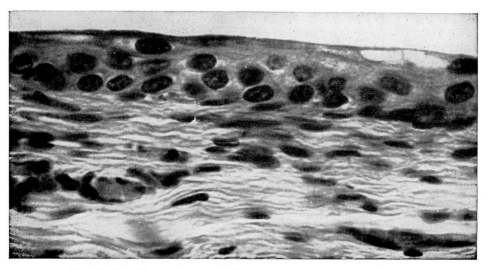

FIG. 15–13. Photomicrograph of section of urinary bladder showing stretched transitional epithelium.

consists of circular fibers continuous with those of the bladder. Where the urethra passes through the **urogenital diaphragm** striated (*voluntary*) circular muscle forms an **external sphincter.**

The **male urethra** (Fig. 15–10) is about 20 cm. (8 inches) long. It passes downward from the bladder, then forward for a short distance and finally downward again unless the penis is erect. It is divided into three parts—the **prostatic, membranous,** and **cavernous urethras.**

The **prostatic urethra** is about 3 cm. (1 inch) in length and at its superior end joins the urinary bladder (Fig. 15–10). At this point is found an involuntary **internal sphincter.** The prostatic urethra, as the name suggests, runs through the **prostate gland** to the **pelvic floor.** This part of the urethra is dilatable and of greater diameter than the remaining portions. Into its posterior wall empty the two ejaculatory ducts of the reproductive system. Between the two openings of the **ejaculatory ducts** is a raised area, the **urethral crest.** In the top of the crest is a small recess, the **prostatic utricle,** which is homologous to the uterus of the female. Sometimes it is called the **uterus masculinus.** The prostatic urethra also receives on its posterior wall the numerous ducts of the prostate gland.

The **membranous urethra,** about 1 to 2 cm. ($\frac{1}{2}$ inch) long, penetrates the region of the pelvic floor below the symphysis pubis, called the **urogenital triangle** (Fig. 15–10). This floor consists of an inner delicate fibrous membrane, below it is a transverse layer of voluntary muscle, the **urogenital diaphragm,** and externally a stout and fibrous **perineal membrane.** The voluntary muscle forms the **external sphincter** of the urethra.

Lying within the pelvic floor and to either side of the membranous urethra are the tiny **bulbourethral** (*Cowper's*) **glands.** They open into the cavernous urethra.

The **cavernous urethra,** the longest (15 cm.) of the three parts, lies in the central one of the three cavernous bodies of the penis, the **corpus spongiosum penis** (*cavernosum urethra*) (Fig. 15–10). It opens at the **urethral orifice,** a vertical slit, at the end of the penis.

Microscopic Anatomy. The male urethra is lined at its superior end with a **transitional epithelium** like that of the urinary bladder. The membranous and cavernous portions having a lining of **stratified** or **pseudostratified columnar epithelium** changing to **stratified squamous epithelium** near the urethral orifice. Occasional mucous goblet cells are found in the surface epithelium and there are many recesses (*lacunae of Morgagni*). These recesses continue into deeper, branching tubules known as the **glands of Littré.** Some of these, mostly on the dorsal surface of the cavernous urethra, extend into the **lamina propria** or even into the cavernous body.

The **lamina propria** of the mucosa is composed of loose connective tissue with numerous elastic fibers. It also contains a few smooth muscle bundles. There is no distinct **submucosa.**

Urination. The kidneys excrete the urine and the ureters, by peristalsis, force it into the urinary bladder where it is stored. The emptying of the bladder through the urethra is called **urination** (*micturition*).

As the urine accumulates, the bladder wall stretches under the influence of the rising pressure. When about three hundred cubic centimeters of urine are in the bladder the pressure is sufficient to stimulate afferent nerves which carry impulses to the central nervous system. Efferent nerves, in turn, bring impulses to the bladder wall and the sphincter muscles. The bladder contracts, the sphincters relax, and the urine is eliminated. This is essentially a reflex action, as any parent should know. However, the child can be "trained," as the signal for urination is felt, to retain the urine by keeping the **voluntary external sphincter** contracted. Voluntary contractions of abdominal muscles, while holding the breath, will increase intra-abdominal pressure and aid urination.

EMBRYOLOGY OF THE URINARY SYSTEM

The urinary organs, like those of reproduction, develop initially from a **urogenital ridge** which arises from the intermediate cell mass or nephrotome of the mesoderm (Fig. 4–10). Their ducts at first empty into a common chamber at the inferior end of the digestive tube, the **cloaca.** When the cloaca later divides into urogenital sinus and rectum, the urogenital ducts empty into the urogenital sinus (Fig. 15–14).

KIDNEYS

Man produces in his development a linear sequence of three kidneys—the **pronephros, mesonephros,** and **metanephros** (Fig. 15–14). Only the last remains as the functional kidney of the adult. This is essentially a recapitulation of the evolution of vertebrate kidneys. The pronephros, or "head kidney," is the earliest and simplest and is functional in the adults of only a few of the early chordates, such as the hagfishes, and in the developmental stages of fishes and amphibians. The adult kidney of fishes and amphibians is the mesonephros, or "middle kidney." It, in turn, gives way in reptiles, birds, and mammals to the metanephros, the "hind kidney."

All three of these kidney types are made up of **uriniferous tubules** which have a common origin in the urogenital ridge. The uriniferous tubules lead into a common longitudinal excretory duct which empties into the cloaca. Unlike other exocrine glands, where the glands are direct outgrowths of their ducts, the

uriniferous tubules of the kidneys develop independently, and secondarily join with their ducts.

Pronephros and **Mesonephros** (Fig 15–14). The uriniferous tubules of the **pronephros** are simple and short-lived. Their common duct, the **pronephric duct,** however, which ultimately reaches the cloaca, remains as the duct of the mesonephros, at which time it is called the **mesonephric** (*Wolffian*) **duct.** The tubules of the **mesonephros** are more numerous and complex than those in the pronephros. As the mesonephros degenerates to give way to the metanephros, some of its tubules remain to become a part of the male reproductive system.

Metanephros and Ureter (Fig. 15–14). At about the fifth week of embryonic life, a bud grows out from the lower end of the mesonephric duct, near the cloaca, and grows cranially to the caudal part of the urogenital fold where the **nephrogenic** tubules of the **metanephros** are developing. The end of the bud widens out to form the **pelvis** of the kidney with its **calyces** and associated **collecting tubules;** its unexpanded portion becomes the **ureter.** The collecting tubules become continuous with the nephrons (*glomerular capsules, convoluted tubules, and Henle's loop*) formed in the nephrogenic mass. The metanephros thus has a double origin—the nephrons, making up the cortex of the kidney, coming from the caudal end of the nephrogenic mass, the **medulla** and **renal pelvis** from the bud of the mesonephric duct.

URINARY BLADDER

As stated earlier, the cloaca divides into a urogenital sinus into which urinary and genital ducts empty, and a rectum, the terminal organ of the digestive system. The part of the urogenital sinus which is to become the bladder enlarges and becomes membranous and tapers

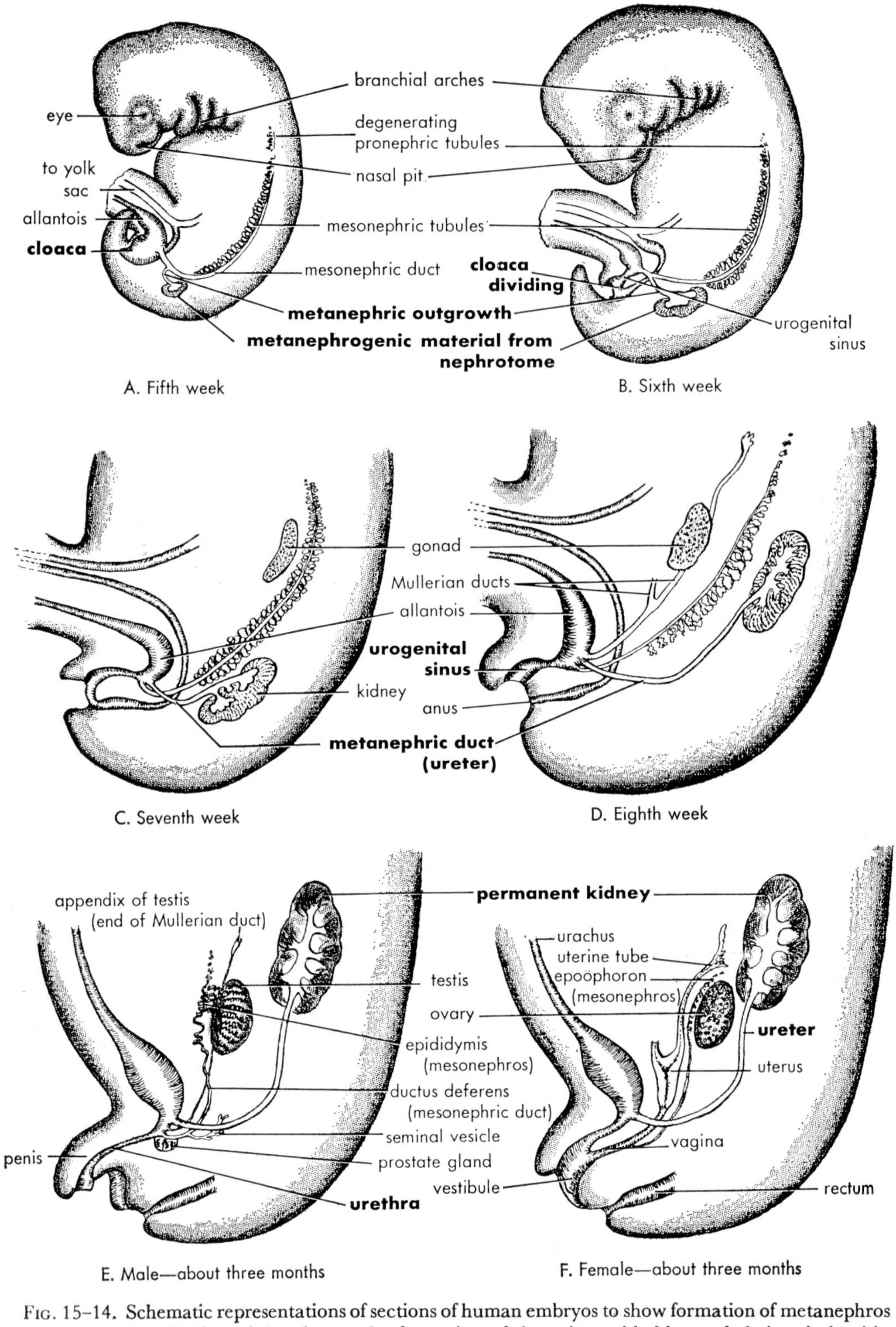

A. Fifth week

B. Sixth week

C. Seventh week

D. Eighth week

E. Male—about three months

F. Female—about three months

FIG. 15–14. Schematic representations of sections of human embryos to show formation of metanephros and its ducts, the division of the cloaca, the formation of the urinary bladder and their relationship to the organs of male and female reproductive systems. (Modified from Patten.)

upward into a tube, the **urachus,** which is, in turn, continuous with the **allantois** at the **umbilicus.** The urachus closes and becomes the **middle umbilical ligament** after birth. At its lower end, the bladder receives on each side the common duct formed from the mesonephric duct and ureter. As growth of the bladder continues, the joined ends of these ducts are absorbed and the ureters gain separate openings into the bladder; the mesonephric ducts shift backward and mediad to enter the pelvic portion of the urogenital sinus (Fig. 15–14).

Urethra and Urogenital Sinus

In the female, the **urethra** is derived by a lengthening of the short duct which leads from the urinary bladder into the urogenital sinus. The pelvic and phallic portions of the sinus form the vestibule into which the urinary and genital ducts empty.

The male urethra is longer and more complicated. Only the upper prostatic portion is equal to that of the female. The pelvic portion of the urogenital sinus becomes the rest of the prostatic and all of the membranous urethra. The phallic portion becomes the cavernous (*spongy*) urethra which is in the penis. It should be noted that most of the male urethra is really a urogenital canal comparable to the vestibule of the female.

The **prostate, bulbourethral,** and **urethral glands** of the male and the corresponding and homologous counterparts in the female, the **urethral, major vestibular,** and **minor vestibular glands,** are endodermal outgrowths of the urethra or vestibule (Fig. 15–10). Functionally, they serve the reproductive system (Fig. 16–5).

The **seminal vesicles** are derived from the mesonephric ducts (*ductus deferens*) and are therefore of mesodermal origin (Fig. 15–14*E*).

QUESTIONS

1. Describe the location of the kidneys in reference to the coelom and peritoneum.
2. What secures the kidneys in position?
3. Name some of the organs which lie in contact with the anterior surface of the kidneys.
4. What structures enter the hilus of the kidney?
5. What parts of the nephron are in the cortex of the kidney?
6. What structures constitute the pyramids of the medulla?
7. Discuss the sequence of kidneys in human development.
8. Describe some of the likenesses and differences between the male and female urethra indicating homologous structures.
9. Name some ways in which the excretory system of the male is related to the reproductive system.
10. Name and discuss the processes by which urine is formed in the kidneys and show how the anatomical relationships serve these processes.
11. Compare the histology of the walls of the ureters and the urinary bladder.
12. What is the trigone?
13. How is urine in the bladder kept from flowing back into the ureters?
14. Describe the mechanism involved in the emptying of the bladder.
15. Discuss the evidence for an evolution of vertebrate kidneys.

*"Owing to the imperfection of language the offspring is termed a new animal,
but is in truth a branch or elongation of the parent."*

—Erasmus Darwin

Chapter 16

The Reproductive System

INTRODUCTION

The Nature of Reproduction. The systems of the body which we have studied thus far have had to do with the support, movement, and maintenance of the **individual.** In reproduction while it deeply involves the individual, there is also an intimate **interpersonal** relationship necessary to perpetuate the species. As such, it leads to the establishment of the family unit. Further, societies of men have evolved very definite attitudes and laws relative to it. Organized religion, codes of morality, and of ethics include guidelines for man's reproductive be-

havior. Many taboos have appeared to confuse an already complicated sequence of events necessary to continue our kind.

Some of these attitudes, guidelines, and laws have not been in keeping with the biological facts and necessities of reproduction. Nor is there appreciable effort on the part of civilized man to insure a reasonable relationship between the time of reproductive maturity and economic self-sufficiency. This problem is further aggravated and enlarged by our attitudes toward material wealth as a mark of success and of status.

However, it is not the purpose of this

book to do more than point out some of the implications of the reproductive process in man. The anatomy of the reproductive system and its general physiology are sufficient to engage our full attention. And to understand these structures and processes is perhaps the best approach to successful use of the system under any circumstances.

Recall that the reproductive and urinary systems have a very close relationship during phylogeny and ontogeny (Fig. 15–2). Even in the adult male of the human species, organs like the urethra are used by both systems. Only in the adult females of primate mammals are the systems well separated (Figs. 16–1, 2).

Of interest too is the fact that while the anatomy of the sexes is distinct through most of the life of the individual, it is not so during the first two months of embryological development. In this "indifferent stage" the individual has all the building materials for either a male or a female system (Figs. 16–1, 2). Sex, being determined basically at the time of conception, will begin to manifest itself anatomically at about the eighth week. It may be further influenced by hormones, temperatures, and metabolic rates. It is, therefore, important to understand that each organ of one sex has a counterpart in the indifferent stage and also in the opposite sex. Such structures are called **homologues**. As an example the **penis** of the male and the much smaller **clitoris** of the female both come from the embryonic **phallus**, hence are homologous.

The Reproductive Organs (Figs. 16–3, 4, 5). The **essential** organs of reproduction are the **testes** of the male and the **ovaries** of the female. The indifferent organ from which these derive is the **gonad**. The testes and ovaries produce the **germ cells; the spermatozoa** and **ova** respectively. They also produce **hormones** which are influential in the devel-

29

opment of **secondary sexual characteristics** and in the regulation of the reproductive cycle. The female reproductive cycle is particularly complex and provides a splendid example of the interactions of endocrine glands. Because of the double function of testes and ovaries they are classified as both reproductive and endocrine organs.

The **accessory organs** of reproduction are those which serve the essential organs. In the male they consist of a scrotum which houses the testes; a system of ducts, the **efferent ductules, epididymis, ductus deferens** and **ejaculatory duct** which convey the sperms from the testis to the urethra; a number of glands, the **seminal vesicles, prostate,** and **bulbourethral** (*Cowper's*) which secrete fluids; and a copulatory organ, the **penis,** for transmission of spermatozoa into the female reproductive tract.

In the female the accessory organs are the **uterine** (*Fallopian*) **tubes** which receive and transmit the ova; the **uterus** in which embryo and fetus is housed and nourished during development, and the **vagina,** which opens to the outside serving for the passage of material from the uterus, including the fetus at term, and for the reception of the penis in copulation. External structures are collectively called the **vulva** or **pudendum** and consist of the **mons pubis, clitoris,** the **labia majora** and **minora,** and a thin membrane, the **hymen,** which partly closes the orifice of the vagina. The **greater vestibular glands** (*glands of Bartholin*) open near the vaginal orifice.

The **mammary glands,** though products of the skin, belong functionally to the reproductive system and are controlled, in part, by it (Fig. 16–19).

The Perineum. The perineum is the region of the outlet of the pelvis. It includes all of the structures between the symphysis pubis and the coccyx. The anterior triangular portion, containing the external urinary and reproductive

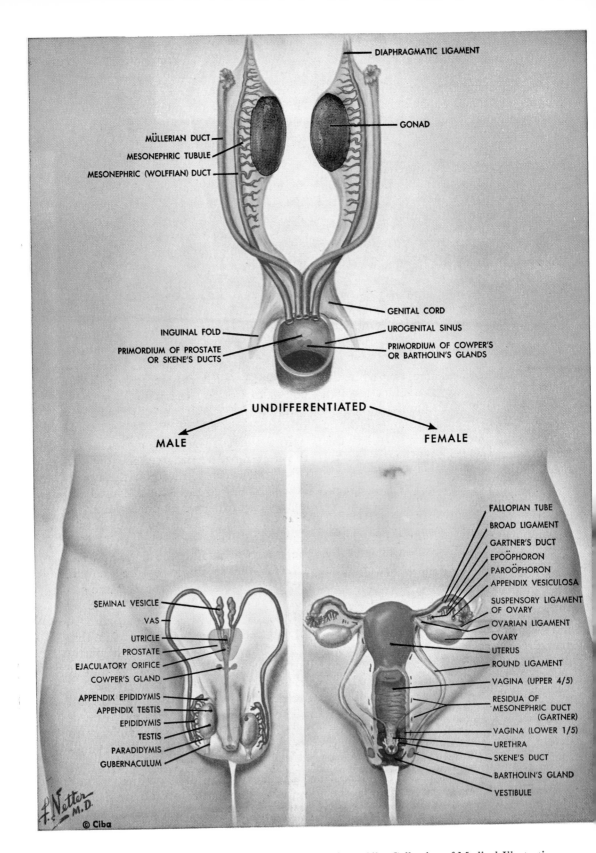

FIG. 16–1. Homologues of internal genitalia. (The Ciba Collection of Medical Illustrations, Vol. II, Reproductive System, Dr. Frank H. Netter.)

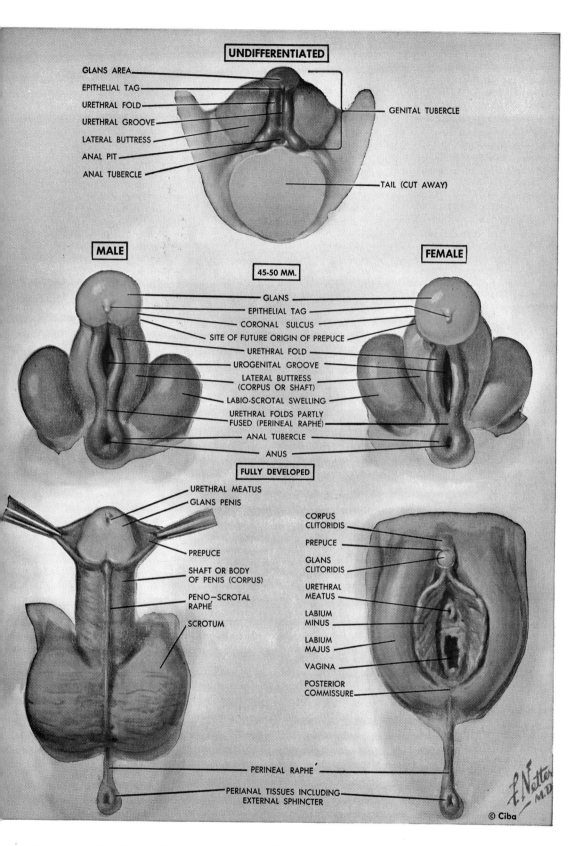

UNDIFFERENTIATED

GLANS AREA
EPITHELIAL TAG
URETHRAL FOLD
URETHRAL GROOVE
LATERAL BUTTRESS
ANAL PIT
ANAL TUBERCLE
GENITAL TUBERCLE
TAIL (CUT AWAY)

MALE **FEMALE**

45-50 MM.

GLANS
EPITHELIAL TAG
CORONAL SULCUS
SITE OF FUTURE ORIGIN OF PREPUCE
URETHRAL FOLD
UROGENITAL GROOVE
LATERAL BUTTRESS (CORPUS OR SHAFT)
LABIO-SCROTAL SWELLING
URETHRAL FOLDS PARTLY FUSED (PERINEAL RAPHÉ)
ANAL TUBERCLE
ANUS

FULLY DEVELOPED

URETHRAL MEATUS
GLANS PENIS
PREPUCE
SHAFT OR BODY OF PENIS (CORPUS)
PENO-SCROTAL RAPHÉ
SCROTUM

CORPUS CLITORIDIS
PREPUCE
GLANS CLITORIDIS
URETHRAL MEATUS
LABIUM MINUS
LABIUM MAJUS
VAGINA
POSTERIOR COMMISSURE

PERINEAL RAPHÉ
PERIANAL TISSUES INCLUDING EXTERNAL SPHINCTER

© Ciba

FIG. 16–2. Homologues of external genitalia. (The Ciba Collection of Medical Illustrations, Vol. II, Reproductive System, Dr. Frank H. Netter.)

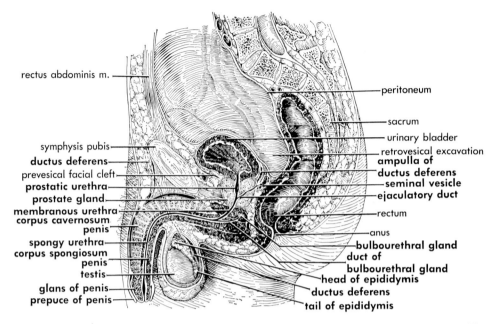

FIG. 16–3. Median sagittal section of male pelvis. (From Crouch, J. E.: Introduction to Human Anatomy: A Lab Manual. National Press Books, Palo Alto, 1966.)

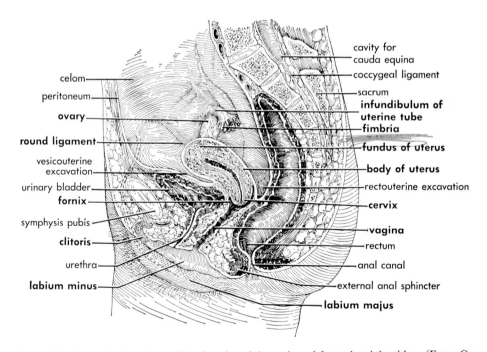

FIG. 16–4. Median sagittal section of the female pelvis as viewed from the right side. (From Crouch, J. E.: Introduction to Human Anatomy: A Lab Manual. National Press Books, Palo Alto, 1966.)

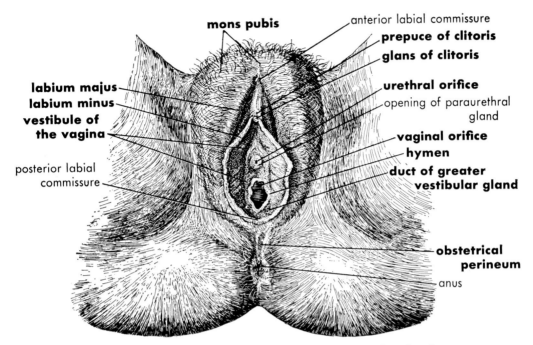

mons pubis

anterior labial commissure

prepuce of clitoris

glans of clitoris

labium majus

labium minus

**vestibule of
the vagina**

urethral orifice

opening of paraurethral
gland

vaginal orifice

hymen

**duct of greater
vestibular gland**

posterior labial
commissure

**obstetrical
perineum**

anus

**A. External genital organs of the female (pudendum).
The labia minora have been parted to reveal the
vestibule and the orifices of the urethra and vagina.**

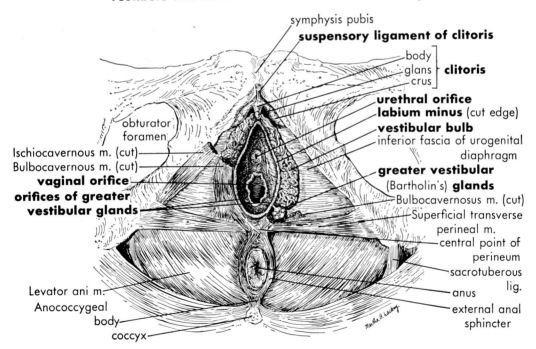

symphysis pubis

suspensory ligament of clitoris

body

glans   **clitoris**

crus

urethral orifice

labium minus (cut edge)

vestibular bulb

inferior fascia of urogenital
diaphragm

greater vestibular

(Bartholin's) **glands**

Bulbocavernosus m. (cut)

Superficial transverse
perineal m.

central point of
perineum

sacrotuberous
lig.

anus

external anal
sphincter

obturator
foramen

Ischiocavernous m. (cut)

Bulbocavernous m. (cut)

vaginal orifice

**orifices of greater
vestibular glands**

Levator ani m.

Anococcygeal
body

coccyx

**B. Dissection of the female genital organs.
Dissection on the left side is deeper than
that on the right.**

FIG. 16–5.   The external genital organs of the human female.

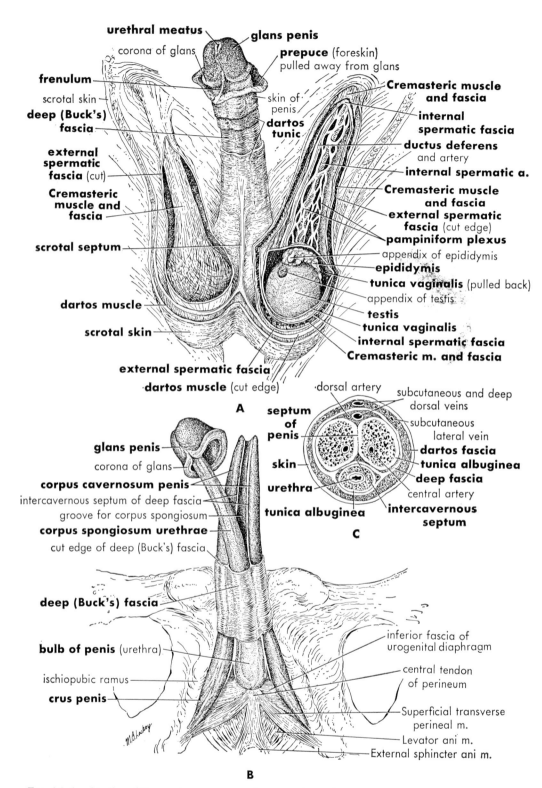

FIG. 16–6. Studies of the penis, scrotum, and spermatic cord. *A*, Drawing of superficial dissection of scrotum and penis. To the right is a deep dissection of the scrotum and spermatic cord. *B*, Deep dissection of the penis. All of the deep fascia is removed except one small segment. *C*, Cross-section of penis.

structures, is the **urogenital triangle;** the posterior portion, containing the anus, is the **anal triangle.** The region between the vaginal orifice and the anus is called the **obstetrical perineum,** because it is often torn during childbirth. To prevent this tearing of the perineum, it is now routine to make a clean incision in this area prior to delivery of the fetus. A clean cut is much more easily and neatly sutured than a ragged wound.

MALE REPRODUCTIVE SYSTEM
(Fig. 16–3)

Scrotum, Testis and Epididymis
(Figs. 16–6, 7)

The Scrotum. The **scrotum** is a medial, pendant pouch of loose skin and superficial fascia evaginated from the lower part of the anterior abdominal wall. A ridge or **raphe** divides it superficially into two lateral portions. The left lateral portion hangs somewhat lower than the right because the spermatic cord, holding the testis, is longer on that side. Involuntary muscle fibers, the **dartos,** lie within the superficial fascia of the scrotum. The **dartos** is subject to temperature conditions, to the extent that cold causes it to contract and the scrotum to pull up and wrinkle; warmth allows it to become elongate and flaccid.

The dartos and accompanying superficial fascia divide the scrotum internally into right and left compartments, each of which houses a testis, epididymis, and associated structures. Each compartment contains also a detached portion of the coelom, the serous membrane of which forms the **tunica vaginalis** of the testis (Fig. 16–6, 7).

The Testis and Epididymis (Figs. 16–6, 7). The **testes** are suspended in the scrotum by the spermatic cords. They average from 4 to 5 cm. in length, and weigh from 10.5 to 14 grams. They are oval in form and compressed laterally. Early in fetal life the testes are contained in the abdominal cavity behind the peritoneum. Before birth they descend and pass through the inguinal canal to reach the scrotum. During their descent they become invested by various coverings derived from the layers of the body wall. These will be discussed in greater detail on page 442.

The tunica vaginalis does not entirely cover the posterior surface of the testis and on this surface is the elongated, flattened **epididymis.** It has an enlarged **head** at the upper extremity, a central portion or **body,** and a pointed lower extremity, the **tail.** The tail is continuous with the duct of the testis, the **ductus deferens.** The head of the epididymis receives the **efferent ductules** which carry the semen and spermatozoa from the testis. The epididymis consists of a tortuous duct which, when straightened out, has a length equal to that of the small intestine, about 6 to 7 meters.

Appendages of the Testis and Epididymis (Fig. 16–7). At the upper end of the testis beneath the head of the epididymis is a small round body, the **appendix of the testis** (*hydatid of Morgagni*), which represents a remnant of the superior end of the **oviduct** (*Müllerian duct*). A similarly shaped body on the head of the epididymis, the **appendix of the epididymis,** is probably a detached and modified efferent duct.

Spermatic arteries enter and veins leave the testis on its posterior surface. The arteries are few and small, the veins large and tortuous and form a large plexus, the **pampiniform** (*tendril-like*) **plexus.** These structures plus the ductus deferens and nerves pass in the spermatic cord to the inguinal canal (Fig. 16–6*A*).

Microscopic Anatomy (Fig. 16–7, 8). If a testis is sectioned so as to reveal its inner structure, it will be found to have under the tunica vaginalis a tough fibrous coat, the **tunica albuginea.** Extending inward from this tunic are a number of incomplete fibrous septa which divide the

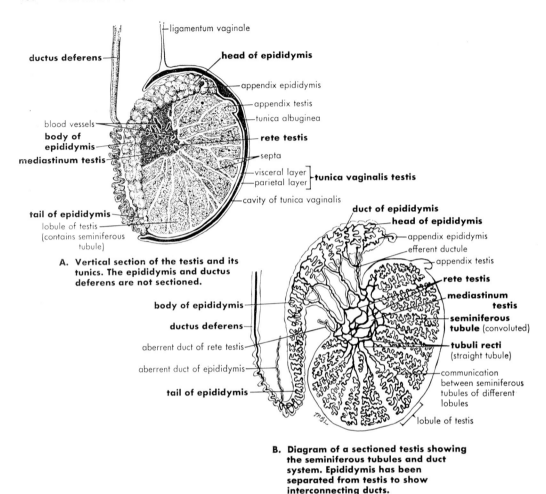

A. **Vertical section of the testis and its tunics. The epididymis and ductus deferens are not sectioned.**

B. **Diagram of a sectioned testis showing the seminiferous tubules and duct system. Epididymis has been separated from testis to show interconnecting ducts.**

Fig. 16–7. The testis, epididymis and ductus deferens showing their relationships.

testis into compartments. Each compartment contains one to three tiny tubules, the **seminiferous tubules** each of which averages about 70 to 80 cm. in length, over a half mile of tubule in each testis. On the posterior side of the testis, in an area colled the **mediastinum testis,** these tubules become straight, the **tubuli recti.** They then form a network of tubes, the **rete testis,** which in turn lead into the efferent ductules, which, as mentioned above, enter the head of the epididymis.

Study of sections of testes taken before sexual maturity reveals solid cords of epithelial cells, only an occasional one showing a lumen within. True lumina appear during puberty and the cords become true seminiferous tubules lined with an unusual stratified epithelium. The cords before puberty contain undifferentiated cells capable of producing both supportive and sex cells of the mature tubules. These sex cells, **spermatogonia,** however, do not proceed beyond this stage of development until puberty.

The intertubular stroma of the testes contain **interstitial cells** before puberty, but these do not produce male sex hormones until puberty.

After sexual maturity, cells of a great variety, representing stages in the development of sperms, and sperms themselves, appear in the walls of the seminiferous tubules. These are stages in spermatogenesis (*meiosis*) which was described on pages 23, 24 in Chapter 4 and illustrated in Figure 4–1.

The supporting cells are called **sustentacular cells** (*of Sertoli*) and are tall

and of irregular shape, and have large vesicular nuclei with prominent nucleoli. They rest upon a basement membrane.

The **interstitial cells** of the mature testis are cells of large size occurring individually or in clumps in the stroma among the seminiferous tubules. They produce the hormone, **testosterone,** which

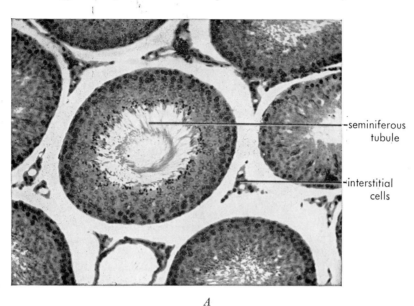

—seminiferous tubule

—interstitial cells

A

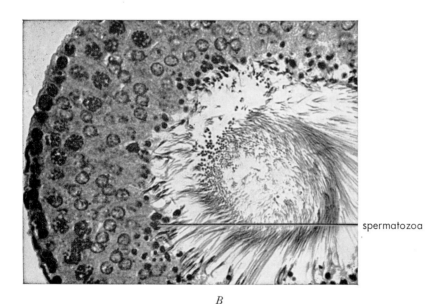

spermatozoa

B

FIG. 16–8.   Photomicrographs of testes of the rat.   *A*, Cross-sections of several tubules.   180×.
B, Cross-section of one tubule showing stages in meiosis.   720×.

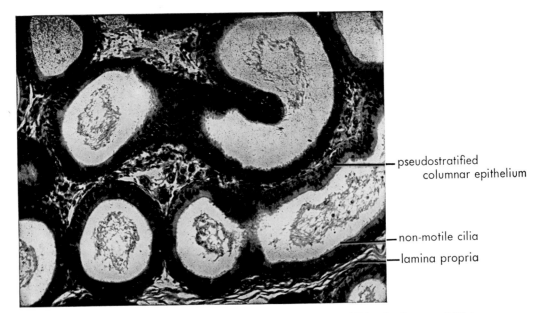

FIG. 16–9. Photomicrograph of a cross-section of the epididymis of a rat. 140×.

is responsible for the development and maintenance of secondary sexual characteristics and virility (Fig. 16–8A).

The **tubuli recti** are lined by simple columnar epithelium, while the **rete testis** has a lining of low columnar epithelium with a few cells each having a single flagellum projecting from its surface.

The **efferent ductules** have a variety of epithelial arrangements in their linings. There are alternating strips of low and high columnar cells and a few regions with pseudostratified epithelium. The low cells may have flagella, the high ones cilia. A thin layer of smooth muscle cells is seen in the lamina propria.

The **epididymis,** which is the principal storehouse for sperms, adds an essential secretion to the fluid in which the spermatozoa are activated and stored. It is lined with tall pseudostratified columnar epithelium with nonmotile cilia (Fig. 16–9). The epithelium rests on a distinct basement membrane. The fibrous lamina propria contains a thin layer of smooth muscle. Contraction of this muscle during ejaculation forces the spermatozoa into the ductus deferens.

DUCTUS DEFERENS AND SEMINAL VESICLE
(Figs. 16–5, 7)

The **ductus** (*vas*) **deferens** is a continuation of the duct of the epididymis. It ascends on the medial side of the epididymis to the groin where it passes through the inguinal canal (page 443). Entering the peritoneal cavity it crosses the lateral wall of the pelvis, passes medially, and crosses the ureter to reach the posterior side of the urinary bladder. Here it joins the duct of the **seminal vesicle** to form the **ejaculatory duct** which empties into the **prostatic urethra** (Fig. 15–10).

The **seminal vesicles** lie lateral to the ductus deferens on the posterior side of the fundus of the bladder (Fig. 16–3). They are coiled membranous pouches which secrete a fluid to be added to the secretion of the testes. They do not, as was once thought, store spermatozoa.

Microscopic Anatomy (Fig. 16–10). The wall of the ductus deferens contains three layers: (1) an outer or **areolar layer;** (2) a middle **smooth muscular layer** with inner circular and outer longitudinal fibers, except at the testicular end where there are both outer and inner longitudinal layers; and (3) an internal **mucous layer** thrown into longitudinal folds and covered with pseudostratified columnar epithelium which is nonciliated through most of the ductus, except at the testicular end.

The walls of the seminal vesicles have layers similar to those of the ductus deferens. The muscular wall is thinner, and the epithelium of the internal wall contains some goblet cells whose secre-

tions contribute to the bulk of the seminal fluid.

SPERMATIC CORD

The **spermatic cord** consists of the ductus deferens and all of the arteries, veins, lymphatics, and nerves that travel along with it. The coverings of the cord and testis are derived from the abdominal wall during the formation of the scrotum and the descent of the testis. They are

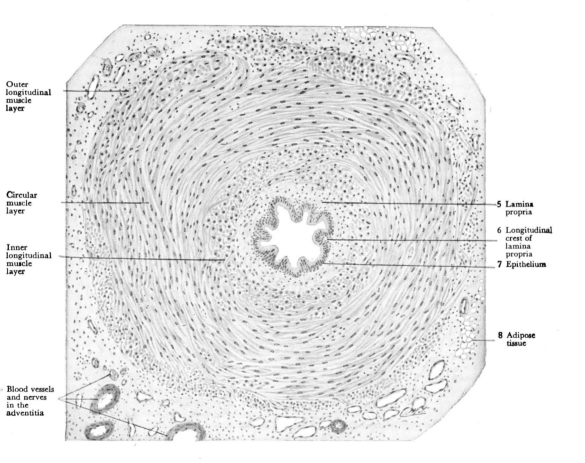

Outer longitudinal muscle layer

Circular muscle layer

Inner longitudinal muscle layer

Blood vessels and nerves in the adventitia

5 Lamina propria

6 Longitudinal crest of lamina propria

7 Epithelium

8 Adipose tissue

Stain: hematoxylin-eosin. 40×.

FIG. 16–10. Transverse section of ductus deferens.
(di Fiore, *An Atlas of Human Histology*, Lea & Febiger.)

listed in Table 16–1 below and illustrated in Figures 16–6 and 11*C*.

Formation of Scrotum and Descent of Testis (Fig. 16–11). To clarify the above relationships the early development of these parts is briefly described. The reader should also review the section pertaining to the inguinal ligament and canal on page 215.

Scrotal swellings appear in the skin in the lower ventrolateral abdominal wall about the seventh week of embryonic life. Corresponding evaginations of the peritoneum push from the inside into the

TABLE 16–1. COVERINGS OF THE SPERMATIC CORD AND THEIR ORIGIN

Coverings	Derivation
1. external spermatic fascia	1. aponeurosis of external oblique abdominal muscle
2. cremaster muscle and fascia	2. internal oblique abdominal muscle
3. internal spermatic fascia	3. transversalis fascia
4. loose connective tissue	4. subserous fascia
5. tunica vaginalis	5. peritoneum (*processus vaginalis*)

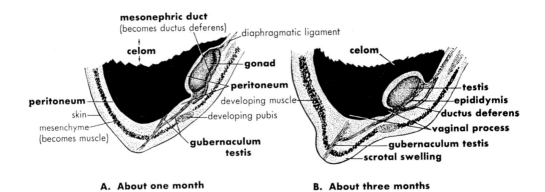

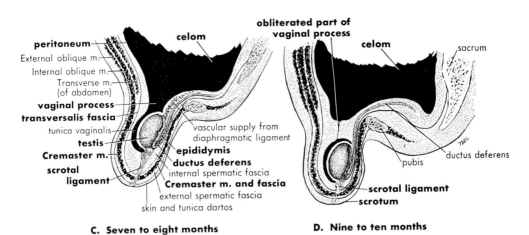

FIG. 16–11.   Sequence of events in the descent of the testis, *A–D*. The lower part of the celom and the vaginal process are indicated in black.

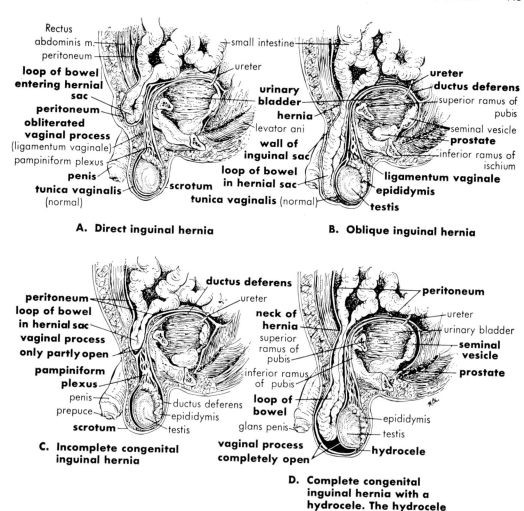

FIG. 16–12. Studies of different types of inguinal hernias.

developing scrotal swellings. Each is a **processus vaginalis.** In between the scrotum and processus vaginalis on each side, the various layers of abdominal muscles and fascia are carried downward also into the developing tube which is the **inguinal canal.**

Meanwhile the testis has been developing in the tissue of the urogenital ridge outside of the peritoneum on the posterior abdominal wall, and in close relationship to the mesonephros (Fig. 15–14). A fibrous connective tissue and smooth muscle structure, the **gubernaculum testis** forms and connects the testis and

epididymis to the inside of the scrotal swelling. As the scrotum grows, the gubernaculum, and hence the testis, are pulled backward. Also the gubernaculum slowly thickens, broadens, and shortens as the overall length of the body increases. As a result, the testis moves downward behind the peritoneum, guided by, and perhaps pulled by the gubernaculum, through the inguinal canal to its position in the scrotal sac. The gubernaculum testis, now much shortened, but still attached to the scrotal wall and testis, becomes the **scrotal ligament.** The testis as it descends drags the ductus

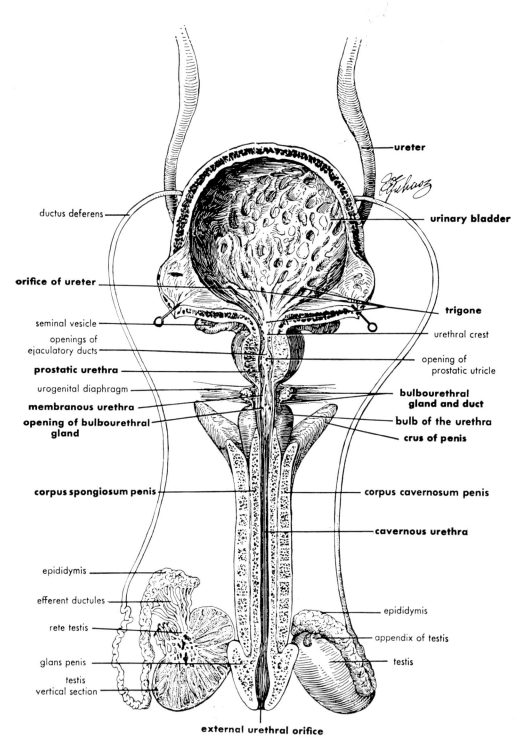

ureter

ductus deferens

urinary bladder

orifice of ureter

trigone

seminal vesicle

urethral crest

openings of
ejaculatory ducts

opening of
prostatic utricle

prostatic urethra

urogenital diaphragm

**bulbourethral
gland and duct**

membranous urethra

bulb of the urethra

**opening of bulbourethral
gland**

crus of penis

corpus spongiosum penis

corpus cavernosum penis

cavernous urethra

epididymis

efferent ductules

epididymis

rete testis

appendix of testis

glans penis

testis

testis
vertical section

external urethral orifice

FIG. 16–13.   Some urogenital relationships in the human male—anterior view.

deferens, blood vessels and nerves behind it, and these become the constituents of the spermatic cord which in turn are covered by the muscular and fascial layers of the abdominal wall.

The processus vaginalis having reached the bottom of the scrotal sac is gradually pinched off from the peritoneal cavity of the abdomen as the inguinal canal constricts. This leaves the distal part of the processus vaginalis isolated in the scrotum, enclosing a part of the true celom. It is now called the **tunica vaginalis testis.** The inner or **visceral layer** of the tunica vaginalis covers closely most of the testis and epididymis, but reflects back on itself as the **parietal layer** to leave the posterior border of the testis uncovered (Fig. 16–7). It is here at the **hilus** or **mediastinum** of the testis that the blood vessels, nerves, and ductus deferens enter from the spermatic cord.

Practical Implications (Fig. 16–12). Assuming that development has followed a normal course, as outlined above, the region of the inguinal canal and spermatic cord remains a weak point in the abdominal wall. As such, it is frequently the site of **hernia.** A hernia is a rupture or separation of some part of the abdominal wall often resulting in the protrusion of some part of the viscera, commonly the intestine. The situation can be chronic or critical, depending upon the nature of the hernia, but should always be called to the attention of a physician.

The intestine, for example, may push out above the medial portion of the inguinal ligament in which case it is a **direct inguinal hernia,** or it may push through the inguinal canal into the scrotum, an **oblique inguinal hernia.** In other cases a hernia may appear in the thigh below the inguinal ligament, a **femoral hernia,** or even in the umbilical region, another weak point in the abdominal wall.

Development does not always follow its usual normal course and may result in anomalous conditions of serious import. The testes may not descend into the scrotum as they should by about the eighth month of intrauterine life. Instead, they may stop at different points along the route, sometimes in the abdomen, other times in the inguinal canal, a condition known as **cryptorchidism.** When the testis remains in the abdomen, the individual is sterile, presumably because viable spermatozoa do not develop in the slightly higher temperatures there, as compared to those in the scrotum. Occasionally a testis may move from its normal course and end up in the thigh or even in the perineum. These are called **ectopic testes.**

Other anomalies result from the failure of the processus vaginalis to close. It is usually closed by birth or within a month thereafter. Closure usually starts at two points, the top of the epididymis and at the abdominal inguinal ring, and from these points gradually extends until the intervening portion is converted into a fibrous cord. Failing to close or only partially closing results in various types of **congenital inguinal hernia** as illustrated in Figure 16–12.

Hydrocele, the accumulation of fluid in the tunica vaginalis, is another condition which may be congenital.

PROSTATE AND BULBOURETHRAL (COWPER'S) GLANDS (Figs. 16–3, 13)

The **prostate,** previously referred to under the urinary system, is located below the bladder and around the superior end of the urethra. It lies in front of the rectum, through the wall of which it can be felt, providing a convenient way to massage it when it becomes congested. It is conical in shape and measures 4 cm. transversely at the base; 2 cm. anteroposteriorly: 3 cm. vertically. It is a lightly encapsulated organ with fibrous connective tissue and smooth muscle fibers. These tissues also extend inward to subdivide the gland into indistinct lobes. Its glandular elements are 30 to 50 in number and are of the compound tubuloacinous serous type. The lumina of the glands are large and irregular and are lined by simple columnar epithelium. Their secretion is thin, milky, and alkaline and escapes through a number of small ducts into the floor of the prostatic urethra. Recall that the ejaculatory ducts also traverse the prostate to join the urethra.

The prostate gland is incompletely developed before puberty, but the male sex hormone brings about its full growth. It tends to hypertrophy in later life and castration after puberty may cause it to atrophy. It is a frequent site of cancer.

The **bulbourethral glands,** about the size of peas, lie one to either side of the membranous urethra between the two layers of the fascia of the urogenital diaphragm. They open into the cavernous urethra near its upper end. They are compound tubuloacinous mucous glands, each made up of several lobules bound by fibrous connective tissue. The acini are lined with columnar epithelium.

PENIS (Figs. 16–3, 6, 13)

The **penis** is the copulatory organ by which spermatozoa are introduced into the female reproductive tract. As such, the essential parts of the organ are the three cavernous or erectile bodies and the urethra which passes through one of them. These are supported by fibrous connective tissue and the whole covered with a thin, loose skin. The skin at the neck of the penis, just in back of the **glans,** leaves the surface and folds back on itself to form the **prepuce** or **foreskin.** This is the structure which is removed in the surgical procedure called **circumcision.** On the glans itself the skin is devoid of hair, but has on its surface a number of highly sensitive papillae. Both the skin and the fascia of the penis are devoid of fat.

Two of the cavernous bodies are lateral and called the **corpora cavernosa penis.** They are surrounded by a double layer of dense fibrous connective tissue, the **tunica albuginea,** and are incompletely separated by a layer of the same tissue, the **septum penis.** They form part of the shaft of the penis. At the symphysis pubis the corpora diverge from each other to form the **crura.** Each **crus penis** just after separating from its fellow has an enlargement, the **bulb,** and at its termination in front of the tuberosity of the ischium has a bluntly pointed process. Further, each crus is firmly bound to the ramus of the ischium and pubis and is surrounded by fibers of the **Ischiocavernosus muscle** (Fig. 11–30).

The **corpus spongiosum penis** (*corpus cavernosum urethra*) is an erectile mass similar to that of the corpus cavernosum penis, but of finer construction, which surrounds the urethra and is median in position. Its midportion is of uniform thickness and forms a part of the shaft of the penis. Its posterior end is bulbous, the **bulbus penis** (urethrae) and is located at the membranous urethra superficial to the urogenital diaphragm. It is enclosed by the **Bulbocavernosus muscle.** The anterior end is expanded into a structure shaped like the cap of a mushroom and called the **glans penis.** It fits closely over the blunt rounded end of the corpora cavernosa penis and since its periphery is greater than that of the shaft of the penis, it forms a raised, rounded border. The urethra opens by a vertical slit-like orifice at the end of the glans.

The cavernous bodies consist internally of a sponge-like meshwork of endothelium-lined spaces. Sexual excitement results in the dilation of the tortuous arteries supplying the organ and these spaces become turgid with blood, causing a pressure to be set up against the fibrous sheaths of the corpora. The penis thus becomes stiff and erect, the blood being held in the cavernous bodies by compression of the deep veins. After ejaculation in which as many as 400 to 500 million spermatozoa may be emitted, or as sexual excitement subsides, the arteries contract and blood drains from the cavernous bodies and the penis again becomes flaccid.

Vessels and Nerves. The penis is a highly vascular and sensitive organ in keeping with the role it plays in reproduction. Most of its blood supply is from the internal pudendal artery, a branch of the hypogastric. Its venous return is mostly through veins of the same name. The internal circulation of the penis is extremely involved and beyond the scope of this book (Fig. 12–23).

The nerves of the penis are derived from the pudendal nerve, hypogastric and pelvic plexuses. The afferent fibers come from numerous receptors in the skin, prepuce, glans, and urethra and travel via the pudendal nerve. Sympathetic fibers reach the smooth muscle of blood vessels by way of the hypogastric and pelvic plexuses. Parasympathetic fibers carried in the pelvic nerve reach the arteries to cause vasodilation, and therefore, the erection of the penis.

Anomalies. The external urethral orifice is not uncommonly found in positions other than at the end of the penis, in the glans. In the condition known as **hypospadias** it may be located at any point along the underside of the penis or even in the perineum. In some situations it reminds one of the female external genitalia, the bifid arrangement extending to the scrotum, a possible persistence or influence from the indifferent sexual stage. **Epispadias** is the condition in which the urethra opens on the dorsum of the penis. Sometimes the penile portion of the urethra is entirely absent. Other anomalies which involve both the

30

male and female reproductive systems are considered later.

SUMMARY

To make clear the anatomical relationships and the general functions of the male reproductive system a brief summarizing review is in order (Fig. 16–13).

The **testes,** the essential organs of reproduction, have a double function of producing **spermatozoa** and the male sex hormone, **testosterone.** The spermatozoa are produced as a result of spermatogenesis (*meiosis*) in the walls of the convoluted **seminiferous tubules** of the testis. From here they pass through the straight seminiferous tubules, the **tubuli recti,** into the anastomosing tubes of the **rete testis,** in the **mediastinal region** of the testis. The sperms next move through **efferent ductules** into the head of the **epididymis.** The epididymis lying on the posterior side of the testis is made up of a mass of coiled tube and serves as a storage place for the spermatozoa. Its lining cells add a secretion to the seminal fluid and smooth muscle fibers aid in the movement of spermatozoa. From the epididymis the spermatozoa go into the **ductus deferens** by whose contractions they are carried upward to the point where the ductus joins the duct of the **seminal vesicles,** which also add their secretion to the seminal fluid and spermatozoa. The **ejaculatory duct,** extending from the point of union of the ductus deferens and the duct of the seminal vesicles carries the spermatozoa into the **prostatic urethra** where the **prostate gland** adds its milky secretion to the seminal fluid. The spermatozoa now make their way through the **membranous and cavernous urethra,** receiving contributions from the **bulbourethral glands** and are ejected through the **external urethral orifice.**

The above description of the passage of the spermatozoa is based upon the assumption that the individual is in a state of sexual excitement, with the penis erect, and that he reaches the point of crisis or **orgasm.** If not, the spermatozoa remain in the epididymis where after considerable time they die and are absorbed.

Ejaculation is brought about by peristaltic contractions of the excretory ducts, particularly the ductus deferens, by the contractions of the smooth muscle fibers of the prostate, and by the bulbocavernosus muscle. The secretions of the prostate gland are thought to activate the spermatozoa, enabling them to move through the uterus and uterine tubes after being deposited in the vagina.

The **testosterone,** the other product of the testes, being derived from the **interstitial cells,** serves the following functions:

1. Contributes to the development of the secondary sexual characteristics as hair distribution and growth, and voice changes.

2. Contributes to sex urge and sexual behavior.

3. Contributes to the development, maintenance, and functioning of accessory sex organs such as the seminal ducts, seminal vesicles, and prostate.

FEMALE REPRODUCTIVE SYSTEM

OVARIES (Figs. 16–4, 14, 15, 16)

The ovaries are paired, oval, and nodular bodies about 3 to 4 cm. in length by 2 cm. in width and less than 1 cm. thick. They lie to either side of the uterus on the lateral walls of the true pelvic cavity. They are attached to the posterior side of the broad ligament of the uterus by a short fold, the **mesovarium,** between the two layers of which the blood vessels and nerves pass to reach the hilus of the ovary. The ovary is further attached by the **ovarian ligament** to the side of the uterus, and to the pelvic wall by the **suspensory ligament** (Fig. 16–14).

Microscopic Anatomy and Function

(Fig. 16–16). Examination of a section through the mature ovary reveals three distinguishing features. (1) The surface is covered with a low simple columnar epithelium, the **germinal** (*ovarian*) **epithelium;** (2) beneath the epithelium is a connective tissue **stroma;** (3) through the stroma are found a variable number of **ovarian follicles,** each containing an **ovum.**

The **germinal epithelium** is continuous with the simple squamous epithelium or **mesothelium** of the abdominal peritoneum. The germinal epithelium is the source of the ova and follicle cells during fetal life.

The **stroma** is composed of an outer **cortex** of soft connective tissue and numerous spindle-shaped cells which resemble smooth muscle fibers, but they lack cytoplasmic fibrils and are not contractile. These cells make it possible to identify an ovary even when ova and follicles are not present in the cortex. The stroma, just beneath the germinal epithelium, is composed of dense fibrous connective tissue and is called the **tunica albuginea.** At the hilus of the ovary where the mesovarium attaches, the tissue of the stroma is loose and here the blood vessels, lymphatics and nerves enter. This is the **medulla** of the ovary and contains some smooth muscle fibers, but no ova or follicles are present (Fig. 16–16).

The basic functions and importance of the ovary center in the ova and their follicles. Here the maturation (*meiosis*) of the ova is initiated, and the follicles and structures derived from them are the sources of important hormones.

It is estimated that there may be as many as 400,000 follicles in both ovaries at birth. Yet only about 400 of these mature during the reproductive life of a woman. The others, with their ova, degenerate at various stages in their development, both before and after puberty. Follicular degeneration is called

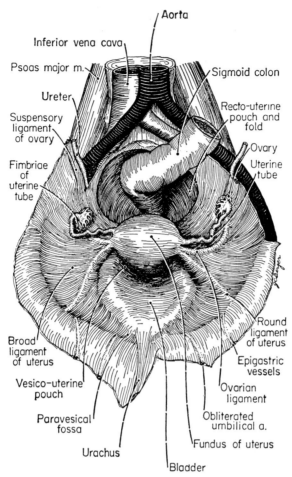

FIG. 16–14.   The peritoneal relations of the pelvic organs in the female, as seen from above.
(Woodburne, *Essentials of Anatomy*, courtesy of Oxford University Press.)

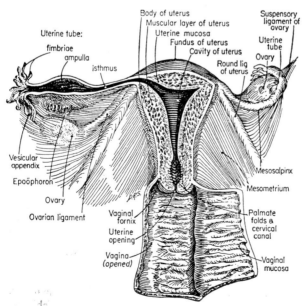

FIG. 16–15.   The internal genital organs in the female, partly in frontal section.
(Woodburne, *Essentials of Anatomy*, courtesy of Oxford University Press.)

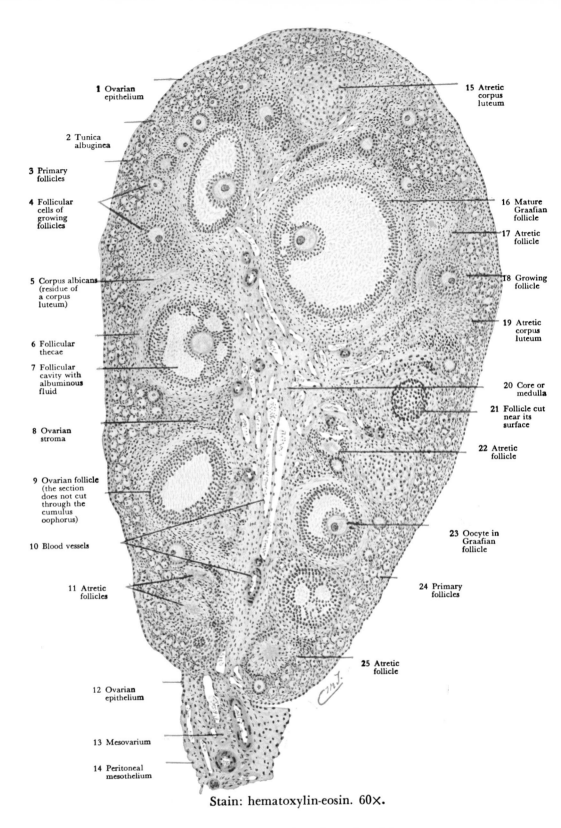

1 Ovarian epithelium

2 Tunica albuginea

3 Primary follicles

4 Follicular cells of growing follicles

5 Corpus albicans (residue of a corpus luteum)

6 Follicular thecae

7 Follicular cavity with albuminous fluid

8 Ovarian stroma

9 Ovarian follicle (the section does not cut through the cumulus oophorus)

10 Blood vessels

11 Atretic follicles

12 Ovarian epithelium

13 Mesovarium

14 Peritoneal mesothelium

15 Atretic corpus luteum

16 Mature Graafian follicle

17 Atretic follicle

18 Growing follicle

19 Atretic corpus luteum

20 Core or medulla

21 Follicle cut near its surface

22 Atretic follicle

23 Oocyte in Graafian follicle

24 Primary follicles

25 Atretic follicle

Stain: hematoxylin-eosin. 60×.

Fig. 16–16.   Panoramic view of a section of the ovary.
(di Fiore, *An Atlas of Human Histology*, Lea & Febiger.)

atresia, signifying an abortive process. After puberty the ovary undergoes cyclic changes relative to its ova and follicles until the menopause, the period when childbearing ceases.

With puberty follicular growth is stimulated by a follicle-stimulating hormone (FSH) from the pars distalis (*anterior lobe*) of the hypophysis (*pituitary*). Some of the cells of primary follicles begin to proliferate and form a stratified layer, while the contained ovum undergoes meiosis. Connective tissue cells form a theca or fibrovascular coat around the outside of the developing follicle. As the follicle cells continue to proliferate and pile up, a small cavity or cavities appear among the cells in which there is a follicular fluid. The cavity ultimately becomes large and the ovum is displaced to one side where it rests in a mass of follicle cells the cumulus oophorus. The lining of the follicular cavity is the membrana granulosa. This is a vesicular (*Graafian*) follicle and its component parts are shown in Figure 16–17.

As the vesicular follicle matures, it moves toward the surface of the ovary where it causes a bulge, and ultimately ruptures through the epithelium and tunica albuginea, casting the ovum with a ring of follicle cells, the corona radiata, into the celom. This is called ovulation and takes place once each twenty-eight days. The rupture of the follicle is accompanied by a little bleeding and sometimes by referred pain, and the basal temperature of the individual is elevated slightly for twelve to fourteen days. This elevated temperature is used by some physicians in helping women who have been unable to conceive. By identifying the time of ovulation, by keeping temperature charts, they can be advised as to the best time for sexual intercourse with the likelihood of conception.

A number of follicles grow during each twenty-eight-day cycle, but only one reaches full maturity and ruptures into the celom. The others undergo atresia and may leave "scars" in the stroma, the corpora atretica. Presumably the ovaries alternate in producing functional ova. Their numbers, one each twenty-eight days, seem trivial compared to the males' continuous contributions of millions of spermatozoa, but the population explosion of today suggests otherwise.

Following ovulation the cells of the ruptured follicle remaining in the ovary undergo rapid increase in size and form a yellow body or corpus luteum. These changes are brought about under the influence of the luteinizing hormones of the pars distalis of the hypophysis.

The status of the corpus luteum depends upon the fate of the ovum. If the ovum is not fertilized, the corpus luteum (*of menstruation*) persists for only about two weeks and then degenerates into a white scar, the corpus albicans. If the ovum is fertilized the corpus luteum (*of pregnancy*) persists for about six months before degeneration starts.

The cells of the developing vesicular follicles and the corpus luteum also produce hormones which have specific effects upon the reproductive life of the individual, and upon the uterus in particular. They also have reciprocal action upon the hypophysis.

The hormone from the follicle is called the female sex hormone or estrogen (other names are theelin, estrin, and folliculin). Its functions are: (1) development of the female secondary sexual characteristics at puberty; (2) stimulation of the lining of the uterus (*endometrium*) in terms of its cyclic proliferation of cells, and in glandular and vascular development; (3) stimulation of the development of the mammary glands; (4) contribution to sexual behavior and sex urge and (5) inhibition of the formation of follicle-stimulating hormones of the hypophysis.

The hormone of the corpus luteum is progesterone which serves the following

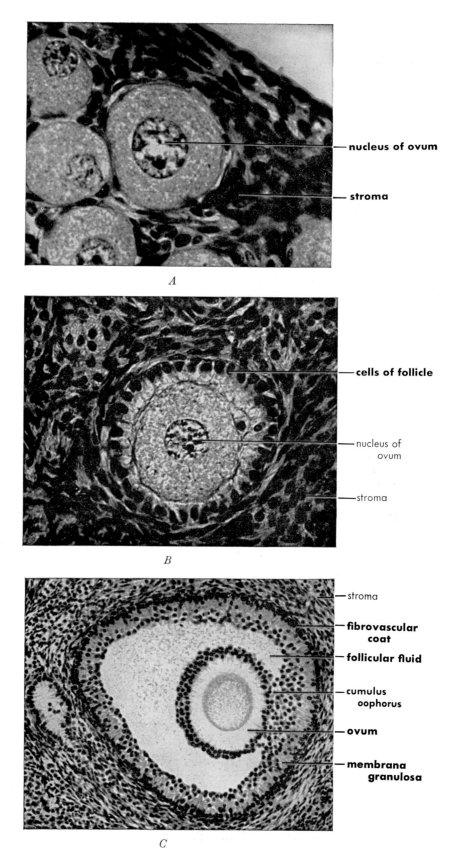

— nucleus of ovum

— stroma

A

— cells of follicle

— nucleus of
  ovum

— stroma

B

— stroma

— fibrovascular
  coat

— follicular fluid

— cumulus
  oophorus

— ovum

— membrana
  granulosa

C

Fig. 16–17.   Photomicrographs of ovary of cat showing growth of follicles.
A, Primary follicles and ova, 700×. *B*, Growing follicle, 360×. *C*, Vesicular (Graafian) follicle, 180×.

functions. It (1) continues the proliferation of the endometrium; (2) continues mammary gland development; (3) inhibits ovulation; (4) makes possible the implantation of the fertilized egg (*embryo*) and the maintenance of pregnancy during the early months, and (5) inhibits uterine contractions. The **placenta** is another source of progesterone which carries on after the corpus luteum has started degeneration.

UTERINE TUBES (Figs. 16–4, 14, 15)

The **uterine** (*Fallopian*) **tubes** are paired, about 10 cm. long, and attach medially to the superolateral angles of the uterus. Each one consists of an **isthmus,** the medial constricted one-third; an **ampulla,** the widest, longest, and most tortuous part of the tube, which reaching to the ovary arches over its pole and turns downward over its free border where it ends in the **infundibulum,** a funnel-like structure formed by many branched processes, the **fimbriae.** The fimbriae spread over most of the medial free surface of the ovary. One of them attaching to the superior pole is called the **ovarian fimbria.** The tubes are contained in the upper free margin of the **mesosalpinx,** which is a part of the broad ligament (Fig. 16–15).

Microscopic Structure and Function. The uterine tube has three coats in its walls. The outer or **serous coat** is furnished by the peritoneum of the mesosalpinx; the middle coat is **muscular** of two poorly delineated layers—the outer, thin and longitudinal; the inner, thicker and with circular or spiral fibers; the inner or **mucous coat** with a simple columnar epithelium and a lamina propria of a cellular, vascular fibrous connective tissue. The mucosa is deeply folded, especially in the ampulla. Some of the epithelial cells are ciliated and the cilia beat toward the uterus and along with contractions of the muscular

coat help to transport the ovum to the uterus. Sperm must, to ascend, move against this "ciliary stream."

Practical Implications. The ovum having been extruded from the ovary into the coelom normally finds its way, possibly by currents set up by the cilia of the fimbria, into the infundibulum of the uterine tube where, we believe, fertilization takes place. The fertilized ovum then is carried into the uterus where it embeds into the endometrium. Since the female reproductive organs form an open pathway between the outside and the celom, it is possible for sperms to enter the celom. It is also possible that ova, since there is no duct between ovary and uterine tube, could, and sometimes do fail to enter the uterine tube. Here they may be fertilized and attach to the peritoneum of the abdominal wall or viscera. Such cases are known, though the embryo thus attached does not obtain adequate nourishment and oxygen and usually dies. A few cases are reported where such **ectopic embryos** have reached term and were delivered surgically. A much more common type of **ectopic gestation** or pregnancy occurs when the fertilized ovum fails to reach the uterus and attaches to the wall of the uterine tube. Such gestation usually terminates when the embryo is extruded into the abdominal cavity through the infundibulum or when the uterine tube ruptures. The latter case demands surgical intervention and should have been detected, prior to rupture, through proper prenatal care.

UTERUS (Figs. 16–3, 14, 15)

The uterus is a single hollow, thick-walled, muscular organ about 7.5 cm. in length, 5 cm. in breadth at its superior end, and about 2.5 cm. in thickness. It is pear-shaped and lies between the urinary bladder in front and the rectum and sigmoid colon behind, and entirely

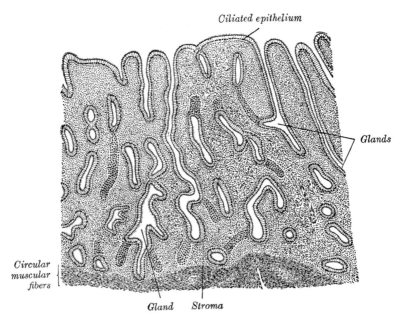

Ciliated epithelium

Glands

Circular muscular fibers

Gland   Stroma

FIG. 16–18.   Vertical section of mucous membrane of human uterus (Sobotta).

within the pelvis. It receives the uterine tubes at its superolateral angles and empties below into the vagina. It may be divided into three parts, the large **body** which lies above a slightly constricted area, the **isthmus.** Below the isthmus is the **cervix.** The part of the body above the points of entrance of the uterine tubes is the **fundus.** The cavity of the body of the uterus is triangular in shape and flattened anteroposteriorly. Its base is formed by the fundus between the orifices of the uterine tubes, the apex by the **internal os** leading into the canal of the cervix. The **canal of the cervix** is broader at the middle than at either end and opens below through the **external os** into the vagina.

The uterus is normally quite movable and is held within limits by its ligaments, but not supported by them. It gets its chief support from the pelvic floor and the surrounding organs.

The uterus has an angle between its body and cervix (*anteflexion*) and also forms an angle of from 100° to 110° with the vagina (*anteversion*). The amount of anteversion or retroversion depends to some extent upon the state of the bladder and rectum. Thus if the bladder is filled, the uterus is moved backward (*retroversion*); if the rectum is full, the uterus is moved forward (*anteversion*). If the organ descends due to a weakened pelvic floor, it is **prolapsed.** Abnormal degrees of anteversion or retroversion, or cases of prolapse, may interfere with circulation in the uterus to the extent that it causes painful menstruation .

The main ligaments holding the uterus in normal position are the **broad ligament,** a fold of peritoneum enclosing the uterus, uterine tubes, **round ligaments,** and ovaries. The round ligaments also run from the body of the uterus near the attachment of the uterine tubes through the inguinal canal to the labia majora. The round ligament and the ligament of the ovary are homologous to the gubernaculum testis, as the scrotum and labia majora are also homologous. Knowing these relationships it is not surprising to realize that the ovary, like the testis, descends along the abdominal wall, usually stopping in the pelvis, but sometimes moving farther down, rarely even into the labium majus.

Microscopic Structure and Function. The function of the uterus is to house and nourish the embryo and fetus, and at term by powerful contractions of its thick muscular walls to expel it through the vagina to the outside (*parturition, birth*). Its activity is closely coordinated with that of the ovary and like it is cyclic, undergoing marked changes during each twenty-eight-day period. The cervix of the uterus differs markedly from the body in that it changes little during the uterine cycle.

The walls of the uterus are made up of three coats similar to those of the uterine tubes with which they are continuous (Fig. 16–18). The inner coat is a mucous membrane, the **endometrium,** and the one subject to the greatest cyclic change. The middle coat, the **myometrium,** is very thick and composed of smooth muscle fibers interspersed with fibrous and elastic connective tissue and is highly vascular. Its muscle fibers attain great length during pregnancy. The outer **serous coat** (*perimetrium*) derived from the peritoneum covers most of the uterine surface except for the cervix (Fig. 16–4, 14).

The **endometrium** has a surface covering of simple columnar epithelium with scattered cilia which beat toward the vagina. Simple and slightly branched tubular glands are abundant and push down into a highly vascular and cellular lamina propria, sometimes called the **endometrial stroma.** Lymphatics are also abundant.

Endometrial Changes During the Menstrual Cycle (Fig. 16–19). The uterine or **menstrual cycle** starts at puberty with the onset of estrogen and progesterone secretion by the ovary. Though twenty-eight days is the usual length given for the cycle, it actually varies widely among women, sometimes even in individuals. It is usually more irregular near puberty and near the menopause. It is interrupted by pregnancy.

It is an interesting fact that considerable information concerning endometrial changes was gained by transplanting pieces of endometrium onto the iris of the eye of a monkey so that changes could be observed through the transparent cornea.

The menstrual cycle is usually divided into four phases in spite of its actually being a continuous process. **Menstruation** (days 1–5) is considered as phase one although it will be described last. It is followed by the following phases in sequence; **repair** (days 4–6), **proliferation** (days 7–15), and **secretion** (days 16–28).

1. **Repair** (days 4–6). Before menstruation has completely ceased, repair begins under the influence of estrogen from the ovary, where follicular development is again under way. Epithelial cells from the uterine glands move out to cover the denuded areas.

2. **Proliferation** (days 7–15). With increased production of estrogen by the ovarian follicles, the growth of the endometrium accelerates. The uterine glands lengthen and produce a thin secretion; connective cells multiply and a new meshwork of reticular fibers appears. The endometrium approaches 2 mm. in thickness. Ovulation takes place.

3. **Secretion** (days 16–28). In this phase estrogen influence gradually gives over to that of progesterone from the corpus luteum. Because of the drop in estrogen, the thickening of the endometrium may be temporarily stopped and there is sometimes intermenstrual bleeding. The endometrium more than doubles in thickness during this period, reaching 4 to 5 mm. Its glands become long, swollen, and tortuous and produce an abundant, thick, mucoid secretion which is rich in glycogen. Convoluted arterioles push into the outer layers of the endo-

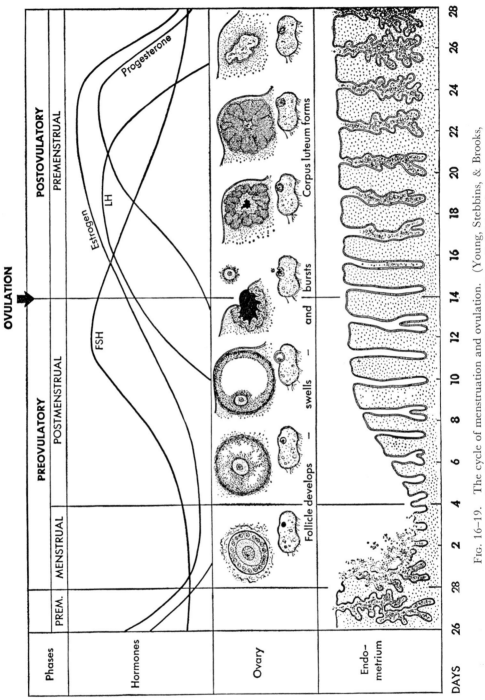

Fig. 16–19.   The cycle of menstruation and ovulation.   (Young, Stebbins, & Brooks, *Introduction to Biological Science*, Harper & Bros., Publishers.)

metrium and develop into **coiled arterioles** which show microscopically longitudinal bands of smooth muscle below their tunica intima. Contraction of these retards the flow of blood. Normal arteriole and capillary relations maintain in the deeper layers of the endometrium. The uterus is now ready to receive an embryo (blastocyst). If one appears and implantation takes place, the endometrium continues its development and the corpus luteum persists. In the absence of a blastocyst, the corpus luteum begins to degenerate, which is the "signal" for the endometrium to break down. The coiled arterioles of the outer endometrium contract and deprive the superficial layers of blood and therefore of oxygen. This initiates the destruction which is characteristic of the next phase.

4. **Menstruation** (days 1–5). Progesterone secretion declines as the corpus luteum begins involution, and the walls of the capillaries and some of the coiled arterioles break down and blood escapes into the stroma of the superficial layer of the endometrium. Pieces of the superficial layer break away and other blood channels are opened. The same contractions of the coiled arterioles, which earlier produced anemia and breakdown, now prevent excessive hemorrhage. The basal portions of the endometrium with remnants of the uterine glands, having a conventional blood supply, remain intact and ready to start repair and replacement of the outer layers. The menstrual flow (**menstruum**) includes sloughed-off superficial endometrium, blood, and glandular secretions.

A dramatic fact is very apparent, and that is the persistence of the reproductive system in attempting to carry out its basic function—to perpetuate the species.

VAGINA (Figs. 16–4, 5, 14, 15)

The **vagina** lies behind the urinary bladder and urethra and in front of the rectum. It extends from the cervix of the uterus to the vestibule and measures about 7.5 cm. along its anterior wall and 9 cm. along its posterior wall. It surrounds the vaginal end of the cervix, its attachment being higher posteriorly than anteriorly. This recessed or moat-like area is called the **fornix** and can be divided into **posterior, anterior** and **lateral fornices** (Fig. 16–4, 15).

Microscopic Anatomy and Function. The **mucosa** of the vagina is covered with a stratified squamous epithelium which rests upon a well-marked papillary zone of lamina propria. The epithelium varies in thickness, being influenced by the estrogenic hormone. There are no glands and the mucosal surface is moistened by secretions from the uterus. Since this secretion is rich in glycogen, bacterial fermentation causes an acid environment especially during the secretion phase of the menstrual cycle. If the seminal fluid were not alkaline, it is possible that the sperm would die in the vagina without reaching the ova.

The lamina propria is highly vascular and is made up of loose tissue and some smooth muscle fibers from the muscular coat. It is **erectile.**

The **muscularis** consists of two layers not clearly distinguished, there being crossing fibers between them. The outer longitudinal layer is stronger than the inner circular. At the lower end, the orifice of the vagina is surrounded by a band of voluntary muscle, the **Bulbocavernosus** (Fig. 11–30).

External to the muscular coats there is a layer of connective tissue which contains a large vascular plexus.

The lower end of the vagina is constricted, the middle is dilated, while the upper end is narrowed. It receives the penis which deposits sperms at its upper end. It is also the channel for carrying the menstrual debris to the outside and for the delivery of the fetus.

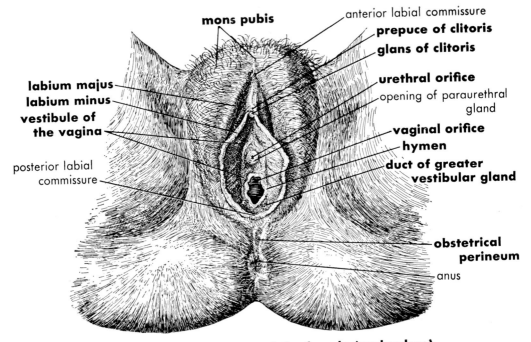

mons pubis

anterior labial commissure

prepuce of clitoris

glans of clitoris

labium majus

labium minus

**vestibule of
the vagina**

urethral orifice

opening of paraurethral
gland

vaginal orifice

hymen

**duct of greater
vestibular gland**

posterior labial
commissure

**obstetrical
perineum**

anus

**A. External genital organs of the female (pudendum).
The labia minora have been parted to reveal the
vestibule and the orifices of the urethra and vagina.**

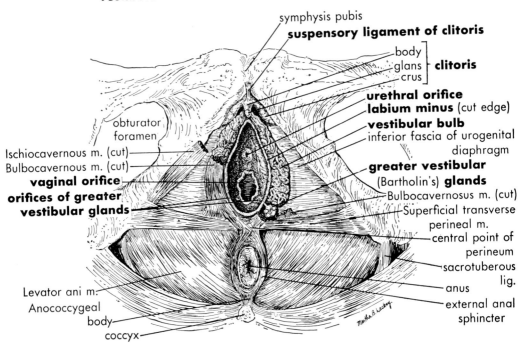

symphysis pubis

suspensory ligament of clitoris

body
glans ⎱ **clitoris**
crus

urethral orifice

labium minus (cut edge)

vestibular bulb

inferior fascia of urogenital
diaphragm

obturator
foramen

Ischiocavernous m. (cut)

Bulbocavernous m. (cut)

vaginal orifice

**orifices of greater
vestibular glands**

**greater vestibular
(Bartholin's) glands**

Bulbocavernosus m. (cut)

Superficial transverse
perineal m.

central point of
perineum

sacrotuberous
lig.

Levator ani m.

Anococcygeal
body

coccyx

anus

external anal
sphincter

**B. Dissection of the female genital organs.
Dissection on the left side is deeper than
that on the right.**

Fig. 16-20.   The external genital organs of the human female.

External Genital Organs
(Figs. 16–4, 20)

The external genital organs of the female are known collectively as the **vulva** or **pudendum**. They are as follows:

The **mons pubis** is a fatty eminence in front of the symphysis pubis which at puberty develops a covering of hair.

The **labia majora** are two large folds of skin which extend backward from the mons pubis on either side of a cleft into which the urethra and vagina open. The outer sides of the labia majora are pigmented and have hair, the inner are smooth and have large sebaceous glands. Inside the labia are areolar tissue, fat and muscle resembling the dartos of the scrotum. The labia majora are **homologous** to the scrotum.

The **labia minora** are two small folds lying between the labia majora. Anteriorly they encircle the clitoris, forming its **preputium clitoridis.**

The **clitoris** is an erectile organ which is homologous to the penis, and which lies in front of the urethral orifice where it is partially hidden by the labia minora. The clitoris is constructed much like the penis having, however, only two **corpora cavernosa** which are enclosed in connective tissue and partly separated by a septum. Posteriorly, they connect to the rami of the pubis and ischium by **crura** and are provided with **Ischiocavernosi muscles.** The free extremity of the clitoris is the **glans clitoridis** which contains erectile tissue and is highly sensitive.

The **vestibule** is the cleft between the labia minora and behind the clitoris. It contains the orifice of the urethra anteriorly, and that of the vagina posteriorly. The latter is provided by a membranous **hymen** of variable shape and size, which determines the size of the vaginal orifice. The hymen may be absent or it may completely close the orifice to which condition the term **imperforate hymen** is applied. It may persist after sexual intercourse. It is not by its presence, therefore, considered a sign of virginity, nor does its absence necessarily mean that sexual intercourse has taken place.

The **bulb of the vestibule** consists of two elongated masses of cavernous tissue placed above the Bulbocavernosus muscle and below the urogenital diaphragm and on either side of the vagina. They unite in front of the urethra as the **pars intermedia.** They are **homologous** to the bulb and corpus spongiosum penis of the male (Fig. 16–2).

The **greater vestibular glands** (Bartholin's glands) (Fig. 16–20). These are situated on either side of the vaginal orifice adjacent to the ends of the bulb of the vestibule. Each has a duct which opens between the hymen and the labium minus. These are the homologues of the bulbourethral glands of the male. They secrete a lubricating substance.

Mammary Glands (Fig. 16–21)

The **mammary glands** structurally and embryologically belong to the integument, but their function makes them accessory to reproduction. They are modified apocrine sweat glands (see page 54) which reach their best development in women during the early childbearing period. They are present in rudimentary form in infants, children, and men.

Each gland consists of about twenty irregular lobes of secreting tissue. Each lobe has one **lactiferous duct** by which it opens onto the nipple of the gland. The lobes pass in a radial direction from the nipple to embed in the connective and adipose tissue of the superficial fascia. The whole organ is covered with skin, that of the **nipple** and the area immediately around it, the **areola,** is very thin. Hair and sweat glands are absent from the nipple and areola, but **areolar** (*sebaceous*) **glands** are present which se-

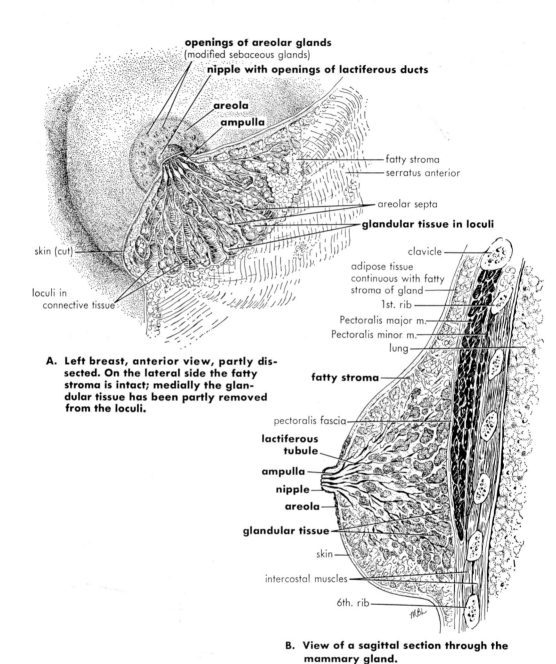

openings of areolar glands
(modified sebaceous glands)
nipple with openings of lactiferous ducts

areola
ampulla

fatty stroma
serratus anterior

areolar septa

glandular tissue in loculi

skin (cut)

loculi in
connective tissue

clavicle
adipose tissue
continuous with fatty
stroma of gland
1st. rib
Pectoralis major m.
Pectoralis minor m.
lung

fatty stroma

pectoralis fascia
**lactiferous
tubule**
ampulla
nipple
areola
glandular tissue
skin
intercostal muscles
6th. rib

A. Left breast, anterior view, partly dissected. On the lateral side the fatty stroma is intact; medially the glandular tissue has been partly removed from the loculi.

B. View of a sagittal section through the mammary gland.

FIG. 16–21.   The mammary gland.

crete a saliva-resisting lubricant to protect the nipple during nursing. The areola has a dark pigment which intensifies during pregnancy and decreases afterwards. The subcutaneous tissue of the areola contains circular and radiating smooth muscle bundles which cause the nipple to erect in response to stimulation.

Hormone Interrelationships. The mamma remains relatively unchanged from infancy to puberty at which time it enlarges by adding glandular tissue, ducts, and adipose tissue. The areola enlarges, develops pigment, and becomes sensitive. Also with the onset of menstruation at puberty there is at each period an increase of glandular tissue and vascular engorgement followed, during the postmenstrual phase, by regression. With pregnancy the glands enlarge starting with the second month; the duct system is developed by the sixth month; the secreting portions by the ninth month. After the birth of the child the secretion of the glands for two or three days is thin and yellowish, but true milk appears by the third or fourth day and continues until nursing ends. Lactation is followed by involution of the glands.

The growth of the mamma at puberty and its participation in the menstrual cycle are the results of stimulation by the ovarian sex hormones, estrogen and progesterone. The increase in size during pregnancy is due to the influence of progesterone. The function of ovarian hormones is taken over by hormones of the placenta after the first part of pregnancy. Lactation is, in part, the result of the influence of the **lactogenic hormone** of the pars distalis of the hypophysis.

EMBRYOLOGY OF THE REPRODUCTIVE SYSTEM
(Figs. 16-1, 2, 13; 15-14)

The **gonads** develop within a restricted part of the genital portion of the uro-genital ridge of the intermediate cell mass. By about the sixth week they appear as bulges into the celom and consist of an external germinal epithelium and an internal blastema of epithelial cells. By about the eighth week they assume characteristics which identify them either as testis or ovary.

The **testis** makes connections with the middle tubules of the mesonephros which persist and become the **efferent ductules.** Other mesonephric ducts may contribute to the **epididymis** while the more caudal ones become the vestige known as the **paradidymis.** The mesonephric duct which developed to carry urine becomes the **duct of the epididymis** and the **ductus deferens.** It gives rise to the **seminal vesicles.**

It is apparent that in the male full use has been made of the "discarded" urinary ducts to transport the spermatozoa. The development of these ducts inferiorly and of the urethra, which also serves to carry spermatozoa, has already been considered under the urinary system. The descent of the testis and its relationship to the development of the scrotum have been described earlier in this chapter.

A pair of **Müllerian ducts** appears in the male, as in the female, but gradually disappears except for an anterior remnant near the testis, the **appendix of the testis,** and a posterior **prostatic utricle** or **uterus masculinus,** which lies within the wall of the prostatic urethra.

In the female the paired **Müllerian ducts** open cranially into the celom as the **ostia** and constitute the **uterine tubes,** while inferiorly they join to form the **uterus** and **vagina.**

The external genitalia, like the internal organs of reproduction remain in an indifferent stage until about the eighth week. At this time there is a median cylindrical elevation, the **phallus** and lateral to it rounded elevations, the **genital tubercles** (Fig. 16-2). The ventral side of the phallus has a lengthwise depression, the **urethral groove.** In the male the phallus becomes the penis and the edges of the urethral grooves close to form the spongy urethra, while the genital tubercles grow and join to form the scrotum. In the female the phallus becomes the clitoris. The urethral groove fails to invade the phallus and to close, hence becomes the open vestibule. Its folds become the labia minora. The genital tubercles, also failing to join, enlarge and form the labia majora.

SUMMARY

Figure 16–22 is an attempt to summarize some of the many and complex controls operative upon the organs involved in human reproduction.

The following table (Table 16–2) gives some of the important homologies of the male and female reproductive systems and their progenitors in the indifferent stages of development. Reference should also be made to Figures 16–1, 2 and to Figure 15–14.

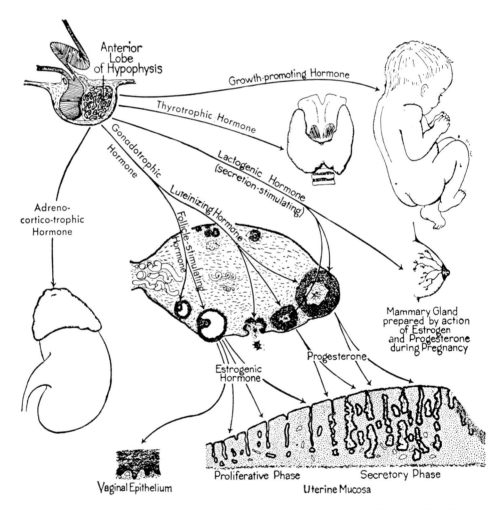

FIG. 16–22. Diagram indicating some of the hormones arising in the anterior lobe of the hypophyses, with special emphasis on those involved in regulating the cyclic activities of the female reproductive organs. (Patten, *Human Embryology*, courtesy of McGraw-Hill Book Co., used by permission.)

TABLE 16–2.   SOME HOMOLOGIES OF THE REPRODUCTIVE SYSTEM
(Adapted from Arey-Developmental Anatomy)

Male	Indifferent Stage	Female
1. Testis	1. Gonad	1. Ovary
2. Mesorchium		2. Mesovarium
3. Gubernaculum testis (caudal part)		3. Round ligament of uterus
4. Efferent ductules and head of epididymis	4. Mesonephric tubules (cranial)	4. Epoöphoron (vestige)
5. Paradidymis (vestige)	5. Mesonephric tubules (caudal)	5. Paroöphoron (vestige)
6. Ductus epididymis	6. Mesonephric duct	6. Duct of epoöphoron (vestige)
7. Ductus deferens	7. Mesonephric duct	7. Gartner's duct (vestige)
8. (Seminal vesicle)	8. Mesonephric duct	8. Gartner's duct (vestige)
9. Ejaculatory duct	9. Mesonephric duct	9. Gartner's duct (vestige)
10. Appendix testis (vestige)	10. Müllerian duct	10. Uterine tube
	11. Müllerian duct	11. Uterus
	12. Müllerian duct	12. Vagina (upper)
13. Upper prostatic urethra		13. Urethra
14. Lower prostatic urethra	14. Urogenital sinus	14. Vestibule (in part)
15. Prostatic utricle	15. Müllerian duct	15. Vagina (lower)
16. Prostate gland		16. Paraurethral ducts, urethral glands
17. Membranous urethra	17. Urogenital sinus	17. Vestibule (middle)
18. Cavernous urethra	18. Urogenital sinus (phallic portion)	18. Vestibule (between labia minora)
19. Bulbourethral glands		19. Vestibular glands
20. Penis	20. Phallus	20. Clitoris
21. Glans penis	21. Glans	21. Glans clitoridis
22. Urethral surface of penis	22. Lips of urogenital groove	22. Labia minora
23. Corpora cavernosa penis	23. Shaft	23. Corpora cavernosa clitoridis
24. Corpora spongiosum penis	24. Shaft	24. Vestibular bulb
25. Scrotum	25. Genital swelling (*tubercles*)	25. Labia majora

QUESTIONS

1. Is it probable that a better understanding of the nature of human reproduction could contribute more to man's welfare and happiness than an understanding of any one other system of the body?
2. Explain your answer to question number one.
3. Describe the organs of the urogenital system in the indifferent stage.
4. At what age does the embryo differentiate sexually?
5. Name the vestiges of urogenital structures in the male and their homologues in the female.
6. Why do we refer to the testes and ovaries as the essential organs of reproduction?
7. Describe the minute anatomy of the testis and the epididymis.
8. Where does the testis originate and how does it reach the scrotum?
9. What is gained by the descent of the testis?
10. Does the ovary descend, and if so to what extent?
11. If the ovary descended to the extent that the testis does, what would be its destination?   Does this ever happen?
12. What is the relationship between the round ligament of the uterus and the gubernaculum testis?

13. What are some implications of the fact that the ovary is separated from its duct by a part of the celom?
14. Define the urogenital and anal triangles of the perineum.
15. What is the relationship of the tunica vaginalis to the testis and epididymis?
16. Where are the interstitial cells of the testis and what are their functions?
17. What tissue lines the epididymis? What is the function of this organ?
18. What and where is the ejaculatory duct? the seminal vesicle?
19. The ductus deferens is well provided with a coat of smooth muscle fibers. What function do these muscle fibers serve?
20. Describe the spermatic cord and its relationship to the inguinal ligament and inguinal canal. What is the processus vaginalis?
21. Discuss some of the anomalies attendant upon the descent of the testis. Define cryptorchidism and ectopic testis.
22. Discuss hernia relative to normal anatomy and congenital involvements.
23. Describe the cavernous bodies of the penis and the glans penis.
24. What causes the erection of the penis?
25. Trace the course that a spermatozoan would follow in moving from its point of origin until it is ejaculated to the outside.
26. Describe the ovary and its ligamentous attachments.
27. Describe the ovary as seen in section and viewed with the microscope.
28. Describe the ovarian cycle.
29. Describe the menstrual cycle, indicating the influence of hormones in its regulation.
30. Describe the gross anatomy of the uterus and its relationship to the vagina.
31. What are the important ligaments and the supporting structures of the uterus?
32. What and where are the fornices?
33. Describe the vulva.
34. Name the parts of the vulva which can be considered homologous to the penis.
35. What are the functions of the greater vestibular glands?
36. Describe the mammary glands, stating their relationships to the skin and to the reproductive system.
37. Name three hormones of the hypophysis which are influential in the control and coordination of reproduction.

The blood more stirs to rouse a lion, than to start a hare.
—Shakespeare, *Henry IV*

Chapter 17

The Endocrine System

INTRODUCTION

Definition. Endocrine glands, in common with glands generally, produce secretions which have general or specific effects upon body functions. Unlike the exocrine glands, which have ducts connecting their secretory cells to the epithelial surfaces from which they were derived, the endocrines are without ducts (*ductless glands*). They empty their secretions, or "chemical messengers," the **hormones,** into the blood and rely upon the circulatory system to transport them. This fact is of major importance, for it means that hormones are broadcast indiscriminately to all parts of the body, making the problem of searching out their exact functions a very difficult one.

The various glands are scattered through the body, associated closely with other systems, and are of diverse embryonic origins, giving little reason to consider them as the parts of one anatomical system (Fig. 17–1). Yet as knowledge of their functions increases ample justification is found for calling them the **endocrine system.** They do form an integrated group of organs, their secretions controlling each other, as well as other tissues and organs of the body.

The matter of assigning organs or tissues to this system is not without its difficulties. Certainly, the **hypophysis** (*pituitary*), **thyroid, parathyroids,** and **suprarenals** are solely endocrine in function. The **pancreas** is both exocrine and endocrine, its **islets** (*of Langerhans*) representing the latter. The **gonads,** though cytogenic, are as obviously endocrine in function. The **placenta** has nutritive and excretory functions; the **duodenum** is concerned in digestion, but both produce hormones. The thymus, too, is a ductless organ and, although no specific hormone has been isolated from it, recent evidence implies a possible hormonal influence by this gland upon immunological processes of the body (Fig. 14–11). The pineal body is ductless, and is often listed as an endocrine gland, but it is here excluded because no hormones have been isolated from it. It is described in other sections of this book.

General Functions. The endocrine system is one of two integrating systems in the body. The **nervous system** is the other. The latter has a structural continuity entirely lacking in the endocrine system. Its actions are more rapid, its targets precise. Among its effectors (*responding structures*) are the voluntary and involuntary muscles, as well as the glands, including some of the endocrine glands. Also at the synapses between neurons, and where efferent neurons reach the effectors, chemical substances are released that make the transfers, and these remind us of hormones though their effects are local. We do, in a sense, run full circle and while the two integrating systems are very different in terms of structure and function, they are not totally so. They are, after all, but parts, interacting parts of one body, and like other systems their functions are meaningful only as identified with the whole body.

More specifically, the functions of the endocrine system may be outlined as follows:

(1) the functional integration of the various tissues and organs of the body, particularly those concerned with metabolic processes.

(2) the development and growth of the body as a whole.

(3) the development, growth, and regulation of the gonads and of the secondary sexual organs.

(4) the regulation of the internal environment in reference to salt and sugar concentration, and the rates of fat, protein and sugar metabolism.

(5) and to be speculative—to regulate

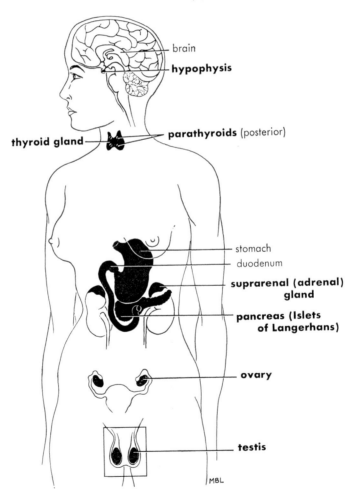

Fig. 17–1.   Endocrine glands of the human body.

the balance of the internal metabolic activities to the external environment and thus to maintain the life of the organism and to determine, in a measure, its character.

Endocrine glands, like other tissues and organs of the body, are subject to a variety of situations which may cause **dysfunction**. These usually result in **hyperfunction** which is overactivity, or in **hypofunction,** underactivity, and result in disturbances in the body often remote from the gland involved. Some of these will be discussed in reference to the individual glands.

HYPOPHYSIS (Fig. 17–2)

The **hypophysis** (*pituitary*) is an organ about the size of a pea which rests in the **hypophyseal fossa** of the sella turcica of the sphenoid bone. Superiorly, it is attached to the hypothalamus of the diencephalon by a stalk, the **infundibu-**

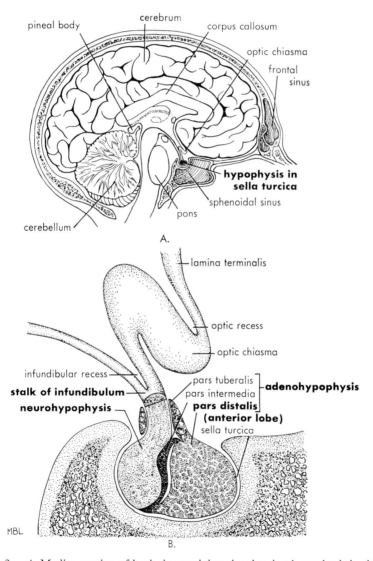

Fig. 17–2. *A,* Median section of brain in cranial cavity showing hypophysis in the sella turcica and related structures. *B,* Hypophysis enlarged to show component parts.

lum. It has a double origin, though from a common germ layer, the ectoderm. Its anterior glandular portion, the **adenohypophysis,** is derived from a diverticulum of ectoderm from the roof of the stomodaeum, known as **Rathke's pouch** (Fig. 13–3). Its posterior **neurohypophysis** (*lobe*) is a diverticulum from the hypothalamus of the brain. The two diverticula come to lie close together. Rathke's pouch loses its connection with the stomodaeum, while the neurohypophysis maintains its connection with the brain. The adenohypophysis sends a small mass of tissue up along the infundibulum which is the **pars tuberalis.** The bulk remains as the **anterior lobe** or **pars distalis** and between it and the neurohypophysis is a **pars intermedia.** The neurohypophysis forms the **posterior lobe.**

The blood supply of the hypophysis is rich. There are eighteen to twenty superior and inferior branches coming into it from the internal carotid, posterior communicating branches, and from the cerebral circle of vessels. The superior arteries go to the stalk and from there to the anterior lobe as a venous portal system; the inferior go to the posterior and anterior lobe. Venous drainage is into neighboring sinuses.

Nerve fibers of the posterior lobe reach it through the stalk by the hypothalamicohypophyseal tract. Fine fibers from the internal carotid plexus have been found in the walls of the vessels, but have not been traced to the secreting epithelial cells of the pars distalis.

Microscopically, the pars distalis is made up of cords of cells which have varied staining reactions by which they can be identified. This is a subject for the student of histology. The nervous part of the hypophysis contains branching, unmyelinated nerve fibers and branching cells called **pituicytes.**

Functions. The pars distalis of the hypophysis produces a number of hormones. Three of these, the **follicle stimulating hormone,** which stimulates the development of not only the follicle cells of the ovary, but the seminiferous tubules of the testis, the **luteinizing hormone,** and the **lactogenic** have already been studied under the reproductive system. In addition, there is the hormone called **adrenocorticotrophic** or in short **ACTH,** which influences the adrenal cortex in the production of its hormones. ACTH has received a great deal of attention in recent years in relationship to the "alarm reaction" and to disease reactions in general and their control (Constantinicles and Carey, 1949; Gray, 1950). A **thyrotrophic** hormone governs the production and release of the thyroxin of the thyroid gland. Finally, there is the growth hormone which, if over-produced in childhood, produces **gigantism,** or in the adult, **acromegaly,** in which the skeleton gradually enlarges, especially in the face, hands, and feet. If there is hypofunction in childhood, it results in **dwarfism.**

The question, of course, arises as to what controls the pars distalis. There is some evidence that the brain, likely the hypothalamus, produces substances which reach the pars distalis through its portal veins, and thus exercises control over it.

The posterior lobe of the hypophysis produces at least two types of hormones. One is **vasopressin,** which causes an elevation of blood pressure except in the kidney where excessive amounts of water are absorbed and the amount of urine secreted is reduced. Hypofunction of the posterior lobe causes a condition marked by excessive excretion of sugar-free urine called **diabetes insipidus.** The other hormone is **oxytocin** (*pitocin*) which induces uterine contractions and may be used artificially to hasten parturition. The posterior lobe, like the pars distalis, is under the influence of the hypothalamus. The **hypothalamico-hypophyseal tract** connects the posterior lobe with nuclei in the hypothalamus and when this tract is cut, diabetes insipidus results.

The hypophysis is often called the **"master gland"** or the **"executive department"** of the endocrine system. This is not due to its superior position in the body, but rather to the wide range of influence which it exerts over other endocrine glands and the body as a whole.

SUPRARENAL (ADRENAL) GLANDS (Fig. 15–1)

The **suprarenal** or **adrenal glands** rest, like cocked hats, on the superior surfaces of the kidneys. They lie behind the peritoneum and are embedded in fat. Each gland is enclosed in a capsule and histologically is made up of an outer **cortex** and an inner **medulla.** The cortex and medulla differ in their embryology, histology and physiology.

The cortex arises from the mesoderm between the intermediate cell mass and the mesentery, near the mesonephric and genital primordia. The medulla, ectodermal in origin, comes from the **neural crest cells** which also are the source of the cells of the **sympathetic ganglia** (Fig. 4–11). Other cells from the neural crests give rise to small groups of cells around the autonomic ganglia and plexuses known as **paraganglia.** Like the cells which form the medulla of the suprarenals, the paraganglia stain brown with chrome salts and are, therefore, called **chromaffin cells.** They all contain **epinephrine,** the hormone of the adrenal medulla.

While in mammals the cortex encloses the medulla to form an adrenal gland, studies in comparative anatomy indicate that this condition has developed only gradually over a long phylogeny. In fishes, these two structures, cortex and medulla, are separate organs; the association in amphibians is closer, and even more intimate in reptiles and birds. The climax is reached in mammals.

Histologically, the cortex is made up of cords of epithelial cells, sinusoidal vessels, and reticular tissue. It shows three distinct layers based upon the arrangement of the epithelial cells, an outer **zona glomerulosa,** a middle **z. fasciculata,** and an **inner z. reticularis** (Fig. 17–3). The medulla is made up of anastomosing cords of granular chromaffin cells bearing little resemblance to the cortex. Sinusoids and venules, which empty into the centrally placed suprarenal vein, are conspicuous. The blood supply to the gland comes from three suprarenal arteries coming from the inferior phrenic, aorta, and renal arteries. The supply is richer than to any other organs relative to their size, with the possible exception of the thyroid glands. The suprarenal vein of the right gland drains into the inferior vena cava, that from the left into the left renal vein.

Many fine medullated preganglionic neurons reach the gland from the celiac plexus and greater splanchnic nerve which end on the secretory cells of the medulla. There may be a few parasympathetic fibers from the vagus. There appear to be no sensory nerves.

Functions. The **cortex** is essential to life. It produces a number of steroid compounds, some of which can sustain life in another animal whose adrenal cortex has been removed. Insufficiency of these hormones produces disturbance in the fluid and electrolyte balance in the body. They have an important influence upon carbohydrate, fat, and protein metabolism. They enable the body to meet conditions of stress such as shock, cold, pain, intense muscular exertion, and emotional excitement. Resistance to infection is also markedly diminished in their absence. A more recent preparation from the cortex called **cortisone** has shown spectacular results in the treatment of rheumatoid arthritis and numerous other disorders. Unfortunately, there are bad side effects and

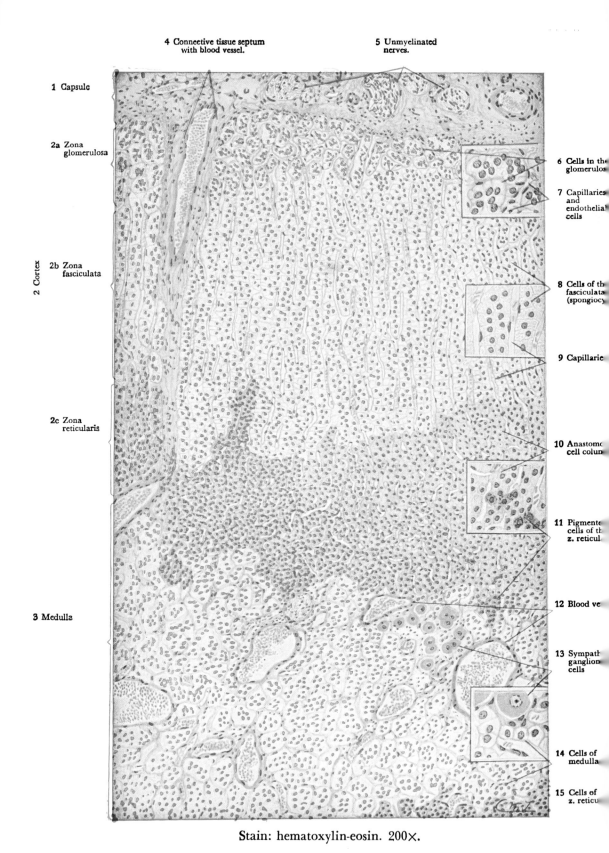

4 Connective tissue septum with blood vessel.

5 Unmyelinated nerves.

1 Capsule

2a Zona glomerulosa

2 Cortex

2b Zona fasciculata

2c Zona reticularis

3 Medulla

6 Cells in the glomerulos

7 Capillaries and endothelia cells

8 Cells of the fasciculata (spongiocy

9 Capillarie

10 Anastomo cell colum

11 Pigmente cells of the z. reticul

12 Blood ve

13 Sympath ganglion cells

14 Cells of medulla

15 Cells of z. reticu

Stain: hematoxylin-eosin. 200×.

FIG. 17–3. Histology of the adrenal gland. (di Fiore, *An Atlas of Human Histology*, Lea & Febiger.)

it does not provide a cure. ACTH, from the pars distalis of the hypophysis, as stated in the previous section, controls the adrenal cortex and evidence indicates some "feed back" or reciprocity of action. This is a fertile field for research.

The **suprarenal medulla** produces the hormones **epinephrine** (*adrenalin*) and **norepinephrine.** Injections of these hormones produce essentially the same body responses as stimulation of the sympathetic or thoracolumbar nervous system and are sometimes called and used as "emergency hormones." They cause the heart to accelerate, the vessels to constrict, the blood pressure to rise. Epinephrine, though not norepinephrine, causes the liver to release sugar and thus raise the blood sugar level. Indeed, considering the common origin of the sympathetic system and the adrenal medulla, we can look upon the latter as a kind of extension of the former.

Implication. As stated earlier in this book embryology, comparative anatomy, histology, and physiology, all lend meaning and provide means for understanding the gross anatomy of the adult human body. The suprarenal glands are a particularly good illustration of this fact, which for interest's sake and for motivation you should always keep before you.

THYROID GLAND (Figs. 17–4, 5, 6)

The **thyroid gland** is the largest of the endocrine glands, weighing from 20 to 30 grams. It lies in the neck region with one lobe on either side of the lower larynx and upper trachea. These lobes are

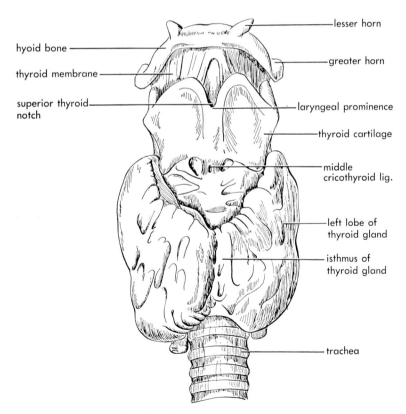

FIG. 17–4. Anterior view of thyroid gland, showing its relationship to the hyoid and trachea.

connected anteriorly across the trachea by the **isthmus** (Fig. 17–4). In about a third of all individuals, a narrow **pyramidal lobe** extends upward from the isthmus in front of the larynx.

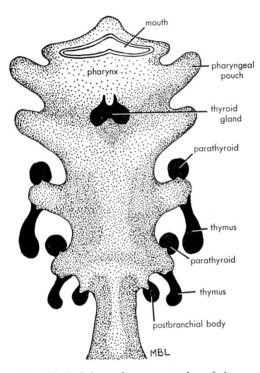

FIG. 17–5. Schematic representation of pharynx to show the origins of some endocrine glands from pharyngeal pouch material.

It rates with the adrenal glands as one of the most vascular glands in the body, receiving its blood from the inferior and superior thyroid arteries and being drained by the inferior, middle and superior thyroid veins. Its nerve supply comes from the cervical part of the sympathetic trunk, its parasympathetic fibers from the vagus.

It originates embryologically as a midventral diverticulum from the pharynx at the level of the second pair of pharyngeal pouches. The diverticulum grows inferiorly, bifurcates and subdivides into a series of cellular cords and forms the lateral lobes and isthmus (Fig. 17–5). The duct, the **thyroglossal duct,** becomes discontinuous and degenerates, leaving at its upper end the **foramen cecum** of the tongue and sometimes at its lower end the pyramidal lobe.

Microscopically, the thyroid consists of a number of follicles intimately associated with the blood and lymph capillaries. The follicle walls are of simple cuboidal or columnar epithelium which rests on a fine, connective tissue layer. The follicles are normally filled with a colloid containing the hormone **thyroxin** (Fig. 17–6).

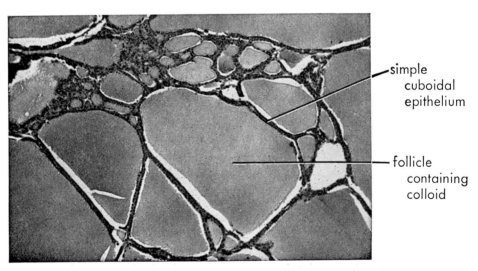

FIG. 17–6.   Photomicrograph of thyroid gland.

Function. The thyroid gland, through its hormone **thyroxin,** regulates metabolism. The "basal metabolism" of an individual is usually measured when the individual is awake, but in a prone position and after a period of fasting. It is determined by finding the rate of oxygen consumption. Also a measure of the glandular function can be obtained by administering radioactive iodine, a tracer substance, and determining the rate of its uptake by the gland. Since iodine is a necessary component of thyroxin, its insufficiency in the diet causes the thyroid gland to enlarge, a condition called **simple** or **endemic goiter.** This may or may not be accompanied by hypofunction, but in either case can be corrected by

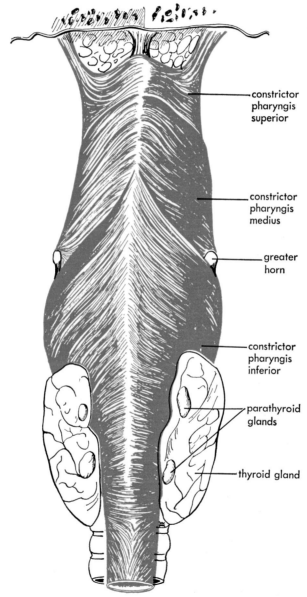

constrictor
pharyngis
superior

constrictor
pharyngis
medius

greater
horn

constrictor
pharyngis
inferior

parathyroid
glands

thyroid gland

FIG. 17–7.   Posterior view of thyroid gland showing parathyroid glands embedded in its tissue.   Relationship to esophagus, larynx and trachea is also shown.

feeding iodine. This is done routinely in the civilized world by placing iodine in the salt which we buy at the stores.

In cases of thyroid hypofunction in an infant, growth, mental, and sexual development are retarded and the condition is known as **cretinism.** Hypofunction in the adult causes **myxedema,** a condition in which there is a great loss of mental and physical vigor, a peculiar thickening of the skin, a loss of hair, and often an increase in weight. Both of these conditions can be treated by feeding thyroxin which, fortunately, is resistant to destruction by the digestive juices.

Hyperthyroidism (*overactive*) produces the condition known as **exophthalmic goiter** (Graves' disease, toxic goiter). The metabolic rate is high, the individual loses weight, is nervous and his eyeballs tend to protrude. The condition can be corrected by surgical removal of part of the thyroid.

The thyroid gland is under the control of the thyrotrophic hormone of the pars distalis of the hypophysis.

PARATHYROID GLANDS
(Figs. 17–5, 7)

As their name suggests, the **parathyroid glands** lie beside the thyroid, specifically on the dorsal surface of the lateral lobes. There are usually four, though there are sometimes more. The blood, lymph, and nerve supply are the same as for the thyroid; their histology and physiology, however, are quite different. They originate as outgrowths from the third and fourth pharyngeal pouches with which they lose all connection as they migrate to the neck area.

Histologically, the parathyroids are composed of interconnecting cords of glandular cells of two types. Chief cells are agranular; oxyphil cells are larger and granular in appearance. These cells are supported by connective tissue containing numerous blood vessels.

Function. The parathyroids secrete a hormone called **parathormone** which is apparently essential to life. It regulates the balance between the calcium in the blood, that in the bone, and that excreted. Since calcium concentrations in the body fluids influence irritability of nerves and muscle tissues, parathormone is a controlling factor in the nervous and muscular systems.

Hyperparathyroidism results in a withdrawal of calcium from the bones and an elevated blood calcium level. This causes a weakening of the bones and they are subject to cyst formation. Calcium deposits may develop in the kidneys and there is often muscular weakness.

Hypoparathyroidism causes a fall in blood calcium, and increased irritability of nerves and muscles, resulting finally in muscular spasms called **tetany.** These are fatal if the condition is not treated either by the feeding or injection of calcium, or by the injection of parathormone. Parathormone is destroyed by the digestive enzymes, so cannot be fed.

ISLETS OF LANGERHANS
(Fig. 17–8)

These are groups of cells irregularly placed among the pancreatic cells. They secrete their hormone, **insulin,** into the blood as true endocrines, while the pancreatic cells secrete into a system of ducts which join to form a common duct or ducts emptying into the intestine. Their secretions serve a digestive function. That a diverticulum of the gut should give rise to cells so different in function as those of the islets of Langerhans and those producing digestive enzymes is amazing only to those who have not been deeply indoctrinated into the ways of living things. Why, indeed, should some cells of a gastric gland produce hydrochloric acid and others pepsin? There is an explanation and man shall find it, as has been true of other "mysteries" of the past. Such is the nature of man!

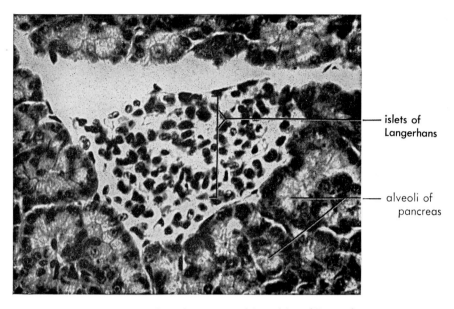

— islets of Langerhans

— alveoli of pancreas

FIG. 17–8.   Section of pancreas with an islet of Langerhans.

Functions. The insulin from the islets determines the level of blood sugar. Their hypofunction results in **diabetes mellitus,** characterized by a high blood sugar (*hyperglycemia*) which regularly "spills over" into the urine. This condition is readily controlled by self-administration of insulin with the advice and direction of a physician. Contrariwise, hypoglycemia may be relieved by administration of glucose.

OTHER ENDOCRINES

As stated at the beginning of this chapter, there are other endocrines. Some have been discussed in previous chapters, notably the **gonads** and the **duodenum.** Others, like the pineal body, seem not to qualify and are taken up in a context where they have more meaning.

In an anatomy text, the consideration of endocrines in particular must seem inadequate, for it is, after all, their functions which justify dignifying them as a system. A textbook of physiology will give wider coverage, but even there one may be left with a sense of frustration—or perhaps of challenge. They belong to the whole body and our understanding of them must ultimately be sought in physiological chemistry.

QUESTIONS

1. What are some of the structural and functional features of the endocrine system which identify it and cause us to call it a system?
2. How does the endocrine system compare with the other controlling and co-ordinating systems of the body? Do their functions overlap?
3. Describe the "double origin"of the hypophysis.
4. What are some of the functions of the pars distalis of the hypophysis (*pituitary*)?
5. Define gigantism, dwarfism, acromegaly in reference to the function of the hypophysis.
6. What is the relationship of the posterior lobe of the hypophysis to kidney and uterine functions?

7. Describe the double nature of the suprarenal gland both in phylogeny and ontogeny.
8. What is the relationship embryologically and physiologically of the medulla of the suprarenal gland and the sympathetic nervous system?
9. The suprarenal glands are retroperitoneal. What does that mean?
10. List some of the functions of the suprarenal cortex.
11. Describe the embryological development of the thyroid gland.
12. Define cretinism, simple goiter, myxedema, and exophthalmic goiter relative to thyroid function.
13. What is the relationship of the thyroids and parathyroids?
14. How do the parathyroids influence calcium metabolism?
15. Where do the islands (*islets*) of Langerhans originate?

Chapter 18

The Nervous System

POINT OF VIEW

Our study of the anatomy of man has thus far dealt mostly with systems which are responsible for protection, support, movement, maintenance, transportation, and reproduction. An effort has been made in dealing with these systems to emphasize their interdependence both in structure and function, and their dependence upon the activating, controlling, and integrating capacities of the endocrine and nervous systems. The endocrine system has been described and its special form of chemical control elucidated. The controlling influence of carbon dioxide and other metabolites and of hydrochloric and other acids has been suggested. But of primary importance is the nervous system and its associated "organs" of general and special sense. Without this system the individual would be very loosely organized and unable to cope with the ever-changing environment, or even to sense the changes.

Consideration of the functional anatomy of the nervous system will be the subject of the remaining chapters of this book. It will complete the story of an integrated organism, a whole man.

FUNCTIONS

The functions of the nervous system may be stated briefly as (1) orientation, (2) coordination, and (3) conceptual thought (*intelligence*). It is evident that these functions, as stated, are overlapping and interdependent. **Orientation** depends initially upon the scope and effectiveness of the organs of general and special sense; upon their capacity to generate nerve impulses in response to changes (*stimuli*) in the **external** and **internal environment** of the body. Impulses so generated travel over sensory nerves to centrally located coordinating areas from which impulses travel outward to the effector organs which respond to the impulses. As a result the organism orients or adjusts itself to the environmental changes. The success, or even the survival, of a man is dependent upon his capacity to orient satisfactorily to an ever-changing environment. **Coordination** is involved in the orientation process, but is a special function of the central organs of the nervous system, which receive impulses, sort them out and direct them to the efferent channels which will result in responses favorable to the organism. **Conceptual thought,** a unique quality in man and a function perhaps of his expanded cerebral cortex, refers to his capacity to record, store, and relate information received and actions taken as a fund of experience to be used in the determination of his future reactions to environmental changes. As men, we do not necessarily respond overtly to new situations at once. We can think them over, consider a number of responses and evaluate their outcomes without committing ourselves to action. Men have the mechanism in their brains for creativity, imagination, calculation, prediction, abstract reasoning, and the control of violence. The extent to which we live up to these capacities is not encouraging, but to seek to do so is perhaps man's greatest challenge and supreme goal. It could be his best chance for survival.

GENERAL PLAN (Fig. 18–1)

The nervous system, for the purpose of discussion and description, is divided into the following parts:

> Central nervous system
> > Brain
> > Spinal cord
> Peripheral nervous system
> > Cranial nerves
> > Spinal nerves
> > Autonomic nervous system
> > > Sympathetic
> > > Parasympathetic
> > Visceral afferent fibers

It should be remembered and understood that this subdividing of the system is arbitrary and that actually there is a functional continuity of nervous tissue from one part to another. The term **voluntary** is sometimes applied to the central nervous system and the spinal and cranial nerves because they provide impulses to the skeletal muscles which we can call into action at will, while the actions induced by the autonomic system are quite beyond willful control and are called **involuntary**. Again this is an arbitrary arrangement, the structures of the two overlapping. There is but one nervous system.

The **brain** and the **spinal cord,** constituting the central nervous system, are continuous one with the other through the foramen magnum of the skull, the brain being in the cranial cavity, the spinal cord in the vertebral canal of the backbone.

In the peripheral system, the **cranial**

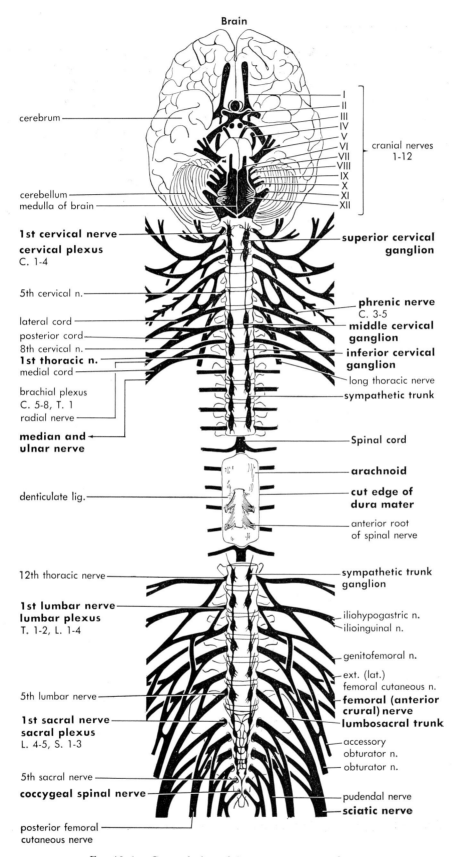

Brain

cerebrum

cranial nerves
1-12

I
II
III
IV
V
VI
VII
VIII
IX
X
XI
XII

cerebellum
medulla of brain

1st cervical nerve
cervical plexus
C. 1-4

5th cervical n.

lateral cord
posterior cord
8th cervical n.
1st thoracic n.
medial cord

brachial plexus
C. 5-8, T. 1
radial nerve

median and
ulnar nerve

denticulate lig.

superior cervical
ganglion

phrenic nerve
C. 3-5
middle cervical
ganglion
inferior cervical
ganglion
long thoracic nerve
sympathetic trunk

Spinal cord

arachnoid

cut edge of
dura mater

anterior root
of spinal nerve

12th thoracic nerve

1st lumbar nerve
lumbar plexus
T. 1-2, L. 1-4

5th lumbar nerve

1st sacral nerve
sacral plexus
L. 4-5, S. 1-3

5th sacral nerve

coccygeal spinal nerve

posterior femoral
cutaneous nerve

sympathetic trunk
ganglion

iliohypogastric n.
ilioinguinal n.

genitofemoral n.

ext. (lat.)
femoral cutaneous n.
femoral (anterior
crural) nerve
lumbosacral trunk

accessory
obturator n.
obturator n.

pudendal nerve
sciatic nerve

FIG. 18–1.   General plan of the nervous system of man.

nerves issue from the inferior side of the brain and supply chiefly the head and neck; the **spinal nerves** come off segmentally from the spinal cord by anterior and posterior roots and, passing through the intervertebral foramina, supply the remainder of the body.

The autonomic nervous system consists of **efferent neurons** and **ganglia** which supply the visceral organs, the blood vessels, and glands. The **sympathetic**

division communicates with spinal nerves and cord from the first thoracic to the third lumbar level. The **parasympathetic** ties in with the midbrain and medulla through four of the cranial nerves, and with the spinal nerves and cord of the sacral region.

The **visceral afferent fibers** come from the viscera and have their cell bodies in cerebrospinal ganglia. They conduct afferent inpulses which convey

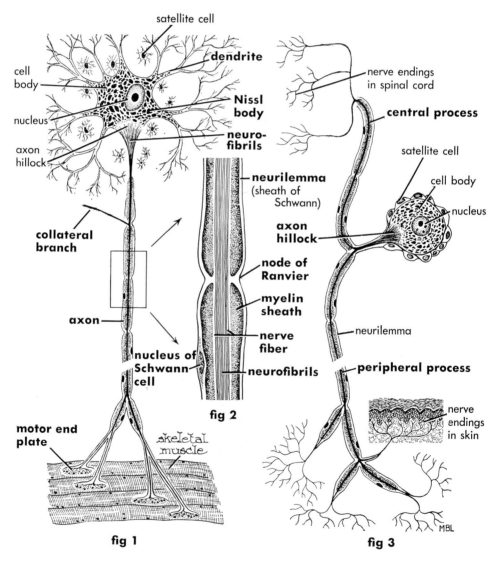

Fig. 18–2.   Types of myelinated neurons.   *A*, Efferent (motor).   *B*, Afferent (sensory).

pain; sensations of sex, nausea, hunger and distension; and some impulses which do not reach the level of consciousness.

NERVOUS TISSUE

Nervous tissue is made up of nerve cells or neurons and supporting cells, the the neuroglia or glial cells.

Neurons (Fig. 18–2, 3). **Neurons** are the basic structural units of the nervous system. They are specialized in the physiological properties of **irritability** and **conductivity.** By irritability we refer to their capacity to respond to

stimulation by a change in structure and activity of the cell at the point which is stimulated. The spread of this activity is called conduction; the activity, a self-propagating physicochemical disturbance, constitutes the **nerve impulse.**

A neuron is composed of a **cell body** (*perikaryon*) and its **processes.** The cell body consists of a mass of protoplasm containing a large and vesicular nucleus with a prominent nucleolus. Within the cytoplasm (*neuroplasm*) are found the usual mitochondria, Golgi bodies, fat droplets, and pigments. In addition, there are the specialized structures of

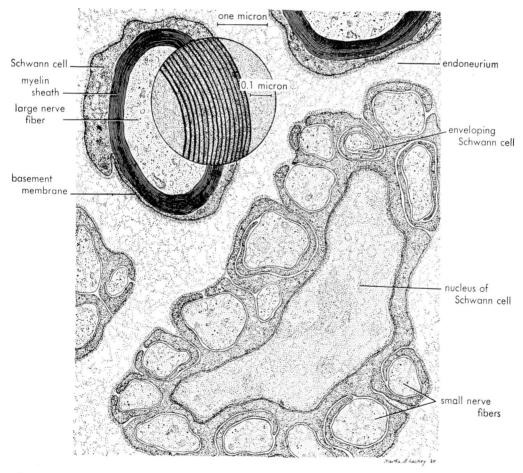

FIG. 18–3. Peripheral nerve fibers and their sheaths. *Lower right*—many small fibers disposed around the nucleus of a single Schwann cell each enveloped by cytoplasmic extensions of the sheath cell. *Upper left*—a large fiber with myelin sheath, the lamellae of which represent successive layers of Schwann cell plasma membrane. Magnification—overall—13,750×—insert in circle 77,750×. (Modified from Porter and Bonneville.)

nerve cell bodies: the chromophil substance in the form of **Nissl bodies** and fine, longitudinally arranged fibrils, the **neurofibrillae.** The processes are thread-like cytoplasmic extensions of two kinds, dendrites and axons. **Dendrites** (*dendron*) receive and conduct impulses toward the cell body. They usually contain all of the structures found in the cell body. There may be one or many dendrites and they vary as to size, shape and the extent of their branching.

Axons (*axis cylinders*) conduct impulses away from the cell body and are commonly called **nerve fibers.** There is only one axon to each neuron, but it usually gives off a number of branches called **collaterals.** Axons are of uniform diameter and contain neurofibrillae and mitochondria but no Nissl bodies or granules. At the point of their origin from the cell body there is an elevated area, the **axon hillock** which, too, is devoid or nearly so of Nissl bodies. At their free ends, axons branch out and either end in effector organs such as muscle tissue or glands, or contact the dendrites and cell body of another neuron to form a synapse, an area of functional continuity between two or more neurons. Since many axons extend from cell bodies in the spinal cord or brain to skeletal muscles at the periphery of the body, in the foot, for example, they must obviously be very long. Others, of course, are short.

Axons may be covered with one or two sheaths: an inner, relatively thick, myelin or medullary sheath and an outer thin, cellular sheath, the neurilemma. The **myelin sheath** is not a continuous covering, but is broken at intervals into separate segments between which are interruptions, the **nodes of Ranvier.** The segments between nodes are called **internodes.** Fibers are referred to as **myelinated** (*medullated*) or **unmyelinated** (*nonmedullated*) on this basis. Myelinated fibers are found in the central nervous system where they give a whitish or yellowish color to the areas in which they are concentrated, hence the white matter. The myelin sheaths of central neurons are more continuous, usually having nodes only where they branch. Myelinated fibers with nodes at regular intervals are found in peripheral nerves. Unmyelinated fibers are found in the autonomic nervous system and some in the gray areas of the central system. Recent studies with the electron microscope have shown that the myelin "sheath" of peripheral nerves is produced by the Schwann cells of the neurilemma as they wind their doubled cell membranes tightly around the axon forming a spiral envelope of several turns. The myelin is therefore a part of the neurilemma and not a separate sheath (Fig. 18–3).

The **neurilemma** (*sheath of Schwann*) is found on fibers of the peripheral system, and probably on those of the central nervous system. It is composed of a single layer of flattened cells, one cell per internode in myelinated fibers and is continuous in the nodes of Ranvier. In unmyelinated fibers the neurilemma is difficult to distinguish from the axon on which it rests, except by the presence of its nuclei. A special staining of the fibers, such as a silver stain which blackens them, will cause the fibers to stand out in contrast to the neurilemma. The neurilemma may contribute to the regeneration of injured nerve fibers.

Classification of Neurons (Fig. 18–4). Neurons are often classed as multipolar, bipolar, and unipolar. **Multipolar neurons** are those with many cell processes. They are the most common and are seen as effector neurons of peripheral nerves, autonomic neurons, and connector or internuncial neurons of the central system. **Bipolar neurons** with two separate processes are found only in the ganglia of the auditory nerve, in the olfactory receptors, and in one layer of the retina of the eye. **Unipolar neurons,**

those with but one process, are found in the sensory ganglia of spinal and certain sensory cranial nerves. Their single process divides into central and peripheral branches very close to the cell body in a T-shaped manner. Some authors argue on the basis of embryological studies that this is really a bipolar cell whose two processes have joined secondarily.

Neuroglia (Fig. 18–5). Neuroglia, a term taken from the Greek, meaning nerve glue, serve as the supporting tissues of the central nervous system. There are three types of cells involved: (1) astro-cytes, (2) oligodendrocytes, and (3) microgliocytes (*microglia*). The first two are often called macroglia and are of ectodermal origin; the microgliocytes are derived from the mesoderm.

The **astrocytes** are relatively large cells with many radiating processes whose terminal portions may have expansions which attach to the pia mater or blood vessels. There are two kinds of astrocytes: (1) fibrous astrocytes found mostly in the white matter and having fibers running through the cytoplasm of their cell bodies; and (2) protoplasmic astrocytes, which occur chiefly in the

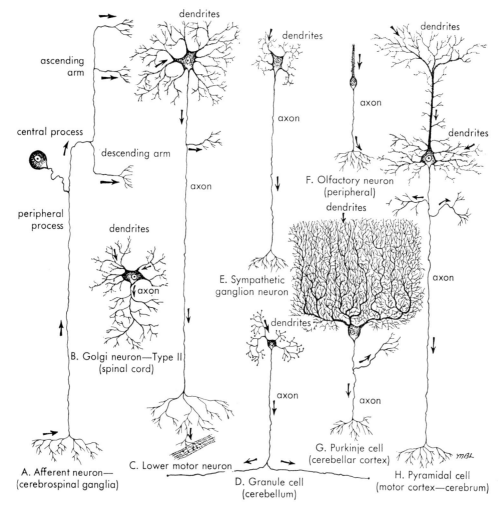

Fig. 18–4.   Schematic diagrams of the principal forms of neurons.
(Modified from Bailey.)

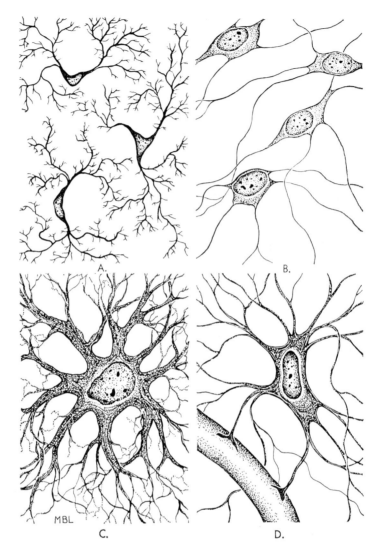

Fig. 18–5. Neuroglial cells of the central nervous system. *A*, Microglia. *B*, Oligodendroglia.
C, Protoplasmic astrocyte. *D*, Fibrous astrocyte.

gray matter.   Astrocytes contribute to the repair processes in the central nervous system and lend support to the nervous tissue.

Oligodendrocytes (*oligodendroglia*) are smaller than astrocytes and have fewer processes.   They are found in close association with small blood vessels and with large nerve cells and in the white matter they lie between bundles of fibers. Their processes frequently clasp the nerve fibers.   Oligodendrocytes are the

"satellite cells" of nerve cells and may serve in the formation and preservation of the myelin sheaths of fibers of the central nervous system, thus assuming the role of the neurolemma of peripheral nerves.

Microgliocytes (*microglia*) are found throughout the gray and white matter of the central system.   They are small cells whose two or more processes are finely branched and feathery.   They are phagocytic and serve to remove the dead or dying parts of tissues and foreign materials.

Glial cells frequently give rise to tumors of the central nervous system. These may be rapidly growing malignancies or slow growing and benign. The latter are often successfully removed surgically.

Nerves (Fig. 18–6). Nerves may be defined as bundles of nerve fibers coursing together outside of the central nervous system. They are held together by well-organized connective tissue sheaths which give protection and strength to what would otherwise be very weak and therefore vulnerable structures. It is because of the connective tissue encasements that tiny axons (*fibers*) can travel great distances in the body to innervate muscles and other structures. A fine sheath of reticular fibers in loose connective tissue cover the individual nerve fibers. These constitute the **endoneurium**. Many of these nerve fibers are collected together into **fasciculi** (*bundles*) under an outer wrapping of areolar connective tissue

called the **perineurium**. A number of bundles, together constituting a nerve, have an outside sheath called the **epineurium**. The connective tissue elements of these various sheaths are continuous one into another. In the sheaths will be found blood vessels and lymphatics appropriate to the size of the nerves.

Nerves vary greatly in size and in the nature of their contained fibers. Spinal and cranial nerves of the peripheral system are large and contain large numbers of myelinated fibers as well as some unmyelinated fibers. Many nerves of the autonomic system have small fibers which are unmyelinated and in sections of these nerves it is sometimes difficult to differentiate between the fibers and the sheath structures.

Tracts. Collections of nerve fibers in the central system are called tracts. They, too, are highly organized and have supporting neuroglial cells (Fig. 18–12).

Ganglia (Fig. 18–1). Collections of

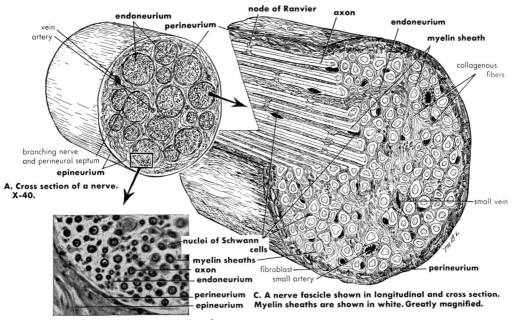

A. Cross section of a nerve. X-40.

B. Photomicrograph of a cross section of a fascicle (bundle) of a myelinated nerve. Osmic acid stain makes the myelin sheath black. x 600.

C. A nerve fascicle shown in longitudinal and cross section. Myelin sheaths are shown in white. Greatly magnified.

Fig. 18–6.   Studies of the microscopic anatomy of a peripheral nerve.

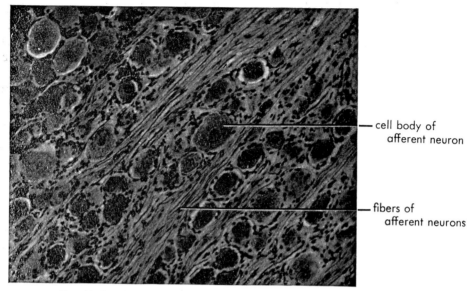

— cell body of
afferent neuron

— fibers of
afferent neurons

FIG. 18–7.   Photomicrograph of a section of a spinal ganglion.   200×.

nerve cell bodies occurring outside of the central nervous system are called ganglia. Some ganglia are large, others very small and in organs of the digestive tract single cells occur. The misconception is common that individual nerve cell bodies occur at intervals along the nerves but they are found only in the gray matter of the central system and in ganglia. Two kinds of ganglia are found in the human body—the sensory and autonomic.

Sensory ganglia are found as spinal ganglia on the dorsal roots of spinal nerves and on the sensory roots of some of the cranial nerves (Fig. 18–7). They are invested with a tough capsule of dense connective tissue which is continuous with the epineurium of the nerve. From the capsule numerous septa penetrate the ganglion carrying blood vessels and lymphatics. The nerve cells are often grouped around the periphery with some centers of deep-lying cells. Fasciculi of fibers run through the ganglion. The nerve cells in spinal ganglia are pear-shaped with a single process which divides near the cell and sends a central branch to the spinal cord and a peripheral branch to the skin or muscle. Capsules

present around the ganglion cells are continuous with the neurilemma of their fibers. Both myelinated and unmyelinated neurons are represented in the spinal ganglia.

Autonomic (*sympathetic*) **ganglia** occur as swellings along the sympathetic nerve trunks which extend through the abdomen, thorax, and neck, one trunk to each side of the vertebral column (Fig. 18–1). They are found also in association with splanchnic nerves in the abdomen. Autonomic ganglion cells are multipolar, being the cell bodies of unmyelinated postganglionic neurons. Myelinated preganglionic neurons synapse with the ganglion cells or pass through the ganglion. Sheath cells form capsules around the ganglion cells.

Nuceli and Nerve Centers. The term **nucleus** is commonly used to designate groups of cell bodies within the central system. They are circumscribed gray areas. A **nerve center** is a functional designation for a nucleus. We speak of respiratory and cardiac centers in the medulla which function in control of breathing and heart action respectively.

Peripheral Nerve Endings. The

endings of the peripheral nerve fibers range from simple naked fibers to those with highly complex accessory structures such as the eye. They may be divided into two functional groups, the receptor (*sensory*) and effector (*motor*) endings. The **receptors** are considered in the next chapter (Figs. 19–1, 2, 3).

Effector nerves are supplied to muscle tissues and glands. Those to **involuntary muscle** (*smooth and cardiac*) and to **glands** are derived from the autonomic system and are composed mostly of unmyelinated fibers. While relatively little is known about the nerve endings to these involuntary structures, it appears that near their terminations, these nerves form numerous branches which communicate and form intimate plexuses near or around the muscles and glands. These plexuses give off tiny branches which in turn break up into fibrillae which run between the involuntary muscles and gland cells to terminate in their surfaces.

Myelinated efferent axons (*fibers*) **to skeletal or voluntary muscles** have their cell bodies in the gray matter of the spinal cord and brain and travel outward in the spinal and cranial nerves. Each axon may supply just one or two fibers in the more critical and precise-acting muscles as those of the fingers or those moving the eye, whereas in the supporting and postural muscles one axon may supply a hundred or more muscle fibers. An axon and the skeletal muscle fibers it innervates constitute a **motor unit.**

As a nerve fiber enters a muscle fasciculus (*bundle*), it branches repeatedly and each terminal branch makes a functional connection with a muscle fiber to form a **myoneural junction** or **motor endplate** (Figs. 18–2, 8). Here the nerve impulse activates the muscle fiber.

The Nerve Impulse

The nature of the nerve impulse is not fully understood, so that any attempt to define it must be considered tentative. The most widely held idea of the nature of the nerve impulse is called the **membrane theory.** According to this the membrane (*cell membrane*) of the neuron is semipermeable and electrically polarized. Specifically, there is a layer of positively charged ions on the outside and a layer of negatively charged ions on the inside of the cell membrane. Presumably, the membrane is impermeable to these ions so polarization is maintained. Polarization is conversely sup-

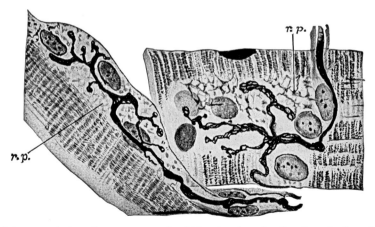

Fig. 18–8. Motor end-plate from tongue of rabbit, showing the "periterminal net" (*r.p.*) of the end-plate. (Redrawn after Ramón y Cajal from Bloom and Fawcett, *A Textbook of Histology*, courtesy of W. B. Saunders Co.)

posed to be in some way important in maintaining the semipermeability of the membrane. A breakdown in one would involve a breakdown in the other. A stimulus applied to the nerve fiber is believed to bring about this breakdown by causing the membrane at the point of stimulation to become permeable. The positive and negative ions move through the permeable gap in the membrane and neutralize one another. This point in the nerve fiber is now electrically negative to adjacent areas and they depolarize, and a wave of depolarization thus moves along the entire fiber. This is the nerve impulse, and may be defined as a physicochemical change in the nerve fibers which, once initiated, is self-propagating. Both direct and indirect evidence support this theory, but it is beyond our scope to consider it. However, as you learn facts about the nerve impulse, see whether you can explain them on the basis of this theory.

Nerve impulses are all alike in kind. The sensation which results, pain, cold, etc., depends on the part of the brain receiving the impulses. The same impulses, if ending in muscle, would cause it to contract, or if ending in a gland, would make it secrete.

The strength of contraction or the intensity of sensation is not dependent upon the magnitude of the nerve impulses, but upon the frequency with which they follow one after the other. The more intense the stimulus, within limits, the greater the frequency of the impulses.

The rate at which impulses travel along nerve fibers depends on the diameter of the fiber; the larger the fiber the greater the rate of conduction, other factors remaining the same. The largest myelinated fibers conduct at a rate of about 120 meters per second, 270 miles per hour; the smallest ones at about 6 meters per second or $13\frac{1}{2}$ miles per hour. Myelinated fibers conduct more rapidly than unmyelinated and with less expenditure of energy.

Nerve fibers also behave according to the all-or-none phenomenon; *i.e.*, if they respond at all to a stimulus, the response is maximal for the condition of the fiber at that time.

There is nothing within a nerve fiber or neuron to determine the direction of passage of the impulse. A fiber stimulated at midpoint would conduct equally well in either direction. Yet, we know that in the nervous system, fibers conduct in one direction only. Where is direction determined?

THE SYNAPSE (Fig. 18–18)

The terminal branches of neurons come into functional contact either with effectors or with other neurons. The area of functional continuity between neurons is the **synapse.** The synapse, among other functions, determines the direction of passage of nerve impulses in the nervous system. What is its nature?

As we understand it, each tip of the many terminal branches of an axon forms a terminal bulb or **bouton terminale** in contact with the dendrites and cell body of the next neuron or neurons in the nerve pathway. This contact is not a point of structural continuity between the neurons, nor does the impulse pass directly from one to another. Rather, the arrival of an impulse at a terminal bulb causes some change or condition which serves to stimulate the next neuron, or make it more, or in some cases less, susceptible to stimulation by the terminal bulbs of other neurons present in the synaptic area. It is likely that the resulting excitation or inhibition is mediated by the release of chemical agents though other means are not excluded. Acetylcholine and adrenalin-like substances have been collected in synaptic areas in the autonomic system as well as at junctions between certain efferent nerves and their effectors.   Recent evidence indicates that these substances are being formed all the time at nerve muscle junctions, and that a nerve impulse does not initiate this secretion, but changes the rate of secretion.   Assuming that the dendrites do not have this capacity to secrete substances into the synapse, whereas axons do, we can understand how synapses serve to establish direction of conduction in the intact nervous system.   We know, too, that conduction is slowed at the synapse and that the synapse is responsible for other important phenomena.   These facts and suggestions should help to encourage your continued interest and study.

SPINAL CORD (Figs. 18–1, 9, 10)

The spinal cord is continuous with the brain at the foramen magnum, passes through the vertebral canal, and ends in the tapered **conus medullaris** at the level between the first and second lumbar vertebrae.   It has a total length of from 43 to 45 cm. (18 inches), an average diameter of about 1 cm.   From the tip of the conus medullaris a narrow, non-nervous thread, the **filum terminale,** extends to the base of the coccyx where it attaches into the coccygeal ligament (Fig. 18–9).   The spinal cord is as long as the vertebral canal until the third month of fetal life, after which the vertebral column lengthens at a more rapid rate than the cord.   At birth, the cord terminates opposite the third lumbar vertebra and gradually recedes during childhood to the adult position.

External Features (Figs. 18–1, 9, 10).   The spinal cord is not of uniform diameter and shape throughout its length.   It has two prominent enlargements which are a structural indication of the complex functions regulated through the spinal nerves of these areas.   The **cervical enlargement** extends from the level of the third cervical to the second thoracic

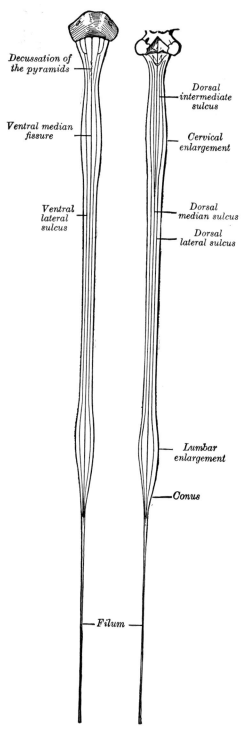

FIG. 18–9.   Diagrams of the spinal cord. (Gray, *Anatomy of the Human Body,* Lea & Febiger.)

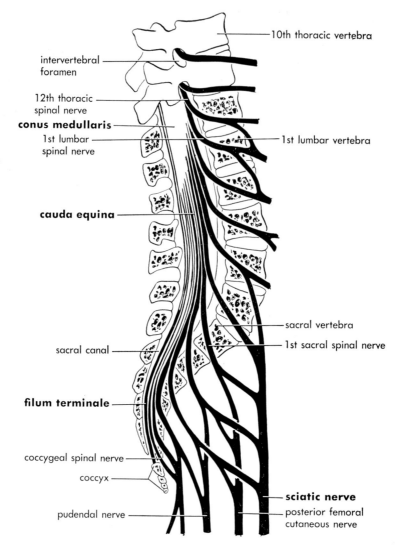

FIG. 18–10.   Inferior portion of spinal cord and vertebral column showing conus medullaris and cauda equina.

vertebra and corresponds to the area supplying the large nerves to the complex musculature and other structures of the upper extremity. The **lumbar enlargement** extends from the ninth thoracic vertebra, reaches its maximum diameter opposite the twelfth thoracic and then tapers rapidly into the conus medullaris. It gives origins to the nerves supplying the lower extremity.

The anterior (*ventral*) and posterior (*dorsal*) surfaces of the spinal cord are marked by the anterior median fissure and the posterior median sulcus, respectively, which divide the cord surface into right and left halves (Fig. 18–9). The **anterior median fissure** extends about 3 mm. into the substance of the cord. The **posterior median sulcus** is a very shallow groove that marks the position of the **posterior median septum,** a sheet of the neuroglia that pushes more than halfway into the substances of the cord.

A **posterior lateral sulcus** is a shallow

longitudinal furrow marking the position of attachment of the posterior roots of the spinal nerves. Between it and the posterior median sulcus in the thoracic and cervical regions of the cord is a **posterior intermediate sulcus** which lies between two important nerve fiber tracts of the cord, the gracilis and cuneatus fasciculi. The **anterior lateral sulcus** is very shallow and inconspicuous and marks the position of attachment of the anterior roots of the spinal nerves.

Regions and Segments of the Cord. There are 31 pairs of spinal nerves which originate from the spinal cord and they are named to correspond to the regions of the vertebral column through which they find passage. The pairs are: 8 cervicals, 12 thoracics, 5 lumbars, 5 sacrals and 1 coccygeal. The cord, in turn, is divided into regions corresponding to the nerves, cervical, thoracic, etc. Further, each part of the spinal cord giving rise to a pair of spinal nerves is called a **spinal segment** or **neuromere** although the nerve roots are the only remaining mark of this embryonic segmentation.

As a result of the unequal lengths of the spinal cord and vertebral column referred to earlier, and the fact that the roots of the spinal nerves emerge through the intervertebral foramina at their appropriate vertebral levels, the nerve roots take a progressively more oblique direction until those in the lumbar and sacral regions are almost vertical and pass below the cord before reaching their foramina. They form there a collection of roots well named the **cauda equina** (*horse's tail*) (Fig. 18–10).

Cross-Section of Spinal Cord (Fig. 18–11). A study of the cross-section of the spinal cord will serve to review some of the features just described and to show their relationships to internal structures. In the middle of the section is a small opening, the **central canal,** a reminder of the dorsal hollow nerve tube

given as a diagnostic feature of all chordate animals in Chapter 1. It is a reminder, too, of embryological development—the rolling-up of the ectoderm to form the neural tube, later to differentiate into cord and brain.

Distinct white and gray areas show in the cut surface of the section. The **gray matter** is centrally located and assumes roughly the form of the letter H. The white matter fills in the areas around the gray. Gray matter is composed of cell bodies, dendrites, unmyelinated nerve fibers, and neuroglia. These nerve fibers are the axons of not only the local neurons but of cells in other parts of the system which synapse with the local cells. Most of them lose their myelin sheaths before they terminate in the gray matter.

The gray matter or substance may be divided into a number of parts which have structural and functional significance. The crossbar of the H is divided by the central canal into **anterior** and **posterior gray commissures** which contain fibers connecting the two halves of the gray substance. The two posterior projections of the H are the **posterior columns.** Central processes of spinal root ganglion cells can sometimes be seen entering the posterior columns whereby the processes send branches into bundles or tracts going to other levels of the cord, or to the brain, or they may synapse with connector neurons, or with the cells in the ventral column at the same level. The **anterior columns** which are the forward-projecting parts of the H contain the large ventral horn cells, easily seen in almost any cross-section. Their axons make up the anterior root fibers of the spinal nerves which innervate skeletal muscles. In the thoracic and lumbar parts of the spinal cord, there are **lateral columns** of gray matter containing cells whose axons go out through the anterior root as preganglionic fibers to sympathetic ganglia of the **thoracolumbar** system. Lateral columns are found also

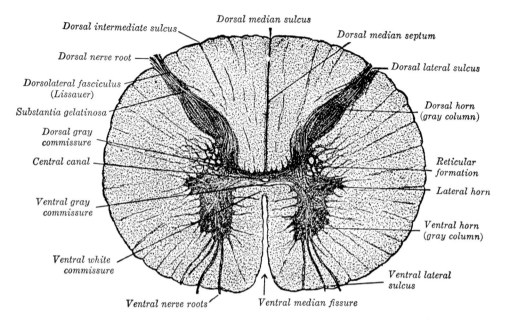

FIG. 18–11.   Transverse section of the spinal cord in the mid-thoracic region.
(*Gray's Anatomy.*)

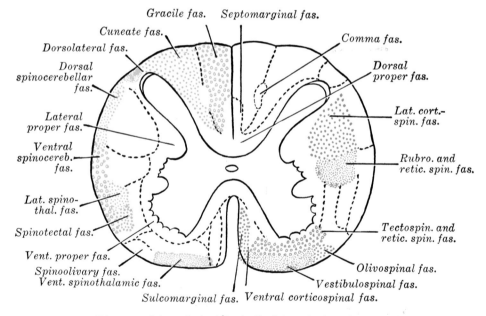

FIG. 18–12.   Diagram of the principal fasciculi of the spinal cord.   Ascending fibers in
blue, descending in red.   (*Gray's Anatomy.*)

in the third and fourth sacral segments of the cord from the cell bodies of which fibers pass through the anterior root to enter the pelvic nerves as a part of the **craniosacral** outflow of the autonomic system. These facts will be further emphasized as we study the components of spinal reflex arcs and the autonomic system.

The columns of gray matter just described, because they are commonly studied in cross-sections may be referred to as **horns,** hence **posterior, anterior** and **lateral horns.** Remember, too, that it is common to use **dorsal** for posterior, **ventral** for anterior. Semantics is an ever-present problem in the study of human anatomy, as in almost any discipline.

The **white matter** or substance of the spinal cord is composed of myelinated and unmyelinated fibers supported by neuroglia. Most of the fibers follow a longitudinal course but many cross from one side to the other in the **white commissure** which lies just anterior to the gray commissure.

The white matter in each *half* of the spinal cord is divided into three portions by the anterior and posterior columns of gray matter. These are the **anterior, lateral,** and **posterior funiculi.** They are made up of ascending and descending **fiber tracts.** Groups of fiber tracts related functionally are sometimes called **fasciculi.** The ascending tracts are associated with the afferent or sensory, the descending with the efferent or motor, parts of the nervous system. They constitute the conduction highways of the spinal cord and are shown diagrammatically in Figures 18–12 and 50 to 54.

The fiber tracts of the spinal cord form a link in the conduction pathways between the receptors of the body and the higher centers in the brain, and between the brain and the effectors. Some tracts merely connect different levels within the spinal cord. Although there are notable exceptions, the names of the tracts are descriptive, giving the location of the cells of origin and the point at which the axons synapse with the next link or neuron in the pathway. For example, the corticospinal tract has its cells of origin in the cortex and its synapses in the spinal cord, whereas the spinothalamic tract has its cells of origin in the spinal cord and its synapses in the thalamus. The spinocerebellar tracts have their cells of origin in the spinal cord, their synapses in the cerebellum. However, the posterior funiculi are made up primarily of the fasciculus gracilis and fasciculus cuneatus tracts and do not follow the rule. The names of many of the tracts are preceded by lateral, dorsal, or ventral, which in most cases places them in the funiculi of the same name. Thus, the lateral spinothalamic tract is in the lateral funiculus; the ventral corticospinal tract is in the ventral or anterior funiculus, but the dorsal and ventral spinocerebellar tracts are both in the lateral funiculus. Nevertheless, these guidelines will be of some help in learning the fiber tracts of the cord.

Some of the fiber tracts of the spinal cord are considered in the context of nerve pathways following the descriptions of the chief landmarks of the whole nervous system (see pages 552–560 and Figs. 18–50 to 54).

MENINGES OF THE SPINAL CORD
(Figs. 18–1, 13)

The spinal cord and brain are enclosed within three membranes, the **meninges.** Named from the outside inward they are the **dura mater, arachnoid,** and **pia mater.** Though very similar in the two regions, our concern now is with their relationship to the spinal cord.

The **dura mater** is a strong, tubular sheath of dense white fibrous and elastic tissue which extends from the foramen magnum to the level of the second sacral vertebra. Between it and the bony wall of the vertebral canal are epidural fat, areolar tissue, and a venous plexus. At the foramen magnum it is continuous with the inner or meningeal dura of the brain and is attached to the occipital bone. It is also connected by fibrous slips to the posterior longitudinal ligament, especially inferiorly. At the second sacral vertebra it tapers abruptly and forms a covering for the filum terminale of the spinal cord, the **coccygeal ligament.** The ligament descends to become continuous with the periosteum of the coccyx. The spinal dura sends separate prolongations outward over the posterior and anterior roots of the spinal nerves which at the posterior root ganglia blend into single sheaths. These sheaths extend over the ganglia to the spinal nerves where they blend with their sheaths. The dura also adheres to the periosteal linings of the intervertebral foramina. The dura tube is lined with mesothelium and is separated from the arachnoid

membrane by a very fine interval, the **subdural space.** This contains enough fluid to moisten the opposed surfaces of the membranes.

The spinal **arachnoid membrane** is a delicate structure continuous with the cranial arachnoid. It follows quite completely the spinal dura mater in all of its ramifications. There is a wide **subarachnoid space** surrounding the cord and its pia mater, being largest at the lower end of the vertebral canal where it is occupied by the nerves of the cauda equina. The subarachnoid space contains the **cerebrospinal fluid,** a fluid which is produced in the ventricles of the brain and circulates through the ventricles, the central canal of the cord, and moving through special foramina in the roof of the fourth ventricle of the brain it reaches the cranial subarachnoid space which is continuous with that of the spinal cord. The combination of structures below the level of the termination of the spinal cord at the second lumbar

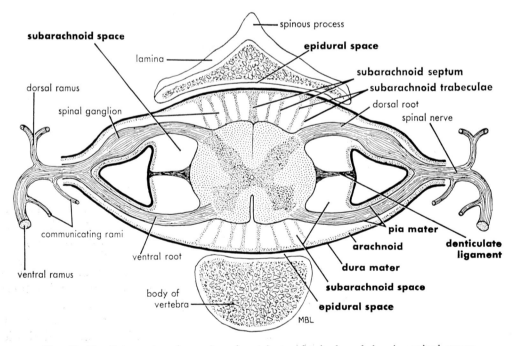

Fig. 18–13.   Schematic cross-section of vertebra and spinal cord showing spinal nerves and meninges (membranes).

vertebra is ideal for the collection of cerebrospinal fluid. A lumbar puncture needle, which is long, thin, and flexible, can be inserted between the separated laminae or spinous processes of the mid-lumbar vertebrae into the large sub-arachnoid sac without danger of puncturing the spinal cord. The nerves of the cauda equina, being round and suspended in fluid, roll out of the way of the needle. Lumbar punctures are used as a means of introducing anesthetics, also as a means of collecting cerebrospinal fluid for examination and diagnosis of certain nervous system disorders.

The **subarachnoid space** is divided by numerous, delicate **trabeculae** coming from both the pia and the arachnoid, and along the posterior median line many of these join to form the subarachnoid septum. The subarachnoid space is also partially divided laterally by the **denticulate ligaments** derived from the pia mater. These pierce the arachnoid and fasten to the dura mater.

The **pia mater** is so intimately related to the spinal cord and to the nerve roots that it cannot be pulled free. It is a thin connective tissue covering with a rich network of blood vessels. It contributes to the trabeculae of the subarachnoid space, to the subarachnoid septum, and its outer longitudinal fibers form the two **denticulate ligaments,** one on either side of the cord. These ligaments help to fix the position of the spinal cord and protect it against sudden displacement, just as the cerebrospinal fluid is an essential buffer material.

SPINAL NERVES
(Figs. 18–1, 6, 13 to 18)

Reference has already been made to the spinal nerves, a part of the peripheral nervous system. They arise as 31 pairs from the segments of the spinal cord in the vertebral canal and pass out through the intervertebral foramina. They are

33

classified as 8 cervicals, 12 thoracics, 5 lumbars, 5 sacrals, and 1 coccygeal. The first cervical spinal nerve leaves the vertebral canal between the occipital bone and the atlas, the eighth cervical between the seventh cervical and first thoracic vertebrae; all the others leave below the vertebrae of the same name and number.

Each spinal nerve arises by a **dorsal** (*posterior*) and a **ventral** (*anterior*) **root** whose fibers join as they pass through the intervertebral foramina (Fig. 18–13, 18). The dorsal roots arise from several rootlets along the posterior lateral sulcus which join to form two bundles which enter the spinal ganglion in the intervertebral foramen. The ventral roots arise in the anterior lateral sulcus from fewer rootlets than the dorsal root. They form two bundles near the intervertebral foramen. The two roots join immediately beyond the spinal ganglion to form the common spinal nerve which emerges from the intervertebral foramen.

The **spinal ganglion** (*posterior root ganglion*) is an aggregate of nerve cell bodies on the dorsal root of the spinal nerve. Most of the spinal ganglia lie within the intervertebral foramina where they are invested with a prolongation of the dura mater. The spinal ganglia of the sacral and coccygeal nerves lie within the vertebral canal and those of the first two cervicals lie on the vertebrae arches of the atlas and axis. The specific composition of the spinal ganglion will become clear as we discuss reflex arcs and the autonomic system.

Sympathetic fibers also contribute to the make-up of the spinal nerves (Fig. 18–13). These are in the form of the **gray rami communicantes** which are made up of **postganglionic unmyelinated fibers** of visceral neurons. They join each spinal nerve just distal to the union of its dorsal and ventral roots and their fibers travel through the dorsal and ventral rami to the smooth muscle tissue

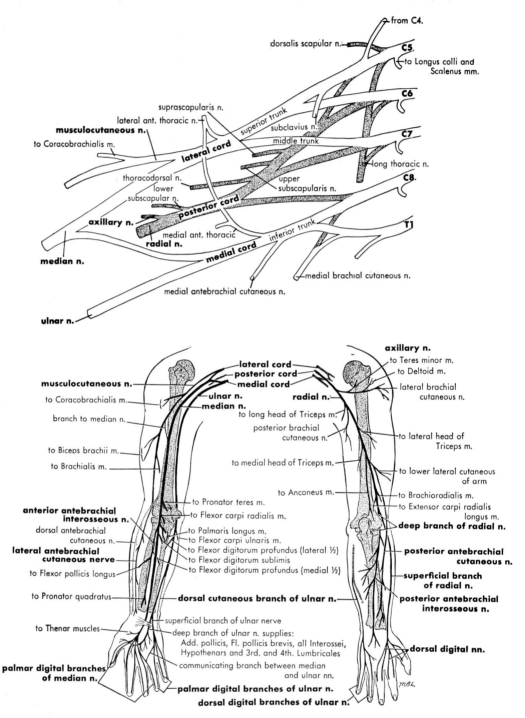

Fig. 18–14.   Right brachial plexus and innervation of the upper limb.   The dorsal division of the plexus is stippled.

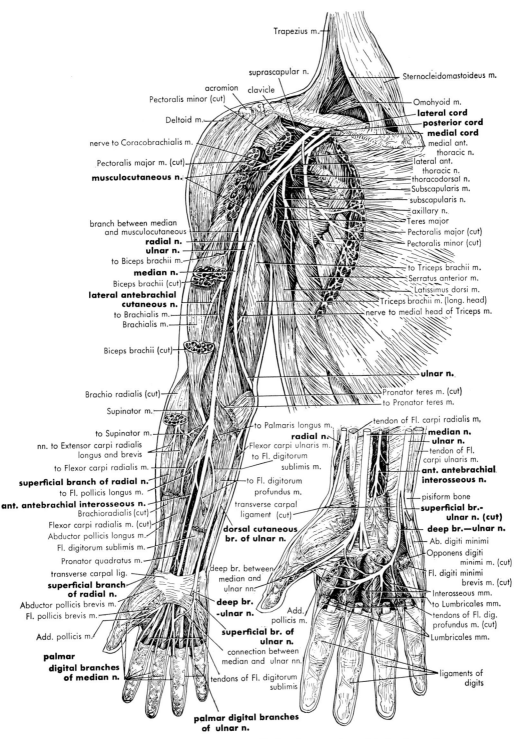

FIG. 18–15.   Nerves of the right upper limb.   Innervation of skeletal muscles
is emphasized.

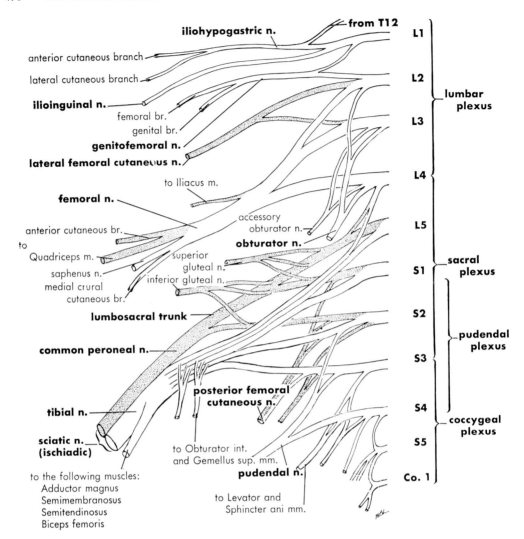

Fig. 18–16. Plan of the lumbar, sacral, pudendal and coccygeal plexuses.
Nerves from the dorsal half of the plexus are stippled.

of blood vessels, Arrectores pilorum and to the glands of the skin.

White rami communicantes made up of **myelinated** fibers of **preganglionic neurons** of the **thoracolumbar system** leave the ventral roots of the twelve thoracic and the first two lumbar spinal nerves just before the ventral roots join the dorsal to form the spinal nerves. The white rami communicantes join the ganglia of the sympathetic chain and therefore can be said to form the roots of these ganglia.

The common spinal nerve is very short. It gives off a small meningeal branch which re-enters the vertebral canal to supply the meninges of the cord. It then almost immediately divides into **dorsal** and **ventral rami** which are predominately, though not entirely, somatic in distribution, supplying the skin and skeletal muscles of the trunk and limbs. Since the common spinal nerve and each ramus contain fibers from both dorsal and ventral roots, they are called "mixed" nerves (Figs. 18–13, 18).

The **dorsal rami** of the spinal nerves pass posteriorly between the transverse processes of adjacent vertebrae and divide into medial and lateral branches, both supplying muscles and one of them sending a cutaneous branch to the skin (Fig. 18–13). The dorsal rami supply the longitudinal muscles which lie between the spinous processes of the vertebrae and the angles of the ribs.

The **ventral rami** of the spinal nerves are larger than the dorsal and appear to be a direct continuation of the spinal nerve (Figs. 18–13). They supply the skin and muscles of the lateral and anterior body walls and of the upper and lower extremities. It should be remembered that some of the appendicular muscles spread to the posterior surface of the body and these, too, are supplied by ventral rami.

The ventral rami of the **thoracic nerves** remain independent of one another and pass out into the intercostal spaces as the **intercostal nerves,** supplying muscles and skin areas of the thorax and upper abdomen.

Plexuses (Fig. 18–1, 14, 16). The distribution of the dorsal and ventral rami described thus far has been largely segmental. In the region of the limbs, however, the **ventral rami** of the spinal nerves undergo considerable regrouping and redistribution of their nerve bundles to form networks of nerves called **plexuses.** These are the **cervical, brachial** and **lumbosacral plexuses.**

The **cervical plexus** is formed by the ventral rami of the first four cervical nerves and supplies muscles and skin in the neck region and makes connections with the last three cranial nerves. One nerve derived from the plexus wanders a long distance from the neck, namely, the **phrenic nerve** which supplies the diaphragm at the lower limit of the thorax. However, embryological studies indicate that the muscle tissue of the diaphragm had its origin in the cervical myotomes and only secondarily moved to this more inferior position (Fig. 18–1).

The **brachial plexus** is derived from ventral rami of the large lower cervical nerves, five through eight, and the equally large first thoracic nerve. These nerves, in turn, come from the brachial enlargement of the cord and supply the complex musculature of the upper limb. The terminal branches of this plexus are the **musculocutaneous, median, ulnar,** and **radial nerves** whose names suggest their distribution (Figs. 18–14, 15).

The **lumbosacral plexus** is often divided into **lumbar** and **sacral plexuses** (Fig. 18–16). The **lumbar plexus** comes from the ventral rami of the first three lumbar nerves and the major portions of the fourth with a minor contribution from thoracic twelve. Among its important branches are the **femoral** which enters the lower limb and in turn gives rise to many branches to the thigh, and its largest and longest branch, the **saphenous,** which continues into the foot. The **sacral plexus** is formed by ventral rami of the fourth and fifth lumbars and the first three sacral nerves uniting to make a large, flattened band, most of which continues into the thigh as the sciatic nerve. The **sciatic nerve** is the largest nerve in the body, supplying the skin of the foot and most of the leg, the muscles of the posterior side of the thigh, and all those of the leg and foot. Its terminal branches are the **tibial** and **common peroneal nerves.** These two great plexuses collectively supply the skin and muscles of the lower limb (Fig. 18–17).

REFLEX ACTIVITY (Fig. 18–18)

The study of spinal reflex arcs will enable us to review and to better understand the structure and functions of the spinal cord and spinal nerves. The spinal cord serves as a center for reflex

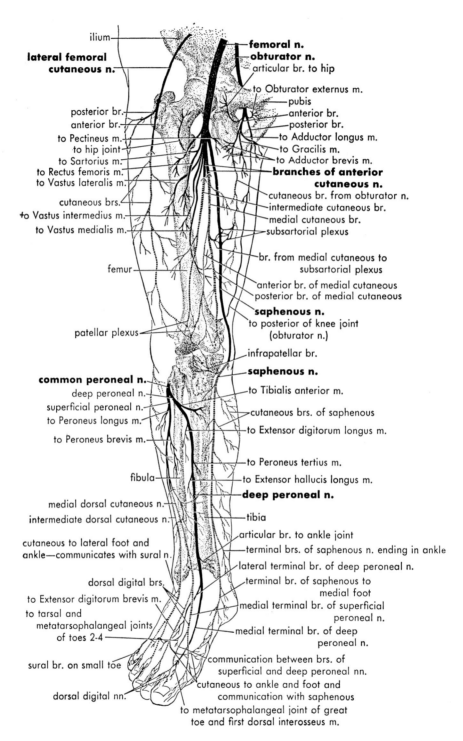

ilium
femoral n.
obturator n.
lateral femoral cutaneous n.
articular br. to hip
to Obturator externus m.
pubis
posterior br.
anterior br.
anterior br.
posterior br.
to Pectineus m.
to hip joint
to Sartorius m.
to Rectus femoris m.
to Vastus lateralis m.
to Adductor longus m.
to Gracilis m.
to Adductor brevis m.
branches of anterior cutaneous n.
cutaneous brs.
to Vastus intermedius m.
to Vastus medialis m.
cutaneous br. from obturator n.
intermediate cutaneous br.
medial cutaneous br.
subsartorial plexus
br. from medial cutaneous to subsartorial plexus
femur
anterior br. of medial cutaneous
posterior br. of medial cutaneous
saphenous n.
to posterior of knee joint (obturator n.)
patellar plexus
infrapatellar br.
saphenous n.
common peroneal n.
deep peroneal n.
superficial peroneal n.
to Peroneus longus m.
to Tibialis anterior m.
cutaneous brs. of saphenous
to Extensor digitorum longus m.
to Peroneus brevis m.
to Peroneus tertius m.
fibula
to Extensor hallucis longus m.
deep peroneal n.
medial dorsal cutaneous n.
intermediate dorsal cutaneous n.
tibia
articular br. to ankle joint
cutaneous to lateral foot and ankle—communicates with sural n.
terminal brs. of saphenous n. ending in ankle
lateral terminal br. of deep peroneal n.
dorsal digital brs.
to Extensor digitorum brevis m.
to tarsal and metatarsophalangeal joints of toes 2-4
terminal br. of saphenous to medial foot
medial terminal br. of superficial peroneal n.
medial terminal br. of deep peroneal n.
sural br. on small toe
communication between brs. of superficial and deep peroneal nn.
dorsal digital nn.
cutaneous to ankle and foot and communication with saphenous
to metatarsophalangeal joint of great toe and first dorsal interosseus m.

Anterior view

Fig. 18–17.

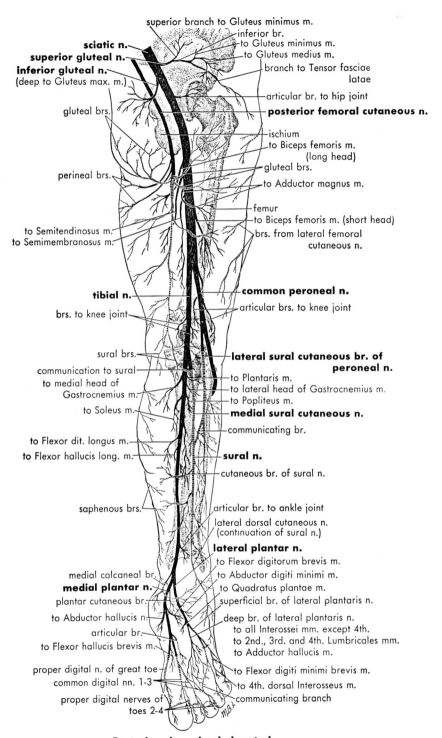

superior branch to Gluteus minimus m.
inferior br.
to Gluteus minimus m.
to Gluteus medius m.
branch to Tensor fasciae
latae
articular br. to hip joint

sciatic n.
superior gluteal n.
inferior gluteal n.
(deep to Gluteus max. m.)

gluteal brs.

posterior femoral cutaneous n.

ischium
to Biceps femoris m.
(long head)
gluteal brs.
to Adductor magnus m.

perineal brs.

femur
to Biceps femoris m. (short head)
brs. from lateral femoral
cutaneous n.

to Semitendinosus m.
to Semimembranosus m.

tibial n.

common peroneal n.
articular brs. to knee joint

brs. to knee joint

sural brs.
communication to sural
to medial head of
Gastrocnemius m.
to Soleus m.

lateral sural cutaneous br. of
peroneal n.
to Plantaris m.
to lateral head of Gastrocnemius m.
to Popliteus m.
medial sural cutaneous n.
communicating br.

to Flexor dit. longus m.
to Flexor hallucis long. m.

sural n.
cutaneous br. of sural n.

saphenous brs.

articular br. to ankle joint
lateral dorsal cutaneous n.
(continuation of sural n.)
lateral plantar n.
to Flexor digitorum brevis m.

medial calcaneal br.
medial plantar n.
plantar cutaneous br.
to Abductor hallucis n.
articular br.
to Flexor hallucis brevis m.

to Abductor digiti minimi m.
to Quadratus plantae m.
superficial br. of lateral plantaris n.

deep br. of lateral plantaris n.
to all Interossei mm. except 4th.
to 2nd., 3rd. and 4th. Lumbricales mm.
to Adductor hallucis m.

proper digital n. of great toe
common digital nn. 1-3
proper digital nerves of
toes 2-4

to Flexor digiti minimi brevis m.
to 4th. dorsal Interosseus m.
communicating branch

Posterior view—heel elevated

FIG. 18–17.   Nerves of the lower limb.   Deep nerves are drawn in
solid black; cutaneous branches in cross-hatch.

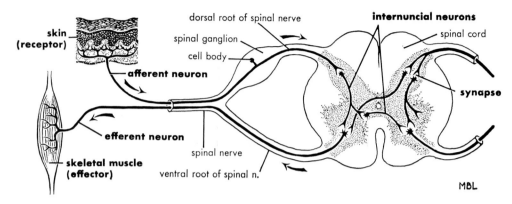

FIG. 18–18.   Diagram of a simple reflex arc.

action as well as a conduction highway for impulses traveling up or down from one spinal segment to another. The spinal nerves establish a functional relationship between the periphery of the body and the cord.

Since the nerve tracts of the cord are continuous with those of the brain, we must sever this connection to study reflex action in the spinal cord without influence from the higher centers. This is easily done on experimental animals, especially cold-blooded ones like the frog. With warm-blooded animals like mammals, the experimental techniques for making and caring for the specimens are much more difficult. Animals treated in this way are called "spinal" animals. Some data have also been made available from human subjects whose spinal cords have been severed by accidental means.

Very shortly after decapitation the "spinal" frog has recovered enough to respond to various stimuli. We know now that the responses which we witness are completely independent of any influence of consciousness, or of any of the inhibiting, augmenting or modifying influences of the higher centers. Before discussing the activity of the "spinal" frog, it will be helpful to look briefly at the mechanism which is involved in its reactions to stimuli—the reflex arc.

REFLEX ARC

Reference to Figure 18–18 will help you to understand the components of a simple reflex arc. They are as follows:

1. A receptor

2. An afferent (*sensory*) neuron—its cell body in the spinal ganglion.

3. An internuncial or connector neuron or neurons in the cord.

4. An efferent (*motor*) neuron—its cell body in the ventral or anterior horn of the gray matter of the cord.

5. An effector organ—a skeletal muscle.

An even simpler reflex arc occurs in the human body, the "knee-jerk" reflex, in which the afferent neuron synapses directly with the efferent neuron in the spinal cord. There is no internuncial neuron.

Most reflex arcs are more complex than the ones mentioned above. There may be many internuncial neurons making connections to different levels in the cord and crossing from one side to the other, as there may be more afferent and efferent neurons synapsing with the internuncial neurons.

The time consumed in the passage of impulses from receptor to effector or from stimulus to response is called **reflex time.** It is the sum of the time involved in nerve conduction and synaptic conduction. Experimentation has shown that conduction over the synapses is slower than that over nerve fibers. Therefore, longer reflex times generally indicate a larger number of synapses in the reflex arc.

Characteristics of Reflex Responses in "Spinal" Animals. Perhaps the outstanding characteristics of reflex responses in "spinal" animals are their apparent *purposefulness* and their *predictability*. They resemble so closely those of the normal animal that they appear to be voluntary acts. If controlled stimuli are repeated, one can accurately predict the responses. They differ from those of the normal animal in that they do not show characteristic variations.

A "spinal" frog suspended with its legs hanging down will show **flexor reflexes** of the legs when "irritating" stimuli are applied to the foot. If a small bit of blotting paper is soaked in a strong acid and applied to the skin of the back, the frog will draw up its hind leg and with the foot, brush the paper off. This demonstrates that the afferent receptor mechanisms show an accurate **local sign.** A "spinal" cat will similarly respond to irritating stimuli like the bending of the hairs, by scratching the site of irritation—the **scratch reflex.** It is as though these animals were conscious, yet these responses can only be mediated through the spinal cord. Indeed, as we know from personal experience these primitive unconscious responses are so well established that when we accidentally touch a hot object or step on a tack, we behave much like the spinal animal and the response is well under way before we are fully conscious of what has taken place. Many such reflexes are protective in nature.

Now if we increase the strength and duration of the stimuli, we can increase the number of impulses arriving in the spinal cord and thus increase the intensity of the response. It results in the spread of nervous activity to other levels of the cord and across the cord and therefore to other effector neurons whose thresholds of stimulation were not attained by the weaker stimuli. This is called **irradiation.** This phenomenon can be shown by the "spinal" frog, which will not only flex the leg on the side on which the stimulus is applied, but will extend the opposite one. Arm and trunk movements may also be initiated. If precise measurements are made, it will be seen that the response lasts after the stimulation has ended, due to increased central reflex time, a phenomenon called **after-discharge.**

The mechanisms of divergence and convergence are also factors in determining the nature of responses. In **divergence** one neuron synapses with two or more neurons and thus can bring about different responses if the initial stimulus is of sufficient strength and duration. In **convergence** the axons of two or more neurons synapse with a single neuron, and by this mechanism we can account for many of the phenomena displayed by the central system.

We must not conclude that all nerve impulses bring about initiation or acceleration of activity. **Inhibition** is an important function of the nervous system. Skeletal muscles, you may recall, are arranged in antagonistic groups, flexors and extensors, abductors and adductors, etc. Also muscles maintained in a state of partial and sustained contraction are an important factor in the maintenance of posture. It is reasonable to suppose that flexion of a part can be facilitated by some inhibition of the opposing extensor muscles and such is the case.

Finally, **summation** plays an important role in reflex conduction or in conduction generally within the nervous system and is again a function of synapses. Stimuli (*subliminal*) which by themselves cause no visible response apparently bring about some change at the synapse which is accumulative or summates and will discharge when it reaches a certain threshold. This is sometimes called **facilitation.** If the subliminal impulses causing facilitation are carried by neurons from different areas of the body (*convergence*), it is called **spatial summation;** if they come from only one neuron or from one area, it is called **temporal summation.**

THE BRAIN

"This Being of mine, whatever it really is, consists of a little flesh, a little breath, and the part which governs."
—Marcus Aurelius, *Meditations II, 2*

The brain is the organ of the central nervous system which occupies the cranial cavity. It is continuous inferiorly with the spinal cord. The diminutive central canal of the spinal cord expands into the ventricles of the brain, and the meninges of the cord continue upward to cover it externally. Twelve pairs of cranial nerves issue from its inferior surface to supply mostly the structures in the head and neck.

The brain is superior not only in anatomic position, but also in function, since it, more than any other part of the system, dominates other parts. The complexity of its integrating areas and its capacity for creativity, prediction, and abstract reason make it man's primary claim to superiority among living organisms.

Early Development (Figs. 18–19, 20). The general form of the brain is best understood through a brief presentation of its early development. In Chapter 4 the origin of the dorsal hollow neural tube from the middorsal ectoderm was briefly outlined (Fig. 4–11). The brain differentiates from the anterior portion of this tube by rapid growth, thickening, flexing, and evagination. Very early the brain shows three enlargements or vesicles, the **forebrain** (*prosencephalon*), **midbrain** (*mesencephalon*), and **hindbrain** (*rhomb-*

encephalon). At about the same time, due to unequal growth, the brain undergoes three successive **flexures,** the **cephalic** bringing the forebrain forward; the **cervical** placing the hindbrain at almost a right angle to the spinal cord; and a **pontine flexure** near the anterior end of the hindbrain in the opposite direction from the other two. The lumen of the neural tube expands in these vesicles to form the **ventricles** of the brain.

These three vesicles, the forebrain, midbrain and hindbrain, maintain their identities as the further development of the brain takes place; they are more obvious in lower vertebrates, but still apparent in man. They constitute the **brain stem,** the "old" brain, and in general, relationships within them are more like those in the cord than in those structures derived from them. Further, especially again in lower vertebrates, an important sensory function was associated with each of them: smell with the forebrain, vision with the midbrain, hearing with the hindbrain (Fig. 18–19). These relationships become somewhat modified, too, in mammals. Finally, the **cranial nerves,** at least the typical ones, from three to twelve, attach to the brain stem (Fig. 18–20).

From each of the three vesicles at least one important outgrowth appears,

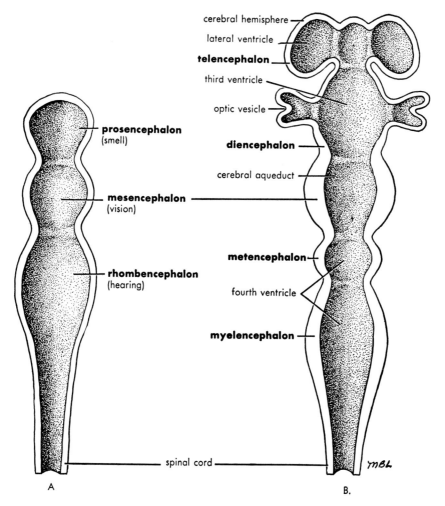

cerebral hemisphere
lateral ventricle
telencephalon
third ventricle
optic vesicle
diencephalon
cerebral aqueduct
metencephalon
fourth ventricle
myelencephalon

prosencephalon
(smell)

mesencephalon
(vision)

rhombencephalon
(hearing)

spinal cord

A

B.

Fig. 18–19.   Schematic presentation of early stages in brain development.   *A*, Three-part stage and its relationship to the senses.   *B*, Five-part stage.

more than one from the forebrain; and the brain of three regions thus develops into one of five regions (Figs. 18–19, 20). The most conspicuous outgrowths of the forebrain are paired hollow pockets which grow forward toward the nasal region. From them develop anteriorly the **olfactory bulbs,** and posteriorly the **cerebral hemispheres.** In mammals and especially in man, the cerebral hemispheres expand to the point where they overshadow all other parts of the brain and their surfaces become so great that they become extensively folded.

Inside they contain the lateral ventricles. These structures constitute the anterior terminal segment of the brain and are called the **telencephalon.** Optic vesicles appear early in development as paired hollow outgrowths of the forebrain. Their ends become expanded into the optic cups which later become the retina of the eye. The stalks become narrowed into the optic stalks and later are invaded with nerve fibers and constitute the optic nerves and tracts.

The remaining part of the forebrain is unpaired and constitutes the **dienceph-**

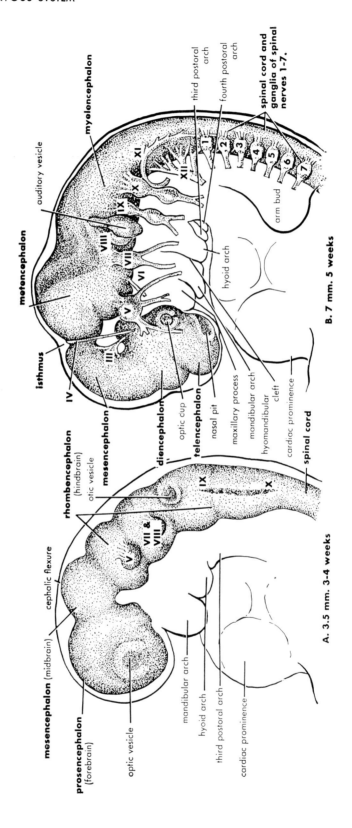

myelencephalon

auditory vesicle

metencephalon

isthmus

IV

III

mesencephalon (midbrain)

cephalic flexure

prosencephalon (forebrain)

optic vesicle

rhombencephalon (hindbrain)

otic vesicle

mesencephalon

diencephalon

optic cup

telencephalon

nasal pit

maxillary process

mandibular arch

hyomandibular cleft

cardiac prominence

spinal cord

third postoral arch

fourth postoral arch

spinal cord and ganglia of spinal nerves 1-7.

arm bud

hyoid arch

B. 7 mm. 5 weeks

A. 3.5 mm. 3-4 weeks

mandibular arch

hyoid arch

third postoral arch

cardiac prominence

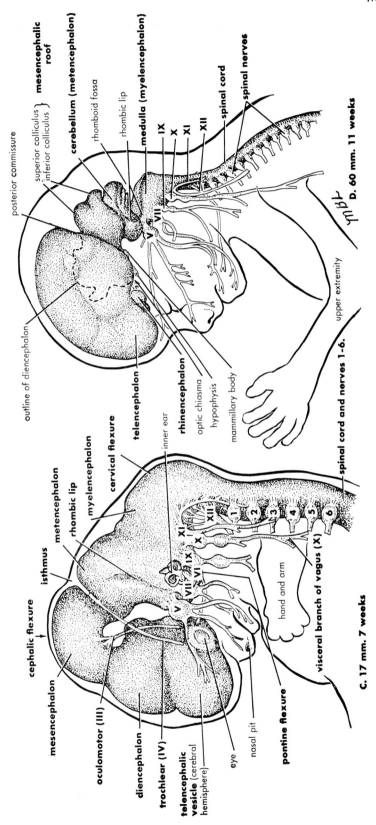

Fig. 18–20.   Stages in the early development of the brain and cranial nerves of man.   The size of the outlines is constant in all of the drawings and the nervous system is drawn in proportion to the outlines.   The actual size of the living brain is shown under each drawing.

alon. From its roof the **pineal body** is formed and from its floor the **infundibulum** joins with a pouch from the primitive mouth and contributes to the hypophysis or **pituitary gland.**

The midbrain or **mesencephalon** shows swellings in the dorsal region, the tectum. From these in the brains of lower vertebrates the optic lobes develop, but in mammals there are four small lobes, the **corpora quadrigemina,** associated with visual and auditory functions.

The hindbrain in its forward portion gives rise to a solid dorsal outgrowth which forms the much-expanded and folded **cerebellum,** and on the ventral side to a prominent transverse bridge of fibers, the **pons.** In between these two is a part of the **medulla oblongata** containing the fourth ventricle. These structures constitute the division called the **metencephalon.** The lower part of the hindbrain is the remainder of the medulla oblongata with a continuation of the fourth ventricle and is the **myelencephalon** of the brain.

The derivatives of the three primary divisions of the embryonic brain are summarized in the outline below.

The structures discussed above should be located on Figures 18–19, 20, 21 to be sure that the gross features of the brain are understood. Also one should realize that as the gross modifications just described are taking place, many internal changes are also in progress.

These involve the microscopic features of brain structure—the organization of fiber tracts and nuclei. Knowledge of these structures is necessary to understand the functions of the brain. It is hoped that enough is gained from our brief study of development to serve as a basis for further understanding.

BRAIN STEM

Medulla Oblongata (Figs. 18–20, 21, 22). Since we have studied the spinal cord, we will start with that part of the brain most closely associated with it, the **medulla oblongata** of the **brain stem.** There is no clear line of distinction between the medulla and the spinal cord, but the upper limit of the medulla is placed at the well-defined lower border of the pons. The central canal of the cord is unmodified in the lower half of the medulla, but in the upper half the medulla opens up dorsally at the dorsal median sulcus to form the expanded **fourth ventricle** which continues on into the pons. The ventral or anterior wall of the fourth ventricle is made up of the medulla and pons; its roof in the medulla region is made of a thin membrane, the **inferior velum.** Superiorly, it lies below the cerebellum. The medulla contains most of the ascending and descending pathways represented in the cord. Some of these, such as fasciculi gracilis and cuneatus, end in nuclei in the

Prosencephalon (forebrain)	Telencephalon . . .	Cerebral hemispheres; olfactory bulbs; basal nuclei; lateral ventricles.
	Diencephalon . . .	Epithalamus; thalamus; hypothalamus; infundibulum; pineal body; third ventricle.
Mesencephalon . . . . (midbrain)		Corpora quadrigemina; cerebral peduncles, cerebral aqueduct.
Rhombencephalon (hindbrain)	Metencephalon . . .	Pons (part of medulla); cerebellum; part of fourth ventricle.
	Myelencephalon . .	Medulla oblongata (part of); fourth ventricle.

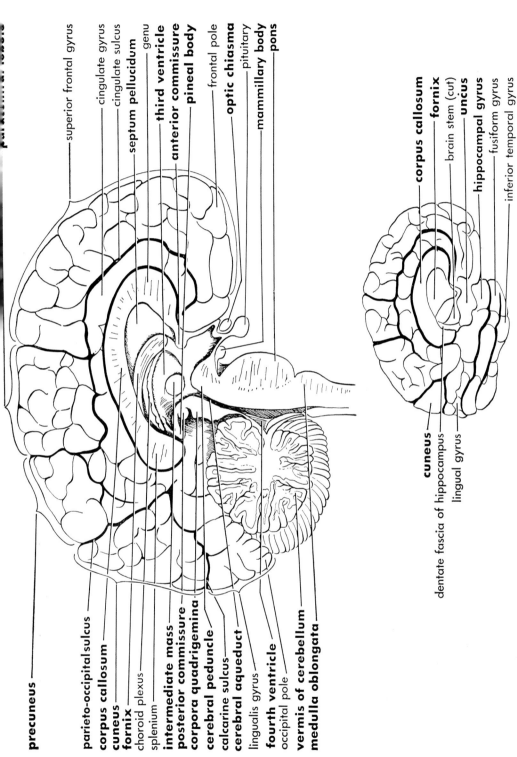

precuneus

superior frontal gyrus
cingulate gyrus
cingulate sulcus
septum pellucidum
genu
third ventricle
anterior commissure
pineal body
frontal pole
optic chiasma
pituitary
mammillary body
pons

parieto-occipital sulcus
corpus callosum
cuneus
fornix
choroid plexus
splenium
intermediate mass
posterior commissure
corpora quadrigemina
cerebral peduncle
calcarine sulcus
cerebral aqueduct
lingualis gyrus
fourth ventricle
occipital pole
vermis of cerebellum
medulla oblongata

corpus callosum
fornix
brain stem (cut)
uncus
hippocampal gyrus
fusiform gyrus
inferior temporal gyrus

cuneus
dentate fascia of hippocampus
lingual gyrus

FIG. 18–21.   Medial section of brain.   Insert shows the medial aspect of the right cerebral hemisphere with the cerebellum and brain stem removed to reveal the lower and medial gyri of the temporal lobe.

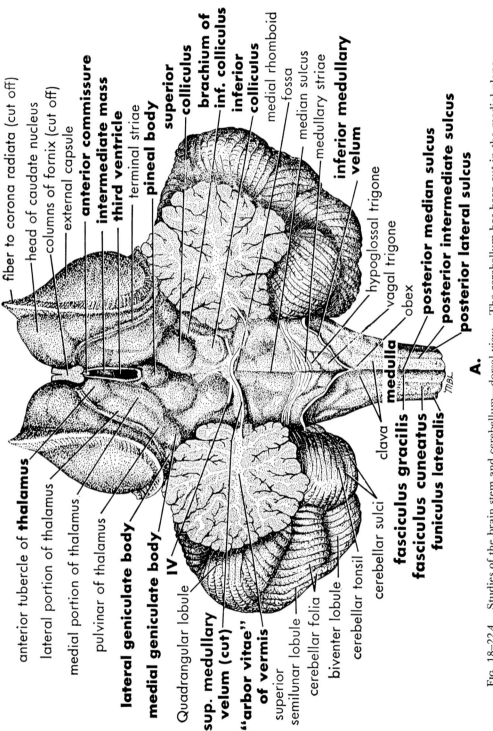

anterior tubercle of **thalamus**

lateral portion of thalamus

medial portion of thalamus

pulvinar of thalamus

lateral geniculate body

medial geniculate body

IV

Quadrangular lobule

sup. medullary velum (cut)

"arbor vitae" of vermis

superior
semilunar lobule

cerebellar folia

biventer lobule

cerebellar tonsil

cerebellar sulci

clava

fasciculus gracilis

fasciculus cuneatus

funiculus lateralis

fiber to corona radiata (cut off)

head of caudate nucleus

columns of fornix (cut off)

external capsule

anterior commissure

intermediate mass

third ventricle

terminal striae

pineal body

superior colliculus

brachium of inf. colliculus

inferior colliculus

medial rhomboid fossa

median sulcus

medullary striae

inferior medullary velum

hypoglossal trigone

vagal trigone

obex

posterior median sulcus

posterior intermediate sulcus

posterior lateral sulcus

medulla

A.

Fig. 18–22A.   Studies of the brain stem and cerebellum—dorsal view.   The cerebellum has been cut in the medial plane and spread to reveal the floor of the fourth ventricle.

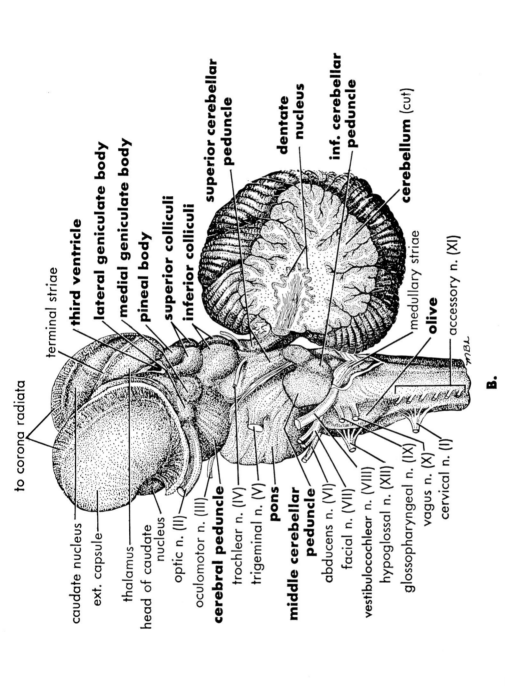

to corona radiata

caudate nucleus
ext. capsule
thalamus
head of caudate nucleus
optic n. (II)
oculomotor n. (III)
cerebral peduncle
trochlear n. (IV)
trigeminal n. (V)
pons
middle cerebellar peduncle
abducens n. (VI)
facial n. (VII)
vestibulocochlear n. (VIII)
hypoglossal n. (XII)
glossopharyngeal n. (IX)
vagus n. (X)
cervical n. (I)

terminal striae
third ventricle
lateral geniculate body
medial geniculate body
pineal body
superior colliculi
inferior colliculi
superior cerebellar peduncle

dentate nucleus
inf. cerebellar peduncle
cerebellum (cut)
medullary striae
olive
accessory n. (XI)

B.

Fig. 18–22*B*.   Studies of the brain stem and cerebellum—dorsolateral view of the brain stem.   Cerebellum cut in parasagittal plane to show the dentate nucleus.

34

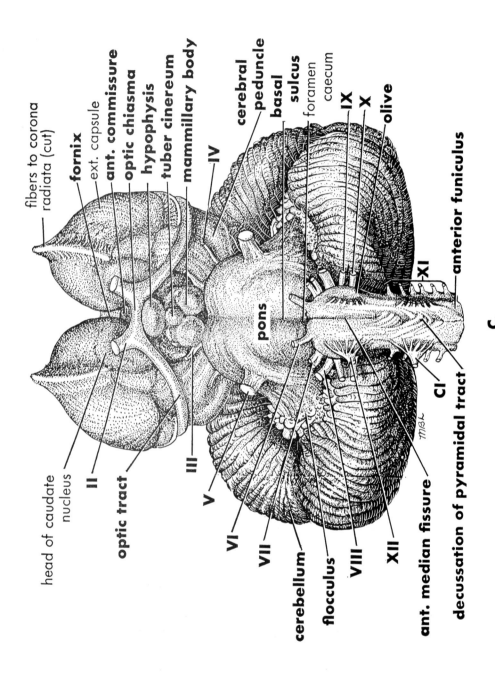

head of caudate
nucleus

fibers to corona
radiata (cut)

fornix
ext. capsule
ant. commissure
optic chiasma
hypophysis
tuber cinereum
mammillary body

IV

**cerebral
peduncle**

**basal
sulcus**
foramen
caecum

IX
X

olive

II

optic tract

III

pons

anterior funiculus

XI

CI

V

VI

VII

cerebellum

flocculus

VIII

XII

ant. median fissure

decussation of pyramidal tract

C.

Fig. 18–22C. Studies of the brain stem and cerebellum—ventral view.

lower part of the medulla on the posterior side. These nuclei receive impulses coming from the cord and relay them to the other side of the medulla where they pass upward to other sensory nuclei and finally to the cerebral cortex. Other pathways pass through the medulla, in some cases crossing over to the other side, as in the descending lateral corticospinal tracts. These crossing or decussating fibers can be seen on the anterior surface where, in crossing through the anterior median fissure, they nearly obliterate it. This is called the **decussation of the pyramids** because these fibers of the corticospinal tract arise in the pyramidal cells of the cortex. They end around the anterior horn cells of the gray matter of the cord. Fibers which do not cross in the brain do cross in the spinal cord. Others, of the ascending tracts, may cross in the cord as do those for pain and temperature. Others will cross as they reach the brain (Figs. 18–50 to 54).

In the medulla there are a number of **centers** (nuclei) for the regulation of vital activities of the body such as the **respiratory, vasomotor,** and **cardiac.** Others regulating deglutition, vomiting,

sweating, and gastric secretion are also present.

The medulla also contains the nuclei of origin of the ninth through the twelfth cranial nerves. Near the upper end of the medulla and laterally there is a swelling formed by a nucleus which, because of its shape, is called the **olive.** It sends a large bundle of fibers across the midline through the substance of the medulla where it enters the cerebellum.

The rest of the substance of the medulla is made up of the **reticular formation,** a minute nerve network, not limited to the medulla but extending through the central part of the brain stem. Until recently relatively little was known about it. Now its great importance is beginning to be realized. J. D. French, in a recent article in Scientific American (*196*:30) called *The Reticular Formation,* sums up its functions in this way, "It awakens the brain to consciousness and keeps it alert; it directs the traffic of messages in the nervous system; it monitors the myriads of stimuli that beat upon our senses, accepting what we need to perceive and rejecting what is irrelevant; it tempers and refines our muscular activity and

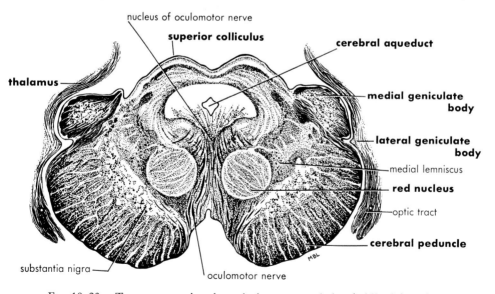

FIG. 18–23.   Transverse section through the mesencephalon (midbrain) at the level of the superior colliculi.

bodily movements. We can go even further and say that it contributes in an important way to the highest mental processes—the focusing of attention, introspection and doubtless all forms of reasoning."

Pons (Figs. 18–20, 21, 22). The **pons** lies on the under surface of the brain between the medulla and the midbrain. As its name suggests it is a "bridge" consisting of large bundles of transverse axons extending from masses of cell bodies, the **pontine nuclei,** into each half of the cerebellum. These fibers constitute the main mass of the **middle cerebellar peduncles.** They relay impulses from the cerebral cortex to the cerebellum.

To the ventrolateral aspect of the pons are attached the motor and sensory roots of the trigeminal nerve (5th cranial) and from the transverse groove between the

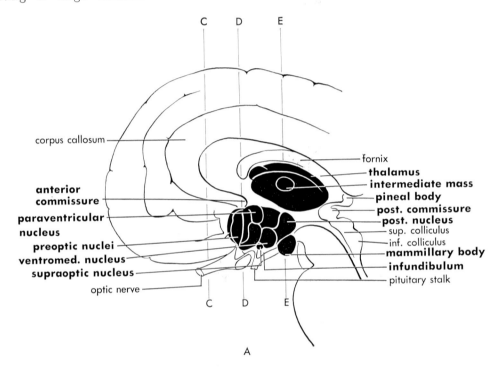

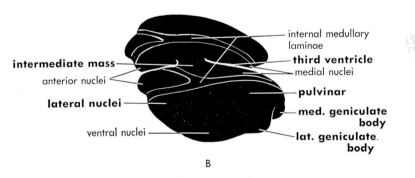

FIG. 18–24.   Legend on opposite page.

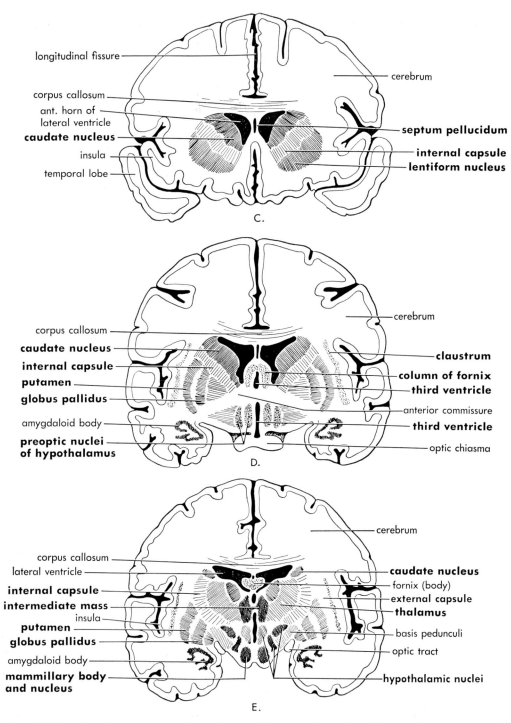

FIG. 18–24. The diencephalon and cerebrum with emphasis on the thalamus, hypothalamus and basal nuclei. *A,* Medial section of brain showing three-dimensional thalamus and hypothalamus. Only some of the main nuclei of the hypothalamus are named. Frontal sections at right side of this illustration are taken from sections cut along lines *C, D* and *E* above. *B,* Schematic presentation of the thalamus. *C,* Frontal section through anterior cornua of lateral ventricles. *D,* Frontal section through anterior commissure. *E,* Frontal section through intermediate mass and mammillary bodies.

pons and the medulla oblongata the abducens (VI), facial (VII), and vestibulocochlear (VIII) cranial nerves emerge. The posterior (*dorsal*) part of the pons forms the upper half of the floor of the fourth ventricle. In this area which is much like the posterior part of the medulla, a number of longitudinal fiber tracts pass through, and here the nuclei of the fifth, sixth, seventh, and eighth cranial nerves are located (Figs. 18-22, 37). The reticular formation is found also in this dorsal area.

Two fiber tracts pass through the substance of the pons, the corticospinal and the corticobulbar; one ends in the pontine nuclei, the corticopontine tract.

Mesencephalon (*midbrain*) (Figs. 18-21 to 24). This constricted portion of the brain stem lies between the pons and the "between" brain or diencephalon. A narrow canal, the **cerebral aqueduct** (*aqueduct of Sylvius*), passes through its center connecting the fourth with the third ventricle. The only parts of the mesencephalon visible from the base of the brain are the **cerebral peduncles** (Figs. 18-22*B*, *C*, 23, 24). Between them issue the oculomotor (III) cranial nerves. Laterally between the cerebral peduncles and the temporal lobes of the cerebrum the trochlear (IV) cranial nerves appear.

The cerebral peduncles are made up mostly of the motor fibers which descend from the cerebrum to the lower parts of the nervous system. Bundles of ascending (*sensory*) fibers passing upward to the thalamus of the diencephalon lie deep in relation to the cerebral peduncles.

The nuclei of the third and fourth cranial nerves, and the anterior part of the nucleus of the fifth nerve are located in the midbrain. The **red nucleus** contains the cells of origin of fibers of the rubrospinal tract and is a relay station for impulses from the cerebellum and higher brain centers (Figs. 18-23, 52). The fibers to the red nucleus from the

cerebellum constitute the superior cerebellar peduncle.

The **corpora quadrigemina** are on the dorsal surface of the midbrain and consist of **superior** and **inferior colliculi** (Figs. 18-22, 23). The superior colliculi connect laterally by their brachia to the lateral geniculate bodies; the inferior colliculi connect superolaterally by their brachia to the medial geniculate bodies. The superior colliculi are reflex centers for visual, auditory, and tactile impulses, while the inferior colliculi are reflex centers for auditory functions.

Diencephalon (Figs. 18-22, 24). Sometimes called the "between" brain the **diencephalon** lies between the midbrain and the "end" brain or telencephalon (*cerebrum*). Its forward limit is at the level of the **interventricular foramina** (*foramen of Monro*) which connect its cavity, the third ventricle, with the lateral ventricles of the cerebral hemispheres. Reference to Figure 18-19 will help to clarify these relationships.

The **third ventricle** is the cavity of the diencephalon and the median portion of the telencephalon. It is deep, but very narrow. Its lateral walls, the thalami, are so thick that they actually come into contact and join at one point in the middle of the third ventricle, forming the **intermediate mass** (Fig. 18-24*D*). The roof of the third ventricle is the epithalamus, the floor the hypothalamus.

The **thalamus** is one of the most important sensory centers of the brain, and is perhaps even more important in the lower vertebrates, where it has not come under the dominance of a large cerebrum, as it has in man. Impulses set up in exteroceptors (*receptors*) reach the lateral nucleus of the thalamus where they are related to the area of general sense in the cerebrum. Its anterior nucleus is a relay station for impulses initiated in the olfactory organ. Proprioceptive pathways relay in the thalamus, and at the posterior end of the thalamus two small elevations, the lateral and medial geniculate bodies, are relay stations for visual and auditory pathways, respectively. As mentioned above in the discussion of the midbrain the geniculate bodies connect medially by their brachia to the superior and inferior colliculi. The thalamus also may serve as an area of crude or uncritical consciousness. Pain may be felt, for example, but not localized through thalamic activity.

The **epithalamus** forms a thin roof for the third ventricle (Fig. 18–21). Internally it has a vascular structure, the **choroid plexus,** which produces cerebrospinal fluid. A small conical structure, the **pineal body,** projects upward from the posterior part of the roof. At the point where the epithalamus becomes continuous with the midbrain it has a thickening, the **posterior commissure.** Some of its fibers may connect the superior colliculi.

The **hypothalamus** forms the floor of the third ventricle (Fig. 18–24). Externally it is marked by the **optic chiasma,** behind which is the **infundibulum.** The infundibulum arises out of an area of gray matter, the **tuber cinereum,** which lies behind the optic chiasma and optic tracts and between the cerebral peduncles. Behind the tuber cinereum is a pair of small rounded bodies, the **mammillary bodies** (Fig. 18–22C). A variety of important functions have been assigned to the hypothalamus. It has centers for the regulation of body temperature—those for heat-loss functions such as sweating are in the anterior, those for preventing loss and for increasing heat production (*shivering*) are in the posterior part. Regulation of water, fat, and carbohydrate metabolism is carried on by the hypothalamus in association with the posterior lobe of the pituitary. Sleep, sexual activity, and emotional control may be influenced by activities in this area.

CEREBELLUM (Figs. 18–20, 21, 22, 25)

The **cerebellum** lies in the posterior cranial fossa below the posterior part of the cerebrum. It is connected below with the brain stem by three bands of fibers, the cerebellar peduncles; the **superior peduncles** connecting it with the midbrain, the **middle peduncle** with the pons, and the **inferior peduncle** (restiform body) with the medulla oblongata. The fourth ventricle intervenes between it and the brain stem.

The cerebellum is a solid mass of nervous tissue consisting of two **cerebellar hemispheres,** separated by a narrow median portion, the **vermis.** The surface of the cerebellum, the gray **cortex,** is thrown up into many more or less parallel ridges, the **folia cerebelli,** which are separated by deep **fissures.** Inside there is a mass of white matter with a few nuclei, among them the **dentate nucleus** whose fibers make up the bulk of the superior cerebellar peduncle mentioned in the section on the midbrain. If the cerebellum is sectioned in the midsagittal plane, the arrangement of the folia and fissures in the vermis form a pattern like a tree in full leaf—hence the name **arbor vitae** or tree of life, given to it by the medieval anatomists (Fig. 18–22A).

The conspicuous and characteristic cells of the cerebellar cortex are the **Purkinje neurons** (Fig. 18–26). Their cell bodies are pear-shaped and they have a multitude of dendrites and a single axon. They occupy the middle layer of the three-layered cortex—an inner granular layer and an outer molecular layer, each made up of neurons whose functions are directed toward taking incoming impulses and spreading them out to reach as many Purkinje dendrites with as many impulses as possible. The Purkinje cells then "fire" a huge volley of impulses back into the white matter to other nuclei and then out to the brain stem through the superior cerebellar peduncle.

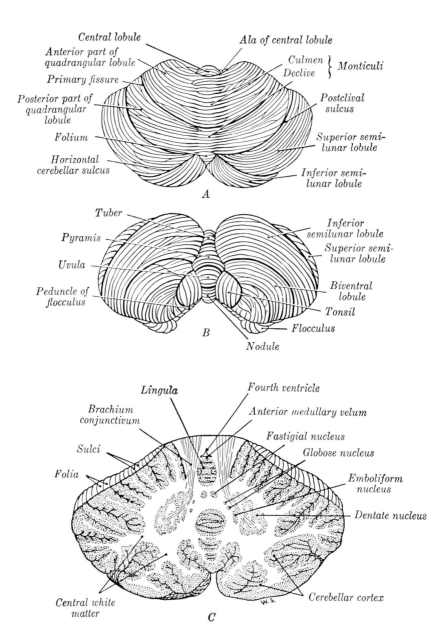

FIG. 18–25.   A, Dorsal surface of the cerebellum; B, ventral surface of the cerebellum so oriented with relation to A as to indicate the continuity of folium and tuber; C, horizontal section of the cerebellum showing the arrangement of the cortical gray matter and the locations of the central nuclei within the white matter (after Sobotta-McMurrich).

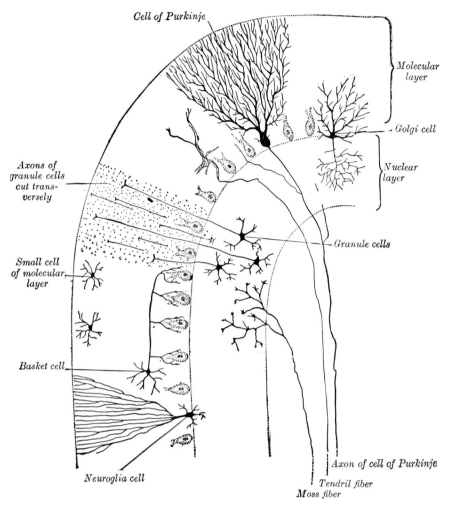

Cell of Purkinje

Molecular
layer

Golgi cell

Nuclear
layer

Axons of
granule cells
cut trans-
versely

Granule cells

Small cell
of molecular
layer

Basket cell

Axon of cell of Purkinje

Neuroglia cell

Tendril fiber

Moss fiber

FIG. 18–26.   Transverse section of a cerebellar folium.
(Diagrammatic, after Cajal and Kölleker.)

The white matter contains also incoming axons from spinal cord and brain areas.

The functions of the cerebellum are to aid in (1) the maintenance of equilibrium, (2) maintenance of muscle tone, (3) maintenance of muscle coordination. It has been shown that the cerebellar cortex is also involved in coordination of the sensations of sight, hearing, and touch—not just those of proprioception.

CEREBRUM

The cerebrum is the largest and most conspicuous single feature of the brain.

By rapid and differential growth in the embryo and fetus it has come to obscure all other structures as seen from above and to a large extent from the lateral view (Figs. 18–20, 21, 27). Only the cerebellum approaches it in size and the brain stem is visible to an appreciable degree only from the ventral side. Like the cerebellum, but unlike the brain stem, the bulk of its gray matter is superficial and is so extensive that it has become greatly convoluted. Only relatively small nuclei occur within its deeper structure.

Surfaces, Poles, and Fissures (Figs.

18–21, 27, 29). The cerebrum is divided by a deep **longitudinal fissure** into right and left hemispheres and is separated from the cerebellum by a **transverse fissure.** At the bottom of the longitudinal fissure the two hemispheres are united by a broad sheet of white matter, the **corpus callosum** (Fig. 18–21). Each hemisphere has a **dorsolateral surface** lying in contact with the cranial vault; an **inferior surface,** the anterior third of which rests in the anterior cranial fossa, the middle third in the middle cranial fossa, and the posterior third on the "roof" of the posterior cranial fossa, above the cerebellum,

and a **medial surface** facing its counterpart in the longitudinal fissure. Three poles of the cerebrum are visible from the lateral view: a full and rounded **frontal pole** at the anterior extremity of the hemisphere, a more pointed posterior extremity, the **occipital pole,** and a **temporal pole** laterally and inferiorly (Fig. 18–27).

Sulci, Gyri, and Lobes. The convolutions of the cerebrum form a characteristic pattern subject to some individual variation. The rounded elevations or convolutions are called **gyri** (gyrus); the intervening furrows are the **sulci** (sulcus).

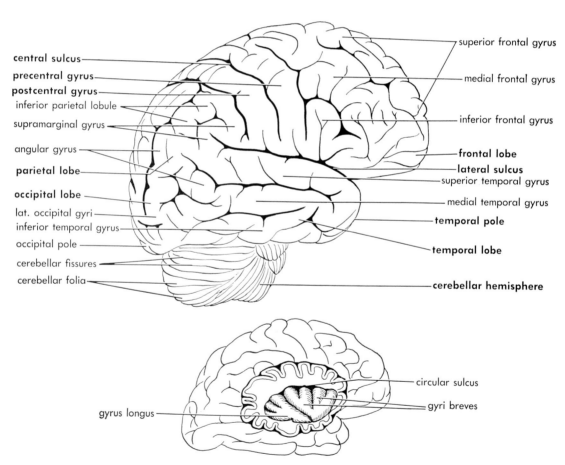

central sulcus
precentral gyrus
postcentral gyrus
inferior parietal lobule
supramarginal gyrus
angular gyrus
parietal lobe
occipital lobe
lat. occipital gyri
inferior temporal gyrus
occipital pole
cerebellar fissures
cerebellar folia

superior frontal gyrus
medial frontal gyrus
inferior frontal gyrus
frontal lobe
lateral sulcus
superior temporal gyrus
medial temporal gyrus
temporal pole
temporal lobe
cerebellar hemisphere

circular sulcus
gyri breves
gyrus longus

Fig. 18–27. Lateral view of the brain showing gyri and sulci of the cerebrum. The insert shows the Insula (island of Reil) revealed by cutting away parts of the frontal, parietal, and temporal lobes along the lateral fissure.

Although these structures are all provided with names, only a few of the more conspicuous ones and those enabling us to delimit important cortical lobes and **functional areas** will be described and named. These perhaps are most easily identified and understood by reference to Figures 18–21, 27, 28.

The **lateral cerebral sulcus** (*fissure of Sylvius*) is a very prominent fissure beginning just above the temporal pole and running obliquely upward on the lateral

surface of the cerebrum. In the depths of its posterior part, the **insula** is hidden. Centers for speech and hearing lie close to this sulcus. It separates the **temporal lobe** from the rest of the cerebrum above. The temporal lobe lies under the temporal bone, hence the name. Other lobes are also named for overlying bones.

The **central sulcus** (*fissure of Rolando*) starts on the medial surface at about midpoint on the superior border and runs downward and slightly forward on the

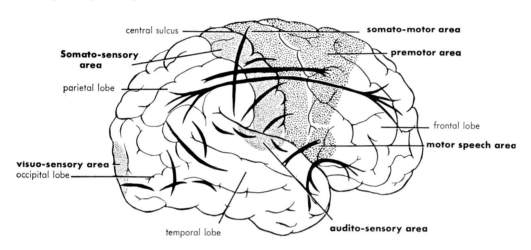

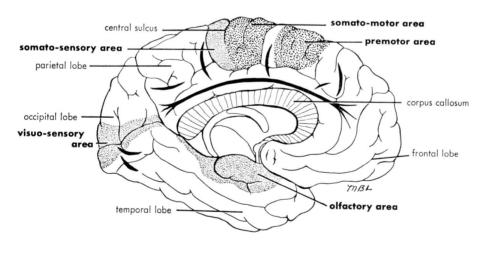

B.

Fig. 18–28. Localization of function and association pathways of the cerebral cortex. Sensory areas are shown in fine stipple; motor areas in coarse stipple. Association fibers connecting the various cortical areas are shown by the broad black lines. *A*, Lateral view; *B*, medial view. (Modified from Netter.)

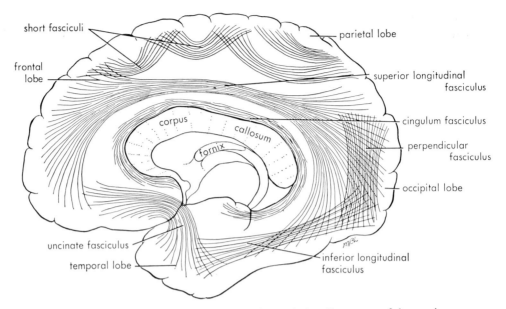

Fig. 18–29. Diagram to show the principal association fiber tracts of the cerebrum.

lateral surface of the cerebrum, stopping just short of the lateral cerebral sulcus. It is the line of demarcation between the **frontal** and **parietal lobes.** In front of it is the **precentral gyrus,** the **motor area;** behind it in the parietal lobe the **postcentral gyrus,** the area of **general sensation.**

The **calcarine sulcus** (Fig. 18–21) is on the medial surface of the posterior part of the hemisphere. It is directed obliquely upward from the occipital pole and joins the **parietooccipital sulcus** which extends upward and backward to the superior border which it just barely crosses, on to the lateral surface. The parietooccipital sulcus divides the **occipital lobe** from the parietal lobe. The **visual area** occupies the walls of the posterior part of the calcarine fissure.

The **sulcus cinguli** is a prominent sulcus on the medial side of the hemisphere which parallels the surface of the corpus callosum. The gyrus between the sulcus and the corpus callosum is the **gyrus cinguli,** the area for **smell association** (Figs. 18–21, 28).

The **collateral sulcus** is on the inferior surface of the hemisphere. It parallels the medial border of the hemisphere from the occipital almost to the temporal pole. The hippocampal gyrus lies medial to it and at its front end turns abruptly in a medial direction to form the **uncus,** the area of **smell appreciation,** possibly also taste (Figs. 18–21, 28).

Study of Figure 18–28 will make one realize that there are still unassigned functional areas on the cerebral cortex, though all of the known ones are not shown on our diagram; some forty-seven have been numbered. These unassigned areas are sometimes called **association areas** or **silent areas.** They suggest the incompleteness of our knowledge of the cerebrum.

White Matter of the Cerebrum. The white matter consists of fiber tracts which fall into three categories: the association, commissural, and projection tracts (Figs. 18–24, 28, 29, 30).

The **association tracts** are those that are limited to one cerebral hemisphere. They serve to bring all parts of the hemisphere into functional relationship. They are divided into **short association fibers**

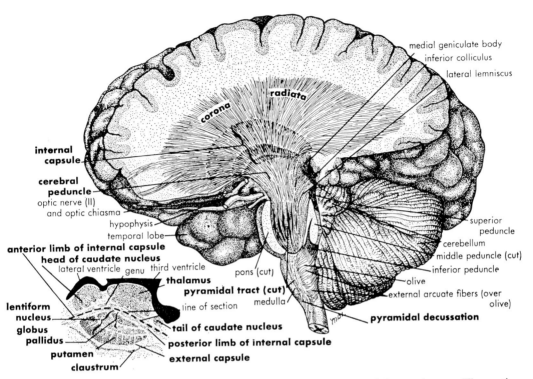

FIG. 18–30. The brain showing the fibers of the projection system of the cerebrum. The section was cut as indicated by the dash line on the insert. Insert shows the relationship of the internal and external capsules to the basal nuclei and thalamus.

which lie immediately beneath the gray matter of the cortex and merely connect one gyrus with the next; and **long association fibers** which run deeper in the cerebrum and connect one region or lobe of the cortex with another. These tracts are shown in Figures 18–28, 29.

The **commissural tracts** cross from one hemisphere to the other, serving to keep all parts of the cortex of one hemisphere in communication with corresponding parts of the other (Figs. 18–21, 28). The **corpus callosum** is made up of these fiber tracts and it roofs over the lateral and third ventricles in a very broad, thick, and compact expanse of tissue.

The **projection tracts** are those by which the cerebral cortex maintains its relationship with the rest of the nervous system and the receptors and effectors of the body. They consist of the **ascending** or **afferent** tracts, the cell bodies of which are located for the most part in the thalamus, and the **descending** or **efferent** tracts, the cell bodies of which are found in the motor area of the cortex. An example of the latter is the corticospinal tract that we have mentioned before. Connecting as they do with almost all parts of the cerebrum, the projection fibers must crowd together to keep within the relatively limited confines of the brain stem. This vast expanse of fibers in the cerebrum is called the **corona radiata** (Fig. 18–30).

Basal Nuclei (Figs. 18–24, 30, 31). The **basal nuclei,** constituting the central gray matter of the cerebrum, lie between the white matter of the cerebrum and the thalamus of the diencephalon. They are the **caudate nucleus, putamen, claustrum,** and **globus pallidus.** The globus pallidus and putamen are often called the **lentiform nucleus.** As the fiber

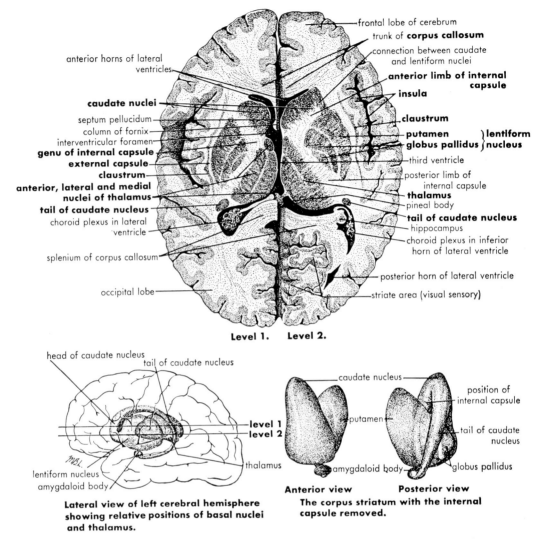

FIG. 18–31. The upper illustration shows horizontal sections of cerebral hemispheres taken at two levels as indicated in the insert to the left. The relationships of the corpus striatum to the thalamus and other cerebral structures are made clear. The insert to the right shows the relationship of the caudate and lentiform nuclei.

tracts from the cerebral cortex approach the brain stem they divide and pass around these nuclei. One part of this divided tract, a broad mass of fibers, called the **internal capsule,** lies between the caudate nucleus and the lentiform nucleus. Some of the gray fibers pass across the internal capsule going between these nuclei, giving the tract a striated appearance. For this reason these structures are often called collectively the **corpus striatum.** The fibers of the internal capsule continue between the thalamus and lentiform nucleus into the midbrain as the cerebral peduncles. The remaining fibers from the cerebrum pass to the outside of the lentiform nucleus as the **external capsule.** The claustrum lies lateral to the external capsule, between it and the insula. The amygdaloid body, a complex of basal nuclei, is located in the roof of the rostral end of the lateral ventricle in close association with the uncus of the gyrus hippocampi. The precise functions of the basal nuclei are not known, but they may have some control over voluntary muscle action because when they are diseased they may cause Parkinson's disease and St. Vitus' dance.

Conditioned Reflexes

In the unconditioned reflexes that we studied earlier the responses are elicited without previous training or experience. The response is specific for the stimulus. They can take place in animals whose cerebral cortex or entire brain is removed. **Conditioned reflexes** are those in which there must be previous training and the stimulus may be neutral or entirely unrelated to the response. The cerebrum is essential to the formation of conditioned reflexes. The classic example, of course, is that of Pavlov's dogs which learned to associate the ringing of a bell with the reception of food until the ringing of the bell alone resulted in salivation. We may salivate when we smell a broiling steak or some other food item which we have associated with dining pleasure. When we drive our cars, we automatically apply the brakes when we see a red light though our thoughts may be on something entirely different. Yet when we were learning to drive, each red light involved us in a conscious effort in the manipulation of a foot and a brake pedal, applying the correct pressure and managing a smooth stop. Through conditioned reflexes we acquire patterns of behavior; habits which facilitate daily living and free our brains for other matters.

Ventricles of the Brain
(Figs. 18–19, 21, 24, 32, 33)

The ventricles have been mentioned briefly in the previous discussion of the brain. They are continuous with the central canal of the spinal cord which extends well into the medulla oblongata before it widens out into the diamond-shaped, shallow **fourth ventricle.** Its floor is the rhomboid fossa of the brain stem; its roof is the cerebellum and its supporting structures. Superiorly, it leads into the cerebral aqueduct of the midbrain. From the roof of the fourth ventricle, the **choroid plexus,** an elongated mass of tortuous blood vessels, mostly of capillary size, projects into and beyond the lateral recesses of the ventricle. The capillaries belong to the pia mater, but they are covered with the epithelial ependymal lining of the ventricle. They secrete cerebrospinal fluid into the ventricle. There are three openings in the roof of the fourth ventricle through which the cerebrospinal fluid can escape into the subarachnoid space. One, the **median aperture** (*foramen of Magendie*), is located at the lower extremity of the ventricle, and one is present in each of the lateral recesses, the **lateral apertures** (*foramina of Luschka*). Closure of these apertures causes cerebrospinal fluid to accumulate in excess, resulting in **hydrocephalus** or water on the brain, a condition sometimes seen in children which may be fatal or at least damaging to brain tissue, due to high pressures it creates.

The **cerebral aqueduct** (*of Sylvius*) of the midbrain leads into the **third ventricle** of the diencephalon. The third ventricle is narrow, but deep and ends anteriorly at the **lamina terminalis** and the **anterior commissure.** The two thalami, which form the bulk of the sidewalls of the ventricle, come together at their medial eminence and an **intermediate mass** connects through the ventricle. It does not function as a commissure, however. In the roof of the third ventricle is a choroid plexus which secretes cerebrospinal fluid (Figs. 18–21, 33). Near its anterior or rostral termina-

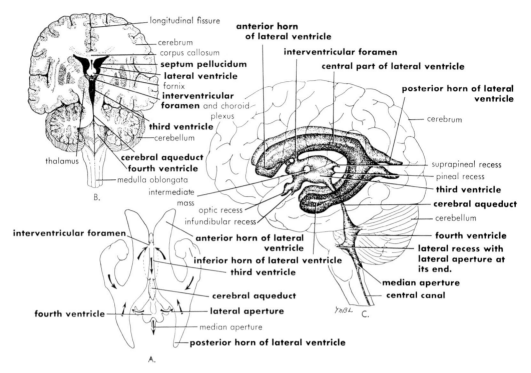

FIG. 18–32.   Ventricles of the brain.   *A*, Ventricular cavities as viewed from above; arrows indicate direction of flow of cerebrospinal fluid.   *B*, Schematic section of brain through ventricles.   *C*, Scheme to show ventricles in relationship to brain surface.   (Modified from Netter and Gray.)

tion the third ventricle opens into the lateral ventricles of the cerebrum by the **interventricular foramina** *(foramen of Monro)*.

The **lateral ventricles**, not called ventricles one and two as might be expected, are medially placed within the cerebral hemispheres and are separated from each other medially by the very thin vertical partition, the **septum pellucidum**.   The lateral ventricles are lined with ependyma and each has a central part and three prolongations: the anterior, posterior, and inferior horns (Figs. 18–24*B*, 32, 33).

The **central part** of the lateral ventricle has in its floor the choroid plexus which is continuous through the interventricular foramen with the choroid plexus of the third ventricle.   It extends also into the inferior horn to its rostral end.   The **anterior horn** of the lateral ventricle

extends into the frontal lobe of the cerebral hemisphere; the **posterior horn** into the occipital lobe; and the **inferior horn** into the temporal lobe.

Cerebrospinal Fluid (Figs. 18–33, 34).   The cerebrospinal fluid is a watery, viscous material similar in chemical composition to that of the lymph and of the aqueous humor of the eye.   It is formed in the choroid plexuses of the brain ventricles and in the central canal of the cord by the ependymal layer.   It is contained in the central canal of the cord, the ventricles of the brain, and in the subarachnoid spaces where it acts as a watery protective cushion around the brain.   It escapes from the brain through the foramina in the roof of the fourth ventricle.   It circulates slowly down the posterior surface of the spinal cord and up the anterior surface of the cord and brain to the dorsal side of the brain.   The

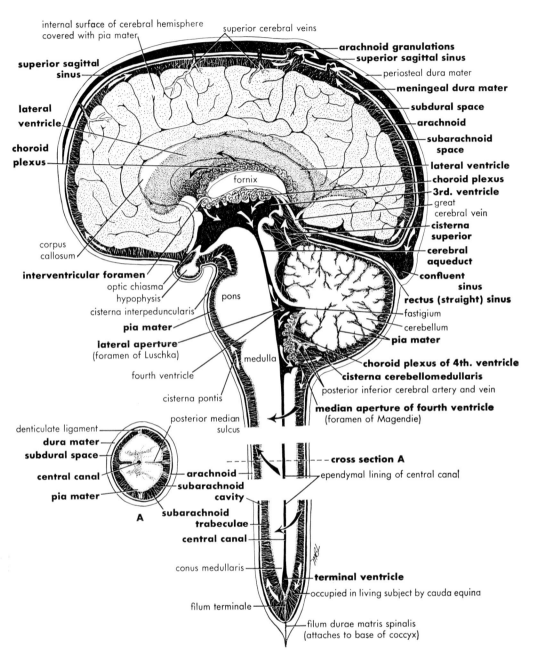

internal surface of cerebral hemisphere
covered with pia mater

superior cerebral veins

arachnoid granulations
superior sagittal sinus

superior sagittal
sinus

periosteal dura mater

meningeal dura mater

lateral
ventricle

subdural space

arachnoid

subarachnoid
space

choroid
plexus

lateral ventricle

choroid plexus

fornix

3rd. ventricle

great
cerebral vein

cisterna
superior

corpus
callosum

cerebral
aqueduct

interventricular foramen

confluent
sinus

optic chiasma

rectus (straight) sinus

hypophysis

pons

fastigium

cisterna interpeduncularis

cerebellum

pia mater

pia mater

lateral aperture
(foramen of Luschka)

medulla

choroid plexus of 4th. ventricle

cisterna cerebellomedullaris

fourth ventricle

posterior inferior cerebral artery and vein

cisterna pontis

median aperture of fourth ventricle
(foramen of Magendie)

posterior median
sulcus

denticulate ligament

dura mater

cross section A

subdural space

central canal

ependymal lining of central canal

arachnoid

pia mater

subarachnoid
cavity

A

subarachnoid
trabeculae

central canal

conus medullaris

terminal ventricle

occupied in living subject by cauda equina

filum terminale

filum durae matris spinalis
(attaches to base of coccyx)

FIG. 18–33. Median section of the brain and spinal cord. The relationships of the meninges, ventricles, and venous sinuses are shown. The arrows indicate the direction of flow of the cerebrospinal fluid. Pia mater shown by light stipple; ventricles and other cavities in black except lateral ventricle which is heavy stipple.

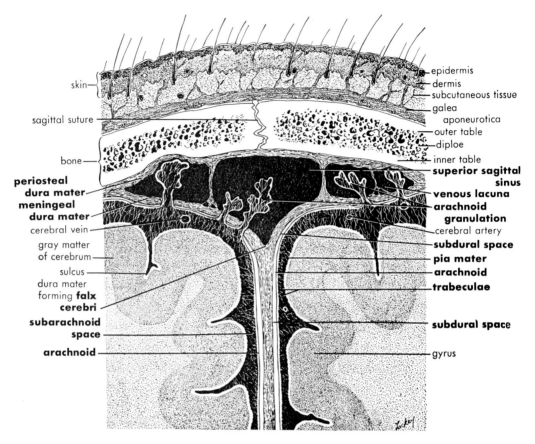

skin

sagittal suture

bone

**periosteal
dura mater**
**meningeal
dura mater**

cerebral vein

gray matter
of cerebrum

sulcus

dura mater
forming **falx
cerebri**

**subarachnoid
space**

arachnoid

epidermis
dermis
subcutaneous tissue
galea
aponeurotica
outer table
diploe
inner table
**superior sagittal
sinus**
venous lacuna
**arachnoid
granulation**
cerebral artery
subdural space
pia mater
arachnoid
trabeculae

subdural space

gyrus

FIG. 18–34.   Frontal section of the scalp, skull, and brain showing relationships of the
meninges, dural venous sinuses, and subarachnoid space.

fluid is resorbed into the superior sagittal sinus of the dura mater of the brain from **arachnoid villi** (*Pacchionian bodies, arachnoid granulations*) which are finger-like processes of the arachnoid meninx which push into the sinus (Fig. 18–34). As suggested in an earlier section of this book, examination of cerebrospinal fluid is often of value in diagnosing diseases of the central nervous system.

MENINGES OF THE BRAIN
(Figs. 18–33, 34, 35)

The **meninges** (*singular, meninx*) of the brain are the same as those around the spinal cord which have been described. There are some variations, however,

which are important to know and understand.

Dura Mater. The dura mater, a single meninx around the cord, becomes double at the foramen magnum, having a periosteal and a meningeal layer. The periosteum of the inner surface of the cranial bones is more closely related to the outer layer of the dura than it is to the bones, therefore it is called the **periosteal dura.** Only at the sutures does it attach more closely to the bones. The **meningeal dura** becomes folded on itself and pushes between parts of the brain to form the falx cerebri, the tentorium cerebelli, the falx cerebelli, and the diaphragma sellae. Also the dural sinuses lie between the meningeal and

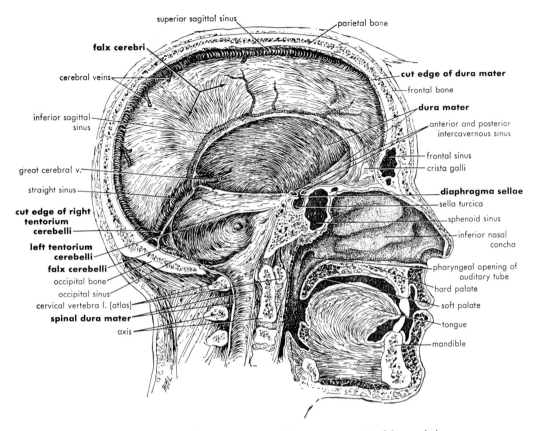

superior sagittal sinus

parietal bone

falx cerebri

cut edge of dura mater

cerebral veins

frontal bone

dura mater

inferior sagittal sinus

anterior and posterior intercavernous sinus

frontal sinus

crista galli

great cerebral v.

straight sinus

diaphragma sellae

sella turcica

cut edge of right tentorium cerebelli

sphenoid sinus

inferior nasal concha

left tentorium cerebelli

falx cerebelli

pharyngeal opening of auditory tube

occipital bone

occipital sinus

hard palate

cervical vertebra I. (atlas)

soft palate

spinal dura mater

tongue

axis

mandible

FIG. 18–35.   Cranial dura mater and its processes—the falx cerebri,
falx cerebelli, and tentorium cerebelli.

periosteal dura or in folds of the meningeal layer.

The **falx cerebri** is a sickle-shaped fold of the meningeal dura which passes into the longitudinal cerebral fissure (Figs. 18–34, 35). In front, it is shallow and attaches to the crista galli of the ethmoid; posteriorly it deepens and becomes continuous over the cerebellum with the tentorium cerebelli. Its superior convex border forms the floor of the superior sagittal sinus and it terminates posteriorly at the internal occipital protuberance. Its inferior border arches over the corpus callosum and houses the inferior sagittal sinus.

The **tentorium cerebelli** (Fig. 18–35) forms a roof over the posterior cranial fossa and is attached peripherally to the edges of the grooves for the transverse sinuses and to the superior angle of the petrous portion of the temporal bone where it leaves a space for the superior petrosal sinus. It attaches anteriorly to the posterior clinoid processes of the sphenoid. Its concave anterior border is free and forms an arch around the brain stem attaching at the anterior clinoid processes. The cerebellum lies below the tentorium, the occipital part of the cerebrum above it. The tentorium arches upward along its median line where it is continuous with the falx cerebri, and there they form the straight venous sinus.

The **falx cerebelli** is a small triangular process whose free border fits into the notch between the cerebellar hemispheres and is attached to the internal occipital crest behind; above, to the under-surface of the tentorium cerebelli. The occipital sinus is contained within the falx cerebelli.

The **diaphragma sellae** is a horizontal fold of the dura mater which connects the clinoid attachments of the tentorium cerebelli, thus forming a roof over the hypophysis (*pituitary*) which rests in the sella turcica. A circular opening in the center of the diaphragm allows for the passage of the infundibulum. Around this opening is the circular sinus.

Pia Mater (Figs. 18–33, 34). The cranial **pia mater** is a delicate vascular membrane which adheres closely to the brain, following faithfully into all its fissures and sulci. In the transverse fissure it forms the tela choroidea of the third ventricle and combining with the ependyma forms the choroid plexuses of the third and lateral ventricles. It forms the tela choroidea and choroid plexus of the fourth ventricle as it passes that area.

Arachnoid (Figs. 18–33, 34). The cranial **arachnoid** is a delicate, web-like, transparent membrane consisting of elastic and collagenous fibers and covered internally by a layer of low cuboidal epithelium. Externally it is covered with a mesothelium and is closely adherent to the dura except for a thin film of fluid in the subdural space. Between the arachnoid and the pia is the **subarachnoid space** into which the arachnoid sends delicate fibrous threads, the **trabeculae,** which connect these two membranes. The arachnoid covers the brain loosely passing over the sulci and entering only the longitudinal and transverse cerebral fissures. Over the gyri the arachnoid and pia are so closely associated that they are called the piarachnoid membrane. Where major segments of the brain join, as at the cerebellar-brain stem juncture, the broader gaps between parts are bridged over by the arachnoid, leaving large subarachnoid spaces, the **cisterns** (Figs. 18–33, 34). The **cisterna**

cerebellomedullaris between cerebellum and medulla oblongata is an example.

The **arachnoid villi** have already been described as finger-like villi which push into the sagittal sinus or its associated venous lacunae and allow the cerebrospinal fluid to be resorbed into the blood (Fig. 18–34).

CRANIAL NERVES (Figs. 18–1, 36, 37)

The cranial and spinal nerves constitute in part the peripheral nervous system. As we have seen, the spinal nerves are paired and form a regular segmental sequence along the spinal cord. They have posterior or afferent roots and anterior or efferent roots. There is a sensory ganglion on the posterior root containing the cells of origin of the afferent neurons; a nucleus made up of the cells of origin of the efferent neurons is in the anterior gray column of the cord.

The cranial nerves issue from the brain. From our understanding of the development of the brain and cord from a common neural tube, we have reason to expect that as the brain bears some resemblance to the cord, the cranial nerves should also bear some resemblance to the spinal nerves. This is indeed the case, but as the brain developed its special features, so did the cranial nerves become more specialized. The student, in studying the cranial nerves, should be alert to these similarities and differences between these two subdivisions of the peripheral nervous system.

Twelve pairs of cranial nerves are commonly recognized. Most of these nerves, like the spinal nerves, have efferent and afferent roots (*mixed nerves*), but a few have only afferent or only efferent roots (*pure nerves*). The afferent or sensory components have their cell bodies in ganglia outside of the brain; the efferent or motor components have their cell bodies in gray nuclei within the brain. Except for the first pair of cranial nerves,

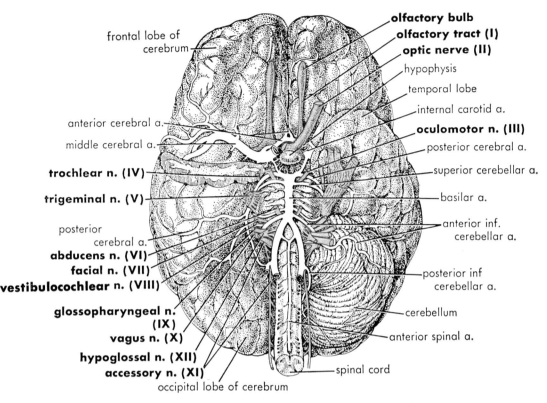

frontal lobe of cerebrum

olfactory bulb
olfactory tract (I)
optic nerve (II)
hypophysis
temporal lobe
internal carotid a.
oculomotor n. (III)
posterior cerebral a.
superior cerebellar a.
basilar a.
anterior inf. cerebellar a.
posterior inf cerebellar a.
cerebellum
anterior spinal a.
spinal cord

anterior cerebral a.
middle cerebral a.
trochlear n. (IV)
trigeminal n. (V)
posterior cerebral a.
abducens n. (VI)
facial n. (VII)
vestibulocochlear n. (VIII)
glossopharyngeal n. (IX)
vagus n. (X)
hypoglossal n. (XII)
accessory n. (XI)
occipital lobe of cerebrum

FIG. 18–36.   Ventral view of the brain showing the cranial nerves (in yellow).

which is atypical, they all attach to the "old" brain stem. Cranial nerves have been give both names and numbers; and Roman numerals should be used. They are as follows:

I. Olfactory	V. Trigeminal	IX. Glossopharyngeal
II. Optic	VI. Abducens	X. Vagus
III. Oculomotor	VII. Facial	XI. Accessory
IV. Trochlear	VIII. Vestibulocochlear	XII. Hypoglossal

Reference to Table 18–1 and to the illustrations will provide basic information about each cranial nerve. The descriptions will be limited except where it is felt that special emphasis is needed.

I. **Olfactory Nerve** (Fig. 19–4). The **olfactory cells** are of the primitive bipolar neuro-epithelial type; their cell bodies are in the olfactory mucosa of the nose, their dendrites pick up the stimuli and pass them on to the central processes or axons, bundles of which pass through the cribriform plate of the ethmoid as the **olfactory nerve** to end in the olfactory bulb at the end of the olfactory tract. The olfactory bulb and tract, though they appear like a cranial nerve, are really a part of the rhinencephalon, a part of the forebrain associated with smell (Figs. 18–36, 38). From the olfactory bulb, fibers of the mitral cells run backward in the olfactory tracts ultimately to reach the areas of olfactory sense in the hippocampus of the cerebrum. The olfactory nerve is pure sensory.

II. **Optic Nerve** (Figs. 18–22, 39, 40). This is a sensory nerve. It is composed

TABLE 18–1. OUTLINE OF THE CRANIAL NERVES

Nerves	Components	Function	Central Connection	Cell Bodies	Peripheral Distribution
I. Olfactory	Afferent Special visceral	Smell	Olfactory bulb and tract	Olfactory epithelial cells	Olfactory nerves
II. Optic	Afferent Special somatic	Vision	Optic nerve and tract	Ganglion cells of retina	Rods and cones of retina
III. Oculomotor	Efferent Somatic	Ocular movement	Nucleus III	Nucleus III	Branches to Levator palpebrae, Rectus superior, medius, inferior, Obliquus inferior
	Efferent General visceral	Contraction of pupil and accommodation	Nucleus of Edinger-Westphal	Nucleus of Edinger-Westphal	Ciliary ganglion; Ciliaris and Sphincter pupillae
	Afferent Proprioceptive	Muscular sensibility	Nucleus mesencephalicus V	Nucleus mesencephalicus V	Sensory endings in ocular muscles
IV. Trochlear	Efferent Somatic	Ocular movement	Nucleus IV	Nucleus IV	Branches to Obliquus superior
	Afferent Proprioceptive	Muscular sensibility	Nucleus mesencephalicus V	Nucleus mesencephalicus V	Sensory endings in Obliquus superior
V. Trigeminal	Afferent General somatic	General sensibility	Trigeminal sensory nucleus	Trigeminal ganglion (Gasserian)	Sensory branches of ophthalmic maxillary and mandibular nerves to skin and mucous membranes of face and head
	Efferent Special visceral	Mastication	Motor V nucleus	Motor V nucleus	Branches to Temporalis, Masseter, Pterygoider, Mylohyoidius, Digastricus, Tensores tympani and palatini
	Afferent Proprioceptive	Muscular sensibility	Nucleus mesencephalicus V	Nucleus mesencephalicus V	Sensory endings in muscles of mastication
VI. Abducens	Efferent Somatic	Ocular movement	Nucleus VI	Nucleus VI	Branches to Rectus lateralis

VII. Facial	Afferent Proprioceptive	Muscular sensibility	Nucleus mesencephalicus V	Nucleus mesencephalicus V	Sensory endings in Rectus lateralis
	Efferent Special visceral	Facial expression	Motor VII nucleus	Motor VII nucleus	Branches to facial muscles, Stapedius, Stylohyoideus, Digastricus
	Efferent General visceral	Glandular secretion	Nucleus salivatorius	Nucleus salivatorius	Greater superficial petrosal nerve, sphenopalatine ganglion, with branches of maxillary V to glands of nasal mucosa. Chorda tympani, lingual nerve, submaxillary ganglion, submaxillary and sublingual glands
	Afferent Special visceral	Taste	Nucleus tractus solitarius	Geniculate ganglion	Chorda tympani, lingual nerve, taste buds, anterior tongue
	Afferent General visceral	Visceral sensibility	Nucleus tractus solitarius	Geniculate ganglion	Great superficial petrosal, chorda tympani and branches
	Afferent General somatic	Cutaneous sensibility	Nucleus spinal tract of V	Geniculate ganglion	With auricular branch of vagus to external ear and mastoid region
VIII. Vestibulocochlear	Afferent Special somatic	Hearing	Cochlear nuclei	Spiral ganglion	Organ of Corti in cochlea
	Afferent Proprioceptive	Sense of equilibrium	Vestibular nuclei	Vestibular ganglion	Semicircular canals, saccule, and utricle
IX. Glossopharyngeal	Afferent Special visceral	Taste	Nucleus tractus solitarius	Inferior ganglion IX	Lingual branches, taste buds, posterior tongue
	Afferent General visceral	Visceral sensibility	Nucleus tractus solitarius	Inferior ganglion IX	Tympanic nerve to middle ear, branches to pharynx and tongue, carotid sinus nerve

TABLE 18-1. OUTLINE OF THE CRANIAL NERVES (Continued)

Nerves	Components	Function	Central Connection	Cell Bodies	Peripheral Distribution
	Efferent General visceral	Glandular secretion	Nucleus salivatorius	Nucleus salivatorius	Tympanic, lesser superficial petrosal nerves, otic ganglion, with auriculo-temporal V to parotid gland
	Efferent Special visceral	Swallowing	Nucleus ambiguus	Nucleus ambiguus	Branch to Stylopharyngeus
X. Vagus	Efferent General visceral	Involuntary muscle and gland control	Dorsal motor nucleus X	Dorsal motor nucleus X	Cardiac nerves and plexus; ganglia on heart. Pulmonary plexus; ganglia respiratory tract. Esophageal, gastric, celiac plexuses; myenteric and submucous plexuses, muscle and glands of digestive tract down to transverse colon
	Efferent Special visceral	Swallowing and phonation	Nucleus ambiguus	Nucleus ambiguus	Pharyngeal branches, superior and inferior laryngeal nerves
	Afferent General visceral	Visceral sensibility	Nucleus tractus solitarius	Ganglion nodosum	Fibers in all cervical, thoracic, and abdominal branches; carotid and aortic bodies
	Afferent Special visceral	Taste	Nucleus tractus solitarius	Ganglion nodosum	Branches to region of epiglottis and taste buds
	Afferent General somatic	Cutaneous sensibility	Nucleus spinal tract V	Jugular ganglion	Auricular branch to external ear and meatus
XI. Accessory	Efferent Special visceral	Swallowing and phonation	Nucleus ambiguus	Nucleus ambiguus	Bulbar portion, communication with vagus, in vagus branches to muscles of pharynx and larynx
	Efferent Special somatic	Movements of shoulder and head	Lateral column of upper cervical spinal cord	Lateral column of upper cervical spinal cord	Spinal portion, branches to Sternocleidomastoideus and Trapezius
XII. Hypoglossal	Efferent General somatic	Movements of tongue	Nucleus XII	Nucleus XII	Branches to extrinsic and intrinsic muscles of tongue

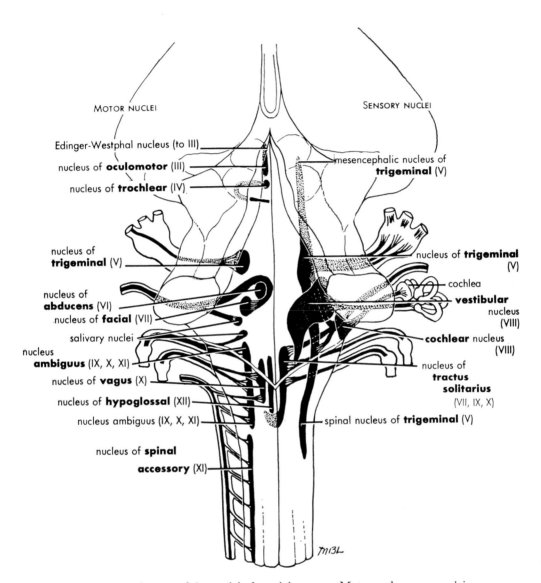

MOTOR NUCLEI

SENSORY NUCLEI

Edinger-Westphal nucleus (to III)

nucleus of **oculomotor** (III)

nucleus of **trochlear** (IV)

mesencephalic nucleus of
trigeminal (V)

nucleus of
trigeminal (V)

nucleus of **trigeminal**
(V)

cochlea

nucleus of
abducens (VI)

nucleus of **facial** (VII)

salivary nuclei

vestibular
nucleus
(VIII)

cochlear nucleus
(VIII)

nucleus
ambiguus (IX, X, XI)

nucleus of **vagus** (X)

nucleus of **hypoglossal** (XII)

nucleus ambiguus (IX, X, XI)

nucleus of
**tractus
solitarius**

(VII, IX, X)

spinal nucleus of **trigeminal** (V)

nucleus of **spinal
accessory** (XI)

FIG. 18–37.   Diagram of the nuclei of cranial nerves.   Motor and sensory nuclei are
represented on opposite sides of the brain.   (Modified from Netter.)

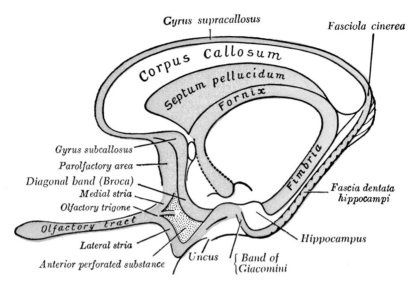

Gyrus supracallosus

Fasciola cinerea

Corpus Callosum

Septum pellucidum

Fornix

Gyrus subcallosus
Parolfactory area
Diagonal band (Broca)
Medial stria
Olfactory trigone

Fimbria

Fascia dentata hippocampi

Olfactory tract

Lateral stria
Anterior perforated substance

Uncus

{ *Band of Giacomini*

Hippocampus

Fig. 18–38.   Schema of rhinencephalon.   (*Gray's Anatomy.*)

of the axons of the ganglion cells of the retina which leave the posterior pole of the eyeball, pass through the optic foramen and enter the opposite side at the **optic chiasma.** Here the fibers from the medial half of each retina cross to the opposite side; those from the lateral half of each retina remain on the same side. From the optic chiasma the fibers pass to the **optic tracts** from which the fibers take different pathways. The majority of them go to the **lateral geniculate body,** a part of the thalamus complex, where they synapse with neurons whose fibers constitute the **geniculo-calcarine tract** which passes as a part of the internal capsule and ultimately to the calcarine cortex of the occipital lobe, the **visual area.** Destruction of an optic tract or of the visual cortex on one side would cause blindness in the medial half of one eye and in the lateral half of the other (Fig. 18–39).

Other fibers of the optic tract may pass to the **superior colliculi** of the midbrain where they synapse with neurons whose fibers either (1) constitute the **tectobulbar fasciculus** distributed to nuclei of cranial nerves III, IV, and VI

to the extrinsic muscles of the eye; or (2) constitute the **tectospinal tract** descending into the cord and synapsing with anterior horn cells of the spinal nerves. Through this **somatic reflex system** widespread motor responses can result from light stimuli.

Finally, some fibers of the optic tracts carry impulses to the **pretectal region** between the thalamus and the midbrain where they synapse with neurons whose axons travel forward to a small part of the oculomotor nucleus. From the cells of this nucleus axons pass out of the brain with the third (oculomotor) nerve to a small **ciliary ganglion** at the back of the eyeball. These last fibers are called preganglionic neurons and belong to the autonomic system. In the ciliary ganglion a final synapse is made with postganglionic neurons whose axons end in the ciliary muscles. These operate on the lens focusing the eye or on the sphincter muscle of the iris which regulates the amount of light reaching the retina. This then is an autonomic reflex—a light reflex (Fig. 18–40). It is apparent that the optic mechanism not only enables us to see objects and to engage in voluntary

VISUAL PATH

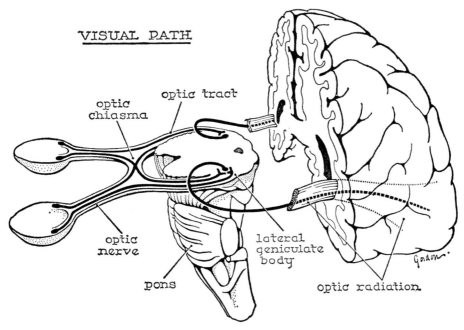

FIG. 18–39.   Scheme of visual pathway from retina to visual area of brain. (From Cates' *Primary Anatomy* by Basmajian, Williams & Wilkins Co.)

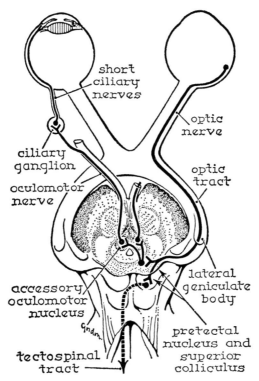

FIG. 18–40.   Scheme of light reflex pathway (from retina, through midbrain, to sphincteric muscle of iris) and somatic reflex pathway. (From Cates' *Primary Anatomy* by Basmajian, Williams & Wilkins Co.)

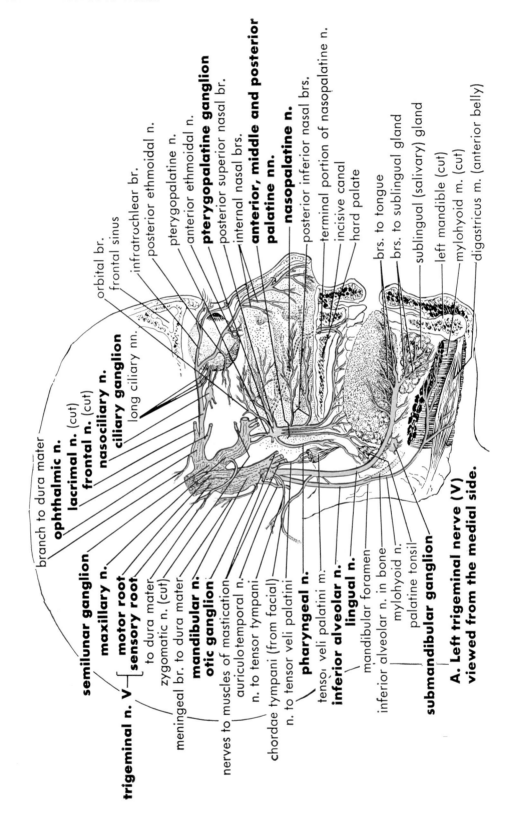

A. Left trigeminal nerve (V) viewed from the medial side.

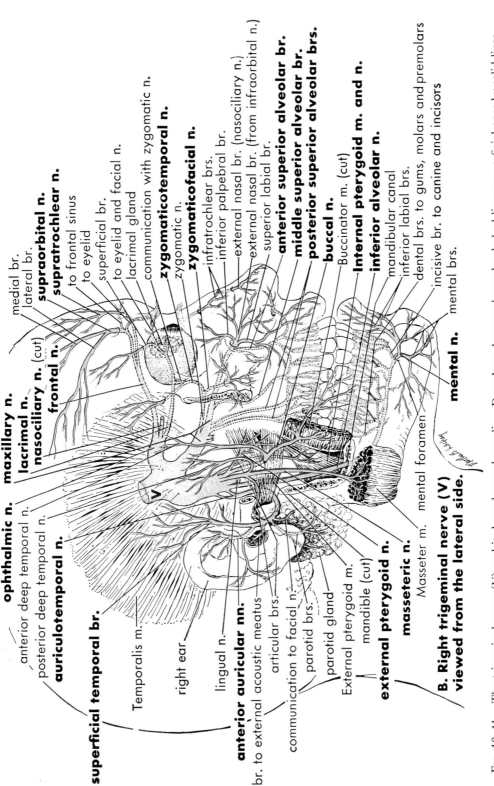

ophthalmic n.
maxillary n.
lacrimal n.
nasociliary n. (cut)
frontal n.

medial br.
lateral br.
supraorbital n.
supratrochlear n.
to frontal sinus
to eyelid
superficial br.
to eyelid and facial n.
lacrimal gland
communication with zygomatic n.
zygomaticotemporal n.
zygomatic n.
zygomaticofacial n.
infratrochlear brs.
inferior palpebral br.
external nasal br. (nasociliary n.)
external nasal br. (from infraorbital n.)
superior labial br.
anterior superior alveolar br.
middle superior alveolar br.
posterior superior alveolar brs.
buccal n.
Buccinator m. (cut)
Internal pterygoid m. and n.
inferior alveolar n.
mandibular canal
inferior labial brs.
dental brs. to gums, molars and premolars
incisive br. to canine and incisors
mental n.
mental brs.

anterior deep temporal n.
posterior deep temporal n.
auriculotemporal n.

superficial temporal br.

Temporalis m.

right ear

lingual n.

anterior auricular nn.
br. to external acoustic meatus
articular brs.
communication to facial n.
parotid brs.
parotid gland
External pterygoid m.
mandible (cut)
external pterygoid n.
masseteric n.

Masseter m. mental foramen

B. Right trigeminal nerve (V)
viewed from the lateral side.

Fig. 18–41. The trigeminal nerve (V) and its branches and ganglia. Deep branches are shown by dashed lines: superficial ones by solid lines.

responses to them, but sets off both somatic and autonomic reflex responses as well.

III. Oculomotor Nerve (Figs. 18–36, 37, 40). This is a mixed nerve. It emerges from the inferior surface of the midbrain just anterior to the pons and medial to the cerebral peduncle. It runs forward, divides into superior and inferior branches and enters the orbit through the superior orbital fissure. The superior branch supplies the Superior rectus and Levator palpebrae muscles; the inferior branch, the Medial and Inferior recti and the Inferior oblique muscles, and gives off a small branch to the ciliary ganglion. These branches also carry afferent fibers from the same muscles. These facts, and reference to the discussion of the optic nerve above, make clear the functions of the oculomotor nerves which may be summarized as follows: they provide for muscle sense and movement in all but two of the extrinsic muscles of the eyeball, constriction of the pupil, and accommodation of the eye.

IV. Trochlear Nerve (Figs. 18–36, 37). This, the smallest cranial nerve, provides for muscle sense and impulses for movement of the Superior oblique muscle of the eye. It arises just behind the inferior colliculus of the midbrain, curves around the lateral surface of the cerebral peduncle just above the pons and passes forward through the superior orbital fissure to the orbit and to the Superior oblique muscle.

V. Trigeminal Nerve (Figs. 18–36, 37, 41). The trigeminal is the largest of the cranial nerves and as its name suggests has three large branches, the ophthalmic, maxillary, and mandibular. It is entirely sensory except for a small **motor branch** which joins with the mandibular branch and supplies the muscles of mastication. By some anatomists this motor branch is designated as another cranial nerve, the **masticatory nerve.** It also carries some proprioceptive sensory fibers.

The trigeminal nerve has two roots placed close together on the ventral lateral surface of the pons: a large sensory root which, as it passes laterally into the middle cranial fossa, has a large **semilunar** (*Gasserian*) **ganglion** and a small motor root. From the semilunar ganglion the three branches radiate forward; the **ophthalmic** branch enters the orbit through the superior orbital fissure; the **maxillary** passes through the foramen rotundum; the **mandibular** through the foramen ovale.

The trigeminal nerve carries impulses from touch, pain, heat, and cold receptors from the face, scalp, and mucous membranes of the head. It also contains special visceral efferent and proprioceptive fibers.

The **ophthalmic branch** receives sensory fibers from the skin of the upper eyelid, side of the nose, forehead, anterior half of scalp, lacrimal gland, cornea, conjunctiva, and eyeball.

The **maxillary branch** receives branches from the upper teeth, upper lip, skin of the cheek, anterior temporal region, mucous membrane of the nose, palate, and adjacent parts of the pharynx.

The **mandibular branch** carries impulses from the skin of the face over the mandible and side of the head in front of the ear and over the chin, the mucous membrane of the floor of the mouth and the anterior two-thirds of the tongue, and from the lower teeth. Since all of the lower teeth on one side are supplied by one branch of the mandibular nerve which enters the mandibular foramen on the inside of the ramus of the mandible, the dentist can anesthetize all of these teeth with one injection.

VI. Abducens Nerve (Figs. 18–36, 37). This nerve supplies afferent and efferent fibers to the Lateral rectus muscles of the eyes, mediating muscle sense and motion respectively. It arises from the brain stem at the lower border of the

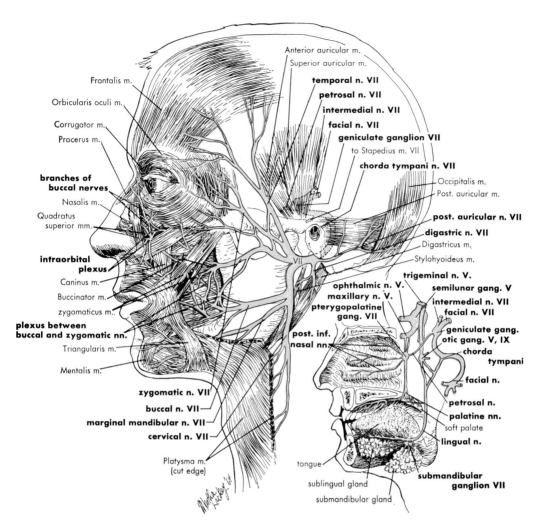

FIG. 18–42.   The facial (VII) nerve and its distribution.   Only those facial muscles that receive their motor nerve supply from the facial nerve are shown.   Insert shows the relationship between the facial (VII) and the trigeminal (V) nerves.

pons and reaches the orbit through the superior orbital fissure.

VII. **Facial Nerve** (Figs. 18–36, 37, 42).   The facial is a mixed nerve. It issues from the brain lateral to the abducens at the lower border of the pons. It enters the petrous portion of the temporal bone through the internal acoustic meatus and by a devious route finally emerges from the stylomastoid foramen. It passes into the parotid gland and divides into numerous branches which reach out to supply the facial and scalp muscles.   This somatic efferent portion makes up the greater part of the nerve. It also, however, contains some visceral efferent fibers supplying the submaxillary, sublingual, and lacrimal glands and also glands of the nose, mouth, palate, and pharynx.

The afferent fibers from the taste buds travel at first in the lingual nerve, a branch of the trigeminal, but later leave it to join the facial nerve by which they reach the geniculate ganglion which lies within the petrous bone.   Their central

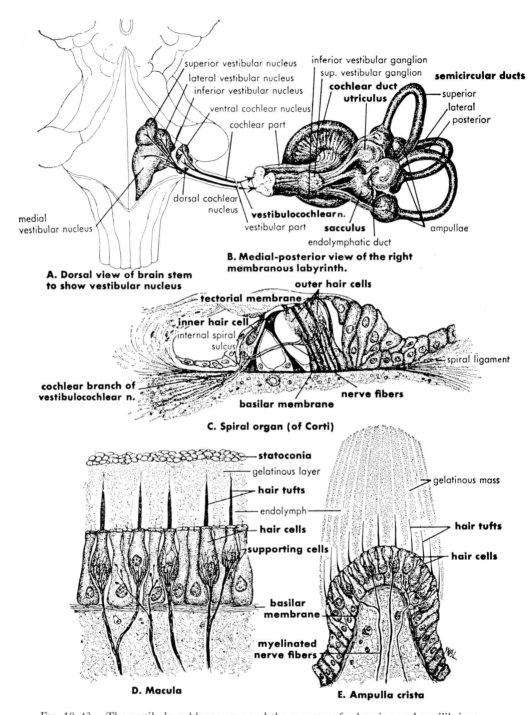

superior vestibular nucleus
lateral vestibular nucleus
inferior vestibular nucleus
ventral cochlear nucleus
cochlear part

inferior vestibular ganglion
sup. vestibular ganglion
semicircular ducts
cochlear duct
utriculus

superior
lateral
posterior

dorsal cochlear nucleus
vestibulocochlear n.
vestibular part   **sacculus**
endolymphatic duct

medial vestibular nucleus

ampullae

A. Dorsal view of brain stem to show vestibular nucleus

B. Medial-posterior view of the right membranous labyrinth.

outer hair cells
tectorial membrane
inner hair cell
internal spiral sulcus
cochlear branch of vestibulocochlear n.
spiral ligament
nerve fibers
basilar membrane

C. Spiral organ (of Corti)

statoconia
gelatinous layer
hair tufts
endolymph
hair cells
supporting cells
basilar membrane
myelinated nerve fibers

gelatinous mass
hair tufts
hair cells

D. Macula

E. Ampulla crista

FIG. 18–43.   The vestibulocochlear nerve and the receptors for hearing and equilibrium.

processes enter the brain stem with the facial nerve and go to their termination in the medulla where they synapse with neurons going to the taste centers in the cerebral cortex.

VIII. **Vestibulocochlear** (*Auditory*) **Nerve** (Figs. 18–36, 37, 43). The vestibulocochlear nerve has two well-defined parts, the cochlear and vestibular nerves—both sensory.

The **cochlear nerve,** the nerve for hearing, arises in the spiral organ of Corti in the cochlea of the internal ear. Its cell bodies lie in the spiral ganglion from which its central fibers pass through the internal acoustic meatus to terminate in the dorsal and ventral cochlear nuclei in the medulla. Synapses with neurons here cross deeply in the medulla and by way of the lateral lemniscus in the midbrain reach the **medial geniculate body** near the posterior end of the thalamus. Here they synapse with the neurons whose axons lead laterally to the hearing center in the temporal cortex. The **inferior colliculus,** like the superior colliculus in its relationship to vision, is a center for somatic reflexes relative to hearing. From it the tectospinal tract carries impulses into the cord to efferent centers in the anterior column.

The second part of the vestibulocochlear nerve, the **vestibular nerve,** arises in the utricle and saccule and semicircular canals and has its **vestibular ganglion** in the petrous bone from which the central processes leave this bone at the internal acoustic meatus to reach vestibular nuclei in the medulla. From here they may go directly to the cerebellar cortex; be relayed to it through the vestibular nuclei; or they may be relayed upward to motor cranial nerves like those of the eye muscles; or downward by the **vestibulospinal tract** to anterior gray column cells. The cerebellum thus receives proprioceptive impulses from the vestibular part of the ear as well as from the tendons, joints, and muscles of

36

the body by way of the **spinocerebellar tracts,** which, with the **vestibulocerebellar** tracts enter through the inferior cerebellar peduncles. By this complicated mechanism the body is kept in equilibrium and the muscles coordinated for efficient functioning.

IX. **Glossopharyngeal Nerve** (Figs. 18–36, 37, 44). The glossopharyngeal nerve is a sensory and motor nerve. Its superficial origin is by several roots on the side of the medulla and it leaves the skull with the vagus and accessory nerves by the jugular foramen. It has two swellings on its trunk, the **superior** and **inferior ganglia,** analogous to dorsal root ganglia. It supplies five branches to the mucous membrane of the pharynx, tonsils, and the taste buds of the posterior third of the tongue. Other branches connect with the carotid sinus. Impulses originating in the peripheral fibers in these structures have their cell bodies in the petrous and superior ganglia and their central fibers go to the brain. Reflex control of blood pressure, taste, and swallowing are provided.

The efferent fibers of the nerve with their cell bodies in the medulla supply the swallowing muscles of the pharynx. This is done through contributions from the vagus and accessory nerves and from the superior cervical ganglion of the sympathetic system, all of which form an extensive pharyngeal plexus. From this plexus the pharyngeal muscles, with a few exceptions such as the Stylopharyngeus, are innervated.

The glossopharyngeal nerve also carries parasympathetic preganglionic secretory fibers to the small **otic ganglion,** situated just below the foramen ovale, from which postganglionic fibers pass to the parotid gland.

X. **Vagus Nerve** (Figs. 18–36, 37, 45). This is the vagrant among the cranial nerves wandering from the head and neck into the thorax and abdomen as far as the transverse colon. Its superficial

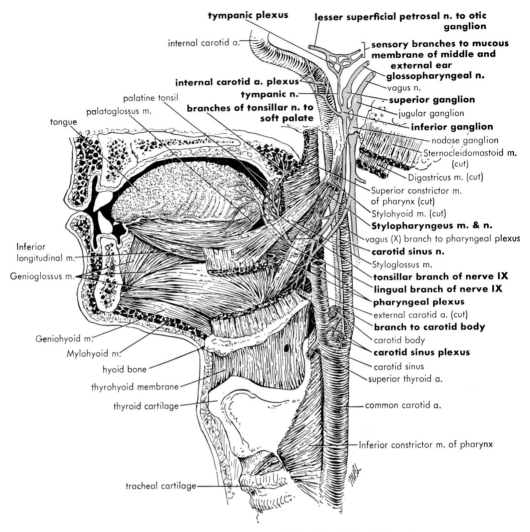

FIG. 18–44. The glossopharyngeal nerve (IX). Its relationship to the vagus nerve (X) is shown.

origin is by rootlets in series with those of the glossopharyngeal nerve on the side of the medulla. It passes through the jugular foramen and just below the skull has two swellings; the smaller upper one is the **jugular** (*superior*) **ganglion,** the lower and larger one the **ganglion nodosum** (*inferior*). They both serve in the same capacity as posterior root ganglia of the spinal nerves. The vagus descends vertically on the side of the pharynx where it is wrapped in the carotid sheath with the common carotid artery and

internal jugular vein. The right and left vagi, as can be seen in Figure 18–45, come together at some points to form plexuses from which certain structures are innervated.

Study of Table 18–1 and Figure 18–45 will make quite clear the functions and distribution of this important nerve. It should be especially noted that it serves as a motor nerve to pharyngeal and laryngeal muscles and that it carries sensory fibers *from* and parasympathetic fibers *to* the heart, digestive tract as far

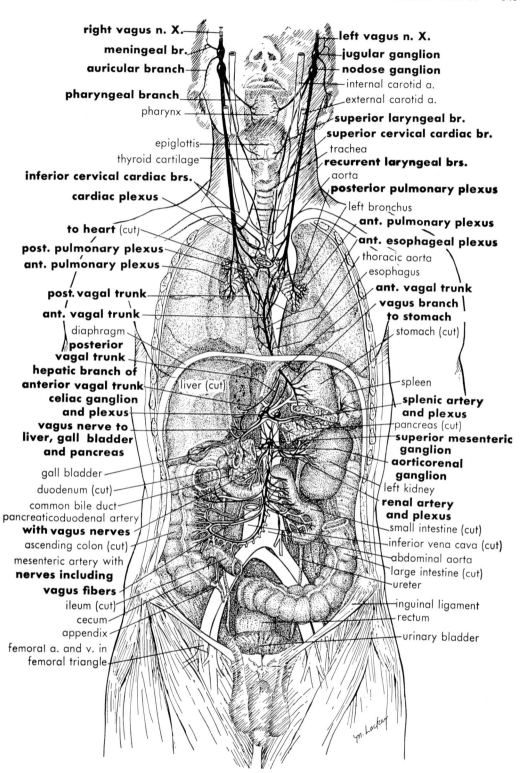

right vagus n. X.
meningeal br.
auricular branch
pharyngeal branch
pharynx
epiglottis
thyroid cartilage
inferior cervical cardiac brs.
cardiac plexus
to heart (cut)
post. pulmonary plexus
ant. pulmonary plexus
post. vagal trunk
ant. vagal trunk
diaphragm
posterior vagal trunk
hepatic branch of anterior vagal trunk
celiac ganglion and plexus
vagus nerve to liver, gall bladder and pancreas
gall bladder
duodenum (cut)
common bile duct
pancreaticoduodenal artery **with vagus nerves**
ascending colon (cut)
mesenteric artery with **nerves including vagus fibers**
ileum (cut)
cecum
appendix
femoral a. and v. in femoral triangle

left vagus n. X.
jugular ganglion
nodose ganglion
internal carotid a.
external carotid a.
superior laryngeal br.
superior cervical cardiac br.
trachea
recurrent laryngeal brs.
aorta
posterior pulmonary plexus
left bronchus
ant. pulmonary plexus
ant. esophageal plexus
thoracic aorta
esophagus
ant. vagal trunk
vagus branch to stomach
stomach (cut)
spleen
splenic artery and plexus
pancreas (cut)
superior mesenteric ganglion
aorticorenal ganglion
left kidney
renal artery and plexus
small intestine (cut)
inferior vena cava (cut)
abdominal aorta
large intestine (cut)
ureter
inguinal ligament
rectum
urinary bladder

liver (cut)

m. Lackey

Fig. 18–45.   The vagus (X) nerve, its branches, and related ganglia.

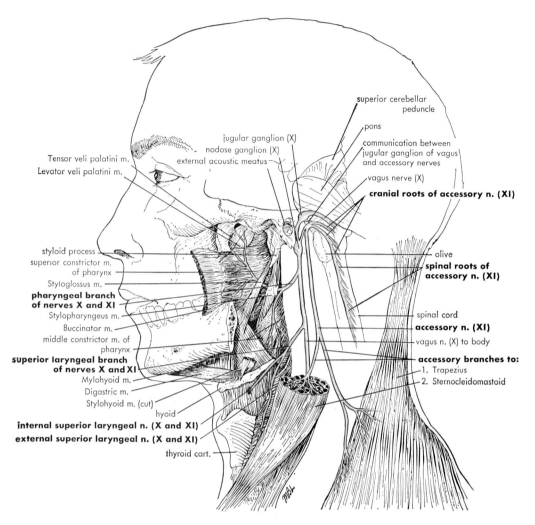

FIG. 18–46. The accessory nerve (XI) showing its distribution and its relationship to the vagus (X).

as the transverse colon, the respiratory tract, kidneys, spleen, liver, and pancreas.

XI. **Accessory Nerve** (Figs. 18–36, 37, 46). The **accessory** nerve has an arrangement different from any other cranial nerves. It is divided into **bulbar** and **spinal** portions. The **bulbar** portion arises from the medulla just behind the vagus and passes through the jugular foramen with that nerve to supply mainly the voluntary muscles of pharynx, larynx, and soft palate. The **spinal** portion consists of nerve fibers from the upper five segments of the spinal cord. These join

to form a common trunk which enters the cranial cavity through the foramen magnum, joins the bulbar portion of the nerve and with it leaves the cavity by way of the jugular foramen. The spinal fibers then supply the Sternocleidomastoid and Trapezius muscles.

XII. **Hypoglossal Nerve** (Figs. 18–36, 37, 47). This nerve arises from the anterior side of the medulla by way of a number of rootlets. It passes through the hypoglossal canal and turns forward to supply both the extrinsic and intrinsic muscles of the tongue.

AUTONOMIC NERVOUS SYSTEM
(Figs. 18–48, 49)

There is a lack of uniformity in the way in which the autonomic nervous system is treated by various authors. Traditionally it has been considered to include only visceral efferent neurons and the ganglia in which they synapse in reaching their effectors which are the smooth and cardiac muscles and the glands. The visceral afferent neurons have been excluded. The terminology used for the autonomic system and its subdivisions has also varied, hence there has been much confusion. The system is often called the involuntary nervous system because the individual has no direct or conscious control over its activities. It has also been called the vegetative, the visceral, and the sympathetic nervous systems.

In this book the nomenclature used is as follows:

    I. Autonomic nervous system
       Visceral efferent neurons
       1. Sympathetic
       2. Parasympathetic
    II. Visceral afferent fibers

VISCERAL EFFERENT SYSTEM
(Figs. 18–48, 49)

This is the part of the system to which the term autonomic is so often applied.

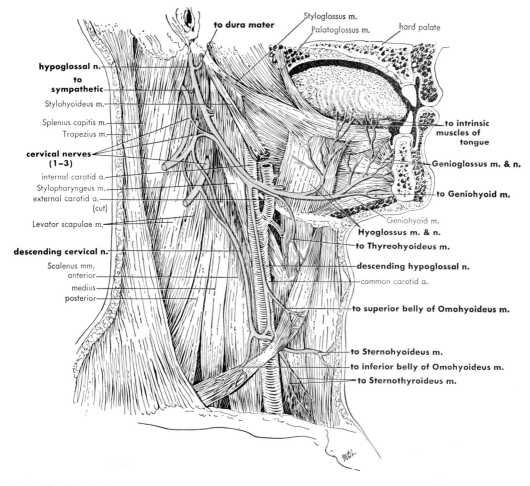

FIG. 18–47. The hypoglossal nerve XII. Its relationship to cervical nerves 1–3 is shown.

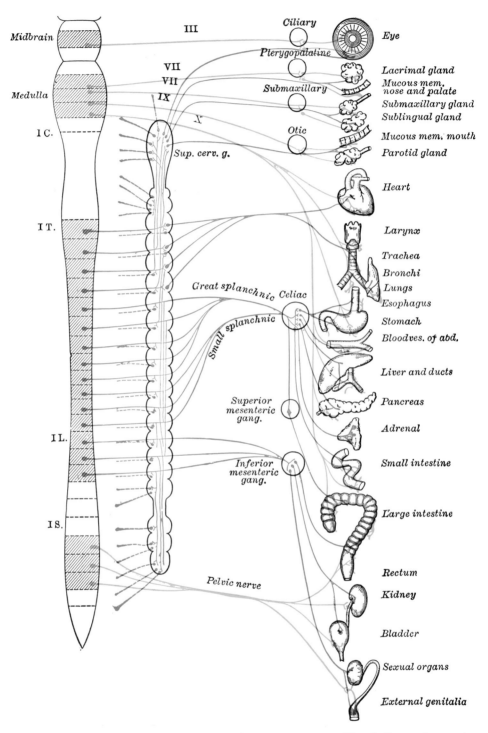

FIG. 18–48.   Diagram of the efferent autonomic nervous system.   Blue indicates the craniosacral outflow of the parasympathetic division; red indicates the thoracolumbar outflow of the sympathetic division.   Postganglionic fibers to spinal and cranial nerves supply vasomotor connections to head, trunk and limbs; motor fibers to smooth muscles of the skin and secretory fibers to the sweat glands. (*Gray's Anatomy*.)

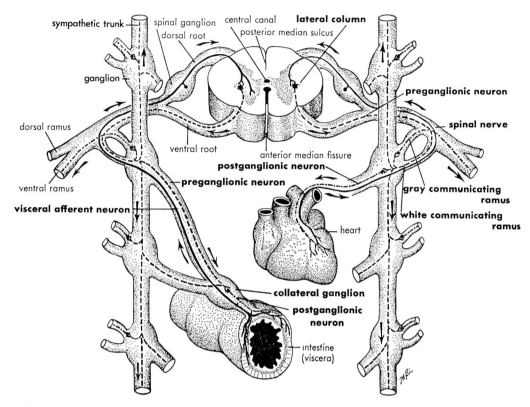

Fig. 18–49.   Sympathetic division of the autonomic nervous system.   A cross section of the thoracic spinal cord with an attached spinal nerve is shown.   Afferent and efferent sympathetic neurons are used to show the relationships between the sympathetics and the spinal cord and nerves.   Afferent neurons in solid lines; preganglionic efferent in dashed lines; postganglionic efferent in dot and dash lines.

This is because this part of the nervous system does have a certain amount of autonomy or independence, but it is still controlled by the central system with which it is connected at different levels. The only part of the autonomic system that may have fairly complete autonomy, *i.e.*, reflexes not involving the central system, is the plexus in the walls of the alimentary tract, the **enteric plexus.** The only important morphological difference between the visceral efferent system and the somatic efferent is that there are two neurons between the central system and the effectors in the former, instead of one, as in the latter (Figs. 18–18, 19). These are called the **preganglionic** and **postganglionic neurons.**

The **visceral efferent system** is divided into two divisions or systems—the **sympathetic** or thoracolumbar, and the **parasympathetic** or craniosacral. They differ in the connections they make with the central nervous system and in the placement of their ganglia. Many of the organs receive innervation from both of these divisions and their physiological effects in each case are antagonistic. In general, the sympathetic system mobilizes the energy of the body for emergency or quick action. The heart beats faster; the pupils dilate; the peripheral blood vessels constrict; blood pressure elevates; the adrenal medulla secretes; glycogen breakdown in the liver is increased; sweating is increased, and the actions of stomach and intestine are inhibited. The parasympathetic system function is directed more toward the conservative and restorative processes. The heart rate slows; the pupils contract; blood pressure drops; the stomach and intestine and their glands become active. It is quite clear now that the hypothalamus has craniosacral and thoracolumbar centers and that the interactions between these two so regulates visceral functions as to maintain the constancy of the internal environment.

Sympathetic Division (Fig. 18–49). The cell bodies of the first or **preganglionic neurons** lie in the **lateral horns** of the gray matter of the spinal cord from the first thoracic to the third lumbar segments. They are often referred to, therefore, as the **thoracolumbar outflow.** The axons of these preganglionic neurons pass from the cord as a part of the anterior roots of the spinal nerves. They quickly separate from the anterior root, however, and join the sympathetic trunk as the **white rami communicantes.** These rami are white because the preganglionic neurons are myelinated.

The sympathetic trunks lie, one on each side of the vertebral column, from the level of the first cervical vertebra to the coccyx. They are chains of ganglia connected by fibers. The ganglia are called **sympathetic trunk** (*vertebral, lateral, central*). The uppermost ganglion of the chain is the **superior cervical ganglion** and here the sympathetic trunk is continued into the head as the internal carotid nerve. At the inferior end the two trunks converge at the coccyx and may join to form a single ganglion, the **ganglion impar** (Fig. 18–1).

The preganglionic fibers, upon reaching the sympathetic trunk, may (1) synapse in the central ganglia with unmyelinated postganglionic neurons, (2) pass up or down the trunk to ganglia at other levels, (3) pass through the ganglia of the trunk to synapse in other outlying ganglia, the **collateral ganglia,** with unmyelinated

postganglionic neurons, or (4) a combination of the above.

The **postganglionic fibers** issuing from the sympathetic trunk constitute its **branches of distribution** and they are of several types. Some go to the spinal nerves, some to cranial nerves, some to individual organs, and some accompany blood vessels or go to the great autonomic plexuses in various parts of the body. Those which return to the spinal nerves constitute the **gray rami communicantes.** The gray rami join all of the spinal nerves, whereas the white rami come only from thoracic and upper lumbar spinal nerves. You should recall that the gray rami were referred to in an earlier section (page 495) as the sympathetic roots of the spinal nerves. Upon entering the spinal nerves, the fibers of the gray rami are distributed by cutaneous branches to the smooth muscle in the walls of peripheral blood vessels, to glands, hair follicles, and to the Arrectores pilorum muscles.

The postganglionic fibers to the head region arise from cells in the superior cervical ganglion and may go quite directly to the cranial nerves or through plexuses along the blood vessels, such as the internal and external carotids, to reach their destinations. Postganglionic fibers to the thoracic and pelvic organs probably reach their destinations through the pulmonary, aortic, and pelvic plexuses.

The **branches of communication** of the sympathetic trunk are those preganglionic neurons which passed through the trunk without synapsing. They form the **splanchnic nerves** which continue to the collateral ganglia. In the collateral ganglia their fibers synapse with postganglionic neurons which supply abdominal and pelvic organs. The important **collateral ganglia** are the **celiac, superior mesenteric** (*aorticorenal*), and **inferior mesenteric** named for the aortic branches near which they lie. The area around the celiac and superior mesenteric ganglia constitutes the "solar" plexus (Figs. 18–45, 48).

The white communicating rami of the sympathetic system, in addition to the small, myelinated preganglionic fibers which they carry, also contain some large and medium-sized myelinated fibers which are believed to be visceral afferent fibers. There are also some unmyelinated fibers.

Parasympathetic Division (Figs. 18–40, 48). The cells of origin of the preganglionic neurons in this division are found in the brain stem and in the second through the fourth sacral segments of the spinal cord; hence the name often applied to it, the **craniosacral outflow.** The cranial outflow is carried in oculomotor (III), facial (VII), glossopharyngeal (IX), and vagus (X) nerves.

In contrast to the sympathetic division the parasympathetic ganglia are **terminal,** lying near or in the organs innervated; hence the preganglionic fibers are long, rather than short; the postganglionic fibers, short, rather than long. The preganglionic fibers of the sympathetic division synapse with far greater numbers of postganglionic neurons than do those of the parasympathetic division, accounting for the greater spread of activity in that division.

The cranial outflow is associated with the following ganglia in which are found the synapses between preganglionic and postganglionic neurons.

(1) The **ciliary ganglion** of the oculomotor nerve has been mentioned in a previous section. From it the ciliary and sphincter (*iris*) muscles of the eye are innervated (Fig. 18–40).

(2) The **submandibular ganglion** of the facial nerve sends fibers to the submandibular and sublingual salivary glands (Fig. 18–42).

(3) The **pterygopalatine ganglion** of the facial nerve gives postganglionic parasympathetic fibers to the lacrimal gland and the glands of the nasal and pharyngeal mucosa (Fig. 18–42).

(4) The **otic ganglion** of the glossopharyngeal nerve as stated previously is the source of postganglionic fibers for the stimulation of the parotid gland (Fig. 18–44).

The parasympathetic influence of the **vagus nerve** is felt all the way from the thorax to the transverse colon as stated in the section under cranial nerves (Fig. 18–45).

The **sacral outflow,** arising in segments two, three, and four of the sacral cord, passes outward with the sacral nerves. It leaves the sacral nerves as their visceral branches which join the pelvic plexus in the subserous fascia of the pelvis. From this plexus the preganglionic fibers reach the ganglia in or near the walls of the pelvic viscera. In the ganglia the postganglionic fibers arise.

Autonomic Reflex Arc (Fig. 18–49). The autonomic reflex arc is built on the same pattern as the somatic reflex. It has a:

(1) Receptor.

(2) Afferent neuron, with cell body in cerebrospinal ganglia.

(3) Synapse with internuncial neuron or directly with a preganglionic neuron.

(4) Preganglionic neuron synapsing in central, collateral, or terminal ganglion with:

(5) Postganglionic neuron carrying impulses from the ganglion to:

(6) Effector organ—smooth muscle, cardiac muscle, or glands.

The possibility of a two-neuron reflex arc does not exist in the autonomic nervous system as it does in the somatic system. The preganglionic nerve, in a way, plays the role of an internuncial or association neuron, yet it is by definition a visceral efferent neuron.

VISCERAL AFFERENT FIBERS

The **visceral afferent fibers** like the somatic afferent neurons have their cell bodies in the sensory ganglia of the cranial and spinal nerves. Their differences lie in their peripheral distribution and in the fact that the impulses they carry have no representation in consciousness or give only vague sensations which we have a very limited capacity to localize. The phenomenon known as **referred pain** is a result of this incapacity to localize the source of stimulation and is well exemplified by "pain" impulses initiated in the heart. They are felt in the axilla, down the ulnar side of the arm, and in the precordial region, rather than in the heart itself. We are not sure of the mechanism that causes this, but there is a suggestion in the fact that such pain is usually referred to those structures from which come the somatic afferent fibers whose central connections are the same as the visceral afferent fibers involved.

Visceral afferent fibers travel not only in the regular autonomic pathways, but in the spinal and cranial nerves. They carry impulses which may initiate either autonomic or somatic reflexes, or may travel upward to the hypothalamus and ultimately to the cortex of the frontal lobe (Fig. 18–49).

CONDUCTION PATHWAYS OF THE NERVOUS SYSTEM

The following account will serve in part as a review of the principal landmarks of the nervous system. It will also aid in bringing these landmarks into meaningful association as parts of a functioning system which orients, controls and coordinates the living body. Only a few of the many nervous pathways known will be described and illustrated but these will be sufficient to demonstrate the general patterns followed by the many. Reference to Figures 18–12, 22, 23, 24, and 28 will be helpful in the descriptions which follow.

The nerve pathways may be divided into **afferent** (*sensory*) and **efferent**

(*motor*). The afferent pathways connect the receptors of the body with the integrating mechanisms of the central nervous system, some extending to the higher centers of the brain, enabling the individual to be conscious of his environment. These generally involve three orders of neurons.

The efferent pathways bring the nerve centers of the central system into communication with the effector tissues—the muscle tissues and glands. Thus the body is able to respond to environmental change. Two orders of neurons are primarily involved, the upper and lower efferent neurons.

Afferent Pathways (Figs. 18–50 to 52). One of the major afferent pathways is that illustrated in Figures 18–50, 51. Impulses carried over it enable us to experience consciousness of touch, pressure, position and movement (*proprioception*). For example, we can know the position of parts of the body without using other senses to locate them. We don't have to watch our feet when we walk or kick. We know when our hands are in our pockets without having to look for them. Our sense of touch enables us to identify objects by their shape and texture and to know when and where objects might be touching the body. Our pressure sense enables us to evaluate the weight of things and thereby to control and coordinate our muscle responses accordingly. The first-order neurons in this pathway have their cells of origin in spinal ganglia. Their peripheral processes are associated with neuromuscular and neurotendinous spindles, receptors in joint capsules and ligaments, Pacinian corpuscles, Meissner's corpuscles, and free nerve networks which are described in the next chapter on organs of general and special sense (Figs. 19–1 to 3). The central processes of these neurons enter the spinal cord through the posterior roots where they divide to form reflex connections, short descending and long and short ascending

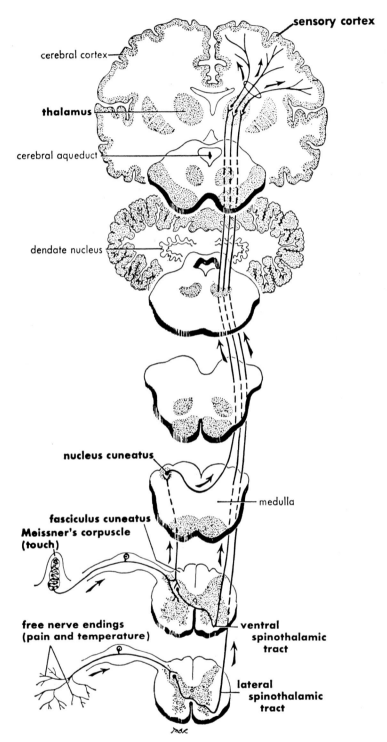

FIG. 18–50. Afferent pathways—sensory mechanisms for pain, temperature, and touch—fasciculus cuneatus, ventral spinothalamic, and lateral spinothalamic tracts. (Schematic.)

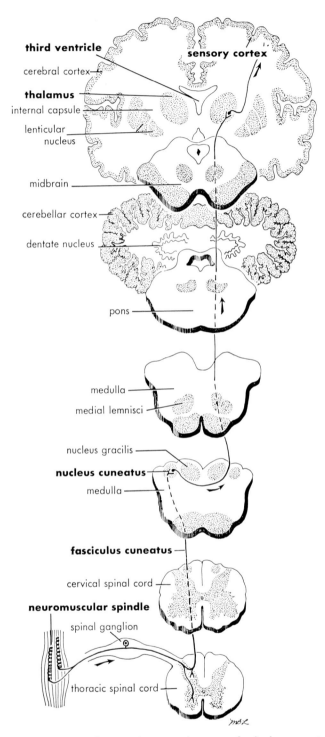

FIG. 18–51. Afferent pathway for conscious muscle sense—fasciculus cuneatus. (Schematic.)

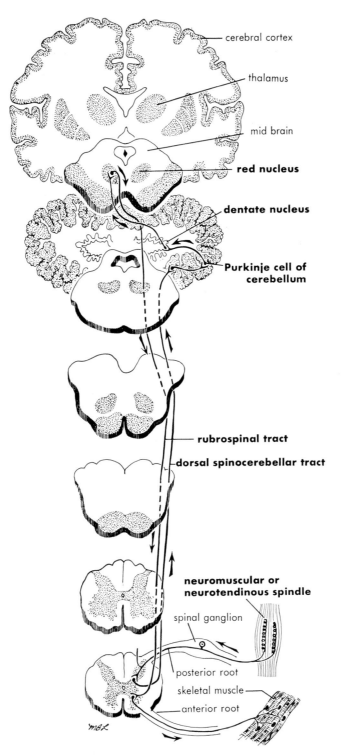

FIG. 18–52.   Afferent and efferent pathways for unconscious muscle movement.
Dorsal spinocerebellar and rubrospinal tracts.   (Schematic.)

branches. The long ascending branches constitute the **fasciculi gracilis** and **cuneatus**—the two major afferent fiber tracts of the spinal cord (Fig. 18–12). These fiber tracts remain on the same side of the spinal cord as their cells of origin and terminate in the **gracilis** and **cuneatus nuclei** of the medulla oblongata. In these nuclei they synapse with second-order neurons whose fibers cross to the opposite side, ascend to the **thalamus** where they synapse with the final or third-order neurons which pass through the internal capsule to reach the cerebral cortex. Hence, stimuli received on one side of the body are represented in the cerebral hemisphere of the opposite side.

A **second afferent pathway** also conveying sensations of pressure, touch, and proprioception is shown in Figure 18–50. The first-order neurons have their cells of origin in the spinal ganglia and send their central fibers into the posterior column of gray matter of the cord. Here they synapse with neurons of the second order whose axons cross to the opposite side of the cord and form the **ventral spinothalamic tract** in the **anterior funiculus** (Figs. 18–12, 50). Some of the first-order neurons send fibers up the cord in the posterior white columns which then give off collaterals which synapse with neurons in the gray matter at a higher segmental level. These in turn cross to the opposite side and join with other fibers of the ventral spinothalamic tract. The fibers of this tract, the axons of the second-order neurons, synapse in the lateral nucleus of the thalamus with the third-order neurons which pass to the cerebral cortex of the same side.

Since the two pathways just described mediate the same functions and because one crosses over in the spinal cord while the other does not, a hemisection of the cord does not result in complete loss of the sensations of touch, pressure, and proprioception on either side of the body.

A **third afferent pathway** (Fig. 18–

50), one for pain and temperature, has its origin in the free nerve endings which are pain receptors and in the end-bulbs of Krause and brushes of Ruffini which are receptors for cold and heat, respectively (Fig. 19–1). The first-order neurons have their cell bodies in the spinal ganglia and their central processes enter the cord and travel upward in the posterior columns as short ascending branches. These branches give off collaterals to the posterior gray columns before their terminal rami also enter the gray columns. Collaterals and terminal rami all synapse with the cells of origin of the **lateral spinothalamic tract** (Figs. 18–12, 50) which are located in the posterior gray columns. The axons of these second-order neurons cross to the opposite side of the cord where they travel upward in the lateral funiculi ultimately to reach the thalamus. The final link in the chain, the third-order neurons, carries the impulses from the thalamus to the cerebral cortex.

A hemisection of the spinal cord would cause loss of pain and temperature sensations on the opposite side of the body below the section since the fibers of the lateral spinothalamic tract are crossed.

The **fourth and final afferent pathway** which we will study is not a truly sensory one since its impulses do not reach consciousness. It arises in the **neuromuscular receptors** and ends in the cerebellum (Fig. 18–52). Its impulses are integrated in the cerebellum and are concerned with coordination of muscles.

The first-order neurons of this pathway have their cell bodies in the spinal ganglia. Their central processes enter the cord and pass to the medial part of the posterior gray column where they synapse with the cells of origin of the second-order neurons. The axons of the second-order neurons pass to the lateral funiculus, mostly on the same side, where they form the **dorsal spinocerebellar tracts** of the cord (Fig. 18–12).

Efferent Pathways (Figs 18–52 to 54).

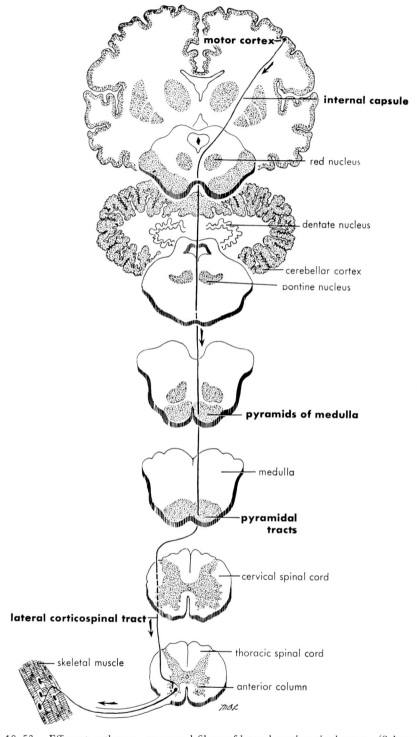

FIG. 18–53.   Efferent pathway—uncrossed fibers of lateral corticospinal tract.   (Schematic.)

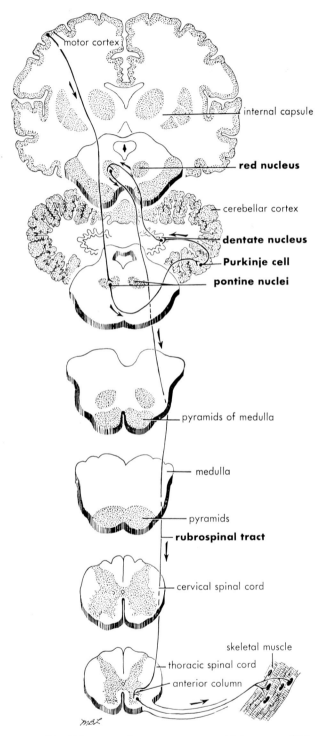

FIG. 18–54.   Efferent pathway—extrapyramidal system. (Schematic.)

The efferent pathways have their cells of origin in the higher centers of the brain and through chains of two or more neurons direct their impulses into the effector muscles. The neurons of these pathways are often called the **upper efferent neuron** or neurons and the **lower efferent neuron.** The cell bodies of the lower efferent neurons are in the anterior gray column. Their axons travel outward through the anterior roots to the spinal nerves. A number of upper efferent neurons from different centers may synapse with one lower efferent neuron; for this reason it is sometimes referred to as the **final common pathway.**

A major efferent pathway involves the two **corticospinal** (*pyramidal*) **tracts** of the spinal cord which have as their cells of origin the large **pyramidal cells** of the motor area of the cerebral cortex (Figs. 18–12, 28, 53). Their axons travel downward through the **corona radiata** and **internal capsule** to leave the cerebrum through the large **cerebral peduncles.** They traverse the midbrain and pons to reach the anterior, vertical, rounded columns of the medulla known as the pyramids (Fig. 18–22C). In the pyramids about four-fifths of the axons cross to the opposite side to the extent that the lower part of the groove between the pyramids is almost obliterated by the crossing fibers. This is often referred to as the **decussation of the pyramids.** These decussating fibers plus a lesser number of the uncrossed fibers make up the **lateral corticospinal tracts** of the lateral funiculi. The remaining uncrossed fibers continue downward to form the **ventral corticospinal tracts** in the ventral funiculus (Fig. 18–12). The fibers of this tract for the most part do cross over, a few at a time, in the anterior white commissure at the various levels of the cord where they terminate. The fibers of the lateral corticospinal tracts supply mostly the lower motor neurons to the muscles of

the extremities; those from the ventral corticospinal tracts supply the muscles of the trunk (Fig. 18–53).

We see in these efferent tracts, as we did in the afferent pathways, that the peripheral structures of one side of the body are represented in the opposite side of the brain—the cerebral cortex in this case. Lesions occurring in the motor cortical areas of the cerebrum or pyramids cause muscular weakness and limitation of precise movements on the opposite side of the body. Lesions of the corticospinal tracts in the cord are more severe since more fibers are involved and paralysis may be more complete. Also lesions in these tracts reduce the inhibitory influence of the higher centers on reflex activity of the postural muscles and such movements become exaggerated. **Spastic paralysis** results.

A second mechanism of efferent pathways, sometimes called the **extrapyramidal system,** influences the anterior horn cells or lower motor neurons, hence the quality of muscular action. Large cell bodies in the cerebral cortex anterior to the motor area and others in the parietal and temporal lobes send axons through the corona radiata, internal capsule, and cerebral peduncles to **pontine nuclei** in the pons. Here they synapse with neurons which pass through the **middle cerebellar peduncles** to the opposite cerebellar cortex where they synapse with the large **Purkinje cells.** The Purkinje cell neurons in turn pass to a nucleus deep in the cerebellum, the **dentate nucleus** where they synapse with a fourth series of neurons whose axons re-enter the brain stem by way of the **superior cerebellar peduncles.** In the brain stem these axons ascend to the midbrain, cross to the opposite side and enter the **red** (*ruber*) **nucleus.** From the red nucleus the **rubrospinal tract** emerges, crosses to the opposite side and descends to the spinal cord to bring its influence to bear upon the anterior

37

column cells of the lower motor neurons (Fig. 18–54).

The anterior gray column cells are acted upon by many other nervous mechanisms too numerous to describe here, but well known to the neurologists.

Enough has been said to show that even the simplest muscular activity requires a great deal of integration for efficient and meaningful performance beyond the initial "commands" given over the corticospinal pathways.

QUESTIONS

1. Compare the roles of the endocrine and nervous systems in the control and integration of the organism.
2. What does the nervous system of man suggest to you concerning his potentialities and his "place in nature"?
3. Diagram a neuron, labeling all of its parts. List some of the variations to be seen among the neurons of the nervous system of man.
4. What might be the functions of the neurilemma and the myelin sheaths?
5. Give an example of a bipolar neuron in the human body; of a unipolar neuron. What type of neuron is most common in man?
6. Name some of the types of supportive cells in the nervous system. What functions do they serve?
7. Describe or diagram and label the structures revealed in a cross section of a nerve.
8. Define a ganglion and name some.
9. What is a nerve tract? A nerve center? A nucleus?
10. Describe a motor end-plate.
11. What is a motor unit?
12. Describe briefly the membrane theory of nerve conduction.
13. Define a synapse. How does a synapse function?
14. What part of the embryo gives rise to the spinal cord?
15. Define filum terminale, cervical enlargement, cauda equina, a spinal segment.
16. Name the parts of the capital H in the central part of the spinal cord. What does each part contain?
17. What constitutes the white matter?
18. Name and describe the meninges of the spinal cord. How do those around the brain differ?
19. Explain from the standpoint of anatomy the reasons for making a spinal puncture in the midlumbar region (between fourth and fifth lumbar).
20. What is the function of the cerebrospinal fluid? Where is it formed and how is it "drained" off?
21. Diagram and label a cross section of the spinal cord with a spinal nerve attached.
22. What structures are supplied by the dorsal rami of the spinal nerves?
23. From what ramus do the appendicular muscles get their nerve supply?
24. What are the plexuses of the spinal nerves and which rami of the spinal nerves are involved?
25. What are the principal nerves to derive from the brachial plexus? The sacral plexus?
26. Diagram, label, and describe a somatic reflex arc and reflex action. Why is this important?
27. What is a subliminal stimulus? An optimal stimulus?
28. What do we mean by refractory period, summation, facilitation?
29. Name the three primary vesicles (divisions) of the brain and the parts which derive from them.
30. Why do we sometimes call the brain stem the "old" brain?
31. Describe the diencephalon.

32. Define the following: Pons, cerebral aqueduct, superior colliculus, cerebral peduncle, red nucleus.
33. What is the reticular formation and why is it important?
34. Is there something in the name of a fiber tract of the brain or cord which tells its destination—such as corticospinal, corticobulbar, rubrospinal, and corticopontine?
35. What are some of the functions assigned to the hypothalamus?
36. How do the Purkinje cells function in relationship to the other cell components of the cerebellum?
37. What is the arbor vitae? The vermis?
38. Name the lobes of a cerebral hemisphere and the fissures or sulci which mark them off.
39. Where are the following—motor area of cortex, area of general sensation, olfactory, visual, and hearing areas?
40. Describe the association and commissural fiber tracts of the cerebrum. Why are they important?
41. What is the importance of the projection system of the cerebrum and what are its two main subdivisions?
42. Describe the functions of the thalamus relative to the projection system of the cerebrum.
43. What are the basal ganglia? The corpus striatum? The internal capsule?
44. Compare unconditioned and conditioned reflexes. Why are conditioned reflexes so important? Give examples.
45. Describe the lateral ventricles of the cerebrum.
46. Where are the foramina of Luschka and Magendie?
47. What are Pacchionian bodies?
48. Describe the tentorium cerebelli.
49. Compare cranial and spinal nerves as to origin, composition, ganglia and nuclei.
50. What is "peculiar" about the olfactory nerves? Are they of advanced or primitive structure?
51. Describe the pathways taken by various fibers of the optic tracts.
52. What happens to vision if one optic tract is destroyed?
53. What do the oculomotor, trochlear, and abducens have in common?
54. What is the largest of the cranial nerves? Is it primarily sensory or motor? What are its three big divisions and its large ganglion?
55. Describe the acoustic nerve in considerable detail.
56. What is the chief motor function of the facial nerve?
57. Describe some of the unique features of the vagus nerve.
58. Describe the (spinal) accessory nerve and why is it so called?
59. Outline the parts of the autonomic nervous system.
60. Make a diagram of a cross section of the thoracic spinal cord with a spinal nerve attached and show how the sympathetic division of the autonomic system ties into this "picture."
61. Comment on the visceral afferent neurons.
62. Compare the sympathetic and parasympathetic systems as to their functions.
63. With what cranial and sacral nerves is the parasympathetic division involved?
64. Describe the sympathetic trunk.
65. Name the collateral ganglia.
66. What are terminal ganglia?

"What can give us more sure knowledge than our senses? How else can we distinguish between true and false?"

—Lucretius

Chapter 19

The Organs of General and Special Sense (Receptors)

POINT OF VIEW

ADJUSTMENT to environment is essential to survival. A critical sensitivity to the environment is prerequisite to adjustment. The receptor "organs" provide this sensitivity and each functions to **generate** nervous impulses in response to selected stimuli. The sensations which arise are the functions of the brain, and the responses that the animal makes are the result of the activity of the nervous and endocrine systems. The sense "organs" are the "scouts" for these integrating systems providing them with vital information.

Receptors or sense "organs" are composed of specialized sensory cells which are highly sensitive to some particular stimulus, the **adequate stimulus,** but are much less sensitive to other stimuli. The adequate stimulus of the retina of the eye is light waves; of the spiral organ of Corti, sound waves; of the taste buds, dissolved substances. The sense "organ" may be only a naked free nerve ending as in those for pain, or it may be as complicated as the eye or ear. The receptors may have accessory structures which protect the more delicate sense cells; which intensify the force of the stimulus, as in the ear; concentrate or bring the stimulus to focus upon the sensory cells, as in the eyes; or transform the character of the stimulus to the end that it may act more effectively upon the sense organ proper, as in the ear.

It should be emphasized that nerve impulses do not necessarily occur in receptor or sense cells. In the photoreceptor cells, the rods and cones, of the vertebrate eye, for example, no one has ever demonstrated a nerve impulse. Yet, when stimulated by light they set up the physicochemical conditions which initiate impulses in the nerve cells associated with them.

As sensitive as our receptors are they operate within definite limitations. These limitations differ among the vertebrate animals and also vary to some extent among individual men. Our eyes are not sensitive to the long infrared or the short ultraviolet rays of light; the human ear detects frequencies between 16 cycles and 20,000 cycles per second, but is insensitive to the high frequency sounds that bats or dogs can detect; our chemical senses, olfaction and taste, are similarly limited. We are, in terms of our sense organs alone, insensitive to vast worlds of experience. As human animals we can compensate for some of this by the development of instruments which can detect and record data from these worlds which to us are otherwise "extrasensory."

We should appreciate also that these limitations are not necessarily unfortunate. Many are indeed a physical necessity. The fact that our ears are least sensitive at low frequencies protects us from hearing our own body sounds. Stick a finger in each ear and by thus closing out air-borne sounds, you will hear the sounds produced by the contracting muscles of your arms and fingers. To have ears more sensitive in the lower frequencies would mean that these and other body sounds would always be present to annoy us.

CLASSIFICATION OF RECEPTORS

Treating the sense "organs" as a system raises the same problems that we experienced with the endocrine "system." They are widely distributed over the body; they are diverse in structure, and there is no physical continuity among them. Yet they do serve a common function in the collecting of data which the nervous and endocrine systems require

in order to perform their orienting and integrating roles. It is more logical perhaps to think of them as parts of the nervous system since with it they are structurally continuous.

Similar problems face us in attempting to classify the receptors. It is common to speak, as the title of this chapter suggests, of the general and the special sense organs. Those are called **general** which have a wide distribution throughout the body such as the receptors for heat, cold, pain, touch and muscle, joint or tendon sense (*proprioception*). The **special** sense organs are found in the head region only and are more advanced and specialized. They are the eye, ear, taste "buds" and olfactory organs, and by many authors are treated separately from the general sense structures.

Perhaps the best way to classify receptors is on the basis of the part of the environment they sample or sense, as follows:

1. **Exteroceptors**
   a. receive stimuli from external environment
   b. located in skin and its apertures
   c. include organs of special senses except taste; also those for pain, temperature, touch and pressure.
2. **Interoceptors**
   a. receive stimuli from "internal environment"
   b. lie within the walls of the digestive tube or its derivations and in walls of other internal organs
   c. includes sense organs of taste, pain, pressure, etc.
3. **Proprioceptors**
   a. receive stimuli from the true internal environment
   b. lie within the muscles and tendons and around joints.

Classified on the basis of location, receptors fall into four groups: 1. **General somatic sensory**—the widely distributed sense organs of skin and skeletal muscles; 2. **special somatic sensory**—those of

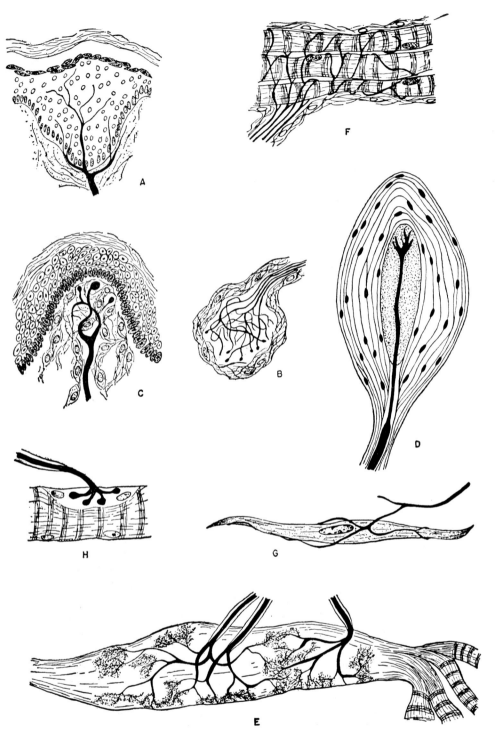

FIG. 19–1. Showing some varieties of peripheral terminations of afferent and efferent nerve fibers. (A) "Free termination" in epithelium (after Retzius). (B) Krause's corpuscle from conjunctiva (after Dogiel). (C) Meissner's corpuscle from skin (after Dogiel). (D) Pacinian corpuscle (after Ruffini). (E) Termination upon tendon sheath (Huber and DeWitt). (F) Neuromuscular spindle (after Dogiel). (G) Motor termination upon smooth muscle fiber. (H) Motor "end-plate" on skeletal muscle fiber (after Böhm and von Davidoff). (Edwards, *Concise Anatomy*, courtesy of McGraw-Hill Book Co., used by permission.)

limited distribution at body surface—the eyes, ears, olfactory organs; 3. **general visceral sensory**—widely distributed in organs of the digestive tract and other viscera; 4. **special visceral sensory**—sense organs of limited distribution in digestive tract, etc.—as the taste buds.

ORGANS OF GENERAL SENSE
(Figs. 19–1, 2, 3)

The structure of these receptors can best be learned and understood by reference to the photomicrographs and diagrams. They range from simple free nerve endings to organs of quite complex design.

Free Nerve Endings (Fig. 19–1A). These consist of dendrites which have lost their myelin and neurilemma and whose fibrillae anastomose and end in knobs or discs between epithelial cells. All the **pain** receptors and some of those for touch, temperature, and muscle sense are of this type.

End-Bulbs of Krause (Fig. 19–1B). They occur quite widely over the body and vary in shape from cylindrical to oval. They consist of a capsule from the connec-

tive tissue sheath of a myelinated nerve fiber. Inside they contain a soft material in which the nerve fiber ends in either a bulb or coiled mass. It has been suggested that they are receptors for **cold.**

Brushes of Ruffini. These were described by Ruffini from the subcutaneous tissue of the human finger and are considered to be **heat** receptors. They are oval in shape and have a tough connective tissue sheath. Inside the nerve fibers branch and end in small free knobs.

Tactile Corpuscles of Meissner (Fig. 19–1C). These are receptors for **light touch** and occur in the papillae of the corium in many areas such as the hands, feet, lips, mucous membranes of tongue, skin of mammary papillae, and the front of the forearm. They are small and oval with a connective tissue sheath and what appear to be tiny plates placed one above the other. The nerve fiber penetrates the capsule and forms a spiral arrangement ending in globular structures among the plates.

Pacinian (lamellated) Corpuscles (Figs. 19–1D, 2). These are large receptors visible to the naked eye, from 2 to 4 mm. in diameter. Each corpuscle is at the end

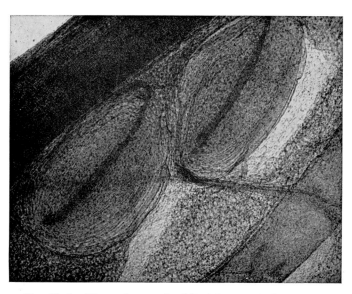

FIG. 19–2. Photomicrograph of Pacinian corpuscles, 30×.

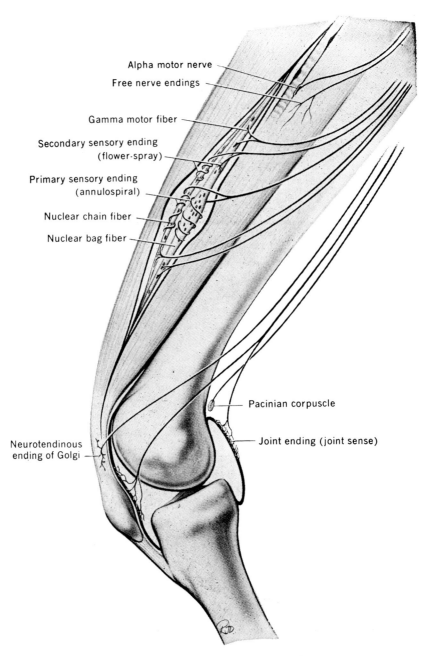

Alpha motor nerve
Free nerve endings
Gamma motor fiber
Secondary sensory ending
(flower-spray)
Primary sensory ending
(annulospiral)
Nuclear chain fiber
Nuclear bag fiber
Pacinian corpuscle
Joint ending (joint sense)
Neurotendinous
ending of Golgi

FIG. 19–3. Nerve endings in voluntary muscles, tendons, and joints. The neuromuscular spindle is disproportionately enlarged and shows only two intrafusal fibers; one nuclear bag fiber and one nuclear chain fiber. (Noback, *The Human Nervous System*, courtesy of McGraw-Hill Book Co.)

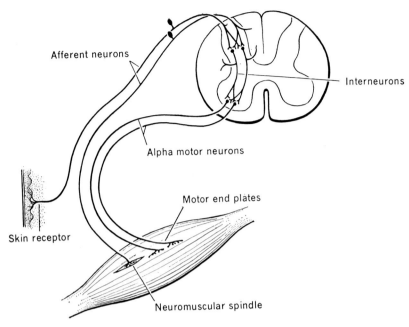

FIG. 19–4.   The flexor reflex loop.   This three-neuron, disynaptic, ipsilateral, intersegmental reflex is comprised of a sequence of sensory receptor in the skin, afferent neuron, spinal intersegmental neurons, alpha motor neurons, and voluntary muscles.   This reflex can be facilitated by the sequence of the "secondary sensory ending" (flower spray ending) of neuromuscular spindle, afferent neuron, spinal interneurons, alpha motor neurons, and volunary muscles.   (Noback, *The Human Nervous System*, courtesy of McGraw-Hill Book Co.)

of a nerve fiber and its structure consists of a number of concentric lamellae reminding one of a section of an onion. The nerve fiber passes into the central part of the corpuscle, losing first its myelin sheath and then the neurilemma as it passes among the specialized layers. The Pacinian corpuscles are **deep pressure** receptors found in the subcutaneous, submucous and subserous connective tissue, being especially common in the palm of the hand, sole of the foot, genital organs, around the joints and in the mesentery around the pancreas.

Neuromuscular Spindles (Figs. 19–1*F*, 3,4). These receptors occur in almost all skeletal muscles and are more numerous in the muscles of the limbs than in the trunk region. Those of the higher vertebrates are complicated structures and the description which follows applies mostly to mammals. In the muscles of man the spindles vary greatly in length from 0.05 to 13 mm. with the more common lengths ranging from 2 to 4 mm. They are arranged parallel to the extrafusal or ordinary muscle fibers.

Muscles spindles are enclosed in lamellar capsules which contain also a viscous fluid. The muscle fibers of the spindle (*intrafusal*) are of two types; those confined to the capsule which are small in diameter, the **intracapsular fibers,** and those of larger diameter which extend beyond the capsule, the **percapsular fibers.** The regular muscle fibers (*extrafusal or extracapsular*) are larger in diameter than any of those of the muscle spindle.

Near the middle of each of the muscle fibers in the spindle one finds an area lacking in striations and in which numerous nuclei accumulate. In the larger percapsular fibers these nuclei clump together forming a nuclear bag, hence the name **nuclear bag fibers.** In the smaller intracapsular fibers the nuclei line up along the fiber and hence are called **nuclear chain fibers.** These areas, and to some extent the adjacent striated areas, are provided with sensory nerve fibers some of which form terminal branches which spiral around the nuclear bags and nuclear chains, others form flower-like or ivy-like branches. The former are called **primary or annulospiral endings,** the latter are **secondary endings** or "**flower sprays**" **of Ruffini.** The secondary endings may be placed poleward from the primary endings and are more common on the smaller intrafusal muscle fibers.

Neuromuscular spindles are also provided with motor nerves of small diameter, the **gamma fibers,** which come from small **gamma motor neurons** in the central nervous system. These are in contrast to the **alpha motor neurons** with fibers of larger diameter, which supply the muscle fibers outside of the spindle. The gamma fibers to the spindles are of two sizes. According to some workers the larger nuclear bag fibers are innervated by gamma efferents which terminate in small motor end-plates along the striated poles of the fiber, while the nuclear chain fibers are innervated by smaller gamma efferents ending in more delicate, narrow, and elongate structures linked by very fine axon branches.

Neuromuscular spindles were for a long time considered to be receptors for muscle sense or **proprioception** (*kinesthesis*) by which one was made aware of the position of the body in space and the relative position of its parts without visual help. Recent investigations have shown this concept to be false and that the sense of position and movement of joints is dependent upon receptors associated with the joints themselves (Fig. 19-3)

Neuromuscular spindles give feedback on data indicating the status of a muscle which is involved in the coordination and efficiency of muscle action. The spindles' sensitivity is under central system influence through its gamma motorneurons.

The simple **knee-jerk reflex** will serve as an example of how a neuromuscular spindle works. This reflex is initiated by tapping the tendon of the Quadriceps femoris muscle. The tap stretches the muscle and the nuclear bag region of the spindles within it. This excites the annulospiral afferent nerve endings which discharge through the afferent neuron to the dorsal root of the spinal nerve and to a synapse with an alpha motor neuron in the anterior gray column of the spinal cord. The alpha motor neuron discharges impulses into the Quadriceps muscle causing it to contract; the knee-jerk. During this contraction the spindles are shortened. This two-neuron reflex arc is known as an alpha reflex loop.

A related example will indicate the role of the gamma motor fibers of the neuromuscular spindle. After the muscle contracts in the preceding example the tension on the spindle fibers decreases resulting in decreased firing of the alpha motor neurons and the inability of the muscle to maintain itself in the continuous stretch necessary to maintain stance. To maintain this stretch a **gamma reflex loop** is involved. It consists of a gamma motor neuron to a neuromuscular spindle, an afferent neuron, an alpha motor neuron and a skeletal muscle. The system operates by stimulation of the gamma motor neuron by an upper motor neuron from the brain. The impulses over the gamma motor neuron cause the intrafusal fibers of the spindle to contract and to stretch their bag region sufficiently to maintain a constant firing of the afferent neurons to the alpha motor neurons which maintain the contraction of the extensor muscles. The central nervous system acts in this situation largely through the gamma loop (Fig. 19–3). Other influences on it are from the receptors in the skin and joints.

Muscle spindles may also play a role in flexor reflexes. A protective three-neuron reflex such as occurs when one pricks or burns a finger may be facilitated by a three-neuron reflex arc consisting of the flower-spray endings of a neuromuscular spindle, their afferent neuron, spinal association or interneuron and alpha motor fibers to the skeletal, voluntary flexor muscle (Fig. 19–4).

Neurotendinous Spindles (organs of Golgi) (Fig. 19–3). These are receptor structures in a tendon at the junction of tendon and muscle or sometimes within a muscle sheath. They consist of branching fibers with leaf-like expansions at their ends. They have no motor control and serve to provide information relative to the static tension and tension changes in the muscle itself.

General Visceral Receptors. Some of the receptors just described, the Pacinian corpuscles, for example, are found also in the viscera. Many visceral receptors consist of free nerve endings. The impulses from visceral receptors do not ordinarily reach the level of consciousness; some, of course, do. Impulses from visceral receptors in the thorax and abdomen travel over sympathetic nerves and rami communicantes to the spinal ganglia. Others by way of the vagus and glossopharyngeal nerves reach their sensory ganglia, while still others travel in the pelvic nerves to spinal ganglia in the sacral area. These relationships will become clearer if the autonomic nervous system is reviewed.

Special Visceral Receptors (Fig. 18–45). These are the receptors for reflex control of respiration and circulation. The nerve fibers are carried in vagus and glossopharyngeal nerves. The carotid and aortic bodies and the carotid sinus mentioned under the circulatory system are among the special visceral receptors. The **carotid body** and **aortic body** are influenced by the concentration of carbon dioxide in the blood; the **carotid sinus,** an enlargement at the beginning of the internal carotid artery, is sensitive to changes in blood pressure and its receptors initiate reflex controls over circulation (Fig. 18–44).

ORGANS OF SPECIAL SENSE

CHEMORECEPTORS

The receptor organs of **taste** and **smell** are called **chemoreceptors** because they are normally stimulated by chemicals. To be effective the chemicals must be in solution and the mouth and nasal cavities are provided with glands to insure this. These receptors have very low thresholds, *i.e.*, they are responsive to chemicals in very great dilution. Musk in dilutions of one part in eight million will activate the olfactory receptors, and one part of quinine in two million parts of water can be tasted.

Taste and smell are often confused. This is partly due to the fact that the mouth and nose communicate freely through the pharynx and also because olfactory and taste receptors are frequently stimulated by the same substances at the same time. Many sensations which we call taste are really odors as is apparent when we have a head cold and say our food is tasteless. It is only because our olfactory receptors are clogged with mucus.

Olfactory Organs (Figs. 14–2, 18–4, 19–5). The **olfactory organs** are located in the uppermost part of the nasal fossae in the mucous membrane covering the superior nasal conchae and the adjacent septum. Branched tubular glands in the subepithelial tissues keep the surface of the mucous membrane protected with moisture.

The **olfactory cells** are the least specialized of the organs of special sense.

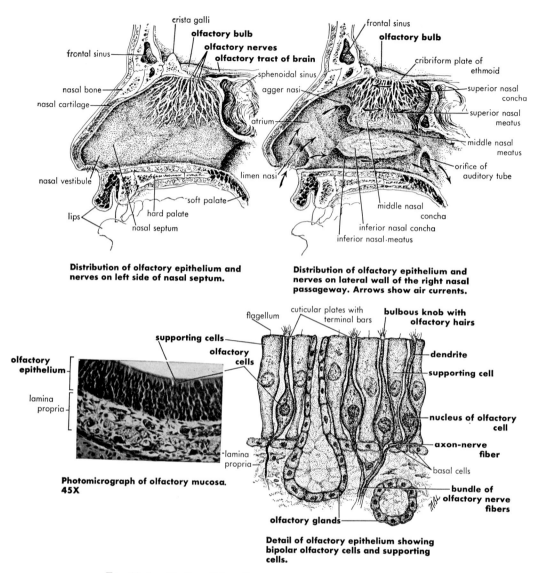

Distribution of olfactory epithelium and
nerves on left side of nasal septum.

Distribution of olfactory epithelium and
nerves on lateral wall of the right nasal
passageway. Arrows show air currents.

Photomicrograph of olfactory mucosa.
45X

Detail of olfactory epithelium showing
bipolar olfactory cells and supporting
cells.

FIG. 19-5. Studies of the olfactory organ and its relationships.

They are modified epithelial cells scattered among the columnar epithelium of the mucous membrane. The columnar cells constitute the **supporting** cells of the olfactory organ. The olfactory cells are **bipolar** with a small amount of cytoplasm and a large nucleus. A slender peripheral process extends from each olfactory cell to the free surface of the mucous membrane and sends out beyond the surface a tuft of very fine processes, the **olfactory hairs.** The central process

passes through the basement membrane and joins with other central processes to form bundles of unmyelinated fibers of **olfactory nerves.** These bundles form a plexus in the submucosa from which about twenty olfactory nerves emerge, pass through the foramina in the cribriform plate and end by synapsing with the **mitral cells** in the glomeruli of the **olfactory bulb** of the brain.

Taste Organs (Fig. 19-6). The organs of taste are the **taste buds** distributed

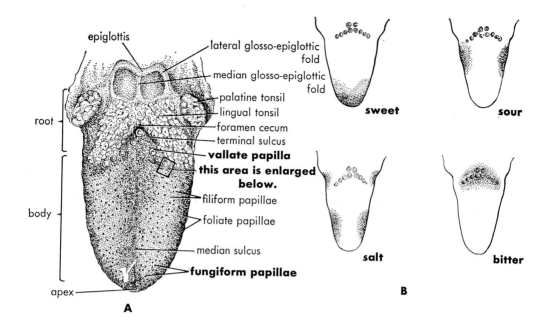

epiglottis

lateral glosso-epiglottic fold

median glosso-epiglottic fold

palatine tonsil

lingual tonsil

foramen cecum

terminal sulcus

vallate papilla

this area is enlarged below.

filiform papillae

foliate papillae

median sulcus

fungiform papillae

root

body

apex

A

sweet

sour

salt

bitter

B

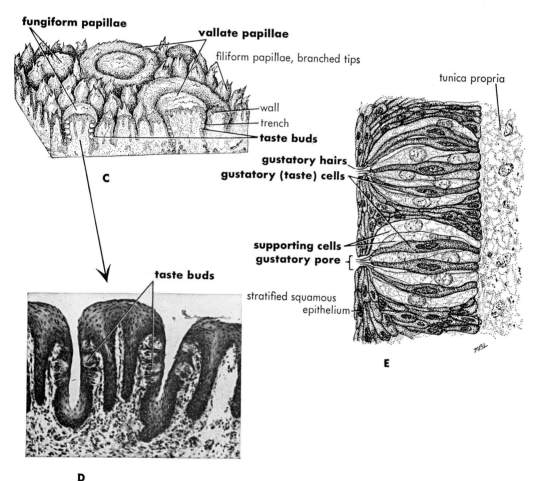

fungiform papillae

vallate papillae

filiform papillae, branched tips

tunica propria

wall

trench

taste buds

gustatory hairs

gustatory (taste) cells

supporting cells

gustatory pore

stratified squamous epithelium

C

taste buds

D

E

Fig. 19–6. Studies of the gustatory (taste) organs showing their location and structure. *A,* The tongue and its papillae—superior surface. *B,* Distribution of the four qualities of taste on the tongue surface. *C,* Schematic section of tongue surface showing vallate and fungiform papillae and taste buds. About 8× (Braus). *D,* Photomicrograph of fungiform papillae and taste buds of rabbit. About 130×. *E,* Detailed drawing of gustatory (taste) buds. Highly magnified.

mainly over the tongue, but present also on the roof of the mouth, in the pharynx and on the epiglottis. They are embedded in the stratified squamous epithelium and are present in large numbers on the vallate papillae, over the sides and back of the tongue and especially on the fungiform papillae.

The taste buds are spherical or ovoid groups of cells occupying pockets which extend through the tongue epithelium and open on the free surface (**gustatory pore**) and through the basement membrane. The bud cells are of two kinds, **gustatory** and **supporting**. The supporting cells form the walls of the pocket and are also scattered between the gustatory cells of the bud. The gustatory cells occupy the middle of the bud, and are spindle-shaped and have a large nucleus. Each gustatory cell has at its peripheral end a delicate hair-like process, the **gustatory hair,** which pushes through the gustatory pore on the surface of the tongue. The central end of the gustatory

cell remains within the taste bud where it comes into intimate contact with many fine terminations of nerves which enter the taste bud through its opening in the basement membrane. These are myelinated nerves but lose their sheaths as they enter the taste bud.

The taste buds on the posterior third of the tongue are supplied by afferent neurons of the glossopharyngeal or ninth cranial nerve, those on the anterior two-thirds of the tongue by the afferent neurons of the facial or seventh cranial nerve. The axons of these nerves enter the brain stem and by internuncial nervous impulses reach the sensory area in the hippocampal gyrus in the temporal lobe.

Only four qualities of taste are recognized: bitter, sour, salty, and sweet. Supposedly, these are associated with four different kinds of taste buds, and there is supporting evidence for this in the distribution of taste sensations over the tongue surface. The many tastes

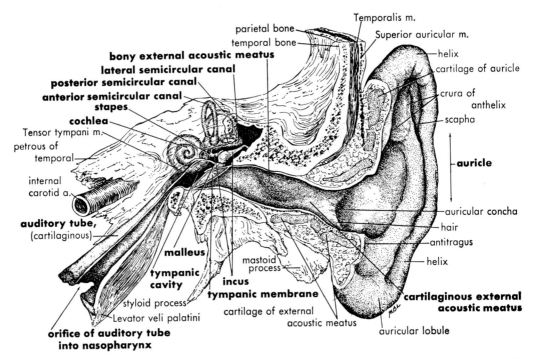

Fig. 19–7. General view of left ear and auditory tube. The external and middle ear are opened; the inner ear is shown as if the surrounding bone were transparent.

which we experience are the result of the blending of the four fundamental qualities, the addition of other sensations by the tongue, and the confusion of smell and taste.

THE EAR (*Vestibulocochlear organ*)

The ear is often considered the most complex of the organs of special sense. It serves two functions, **equilibrium** and **hearing.** And strange as it may seem, hearing developed as a function of the ear in relatively recent geological time, as a secondary function of the ear. The "ears" of fish were not for hearing, but for maintaining equilibrium. The ear is divisible into three parts—an external, middle, and internal ear (Fig. 19–7).

External Ear. The external ear consists of an expanded portion projecting from the side of the head, the **auricle** (*pinna*), and a tube, the **external acoustic meatus,** leading inward to the **tympanic membrane** (*ear drum*). The auricle is directed mostly upward and backward and is of irregular shape. It is supported by an elastic cartilage and covered with thin skin. Its function is to direct sound waves into the external acoustic meatus, but it is relatively ineffective as indicated by man's tendency to cup his ears when hearing gets a little difficult.

The external acoustic meatus is an S-shaped canal about 2.5 cm. long leading from the "funnel" of the auricle inward to the tympanic membrane. Its outer part is supported by cartilage continuous with that of the auricle; its inner portion, about 1.6 cm., is supported by bone. It is lined with skin which has many fine hairs and sebaceous glands near its orifice. In its upper wall are modified sweat glands, the **ceruminous**

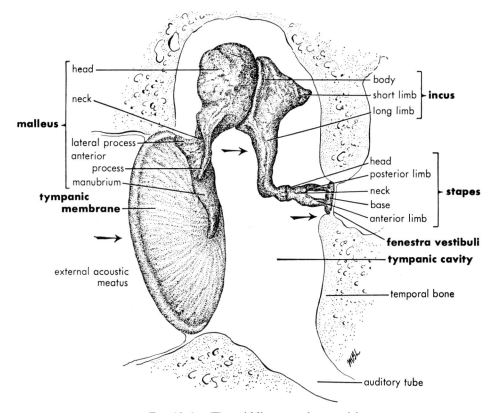

Fig. 19–8. The middle ear and ear ossicles.

glands which secrete cerumen or ear wax. The cerumen and hairs help to guard the ear against the entrance of foreign materials such as insects.

Middle Ear (Figs. 19–7, 8). The **middle ear** or **tympanic cavity** is a laterally compressed space in the petrous portion of the temporal bone. It is separated from the external acoustic meatus by the obliquely positioned **tympanic membrane**, and from the internal ear by a bony wall in which are found two small windows covered by membrane, the oval window or **fenestra vestibuli** above and the round window or **fenestra cochleae** below. The posterior wall of

the tympanic cavity has a large opening which leads into a large air space, the **mastoid antrum**, which in turn communicates with the **mastoid air cells**. Anteriorly the **auditory tube** (*Eustachian tube*) connects the tympanic cavity with the nasopharynx, which enables one to keep the air pressure within the middle ear equalized with atmospheric pressure. Swallowing facilitates this and for this reason the commercial air lines prior to the time of pressurized cabins, provided their passengers with mints or chewing gum at the time of take-off or landing to promote salivation and swallowing.

Three small bones or **ossicles** bridge

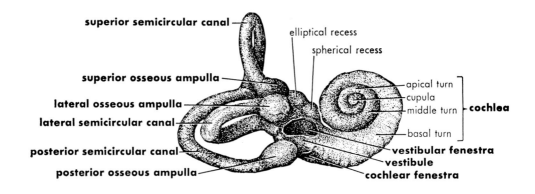

Antero-lateral view of right bony labyrinth

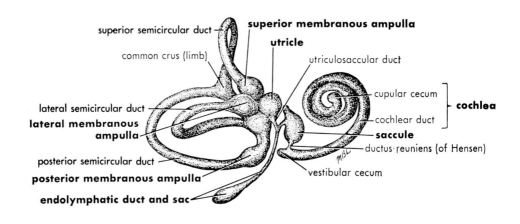

Antero-lateral view of right membranous labyrinth

FIG. 19–9.   Right bony and membranous labyrinths.

the tympanic cavity from the tympanic membrane to the fenestra vestibuli (*oval window*). The **malleus** (*hammer*) attaches to the upper part of the tympanic membrane; the **stapes** (*stirrup*) base fits into the fenestra vestibuli; and the **incus** (*anvil*) lies between and has synovial articulations with the other two ossicles.

Mucous membrane continuous with that of the nasal passageways and pharynx lines the auditory tube, the tympanic cavity, mastoid antrum, and mastoid cells. It forms the medial layer of the tympanic membrane and the lateral layer of the membrane of the fenestra cochleae, called the **secondary tympanic membrane,** and it covers the ear ossicles, muscles, and nerves of the cavity.

The continuity of the mucous membranes from nasal passageways and pharynx is a natural pathway for the spread of infections of nose and throat to middle ear and mastoid cells. Such infections are always troublesome, often resulting in mastoiditis, and may impair hearing temporarily or permanently unless prop-

erly cared for. These are some of the reasons why even the common cold should be taken seriously.

Internal Ear (Figs. 19–9 to 12). The internal ear is the organ of hearing and equilibrium. It consists of an **osseous** and a **membranous labyrinth.** The osseous labyrinth is hollowed out of the substance of the petrous portion of the temporal bone and consists of a number of chambers and canals. It is lined with a thin fibro-serous membrane which secretes a fluid called **perilymph.** The parts of the osseous labyrinth are the **vestibule,** the **semicircular canals** and the **cochlea.**

The **vestibule** is a chamber just medial to the tympanic cavity, separated from it by a thin partition of bone in which is found the **fenestra vestibuli,** which in the living state is closed by the base of the stapes and its annular ligament. The vestibule contains a number of fossae in which are found small openings or foramina for the sensory nerve fibers which have their origins in the inter-

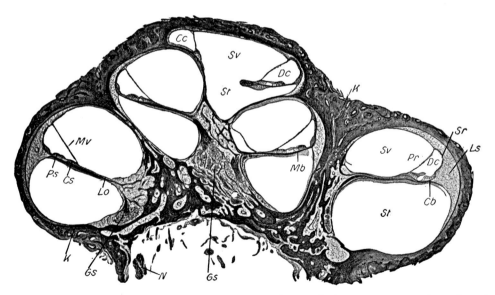

Fig. 19–10. Axial section of cochlea of a man. *Cb,* crista basilaris; *Cc,* cecum cupulare; *Cs,* crista spiralis; *Dc,* ductus cochlearis; *Gs,* ganglion spirale; *K,* bony wall of the cochlea; *Lo,* lamina spiralis ossea; *Ls,* ligamentum spirale; *Mb,* membrana basilaris; *Mv,* membrana vestibularis; *N,* cochlear nerve; *Pr,* prominentia spiralis; *Ps,* organ of Corti; *Sr,* stria vascularis; *St,* scala tympani; *Sv,* scala vestibuli. ×16. (After Schaffer.) (Bloom and Fawcett's *A Textbook of Histology,* courtesy of W. B. Saunders Co.)

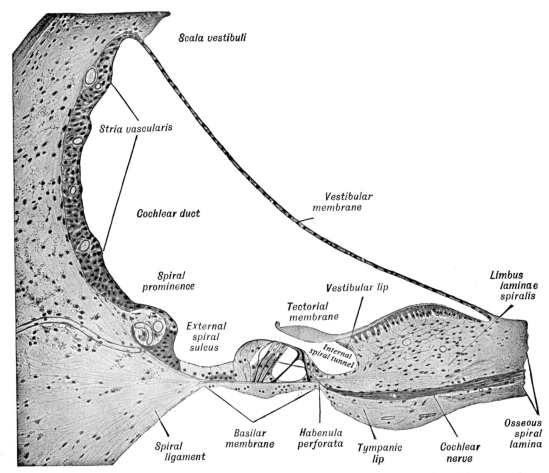

FIG. 19–11.   Radial transection of cochlear duct in the first coil of human cochlea.   The separation of the free end of the tectorial membrane from the organ of Corti is due to improper fixation (after Held). (Bloom and Fawcett's *A Textbook of Histology*, courtesy of W. B. Saunders Co.)

nal ear. Posteriorly, there are five orifices which communicate with the semicircular canals; anteriorly, there is one orifice for communication with the cochlea.

The three bony **semicircular canals** lie posterior to the vestibule and communicate with it.   One canal lies in each of the three planes of space and they are called **superior, posterior,** and **lateral.**   Each one is provided with a swelling, the ampulla, at the end of one of its arms.

The bony **cochlea** is shaped like the shell of a snail consisting of two and three quarter spiral coils.   The basal coil of the cochlea is broad and it tapers as it

spirals to a narrow apex. It has a conical-shaped central axis called the **modiolus,** from which projects into the canal an **osseous spiral lamina.**   This lamina, following the windings of the canal, partially divides the canal into two parts. In the living subject this partial division is completed by a **basilar membrane.** The two parts of the canal are now separate except at the apex of the cochlea where they are continuous.   The upper part of the divided canal is called the **scala vestibuli;** the lower part the **scala tympani.**   The scala vestibuli enters the vestibule at its basal end; the scala tympani ends at the **fenestra cochlea**

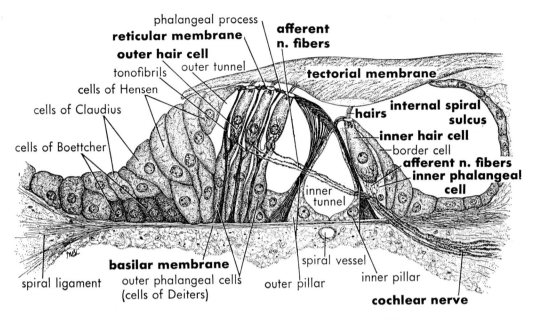

phalangeal process
reticular membrane
outer hair cell
tonofibrils   outer tunnel
cells of Hensen
cells of Claudius
cells of Boettcher
spiral ligament
basilar membrane
outer phalangeal cells
(cells of Deiters)
**afferent
n. fibers**
tectorial membrane
hairs
**internal spiral
sulcus**
inner hair cell
border cell
**afferent n. fibers
inner phalangeal
cell**
inner
tunnel
spiral vessel
outer pillar
inner pillar
cochlear nerve

FIG. 19–12. Radial transection of the spiral organ of Corti.

which in the living state is closed by the secondary tympanic membrane. It also communicates with the subarachnoid space by the cochlear aqueduct.

The **membranous labyrinth** lies within the osseous labyrinth and has much the same form except that the vestibular portion is divided into two chambers, the **utricle** and **saccule.** It is smaller, however, than the osseous labyrinth and the space between it and the bony walls is filled with **perilymph.** Inside the membranous labyrinth is a fluid, the **endolymph,** and in its walls the fibers of the vestibulocochlear nerve are distributed.

The **utricle** is larger than the saccule and is in the upper back part of the vestibule. It communicates with the five openings of the **semicircular ducts** and has in thickened areas in its walls the utricular filaments of the vestibular nerve (*macula*). It communicates with the endolymphatic duct (Fig. 18–43*D*).

The **saccule** also has a thickened area in its wall provided with saccular filaments of the vestibular nerve (*macula*). It communicates with the cochlear duct

through a small canal from its lower side and with the endolymphatic duct from its posterior wall. The **endolymphatic duct** is joined by a small duct from the utricle after which it continues along the vestibular aqueduct ending in a blind pouch on the posterior surface of the petrous temporal bone where it lies against the dura mater (Fig. 19–9).

Suspended in the endolymph in contact with the free hairs of the sensory cells of utricle and saccule are two small bodies made up of grains of carbonate of lime enmeshed in a gelatinous substance. These are called **statoconia** (*otolith*) (Fig. 18–43).

The **semicircular ducts** follow in pattern the bony semicircular canals but are only about one-fourth their diameter. Each is provided at one end with an **ampulla** which contains an elevated transverse septum in which the sensory cells are located. They send their hair-like filaments into the ampullar cavity (*crista*) (Fig. 18–43*D*). The filaments of the vestibular nerve, having lost their myelin sheaths, enter and penetrate among the hair cells.

The **cochlear duct** lies between the scala vestibuli above and the scala tympani below. It rests on the **basilar membrane** and is limited above by the thin **vestibular membrane.** It is filled with endolymph and contains the spiral organ of Corti, the essential receptor organ of hearing (Figs. 19–10, 11, 12).

The **spiral organ of Corti** consists of a series of epithelial structures placed upon the inner surface of the basilar membrane. The hair cells which are believed to be the receptors are arranged in rows. Medially, there is a single row of about 3500 **inner hair cells** extending the length of the coiled cochlea, and lateral to these about 12,000 **outer hair cells** arranged in three rows in the basal coil of the cochlea, and in four rows in the apical coil. These cells have long hair-like processes at their free ends and large basal nuclei. The terminal filaments of the cochlear branch of the vestibulocochlear nerve are in contact with their deep ends. The outer hair cells are about twice as long as the inner ones. Some columnar cells serve to support the inner hair cells and special cells of Deiters support the outer hair cells. Also, a reticular framework supported by specialized rods "fits over" the free ends of the outer hair cells to give them additional support.

The spiral organ is covered by a very delicate and flexible membrane, the **tectorial membrane.** It attaches medially and extends roof-like over the hair cells, contacting their hair-like processes.

The vestibulocochlear nerve, or the eighth cranial, forms two branches, the **vestibular** and **cochlear.** The vestibular branch supplies the ampullae of the semicircular ducts, the utricle and saccule. Within the internal acoustic meatus, on the trunk of the nerve is the **vestibular ganglion** where the cells of origin of the fibers of this nerve are located. The distal fibers of the ganglion split into three branches to serve the vestibular structures.

The **spiral ganglion** of the cochlear nerve occupies the spiral canal of the modiolus. It is made up of **bipolar** nerve cells, the cells of origin of this nerve.

Equilibrium. The utricle, saccule, and the ampullae of the semicircular ducts, as we have seen, house sensory nerve endings from the vestibular branches of the vestibulocochlear cranial nerve. The saccule and utricle both have sensitive areas in their walls, the **maculae,** and attached to the hairs of the sensory cells are the **statoconia** (Fig. 18–43). Any disturbance by an altered position of the head causes the statoconia to exert a pull on the hair cells and stimulation results. The information received in this way from all four of the maculae, together with supplemental information from the eyes and proprioceptive organs in the neck, enables us to maintain the normal head position. This does not mean that we need to be conscious of these changes. They will take place in mammals in which all pathways to the conscious centers of the brain are cut off. The utricle and saccule seem therefore to be involved in maintaining static equilibrium —the relationship of the head to the pull of gravity.

The ampullae of the semicircular ducts are provided with patches of hair cells, the **cristae.** Their hair-like filaments project into the endolymph. Since the ducts occur in all planes, any movement of the head will cause a pull on the hairs of one crista or another, due to the inertia of the fluid. The bending or pull on the filaments sets up impulses in the sensory cells of the crista; impulses travel to the brain and reflex adjustments are made. These impulses do not necessarily give rise to conscious sensations. They do involve reflex movements of the eyes and the muscles that maintain equilibrium. This is a dynamic equilibrium involving the whole organism.

Hearing. You are probably asking yourselves by now, "Why must the hearing mechanism be so complex?" A part of the answer would be that it does so much for us in terms of its great sensitivity, its high capacity for selectivity, and at the same time its ability to shut out unwanted sounds. Specifically, it has to solve the very difficult mechanical problem of getting the maximum energy out of the sound waves striking the eardrum, to keep them from being lost by reflection. Sound waves in the air, of large amplitude, must be converted into more forceful vibrations of smaller amplitude. The ear does this by using the principle of a hydraulic press—it takes the small pressure on the large area of the eardrum and concentrates it upon the much smaller area of the foot plate of the stapes (*stirrup*) increasing 22-fold the pressure on the perilymph fluid of the internal ear (Fig. 19–8).

It is apparent that the external ear collects sound waves and it is not particularly efficient; a larger auricle would collect more, but it might create other problems. The burden then falls upon the eardrum and middle ear ossicles to use effectively the sound waves available. It has been stated above how this is carried out. Through the arrangement of the ossicles, the piston-like movement of the eardrum is changed to the rocking movement of the stapes, thus reducing the amplitude of the excursion. The vibrations of the stapes are exerted through the fenestra vestibuli to the whole fluid system of the internal ear, setting it into motion. The vibrations travel through the scala vestibuli, through the small opening at the apex of the cochlea leading into the scala tympani, and descend to the base of the cochlea where they expend themselves against the secondary tympanic membrane of the fenestra cochleae (*round window*).

The next question is, how are these vibrations in the internal ear fluid registered by the spiral organ of Corti? This is a complex problem and space allows only a few comments. The **basilar membrane** appears to play a very important role. This membrane, you may recall, is the one on which the organ of Corti rests. It is narrowest and possesses its shortest transverse fibers in the lowest coil of the cochlea and it widens and its fibers lengthen as it ascends to the apex. Sound causing vibration in the internal ear fluid sets a portion of this membrane into motion, and as the structure of the membrane would suggest, the various tones produce maximal vibrations in different parts of the membrane. The basilar membrane also makes a frequency analysis for the determination of pitch. These vibrations are faithfully transmitted by the hair cells in contact with the tectorial membrane which set up impulses in the cochlear nerve. These impulses are carried to the hearing area in the temporal lobe of the cerebral cortex and sound is perceived.

This knowledge, so briefly stated, represents the outcome of a great deal of experimentation, and the use of the latest of electronic recording devices by which the electrical phenomena in nervous structures can be followed and analyzed. Using such instruments one can show objectively that stimulation of certain spots on the basilar membrane does project to certain spots in the temporal lobe of the brain. We now need to know how the auditory mechanism sharpens this crude analysis of the basilar membrane into the pure tones we actually hear.

Deafness. At one time before the day of the so-called "miracle drugs" or antibiotics, infections of the ear were the cause of the majority of cases of deafness. Now this is rare.

Vestibulocochlear nerve destruction and otosclerosis remain the common causes of deafness. There is no cure for nerve deafness. Otosclerosis is a painless tumorous bone growth in the temporal bone. If it does not involve the sound transmission part of the ear, no harm is done. If it happens to involve the footplate of the stapes, it may reduce or freeze its action and sound conduction stops at this point—except, of course, that one can hear by bone conduction. Hearing aids are available which help such cases, but they exert pressures and create other physical and psychological problems causing discomfort or embarrassment to the user. Therefore, many now resort to surgery—the so-called fenestration operation—whereby a small hole is drilled in the lateral wall of the internal ear near the stapes and then covered with a flap of skin giving a means of transmitting vibrations from the middle ear to the fluids of the internal ear.

Most important of all, perhaps, is to learn to avoid situations which may cause nerve deafness. Intense noise, particularly of high frequency, can in time cause deafness and the use of certain drugs such as streptomycin can be dangerous. Disease processes and age may also deteriorate the nervous mechanisms of hearing.

Comparative Anatomy. Reference was made at the beginning of this chapter to the relatively recent development of hearing as a function of the ear. In fish only the parts of the internal ear concerned with detection of position and movement were present. All the other most complicated parts of the internal ear and the whole middle and external ear have been added during the course of vertebrate evolution. It is only in mammals that the complete mechanism just described is found. Birds have only one ossicle in the middle ear and do very well without the other two, for as is well known, they have a keen sense of hearing.

Where did these hearing structures come from? A part of the answer was given in Chapter 1 of this book where chordate characteristics were defined. You might review that section now. It tells how the middle ear cavity and the auditory tube are derived from a pharyngeal or gill pouch and the external auditory meatus from the corresponding invagination of the body wall. The tympanic membrane is the structure separating these two which in fish would break through to form a gill slit. The ear ossicles can be shown, through studies in embryology and comparative anatomy, to have come from parts of the hyoid and mandibular (*jaw*) arches. In mammals therefore, the jaw articulation is of a different type than that found in other vertebrate classes. It lies between the mandible and temporal bones. The hearing structure of the internal ear, the cochlea, represents an outgrowth of the saccule. It would appear that in evolution "new" structures represent outgrowths and modifications of the old.

It is hoped that the frequent references to comparative anatomy and embryology have added interest and understanding to the study of the anatomy of man. Certainly they must to those who are interested in the nature of man and his place in the natural world.

The Eye (Figs. 19-13 to 16)

The eye, unlike the ear, appears fully developed in vertebrate animals from fish to man, and while changes in its structure and function do occur, they are relatively minor adaptations to the way of life of the animal possessing them. It is interesting, too, that in the history of vertebrates we do find one and sometimes two unpaired eyes on the top of the head, one of them associated with the pineal body. None of these single eyes functions, however, in present-day vertebrates.

The eyes like the ears are distance receptors (*telereceptor*), informing us of objects and events far removed from the body as well as those close at hand. Light rays from the sun or other distant stars, and that produced artificially,

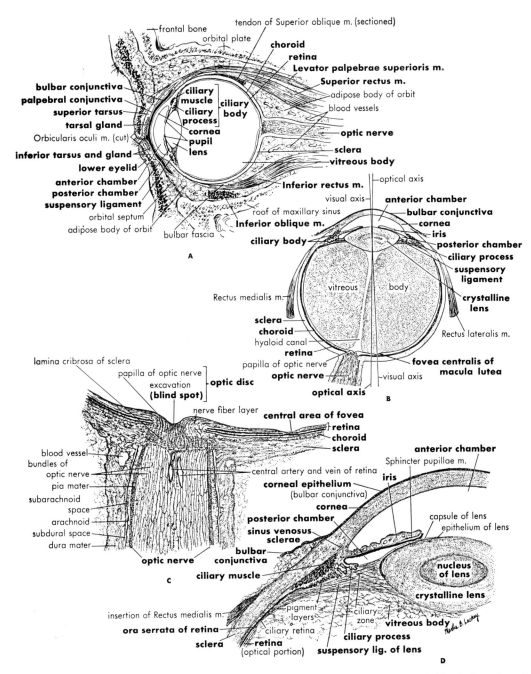

Fig. 19–13. Studies of the structure of the human eye and accessory parts. *A*, Vertical section of orbital cavity containing eye and related structures. *B*, Schematic section through the right eye—horizontal plane. *C*, Enlargement from *B* of the optic nerve and adjacent retina. *D*, Enlargement of anterior part of eye from drawing *B*—horizontal plane.

though from the same ultimate sources, constitute the adequate stimulus for this receptor organ. The eye and ear initiate far more impulses to be carried to the brain than all the other sense organs combined.

Eye and Camera. It appears to be an honored tradition to compare eye and camera and the analogy is good, up to a point. One wonders if the photographer in explaining the "anatomy and physiology" of the camera compares it to the eye. Or does modern man know more about the relatively simple man-made camera, and perhaps have more interest in it, than he does in the truly remarkable living cameras with which he is born, and on which he depends all his life.

The eye and the camera have their more delicate parts protected, the one in a box of some kind, the other in a sphere, the eyeball. Each has an optical system through which light passes and can be bent or focused on a light-sensitive layer, the film or the retina, though they do their focusing in different ways. Each also has a diaphragm or iris by which the amount of light entering can be regulated. But the human eye has the advantage of being more automatic than even the latest automatic camera and perhaps this is why most of us know little about it. When we come to the development of the film, the analogy breaks down. The film in the eye, the retina, remains there to be used again and

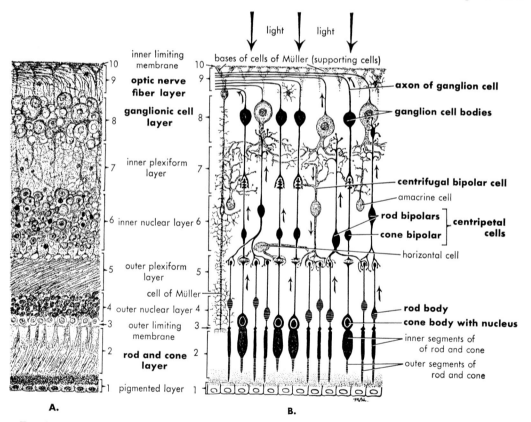

Fig. 19–14. Layers of the adult human retina. *A*, Retina stained routinely and highly magnified. *B*, Schematic presentation of retina to show types of nerve cells and their relationships. Large arrows at top indicate light coming to retina; small arrows show the direction of passage of nerve impulses. Note that the light must penetrate most of the retina to reach the light sensitive rods and cones. The impulses then pass inward to the retina surface and to the optic nerve.

again, and only the impulses initiated in it go to the "developers" in the brain where we see our picture within fractions of a second. We see with the brain, and two perfect eyes would be useless to us without it.

Structure of the Eyeball (Fig. 9–13). The eyeball is a sphere about 2.5 cm. in diameter, modified by an anteriorly bulging, smaller segment, the transparent cornea. It is contained in the cavity of the orbit where it is embedded in a thick layer of fascia and fat and is protected from injury by bone and a variety of accessory structures to be described later. It is moved by a number of muscles which attach to its outer surface and nearby bone. A line drawn from the center point of the cornea to the center point of the larger sphere posteriorly constitutes the **optic axis.** The optic axes of the two eyes are about parallel.

Tunics of the Eyeball (*bulb*). There are three tunics of the eye named from the outermost inward, the: (1) **fibrous tunic,** consisting of the sclera and cornea; (2) **vascular tunic** consisting, from back to front, of the choroid, ciliary body and iris; and (3) **nervous tunic,** the retina (Fig. 9–13).

The **sclera** constitutes the posterior five-sixths of the eyeball or bulb. It consists of a dense, hard, smooth, unyielding membrane which serves to maintain the form of the bulb. Just medial to the posterior pole it is pierced by the fibers of the optic nerve and is continuous through the sheath of this nerve with the dura mater of the brain. Around the optic nerve entrance there are other small apertures for the entrance of ciliary vessels and nerves. The central artery and vein of the retina enter with the optic nerve. The sclera is continuous in front with the cornea at the **sclerocorneal junction** in which is found an encircling sinus, the **sinus venosus sclerae** (*canal of Schlemm*).

The transparent **cornea** makes up the anterior one-sixth of the fibrous tunic of the bulb. Being a segment of a smaller sphere than the sclera it bulges forward from it. Its anterior surface is covered with a stratified squamous epithelium or **bulbar conjunctiva.** It is nonvascular but richly supplied with sensory nerves. It gets its nourishment by diffusion from the capillaries at the junction of the corneal and conjunctival epithelium. The cornea is an important part of the refracting system of the eye. Unequal curvature of the corneal surface causes a blurring of vision called astigmatism.

The **choroid** layer of the vascular tunic occupies the posterior five-sixths of the bulb. It is a thin, highly vascular membrane, dark brown in color due to the pigment cells in its outermost layer. It is loosely joined to the sclera except at the point where the optic nerve pierces it, where it is firmly attached. Internally, the choroid relates intimately to the pigment layer of the retina. It is analogous to the pia-arachnoid of the brain.

The **ciliary body** is the thickest portion of the vascular coat of the eye, extending forward from the **ora serrata** of the retina to a point just behind the sclerocorneal junction (Fig. 19–13*D*). It forms a ring around this part of the eye to which the **suspensory ligament** of the lens attaches. It consists of the ciliary muscle and ciliary process. The inner surface of the ciliary body is covered by an epithelium which is continuous with the retina. This area is thrown into 70 or 80 radiating folds, especially prominent at the free mesial edge of the ciliary body, which constitute the **ciliary processes.** The **ciliary muscle** consists of smooth muscle fibers arranged in meridional, radial, and circular fashion. The meridional fibers run from the sclera in front to the choroid behind and by contraction tense the choroid. The radial fibers run from the sclera into the ciliary body; the circular fibers form a sphincter-like muscle near the base of the iris. The ciliary muscle functions in the **accommodation** (focusing) of the eye through its action on the lens.

The **iris,** the colored part of the eye, is a thin circular, muscular diaphragm which is attached at its periphery to the ciliary body and has at its center a circular aperture, the **pupil.** It lies in front of the lens and divides the chamber between the front of the lens and the cornea into the **anterior** and **posterior chambers** of the eye. The posterior chamber is limited to a small peripheral space behind the iris and in front of the suspensory ligament and ciliary process.

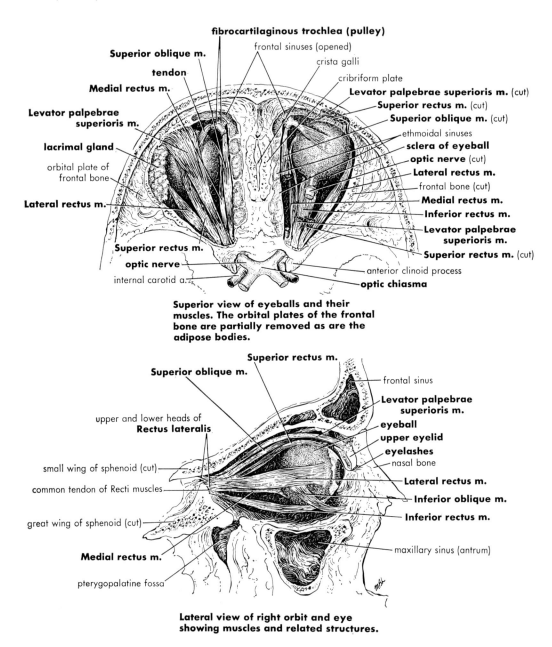

Superior view of eyeballs and their muscles. The orbital plates of the frontal bone are partially removed as are the adipose bodies.

Lateral view of right orbit and eye showing muscles and related structures.

FIG. 19–15.   Studies of the eyes and their extrinsic muscles.

The muscle fibers of the iris are in two groups, circular and radiating. When the circular fibers contract, the pupil becomes smaller; when the radiating fibers contract, it is made larger. These muscles and the ciliary muscles are supplied by the oculomotor or third cranial nerve.

The **retina** is a delicate nervous coat and forms the lining layer of the eyeball. It is continuous behind with the optic nerve. It diminishes in thickness from back to front and near the ciliary body appears to end in a jagged margin, the **ora serrata.** The nervous tissue does end at this point, but a thin pigmented portion of the membrane continues forward on to the back of the ciliary body and iris. In the center of the posterior part of the retina is an oval yellowish area where the retina has thickened, the **macula lutea.** In the center of the macula is a depression in which the retina is exceedingly thin and only the cone cells remain, the receptors for bright light and for color vision. The cone cells here are more numerous and longer than elsewhere in the retina. This is the **fovea centralis** the point of greatest visual acuity (Fig. 19–13C). Everything which is viewed closely and critically comes to focus on it. Just to the nasal side of the fovea is the exit of the optic nerve, the **optic disc,** the margins of which are elevated. In its center is the central artery of the retina. This part of the retina has no light-sensitive cells. It is called the **blind spot.**

Microscopic Anatomy of the Retina (Fig. 19–14). We can mention only briefly some of the important features of this complex light-sensitive layer. It consists of an outer **pigmented layer,** one cell thick, and a nervous layer, the **retina proper.** The retina proper, when sectioned perpendicular to its surface, shows seven layers of supporting and of nervous structures. Excluding the outer pigmented layer and the synaptic and fiber layers, three layers of cell bodies belonging to visual cells (*rod and cones*), bipolar neurons, and optic (*ganglion*) neurons are seen. Other nuclei are present, but the above are the most important. The nonnucleated layers in the retina are made up of the synapses between visual cells and bipolar neurons and between the latter and the optic neurons.

The visual cells, the **rods** and **cones,** are the photoreceptors of the retina. Their adequate stimulus is the radiant energy of the visible spectrum, a relatively narrow band between the longer infrared and the short ultraviolet waves. The **rods** are the dim light receptors (*scotopic vision*). They enable us to see fairly well at night once the eyes have adapted to the darkness. Night vision is not discriminating. It gives us vague outlines of objects which are more readily detected when they move. Under higher illumination rod vision is gradually lost and the **cones** come into action. The cones are **photopic** or discriminative and enable us to see details of form, structure, and color. Reference to the drawings of rods and cones in Figure 19–14 will help you to understand the structure of these specialized cells and will indicate their relative position in the retina proper.

Nerve impulses are initiated in the **bipolar neurons** by the rod and cone cells. The axons of bipolar cells in turn synapse with the dendrites of multipolar ganglion cells. The rods and cones are often referred to as first order, the bipolar as second order and the ganglion cells as third-order neurons. The unmyelinated axons of the ganglion neurons turn and course parallel to the surface of the retina, turn outward at the optic disc and form the bundles of the optic nerve. They acquire myelin sheaths at the optic disc and continue to the brain of which they are actually a part, since in development the material which produces the retina is an outgrowth of the brain

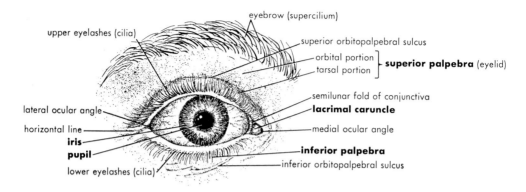

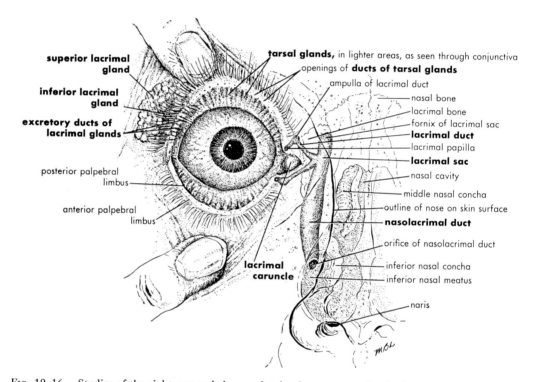

FIG. 19–16.   Studies of the right eye and the nasolacrimal apparatus.   In the lower figure the eyelids are spread to show the relationship of the lacrimal apparatus.

tissue (Fig. 4–21).   Each optic nerve contains about a million fibers.

Refracting Media of the Eye (Fig. 19–13, 16).   In addition to the cornea, which has been considered, the refracting media of the eye are the crystalline lens, the aqueous humor, and vitreous body.

The **crystalline lens** lies immediately behind the pupil in front of the vitreous. It is enclosed in a transparent capsule

and held in position by the **suspensory ligament** (*zonula ciliaris*) of the ciliary body.   It is a transparent, biconvex, elastic structure, its anterior convexity being less than its posterior.   The lens is one of the refracting structures of the eye and is important in accommodation (*focusing*).   When the ciliary muscle contracts, the choroid is pulled forward causing a release of tension on the lens.

Since the lens is elastic, it becomes more spherical, thus increasing its index of refraction. In this way close objects can be brought into focus on the retina. Since this action involves muscular contraction, long hours of close work may cause one to feel a sense of "eye strain." No accommodation is needed in the **normal** (*emmetropic*) eye for distant vision (*20 feet and beyond*). In a **nearsighted** (*myopic*) person, however, where the rays come into focus in front of the retina, or a **farsighted** (*hypermetropic*) individual where the rays focus behind the retina, lenses must be used to make the correction. People with myopic eyes need lenses which diverge the light rays, those with hypermetropic eyes need lenses which converge the rays.

The **aqueous humor** is a dilute alkaline solution which fills the anterior and posterior chambers of the eye. It is secreted by the ciliary process into the posterior chamber of the eye from which it circulates through the pupil into the anterior chamber and is finally filtered into the sinus venosus sclerae or resorbed by the blood vessels of the iris and other veins.

The **vitreous body** is transparent and semigelatinous, consisting mainly of water with some salt and albumen. It fills the large cavity of the eye behind the lens and adheres to the retina, especially at the ora serrata.

Accessory Organs of the Eye (Figs. 19–13, 15, 16). In addition to the fascia around the eyeball and the bony framework of the orbit mentioned earlier, the accessory organs of the eye include the ocular muscles, eyebrows, eyelids, conjunctiva, and lacrimal apparatus.

The **ocular muscles** are the extrinsic muscles of the eyeball which aid in holding it in place and control its movements. They are as follows (Figs. 19–13, 15):

Rectus superior	**Rectus lateralis**
Rectus inferior	**Obliquus superior**
Rectus medialis	**Obliquus inferior**

All these muscles except the Obliquus inferior arise from the apex of the orbit near the optic foramen. The Obliquus inferior arises near the anterior margin of the orbit on the medial side. They insert into the outer coat of the eyeball. Study of Figure 19–15 will help you to understand the arrangement and action of these muscles. Note especially the attachment of the Oblique muscles relative to the Rectus group and also the fact that the tendon of the Obliquus superior muscle passes through a fibrocartilaginous pulley on the medial side of the orbit. Acting together, these muscles move the two eyes so that they focus upon a single field of vision.

The **eyebrows** are thickened areas of the integument which arch over the orbits and support a growth of obliquely directed hairs (Figs. 19–13, 15, 16). They are connected with the Orbicularis oculi, Corrugator, and Frontalis muscles and hence capable of limited movement. Their function would appear to be like that of the facial region generally—a means to give outward expression to the inner feelings, the emotions.

The **eyelids,** upper and lower, are made up of dense fibrous connective tissue structures which give form to the lids, the **tarsal plates.** These plates are covered externally by thin skin which turns inward over their edges and on to the inside of the lids as a mucous membrane, the **conjunctiva.** In the inner surface of the eyelids, between the tarsi and the conjunctiva, are numerous modified sebaceous glands, the **tarsal glands.** Their secretion, an oily liquid, helps to keep the eyelids from adhering. In the outer surface of the eyelids, between the skin and tarsi, are the fibers of the Orbicularis oculi whose contraction closes the lids. Inserted into the upper lid is the aponeurosis of the Levator palpebrae which raises the upper lid. In the margins of the eyelids are modified hairs, the **eyelashes.** They are long and turn upward in the upper lid, shorter and turn downward in the lower lid (Fig. 19–13A).

The conjunctiva is divisible into **palpebral** and **bulbar portions** (Figs. 19–13, 14). The **palpebral portion,** mentioned above, is on the backs of the eyelids and is thick and vascular and contains papillae and some lymphoid tissue in its deeper portions. The **bulbar portion** of the conjunctiva is a continuation of the palpebral portion at the base of the eyelids where it is reflected over the anterior part of the sclera and the cornea. It is thin and transparent on the cornea consisting only of stratified squamous epithelium.

At the medial corner of the eye there is a small, pinkish elevation, the **caruncula,** containing sebaceous and sudoriferous glands. It secretes the whitish material which collects in this region (Fig. 19–16).

The **lacrimal apparatus** consists of the lacrimal glands, lacrimal ducts, lacrimal sac, and nasolacrimal duct (Fig. 19–16).

The **lacrimal gland** which produces the tears lies in the upper lateral side of the orbit and opens by six to twelve ducts from the lateral and upper part of the palpebral conjunctiva. In this way the tears move in a constant protective stream over the anterior surface of the cornea and drain to the medial side of the eye where they empty into two small **lacrimal ducts** leading into the **lacrimal sac.** The lacrimal sac fits into a deep groove formed by the lacrimal bone and frontal process of the maxilla. It drains into the **nasolacrimal duct** which empties into the inferior meatus of the nose. When one cries and increases the secretion of tears, his nose tends to run and as he further overtaxes this system the tears spill down the cheeks.

QUESTIONS

1. What is the role of the organs of general and special sense relative to that of the nervous system?
2. On what basis do the sense organs constitute a system?
3. Define adequate stimulus, receptor, exteroceptor, interoceptors, and proprioceptors. Give examples of each.
4. Describe the distribution of the Pacinian corpuscles.
5. Describe the function of the neuromuscular spindles.
6. What are chemoreceptors? Give examples.
7. Describe the olfactory organ.
8. What is man's capacity for taste?
9. What are the functions of the ear? Which of these is older in terms of vertebrate evolution?
10. What part does the middle ear play in the function of hearing?
11. What is the function of the auditory tube and what does it connect? How do these facts relate to mastoid infections?
12. Describe the parts of the bony labyrinth of the internal ear. What is its relationship to the membranous labyrinth?
13. What functions do the perilymph and endolymph perform?
14. Describe the spiral organ (of Corti), and state briefly how it functions.
15. Explain the differences in the receptors within the saccule and utricle and those in the semicircular canals.
16. Name the two branches of the auditory nerve which serve the two functions of the internal ear.
17. Where did the ear ossicles come from?
18. Compare the sclera and cornea as to structure and function.
19. Name the parts of the middle tunic of the eyeball.

20. Name the refracting media of the eye.
21. Describe the focusing mechanism of the eye.
22. What are the boundaries of the anterior and posterior chambers of the eye? What do these chambers contain?
23. How does the eye regulate the amount of light which enters it?
24. What are myopia and hypermetropia and how may they be corrected?
25. What is astigmatism?
26. What are the principal types of cells in the retina and how are they related one to another in structure and function?
27. Describe the optic nerve and state why it is considered a part of the brain.
28. Describe the structures involved in the production and drainage of "tears."
29. Name the extrinsic eye muscles and state their function.
30. What muscles operate the eyelids?
31. Describe the conjunctiva.

Epilogue

As you come to the end of this book you may have a sense of having accumulated a considerable mass of information and some understanding. You probably realize too that you have learned only a small part of what is known of the human body. If this experience has served to stimulate your imagination and interest to the point that you wish to know more, it has served its purpose. Or, if unable or unwilling to continue your study, you are more cognizant of the complexity of man, more humble in his presence, and have a greater respect for all of life, you have gained something.

But this is not all. It is hoped that man is more to you than structure and function; that now you can see him in the perspective of all of nature—as one especially endowed with a capacity to live, to learn, to plan, and to attain a glorious future. To do this, man must become as objective about his relationship to other men as he has been about the exploration of his physical nature. With each advance in the natural sciences and technology which has benefited man, there has come some threat to his survival. War, never a rational means of solving basic problems, is now inconsistent with survival if pursued with all available weapons. Population is a threat which receives too little consideration. Our wastage of natural resources cannot continue if man is to have an abundant life.

A problem, however, which overrides all others, is man's failure to encourage and promote his potentially creative mind. All are not endowed with great creative powers, but there are always those who are. The rest of mankind must have the wisdom to give such people every opportunity to create, to innovate, and to explore. This means the development of a society in which man has faith in himself; one in which each is encouraged to develop to the fullest his peculiarly human capital of intelligence; one in which the spiritual, moral, and intellectual climate is sympathetic to even the revolutionary discoveries of its highly endowed citizens. It means, finally, that suspicion, fear, self-interest, and materialism cannot be allowed to determine man's direction and his goals. They can only lead to the collapse of our American, and yes, our world dream. Man will have thrown away his opportunity of being master of his own destiny and may thus become one of the least effective of nature's creatures. This need not be the outcome if we learn to realize and to capitalize upon the full potential of the anatomy of man.

References

One should not depend entirely upon only one or two anatomy books as a source of his knowledge. It is refreshing to occasionally look into the writings of other authors for different points of view, organization, and language. Also the writings in related fields are often helpful. The following list of references has been selected to aid the reader in broadening his view and in pursuing special topics which may have interested him.

Nomenclature

Borror, Donald, J. 1960. *Dictionary of Word Roots and Combining Forms.* N–P Publications, Palo Alto.

Blakiston's New Gould Medical Dictionary. 1956. 2nd Ed. McGraw-Hill Book Co., New York.

Dorland's Illustrated Medical Dictionary. 1965. 24th Ed. W. B. Saunders Co., Philadelphia.

Henderson, I. F. and W. D. Henderson. 1963. *A Dictionary of Scientific Terms.* 8th Ed. (by Kenneth, J. H.). D. Van Nostrand Co. Inc., Princeton, N.J.

Jaeger, E. C. 1960. *The Biologist's Handbook of Pronunciations.* Charles C Thomas, Springfield, Ill.

Nomina Anatomica, 3rd Ed., 1966. Revised by the International Anatomical Nomenclature Committee and approved by the 8th International Congress of Anatomists held at Wiesbaden in 1965. V + 112 pages. Excerpta Medica Foundation, Amsterdam.

Gross Anatomy, Textbooks

Anson, Barry J. 1966. *Morris' Human Anatomy.* 12th Ed. McGraw-Hill Book Co., New York.

Cunningham, D. J. 1964. *Textbook of Anatomy,* edited by G. J. Romanes. 10th Ed. Oxford University Press, London.

Gardner, E., D. J. Gray, and R. O'Rahilly. 1969. *Anatomy,* 3rd Ed. W. B. Saunders Co., Philadelphia.

Grant, J. C. B. and J. V. Basmajian. 1965. *A Method of Anatomy.* 7th Ed. The Williams & Wilkins Co., Baltimore.

Gray, H. 1966. *Anatomy of the Human Body,* edited by C. M. Goss. 28th Ed. Lea & Febiger, Philadelphia.

Lockhart, R. D., G. F. Hamilton, and F. W. Fyfe. 1965. *Anatomy of the Human Body.* 2nd Ed. J. B. Lippincott Co., Philadelphia.

Marshall, Clyde and E. L. Lazier. 1955. *An Introduction to Human Anatomy.* 4th Ed. W. B. Saunders Co., Philadelphia.

Woodburne, R. T. 1965. *Essentials of Human Anatomy.* 3rd Ed. Oxford University Press, New York.

Gross Anatomy, Atlases

Anson, B. J. 1963. *Atlas of Human Anatomy.* 2nd Ed. W. B. Saunders Co., Philadelphia.

Grant, J. C. B. 1962. *An Atlas of Anatomy.* 5th Ed. The Williams & Wilkins Co., Baltimore.

Jamieson, E. B. 1959. *Illustrations of Regional Anatomy.* 8th Ed. The Williams & Wilkins Co., Baltimore.

Sobotta, J., and Figge, Frank, H. J. 1963. *Atlas of Descriptive Human Anatomy.* 8th Ed. Hafner Publishing Co., New York.

Spalteholz, W. 1943. *Hand Atlas of Human Anatomy,* edited by L. F. Barker. 7th Ed. J. B. Lippincott Co., Philadelphia.

Anatomy and Physiology

Kimber, D. C. and Carolyn E. Gray, Caroline E. Stackpole and Lutie C. Leavell. 1966. *Anatomy and Physiology.* 15th Ed. The Macmillan Co., New York.

King, B. G. and M. J. Showers. 1969. *Human Anatomy and Physiology.* 6th Ed. W. B. Saunders Co., Philadelphia.

Langley, L. L. and E. Cheraskin and R. Sleeper. 1963. *Dynamic Anatomy and Physiology.* 2nd Ed. McGraw-Hill Book Co. (Blakiston Division), New York.

Steen, E. and A. Montagu. 1959. *Anatomy and Physiology.* Vols. I, II, College Outline Series. Barnes & Noble, Inc., New York.

Physiology

Best, C. H. and N. B. Taylor. 1966. *Physiological Basis of Medical Practice.* 8th Ed. The Williams & Wilkins Co., Baltimore.

Carlson, A. and V. Johnson. 1961. *Machinery of the Body.* 5th Ed. University of Chicago Press, Chicago.

39

Eldred, Earl. 1965. *The Dual Sensory Role of Muscle Spindles*. Children's Bureau Publication Number 432—1965. U. S. Department of Health, Education, and Welfare. pp. 8–31.

Gerard, R. W. 1944. *Unresting Cells*. Harper & Bros., New York.

Giese, A. C. 1968. *Cell Physiology*. 3rd Ed. W. B. Saunders Co., Philadelphia.

Harvey, W. 1949. *Anatomical Studies on the Motion of the Heart and Blood*. 3rd Ed. Charles C Thomas, Springfield, Ill.

Rose, Jerzy E. and Vernon B. Mountcastle, Touch and kinesthesis. 1959. *Handbook of Physiology*. Section 1: Neurophysiology, Vol. 1, pp. 387–429. American Physiological Society, Washington, D. C.

Tuttle, W. W. and B. A. Schottelius. 1969. *Textbook of Physiology*. 16th Ed. C. V. Mosby Co., St. Louis.

Endocrinology

Hall, P. F. 1959. *Functions of the Endocrine Glands*. W. B. Saunders Co., Philadelphia.

Soffer, Louis J. 1956. *Diseases of the Endocrine Glands*. 2nd Ed. Lea & Febiger, Philadelphia.

Turner, C. D. 1966. *General Endocrinology*. 4th Ed. W. B. Saunders Co., Philadelphia.

Developmental Anatomy

Arey, L. B. 1965. *Developmental Anatomy*. 7th Ed. W. B. Saunders Co., Philadelphia.

Balinsky, B. I. 1970. *Introduction to Embryology*. 3rd Ed. W. B. Saunders Co., Philadelphia.

Patten, B. M. 1964. *Foundations of Embryology*. 2nd Ed. McGraw-Hill Book Co., New York.

Patten, B. M. 1968. *Human Embryology*. 3rd Ed. McGraw-Hill Book Co. (Blakiston Division), New York.

Comparative Anatomy

Crouch, James E. 1969. *Text-Atlas of Cat Anatomy*. Lea & Febiger, Philadelphia.

Romer, A. S. 1970. *The Vertebrate Body*. 4th Ed. W. B. Saunders Co., Philadelphia.

Smith, H. M. 1960. *Evolution of Chordate Structure (An Introduction to Comparative Anatomy)*. Holt, Rinehart, & Winston, Inc., New York.

Microscopic Anatomy

Bloom, W. and D. W. Fawcett. 1968. *A Textbook of Histology*. 9th Ed. W. B. Saunders Co., Philadelphia.

Copenhaver, W. M. and D. D. Johnson. 1964. *Bailey's Textbook of Histology*. 15th Ed. The Williams & Wilkins Co., Baltimore.

Di Fiore, M. S. H. 1967. *Atlas of Human Histology*. 3rd Ed. Lea & Febiger, Philadelphia.

Finerty, J. C. and E. V. Cowdry. 1960. *Textbook of Histology*. 5th Ed. Lea & Febiger, Philadelphia.

Ham, A. W. 1969. *Histology*. 6th Ed. J. B. Lippincott Co., Philadelphia.

Jordan, H. E. 1952. *Textbook of Histology*. 9th Ed. Appleton-Century-Crofts, Inc., New York.

Porter, Keith R. and Mary A. Bonneville. 1968. *An Introduction to the Fine Structure of Cells and Tissues*. 3rd Ed. Lea & Febiger, Philadelphia.

Siles, Karl A. 1968. *Handbook of Histology*. 5th Ed. McGraw-Hill Book Co., New York.

Windle, W. F. 1960. *Textbook of Histology*. 3rd Ed. McGraw-Hill Book Co., New York.

Neuroanatomy

Crosby, E. E., T. Humphrey, and E. Lauer. 1962. *Correlative Anatomy of the Nervous System*. The Macmillan Co., New York.

Gardner, E. 1968. *Fundamentals of Neurology*. 5th Ed. W. B. Saunders Co., Philadelphia.

Krieg, W. J. S. 1953. *Functional Neuroanatomy*. 2nd Ed. McGraw-Hill Book Co. (Blakiston Division), New York.

Kuntz, A. 1953. *Autonomic Nervous System*. 4th Ed. Lea & Febiger, Philadelphia.

Noback, Charles R. 1967. *The Human Nervous System*. 1st Ed. McGraw-Hill Book Co., New York.

Ranson, S. W. and S. L. Clarke. 1959. *Anatomy of the Nervous System*. 10th Ed. W. B. Saunders Co., Philadelphia.

Rasmussen, A. T. 1952. *The Principal Nerve Pathways*. 4th Ed. The Macmillan Co., New York.

Truex, Raymond C. and M. B. Carpenter. 1969. Strong and Elwyn's *Human Neuroanatomy*. 5th Ed. The Williams & Wilkins Co., Baltimore.

Special Anatomy

Duvall, E. N. 1959. *Kinesiology—The Anatomy of Motion*. Prentice-Hall, Inc., Englewood Cliffs, New Jersey.

Hollinshead, W. H. 1969. *Functional Anatomy of the Limbs and Back*. 3rd Ed. W. B. Saunders Co., Philadelphia.

Kronfeld, P. C. 1943. *The Human Eye in Anatomical Transparencies*. Bausch and Lomb Press, Rochester, New York.

Rasch, Philip J. and Burke, Roger K. 1971. *Kinesiology and Applied Anatomy*. 4th Ed. Lea & Febiger, Philadelphia.

Wells, Katherine F. 1966. *Kinesiology*. 4th Ed. W. B. Saunders Co., Philadelphia.

Biology

Burnet, Macfarlane. 1962. The Thymus Gland. Sci. Am., *207*, 5, 50–57.

French, J. D. 1957. The Reticular Formation. Sci. Am., *196*, 30, 54–60.

Korzybski, A. 1921. *Manhood of Humanity*. E. P. Dutton & Co., New York.

Siekevitz, P. 1957. Powerhouse of the Cell. Sci. Am., *197*, 131–4.

Storer, T. I. and R. L. Usinger. 1965. *General Zoology*. 4th Ed. McGraw-Hill Book Co., New York.

Swanson, C. P., *et al*. 1960–61. *Foundations of Modern Biology Series*. Prentice-Hall, Inc., Englewood Cliffs, N.J.

Teilard de Chardin, P. 1959. *The Phenomenon of Man*. Harper & Bros., New York.

Weisz, P. B. 1967. *The Science of Biology*. 3rd Ed. McGraw-Hill Book Co., New York.

Laboratory Manuals

Brash, James C. 1957. *Cunningham's Manual of Practical Anatomy*. 12th Ed., 3 vol. Oxford University Press, London.

Crouch, J. E. 1972. *Introduction to Human Anatomy, a Laboratory Manual*. 4th Ed. N-P Publications, Palo Alto. A manual for use in courses designed for nursing and physical education students.

Davenport, H. A. 1963. *A Dissector's Manual of Human Anatomy*. W. B. Saunders Co., Philadelphia.

Grant, J. C. B. 1967. *A Handbook for Dissectors*. 6th Ed. The Williams & Wilkins Co., Baltimore.

Shearer, E. M. 1967. *Manual of Human Dissection*. 5th Ed. McGraw-Hill Book Co., New York.

Glossary

If you cannot find the word you wish in the glossary, check the index for a text reference.

abducens (ăbdū′sĕnz) [L. *abduceu*, to lead away] The sixth cranial nerve.

abduction (ăbdŭkt′shŭn) [L. *abductus*, led away] Movement away from the central axis of the body or part.

absorption (ăbsôrp′shŭn) [L. *absorbere*, to suck in] Passage of materials into or through living cells.

acinus (as′e-nus) [L. *grape*] A small sac-like dilatation.

acetabulum (ăsĕtăb′ūlŭm) [L. *acetabulum*, vinegar-cup] The socket in the pelvic girdle for the head of the femur.

acromegaly (ăk′römĕg′ălĭ) [Gk. *akros*, tip; *megalon*, great] Gigantism due to excessive activity of part of hypophysis, produces abnormal proportions of body in adult.

adduction (ădŭk′shŭn) [L. *ad*, to; *ducere*, to lead] Movement toward the central axis of the body or part.

adenohypophysis (ăd′ĕnōhīpof′ĭsĭs) [Gk. *aden*, gland; *hypo*, under; *physis*, growth] Anterior glandular portion of the hypophysis.

adipose capsule (ăd′ipōs) [L. *adips*, fat] A mass of fat around the kidney giving it protection and support.

adrenal (ădrē′năl) [L. *ad*, to; *renes*, kidneys] An endocrine gland located on the superior surface of the kidney.

adrenocorticotrophic (ădrē′nökôr′tĭkötrŏf′ik) [L. *ad*, to; *renes*, kidneys; *cortex*, bark; Gk. *trophe*, nourishment] Hormone secreted by anterior lobe of hypophysis which influences activity of adrenal cortex.

afferent (ăf′ĕrĕnt) [L. *affere*, to bring] Conveying to.

after-discharge. A phenomenon in which the response lasts after the termination of the stimulus.

allantois (ălăn′tŏĭs) [Gk. *allas*, sausage] A fetal membranous sac arising from posterior part of alimentary canal.

alveolus (ălvē′ölŭs) [L. *alveolus*, small pit] A tooth socket or small depression.

amnion (ăm′nĭŏn) [Gk. *amnion*, fetal membrane] A fetal membrane enclosing amniotic fluid and embryo.

amniotic cavity (ămnĭŏt′ĭk) [Gk. *amnion*, fetal membrane] Pertaining to the amnion. A cavity enclosed in the amnion within the ectoderm cells of the blastocyst.

ampulla (ămpŭl′a) [L. *ampulla*, flask] A membranous vesicle.

anabolism (ănăb′ŏlĭsm) [Gk. *ana*, up; *gale*, throw] The constructive chemical processes in living organisms.

anal glands (ā′năl) [L. *anus*, anus] Large modified sweat glands in the stratified squamous epithelium located above the anal opening.

anaphase (ăn′ăfāz) [Gk. *ana*, up; *phasis*, appearance] The stage in mitosis when paired chromatids move to opposite poles.

anastomosis (ănăs′tömö′sis) [Gk. *ana*, up; *stoma*, mouth] Interconnecting of blood vessels or nerves to form network.

anatomy (ănăt′ŏmĭ) [Gk. *ana*, up; *tome*, cutting] The science which deals with the structure of the body.

anemia (ănē′mĭă) [Gk. *an*, not; *haima*, blood] A condition in which there is a reduced number of erythrocytes or erythrocytes with a reduced amount of hemoglobin.

antagonist (ăntăg′ŏnĭst) [Gk. *antagonistes*, adversary] A muscle acting in opposition to the action produced by a prime mover.

anthropoids (an′thrŏpŏĭd) [Gk. *anthropos*, man; *eidos*, form] The tailless apes, including chimpanzee, gibbon, gorilla, and orangutan.

aortic sinuses (āôr′tĭk si′nŭsĭz) [Gk. *aorte*, the great artery; L. *sinus*, cavity] Dilated pockets between the cusps of the semilunar valves and the aortic wall.

apex (ā′pĕks) [L. *apex*, summit] Tip or summit.

apical foramen (ăp′ĭkăl fŏrā′mĕn) [L. *apex*, summit; *foramen*, opening] The opening of the root of the tooth to the root canal.

apocrine (ap'o-krīn) [Gk. *apo*, from; away from; *krino*, to separate] A type of gland in which the secretions gather at the outer end of the gland cells, which are then pinched off to form the secretion. Example—mammary glands.

aponeurosis (ap'onūro'sĭs) [Gk. *apo*, from; *neuron*, tendon] A white, flattened, sheet-like tendon.

aqueous humor (ā'kwĕŭs hū'mŏr) [L. *aqua*, water; *humor*, moisture] A dilute alkaline solution filling the anterior and posterior chambers of eyes.

arachnoid (ărăk'noid) [Gk. *arachne*, spider; *eidos*, form] The intermediate meninx which is thin, weblike, and transparent.

arbor vitae (âr'bŏr vī'tē) [L. *arbor*, tree; *vita*, life] Arborescent appearance of cerebellum in midsagittal section.

arteriole (ârtē'rĭŏl) [L. *arteriola*, small artery] An artery of under 0.5 mm. in diameter.

arteriosclerosis (ârtē'rĭō'sklĕ'ro'sis) [L. *arteria*, artery; *sclerosi*, hardness] Abnormal thickening and hardening of the arteries.

artery (âr'tĕrĭ) [L. *arteria*, artery] A vessel which carries blood away from the heart.

arthritis (ârthrīt'ĭs) [Gk. *arthron*, joint] Inflammation of a joint.

arthrology (ăr'thrŏl'ŏjī) [Gk. *arthron*, joint; *logos*, discourse] Study of joints.

articulation (ărtĭkūlā'shŭn) [L. *articulus*, joint] A joint by which bones are held together.

aryepiglottic fold (ar'yep'iglot'tic) Upper border of the vestibular membrane found between the epiglottis and the apex of arytenoids.

arytenoid (ăr'ĭtē'noid) [Gk. *arytoina*, pitcher; *eidos*, form] A pair of small cartilages of the larynx articulating with cricoid cartilage.

atresia (ah-tre'ze-ah) [Gk. *a*, not; *tretos*, perforated] Follicular degeneration.

atrium (ā'trĭŭm) [L. *atrium*, chamber] A superior cavity of the heart which acts as the receiving chamber; also a part of the tympanic cavity of the ear.

atrophy (ăt'rŏfĭ) [Gk. *a*, without; *trophi*, nourishment] Disappearance or diminution in size and function.

auricle (ôr'ĭkl) [L. *auricula*, small ear] Any ear-like lobed appendage as related to the atria of the heart.

autonomic (ôt'ŏnŏm'ĭk) [Gk. *autos*, self; *nomos*, province] Self-governing, spontaneous; as the involuntary nervous system.

axon (ak'son) [Gk. *axon*, axle] Nerve cell process limited to one per cell which is involved in conducting away from cell body.

axon hillock. Elevated area of perikaryon at which axon originates, which is nearly devoid of Nissl bodies.

basophil (bā'söfĭl) [Gk. *basis*, base; *philain*, to love] Rare leukocytes often having an S-shaped nucleus and large granules which stain purplish blue with Wright's stain.

bipennate (bipĕn'āt) [L. *bi*, twice; *penna*, contour feather] Muscle fiber arrangement in which the fibers are attached to both sides of a tendon as in a contour feather.

blastocyst (blăs'tösĭst) [Gk. *blastos*, bud; *kytos*, hollow] The germinal vesicle.

blood (blŭd) [A.S. *blod*, blood] The fluid tissue of the vascular system of animals.

bone (bōn) [A.S. *bon*, bone] Connective tissue whose ground substance contains salts of lime.

boutons terminaux (boo-taw' tār-mino-o) Bulb-like enlargements of terminal branches of axons which are in relation to the dendrites and cell bodies of other neurons.

Bowman's capsule [Sir William Bowman, English physician 1816–1892] The vesicle of a renal tubule; capsula glomeruli.

branched acinous (as'e-nus) [L. *acinus*, berry] Type of gland with a single duct and having more than one dilation or acinus.

branchial arch (brăng'kiăl) [Gk. *brangchia*, gills] A cartilaginous or bony arch on the side of the pharynx supporting the gill bars.

branchiomere (brăng'kiŏmēr) [Gk. *brangchia*, gills; *meros*, part] A branchial segment.

bronchiole (brŏng'kiōl) [Gk. *brongchos*, windpipe] One of the finer subdivisions of the bronchiolar tree.

bronchus (brŏng'kŭs) [Gk. *brongchos*, windpipe] Short connecting tube between trachea and lungs.

Brunner's glands [Johann Conrad Brunner, Swiss anatomist, 1653–1727] Compound tubuloalveolar glands found in the submucosa of the duodenum.

brushes of Ruffini [A. Angelo Ruffini, Italian anatomist, 1854–1929] Cylindrical end-bulbs in subcutaneous tissue of finger.

buccal frenula (bŭk'ăl frĕn'ū lah) [L. *bucca*, cheek, *frenum*, bridle] Small folds of membrane between the cheeks and the gums.

buccal glands (bŭk'ăl) [L. *bucca*, cheek] Small glands of the submucosa of the cheeks which secrete mucus into the vestibule.

bulbo-urethral glands (bŭl'bö-u-re'thral) [L. *bulbus*, bulb; Gk. *ourethra*, urethra] Glands lying in the pelvis floor to either side of the membranous urethra of the male; open into cavernous urethra.

bundle of His [W. His, German anatomist, 1863–1934] Band of specialized muscle fibers in the interventricular septum of the heart—a part of the conducting mechanism.

bursa (bŭr'să) [L. *bursa*, purse] A fluid-filled sac-like cavity situated in the tissues at points of friction or pressure—mostly around joints.

calcification (kălsĭfĭkā'shŭn) [L. *calx*, lime; *facere*, to make] The process by which lime salts are deposited in the matrix of bone or cartilage.

calyx (kāl'ĭks) [Gk. *kalyx*, calyx] Cup-like extensions of pelvis of kidney.

canaliculi (kănălĭk'ūlī) [L. *canaliculus*, small channel] Microscopic canals through which processes of the bone cells connect.

cancellous bone (kăn'sĕlŭs) [L. *cancellous*, chambered] Inner, more spongy portion of bony tissue.

cancer (kan'ser) [L. for crab] A malignant tumor, capable of metastasis (spreading through circulation and lymph).

canine (kănīn) [L. *canis*, dog] One of the teeth primarily for tearing, found on either side of the incisors.

capitulum (kăpĭt'ūlūm) [L. *caput*, head] A knob-like swelling at end of a bone.

cardiac glands (kăr'dĭăk) [Gk. *kardia*, heart] One of the three gastric glands found in the immediate area of the cardiac orifice.

carotene (kăr'ötēn) [L. *carota*, carrot] A yellow pigment.

carpus (kăr'pŭs) [L. *carpus*, wrist] Collective term for the eight bones which support the wrist.

cartilage (kăr'tĭlĕj) [L. *cartilago*, cartilage] A form of connective tissue usually bluish-white, firm and elastic; cells placed in groups in spaces called lacunae.

caruncula (kăr-ung'ku-lah) [L. small piece of flesh] Small pinkish elevation at the inner angle of the eye.

catabolism (kătăb'ölĭsm) [Gk. *kata*, down; *bale*, throw] The destructive chemical processes in living organisms.

catalyst (kat'-al-ist) [Gk. *kata*, down; *lysis*, loosing] A substance which accelerates or retards a reaction without entering into the reaction itself.

cauda equina (kô'dă e-kwin'a) [L. *cauda*, tail; *e'quus*, horse] A tail-like collection of spinal nerves at the end of the spinal cord.

caudal (kô'dăl) [L. *cauda*, tail] Of or pertaining to the tail end of the animal.

cavernous body (kăv'ĕr-nus) [L. *cavernosus*, chambered] A structure of the penis and clitoris containing blood spaces; involved in erection of these organs.

cecum (sē'kŭm) [L. *caecus*, blind] A large blind pouch found at the beginning of the large intestine.

celom (sē'lŏm) (also coelom) [Gk. *koiloma*, a hollow, fr. *koilos*, hollow] A cavity formed within the mesoderm and generally lined by mesothelium.

cementum (sēmĕnt'ŭm) [L. *caementum*, mortar] The covering of the root of the tooth which is connected to the alveolar bone.

centriole (sĕn'trĭōl) [L. *centrum*, center] Found in the cytoplasm near the nucleus, and important in mitosis.

cerebellum (sĕr'ĕbĕl'ŭm) [L. dim. of *cerebrum*, brain] A solid mass of nervous tissue consisting of two hemispheres located in the posterior cranial fossa below posterior portion of cerebrum; concerned with coordination and balance.

cerebral aqueduct (sĕr'ĕbrăl' ăk'wĕdŭkt) [L. *cerebrum*, brain; *aqua*, water, *ducere*, to lead] A narrow canal passing through the mesencephalon connecting the third and fourth ventricles.

cerebral peduncle (sĕr'ĕbrăl pĕdŭng'kĕl) [L. *cerebrum*, brain; *pedunculus*, small foot] Large bundles of nerve fibers which form the inferior portion of the midbrain.

cerebrospinal fluid (sĕr'ĕbrŏspī'năl) [L. *cerebrum*, brain; *spina*, spine] A fluid produced in the choroid plexuses of the ventricles of the brain.

ceruminous glands (sĕrū'mĕnŭs) [L. *cera*, wax] Wax glands of the external auditory meatus.

chemoreceptor (kĕm'örĕsĕp'tŏr) [Gk. *chemos*, juice; L. *recipere*, to receive) Receptor organ which responds to chemical stimuli.

choana (kŏ'ănă) [Gk. *choane*, funnel] A funnel-shaped opening; the internal naris.

cholecystitis (kŏl'ē-sĭs-tī-tĭs) [Gk. *chole*, bile; *kystes*, bladder; *itis*, inflammation] Inflammation of the gallbladder.

chondrin (kôn'drĭn) [Gk. *chondros*, cartilage] Gelatinous substance obtained from cartilage.

chondrocranium (kôn'dōkrā'nĭŭm) [Gk. *chondros*, cartilage; *kranim*, skull] The skull when in a cartilaginous condition, either temporary or permanent.

chondrocyte (kôn'drōsīt) [Gk. *chondros*, cartilage; *kytos*, hollow] Cartilage cell.

chordae tendineae (chor'de tendi-neae) [Gk. *chorde*, string; *tendene*, to stretch] Fine tendinous strings connecting the ventricular walls of the heart to the valve cusps or flaps.

chordata (kor-da'-tah) [Gk. *chorde*, cord, string] A phylum which includes the vertebrates and other animals that have a notochord.

chorion (kō'rion) [Gk. *chorion*, skin] An embryonic membrane external to and enclosing the amnion.

choroid (kōr'oid) [Gk. *chorion*, skin-like] The middle layer of the eyeball between the retina and sclera.

choroid plexus (kō'roid plĕk'sŭs) [Gk. *chorion*, skin-like; L. *plexus*, interwoven] Vascular structures in the roofs of the four brain ventricles which produce cerebrospinal fluid.

chromatophores (krō'mătōfōr) [Gk. *chroma*, color; *pherein*, to bear] Branched, pigmented cells as the dermal chromatophores.

chromatid (krō'mătid) [Gk. *chroma*, color] A component of a tetrad in meiosis.

chromatin (krō'matin) [Gk. *chroma*, color] A substance in the nucleus which contains nucleic acid proteids and stains with basic dyes.

chromosome (kro'mosom) [Gk. *chroma*, color; *soma*, body] Cell structures made up of genes.

chyle (kīl) [Gk. *chylos*, juice] Lymph-containing globules of emulsified fat found in the lacteals.

chyme (kīm) [Gk. *chymos*, juice] The mass of partially digested food forced through the pyloric sphincter into small intestines.

ciliary glands (sĭl'e-ere)) [L. *cilium*, eyelid] Glands of the eyelids.

circle of Willis [Thomas Willis, English anatomist and physician, 1621–1675] A circular system of arteries inferior to the brain.

circumcision (ser-kum-sizh'un) [L. *circumcisio*, a cutting around] The surgical procedure by which the prepuce is removed.

circumduction (sër-kŭmdŭk shŭn) [L. *circum*, around; *ductus*, led] An action which involves flexion, extension, abduction, adduction and rotation.

cisterna (sĭs'tërnăh) [L. *cistern*] Any closed space serving as a reservoir, especially one of the enlarged sub-arachnoid spaces.

cleavage (klē'vëj) [A.S.] Early cell divisions.

clitoris (klī'tōrĭs) [Gk. *kleiein*, to enclose] An erectile organ of female which is homologous to the penis.

cochlea (kŏk'lëă) [Gk. *kochlias*, snail] Anterior part of labyrinth of the ear, spirally coiled like a snail shell.

collagenous fibers (kŏl'ăjĕn'ŏus) [Gk. *kolla*, glue; *genos*, offspring] Strong, inelastic fibers composed of many parallel fibrils.

collateral (kŏlăt'ërăl) [L. *cum*, with; *latera*, sides] Fine lateral branches of the axon.

çolon (kō'lŏn) [Gk. *colon*, colon] Portion of the large intestine between the cecum and rectum.

concentric lamellae (kŏnsĕn'trik lămĕl'ae) [L. *cum*, together; *centrum*, center; L. *lamella*, small plate] Concentric circles of bony matrix arranged around Haversian canals.

concha (kŏng'kă) [Gk. *kongche*, shell] A structure resembling a shell as the nasal conchae (turbinates) or the hollow of the external ear.

conductivity (kŏn'dūktĭv'ĭtĭ) [L. *conducere*, to lead together] Power of protoplasm to carry the effect of a stimulation from one part to another.

condyle (kŏn'dīl) [Gk. *kondylos*, knuckle] A rounded process on a bone for articulation.

cone—One of the photopic and color receptors of the retina.

connective tissue (kŏnĕk'tĭv) [L. *cum*, together; *nectere*, to bind] Characterized by cells separated by large amounts of intercellular material.

contractility (kŏn'trăktĭl'ĭtĭ) [L. *cum*, together; *trahere*, to draw] The capacity to change form.

conus arteriosus (kō'nŭs arterio'sus) [L. *conus*, cone; *arte'ria*, to keep air] A structure between ventricle and aorta in fishes and amphibians; the upper and anterior part of the right ventricle of the human heart from which the pulmonary artery arises.

conus medullaris (kō'nŭs mĕd'ŭ'lā'rĭs) [Gk. *konos*, cone; L. *medullaris*, marrow] Terminal, tapering portion of the spinal cord.

convergence (kŏnvĕr'jĕns) [L. *convergere*, to incline together] The process in which the axons of two or more neurons synapse with single neurons.

coordination—A special function of the central organs of the nervous system by which impulses are sorted and channeled for favorable response.

cornea (kôr'nëă) [L. *corneus*, horny] The anterior, transparent, and bulging portion of the outer fibrous coat of the eye.

corniculate cartilage (kôrnĭk'ūlāt) [L. *cornu*, horn] One of two small, conical, elastic cartilages articulating with apex of arytenoids.

coronal (kŏr'ŏnăl) [L. *corona*, crown] (same as frontal) A plane vertical to the median plane which divides the body into anterior and posterior parts.

corpora atretica (kôr'pŏrä ah-tret'ik-ă) [L. *corpus*, body; Gk. *a*, not; *tretos*, perforated] The scars in the stroma of the ovary indicating the death of an ovarian follicle or spurious corpora lutea.

corpora quadrigemina (kor'po-rah kwod-re-jem'i-nah) [L. *corpus*, body; *quad*, four; *gemma*, bud] Four small lobes on dorsal region of mesencephalon associated with visual and auditory functions.

corpus albicans (kôr'pŭs ăl'bĭkănz) [L. *corpus*, body; *albicare*, to grow white] A white scar formed when the corpus luteum degenerates.

corpus callosum (kôr'pŭs kălō-sum) [L. *corpus*, body; *callosus*, hard] Broad sheet of white matter uniting the two cerebral hemispheres below the longitudinal fissure.

corpus luteum (kôr'pŭs lu'teum) [L. *corpus*, body; *luteus*, orange or yellow] The glandular body developed from a Graafian follicle after extrusion of ovum; produces progesterone.

cortex (kôr'těks) [L. *cortex*, bark] Outer or more superficial part of an organ, as the cortex of the adrenal gland or of the cerebrum.

cranial (krā'nĭăl) [Gk. *kranion*, skull] Referring to the head end of the body.

cretinism (krē'tĭn'ĭzm) A condition in which mental and sexual development are retarded in infancy due to hypofunction of thyroid.

cricoid cartilage (krĭk'oid) [Gk. *krikos*, ring; *eidos*, form] Thick ring-like cartilage in larynx, articulating with the thyroid and arytenoid cartilages.

crista galli (krĭs'tă gal'e) [L. *crista*, crest; *gallus*, chicken, cock] A process on the superior surface of the ethmoid.

cryptorchidism (krĭptôr'kĭdĭzm) [Gk. *kryptos*, hidden; *orchis*, testis] A condition in which the testes are abdominal in position; do not descend normally.

crypts of Lieberkühn [Johann Nathaniel Lieberkühn, German anatomist, 1711–1756] Tubular glands of the small intestine.

cuneate (ku'ne'āt) [L. *cuneus*, wedge] A wedge-shaped body.

cuneiform cartilage (kūnē'ĭfôrm) [L. *cuneus*, wedge; *forma*, shape] One of two small, elongated pieces of elastic cartilage found in the aryepiglottic folds.

cutaneous plexus (kūtā'nĕŭs plĕk'sŭs) [L. *cutis*, skin; *plexus*, interwoven] Network of arteries at the interface of the corium and subcutaneous layers.

cuticle (kū'tĭkl) [L. *cutis*, skin] A layer of more or less solid substance secreted by and covering the surface of an epithelium and sharply delimited from the cell surface.

cystic duct (sĭs'tĭk dŭkt) [Gk. *kystis*, bladder; L. *ducere*, to lead] Duct of the gallbladder which empties into the common bile duct.

cytology (sītŏl'ŏjĭ) [Gk. *kytos*, hollow, hollow vessel; *logos*, discourse] The science dealing with the structure, functions, and life history of cells.

cytoplasm (sī'toplazm) [Gk. *kytos*, hollow; *plasma*, form] Living substance of the cell body, excluding the nucleus.

dartos (dâr'tŏs) [Gk. *dartos*, flayed] Involuntary muscle fibers which lie within the superficial fascia of the scrotum.

decidua basalis (dē sĭd'ŭă ba'sal is) [L. *de*, away; *cadere*, to fall] The portion of the endometrium between the embryo and the muscular wall of the uterus.

decidua capsularis (dē sĭd'ŭa kăp'sŭl arĭs) [L. *de*, away; *cadere*, to fall] The portion of the endometrium which closes over the implanted embryo.

deciduous (dē sĭd'ŭŭs) [L. *de*, away; *cadere*, to fall] Falling at end of growth period or at maturity.

deglutition (dē gloot ĭsh'ŭn) [L. *deglutire*, to swallow down] The process of swallowing.

dendrite (děn'drīt) [Gk. *dendron*, tree] Nerve cell process which normally conducts impulses toward cell body.

dentin (děn'tĭn) [L. *dens*, tooth] A hard, elastic substance constituting the greater part of the tooth.

dermatocranium (děr'mă to krā'nĭ ŭm) [Gk. *derma*, skin; *kranion*, skull] Intramembranous bones of the roof and sidewalls of the cranial cavity and bones of the face.

dermis (děrm'ĭs) [Gk. *derma*, skin] A layer of dense connective tissue derived from mesoderm germ layer; the inner skin.

desquamation (děs'kwăm ā'shŭn) [L. *de*, away; *squama*, scale] Shedding of cuticle or epidermis in flakes.

diabetes (dī'a bē'tēz) [Gk. *diabetes*, fr. *diabainein*, to pass through] A deficiency condition marked by habitual discharge of an excessive quantity of urine.

diad (dīăd) [Gk. *di*, two] In meiosis, a structure consisting of two chromatids.

diaphragm (dī'ă frăm) [Gk. *diaphragma*, midriff] A partition partly muscular, partly tendinous, separating cavities of chest from abdominal cavity; a most important organ of breathing.

diaphysis (dī ăf'ĭ sĭs) [Gk. *dia*, through; *phyein*, to bring forth] Shaft of bone.

diastole (di ăs'tōlē) [Gk. *diastole*, difference] Relaxation phase of the heart beat.

diencephalon (dī'ĕn sĕf'ă lŏn) [Gk. *dia*, between; *enakephalos*, brain] Hindpart of forebrain.

differentiation (dĭf'ĕrĕn'shĭā'shŭn) [L. *differre*, to differ] Modifications in structure and functions of the parts of an organism.

digestion (dī jĕs'chŭn) [L. *digestio*, digestion] Mechanical and chemical breakdown of food whereby it may be absorbed.

diploe (dip'loë) [Gk. *diploos*, double] The cancellous layer which lies between the inner and outer tables of compact bone as in the bones of the skull.

distal (dĭs'tăl) [L. *distare*, to stand apart] End of any structure farthest from midline or from point of attachment.

divergence (dī vĕr'jĕns) [L. *diverge*, to bend away] Process by which one neuron synapses with two or more neurons.

dorsal (dôr'săl) [L. *dorsum*, back] Pertaining to or lying near the back.

ductus deferens (dŭk'tŭs def'er ens) [L. *ducere*, to lead; *deferens*, to carry away] The excretory duct of the testis leading from the testis to the ejaculatory duct.

duodenum (dū'ö dē'nŭm) [L. *duodeni*, twelve each] The short upper portion of the small intestine.

dura mater (dū'ră mā'tĕr) [L. *dura*, hard; *mater*, mother] The outermost and toughest meninx.

dysfunction (dis'fungk'shun) [Gr. *dys*, ill, bad or hard] Impaired functioning.

ectoderm (ĕk'tö dĕrm) [Gk *ektos*, outside; *derma*, skin] The outer germ layer of a multicellular animal.

ectopic (ĕk tŏp'ĭk) [Gk *ek*, out of; *topos*, place] Not in normal position.

effectors (ĕf fĕk'tŏrz) [L *efficere*, to carry out] Muscles and glands which respond to impulses carried to them by nerves.

efferent (ĕf'fĕr ĕnt) [L *ex*, out; *ferre*, to carry] Conveying from.

efferent ductules (ĕf'fĕr ĕnt dŭk'tūlz) [L. *ex*, out; *ferre*, to carry; *ducere*, to lead] Tubes from testes to the head of epididymis carrying spermatozoa.

egestion (ē jĕst'shŭn) [L. *ex*, out; *gerere*, to carry] Elimination at the inferior end of the digestive tube.

ejaculation (ë jăk'ū lā'shŭn) [L. *ejaculatus*, thrown out] The process by which the seminal fluid is emitted.

ejaculatory duct (ë jăk'ū lă törĭ dŭkt) [L. *ex*, out; *jacere*, to throw] A continuation of the ductus deferens from the point of entrance of the seminal vesicles to the prostatic urethra.

elastic fibers (e-las'tik) [L. *elasticus*] Long cylindrical threads with no fibrillar structure, which contain elastin.

eleidin (ĕlē'ĭdĭn) [Gk. *elaia*, olive] Substance related to keratin found in the stratum lucidum of the skin.

embryo (ĕm'brĭö) [Gk. *embryon*, embryo] A young organism in early stages of development, before it becomes self-supporting; in human embryology the first weeks of development.

embryology (embrĭŏl'öjĭ) [Gk. *embryon*, embryo; *logos*, discourse] Science of development from egg to birth or hatching.

enamel (ĕnăm'ĕl) [O.F. *esmaillier*, to coat with enamel] The hard material which forms a cap over dentin.

end bulbs of Krause [Wilhelm Johann Friedrich Krause, German anatomist, 1833–1910] A cylindrical or oval capsule derived from the connective tissue sheath of a myelinated nerve fiber; contain nerve fibers; receptors for cold.

endocardium (ĕn'dokar'dium) [Gk. *endon*, within; *kardia*, heart] The inner layer of the heart wall.

endochondral (en'dökôn'drăl) [Gk. *endon*, within; *chondros*, cartilage] Bones formed by the replacement of hyaline cartilage.

endocrine (ĕn'dökrĭn) [Gk. *endon*, within; *krinein*, to separate] A ductless gland which conveys its secretions into the blood for distribution.

endoderm (ĕn dödĕrm) [Gk. *endon*, within; *derma*, skin] The innermost of the three germ layers.

endolymph (ĕn'dölĭmf) [Gk. *endon*, within; L. *lympha*, water] The fluid found inside the membranous labyrinth.

endomysium (ĕn'dömĭz'ĭŭm) [Gk. *endon*, within; *mys*, muscle] Sheath-like covering of connective tissue around each muscle fiber.

endoneurium (ĕn'dönū'rĭŭm) [Gk. *endon*, within; *neuron*, nerve] The delicate connective tissue holding together and supporting nerve fibers within fasciculi.

endoskeleton (ĕn'döskĕl'ĕtŏn) [Gk. *endon*, within, *skeletos*, dried up] The bony and cartilaginous structures of the body, exclusive of that part of dermal origin.

endosteum (ĕndŏs′tēŭm) [Gk. *endon*, within; *osteon*, bone] The internal periosteum lining the cavities of bones.

endothelium (ĕn′dŏthē′lĭŭm) [Gk. *endon*, with; *thele*, nipple] A simple squamous epithelium which lines cavities of the heart, blood and lymphatic vessels.

enzyme (ĕn′zīm) [Gk. *en*, in; *zyme*, leaven] Organic catalysts which act only upon specific substances and under specific conditions.

eosinophils (ē′ōsĭn′ŏfĭl) [Gk. *eos*, dawn; *philein*, to love] Leukocytes, having a two-lobed nucleus and cytoplasmic granules which stain bright red with Wright's stain.

epaxial (ĕpăk′sĭăl) [Gk. *epi*, upon; L. *axis*, axle] Above the axis; dorsal muscle mass.

epicardium (ep′ĭkâr′dĭŭm) [Gk. *epi*, upon; *kardia*, heart] The thin transparent outer layer of the heart wall, also called visceral pericardium.

epicondyle (ĕp′ikŏn dīl) [Gk. *epi*, upon; *kondylos*, knob] A projection above or upon a condyle.

epididymis (ĕp′ĭdĭd′ĭmĭs) [Gk. *epi*, upon; *didymos*, testicle] A convoluted duct found on the posterior surface of the testis.

epigenesis (ĕp ĭjĕn′ĕsĭs) [Gk. *epi*, upon; *genesis*, descent] A theory of generation.

epiglottis (ĕp′ĭglŏt′ĭs) [Gk. *epi*, upon; *glotta*, tongue] A leaf-shaped elastic cartilage, between root of tongue and entrance to larynx.

epinephrine (ĕp inĕf′rĕn) [Gk. *epi*, upon; *nephros*, kidney] The hormone of the adrenal medulla.

epineurium (ĕp′ĭneū′rĭŭm) [Gk. *epi*, upon; *neuron*, nerve] The external connective tissue sheath of a nerve.

epiphysis (ĕpĭf′ĭsĭs) [Gk. *epi*, upon; *phyein*, to grow] Enlarged ends of bones, formed from separate centers of ossification.

epiploic foramen (ĕp′ĭplŏ′ĭk fŏrā′mĕn) [Gk. *epiploon*, caul of entrails; L. *foramen*, opening] The opening above the duodenum from the greater into the lesser peritoneal cavity (Foramen of Winslow).

epispadias (ep-e-spa′de-as) [Gk. *epi*, upon; *spadoros*, no generative power] The condition in which the external urethral opening is on the dorsum of the penis.

epithalamus (ep′ithal′amus) [Gk. *epi*, upon; *thalamos*, chamber] The thin roof of the third ventricle.

epithelial (ĕp′ĭthē′lĭăl) [Gk. *epi*, upon; *thele*, nipple] Characterized by cells closely joined together and found on free surfaces of the body.

eponychium (ĕp′onĭk ĭŭm) [Gk. *epi*, upon; *onynx*, nail] The fold of stratum corneum which overlaps the lunula of nail.

erythrocyte (ĕrĭth′rōsīt) [Gk. *erythros*, red; *kytos*, hollow] A red blood corpuscle.

estrogen (ēs′trōjĕn) [Gk. *oistros*, gadfly; *gennaein*, to produce] Female sex hormone produced by the vesicular (Graafian) follicle.

Eustachian tube (ūstā′kĭăn) [B. Eustachio, Italian physician] Tube connecting the middle ear with the nasopharynx; auditory tube.

eversion (ē′vĕr′shŭn) [L. *evertere*, to turn] Rotation of the foot which turns the sole outward.

evolution (ĕv′ōlū′shŭn) [L. *evolvere*, to unroll] The process of development of organisms from pre-existing forms.

excretion (ĕkskrē′shŭn) [L. *ex*, out; *cernere*, to sift] The passage of waste products from the internal environment through living membranes to the external environment.

exoskeleton (ĕk′söskĕl′ĕtŏn) [Gk. *exo*, without; *skeletos*, hard] A non-living skeleton which must be shed and replaced at growth periods.

expiration (ĕk′spīrā′shŭn) [L. *ex*, out; *spirare*, to breathe] The act of emitting air from lungs.

extension (ĕkstĕn′shŭn) [L. *ex*, out; *tendere*, to stretch] A motion which increases the angle between two bones.

exteroceptor (ĕk′stĕrosep′tŏr) [L. *exter*, outside; *capere*, to take] A receptor which receives stimuli from the external environment.

extraembryonic celom (ĕk′strâĕm brĭŏn′ik) [L. *extra*, outside; Gk. *embryon*, fetus] The cavity between the layers of extraembryonic mesoderm.

extraembryonic mesoderm (ĕk′strâĕm′brĭŏn′ik) [L. *extra*, outside; Gk. *embryon*, fetus] A layer of cells lying between the yolk sac and the trophoblast, outside the embryo proper.

facet (făs′ĕt) [L. *facies*, face] A smooth, flat, or rounded surface for articulation.

family (făm′ĭlĭ) [L. *familia*, household] A group of related genera; families being grouped into orders.

fasciculi (făsĭk′ūli) [L. *fasciculus*, little bundle] Bundles of fibers.

fenestra cochleae (fĕnĕs′tră) [L. *fenestra*, window] A round window below the fenestra vestibuli in the bony wall between the middle ear and the cochlea of the inner ear.

fenestra vestibuli (fĕnĕs′tră) [L. *fenestra*, window] An oval window above the fenestra cochlea in the bony wall between the middle ear and the vestibule of the inner ear.

fertilization (fĕr′tĭlĭza shŭn) [L. *fertilis*, fertile] The union of male and female pronuclei.

fetus (fē′tŭs) [L. *foetus*, offspring] Product of conception after the second month of gestation.

fibril (fī′brĭl) [L. *fibrilla*, small fiber] Fine thread-like structures which give cell stability.

fibroblasts (fī′brōblăst) [L. *fibra*, band; Gk. *blastos*, bud] Connective tissue cell found close to collagenous fibers which give rise to fibers.

fibrocartilage (fī′brö-kâr′tĭlĕj) [L. *fibra*, band; *cartilago*, gristle] A cartilage which is characterized by parallel collagenous bundles within its matrix.

fibrous joints (fībrŭs) [L. *fibra*, band] Joints in which the primitive joint plate develops into fibrous tissue.

filum terminale (fī′lŭm tur′mĭ-nāl′lē) [L. *filum*, a thread; *terminalis*, terminal] The non-nervous terminal thread extending from the conus medullaris of the spinal cord to the coccyx.

fimbria (fĭm′brĭă) [L. *fimbria*, fringe] Any fringe-like structure; as on the infundibulum of the uterine tube.

fissure (fĭsh-ūr) [L. *fissus*, cleft] A cleft, deep groove, or furrow dividing an organ into lobes.

fixators (fiks-a′tors) [L. *fixatio*, to hold] Muscles which maintain the position of the body; which fix one part to support the movement of another.

flexion (flĕk′shŭn) [L. *flexus*, bent] A movement which decreases the angle between two bones.

folia cerebelli [L. *folium*, leaf] More or less parallel ridges of the gray cortex of the surface of cerebellum.

fontanels (fŏn′tănĕl) [F. *fontanelle*, little fountain] Membranous areas of the head of the fetus and infant.

foramen (fŏră′mĕn) [L. *foramen*, opening] A hole to allow passage of blood vessels or nerves.

foramen of Monro [John Cummings Monro, Boston surgeon, 1858–1910] The opening between the third ventricle and a lateral ventricle of cerebral hemisphere.

fornix (fôr′nĭks) [L. *fornix*, vault, arch] A moat-like area around the cervix where it protrudes into the superior end of the vagina.

fossa (fŏs′ă) [L. *fossa*, ditch] A depressed area; usually broad and shallow.

fovea centralis (fō′vĕă cen-tra′lis) [L. *fovea*, pit; *centrum*, center] A depression in the center of the macula lutea which marks point of keenest vision; contains only cone cells.

frontal (frŭn′tăl) [L. *frons*, forehead] A plane, vertical to the median plane which divides the body into anterior and posterior parts; the bone of the forehead.

fundic glands (fŭn′dĭk) [L. *fundus*, bottom] Principal glands of the stomach found throughout body and fundus.

gallbladder (gōl′blăd′ĕr) [A.S. *gealla*, gall; *blaedre*, bag] The structure for storing bile located on the inferior surface of the liver.

gametes (gămēts′) [Gk. *gametes*, spouse] Sexual cells; haploid.

ganglion (găng′glĭŏn) [Gk. *gangglion*, little tumor] Collections of nerve cells occurring outside of central nervous system.

gastric pits (găs′trĭk) [Gk. *gaster*, stomach] Grooves in the mucous membrane of the stomach wall which in cross-section appear as pits.

gene (jēn) [Gk. *genos*, descent] The heredity determining unit of the cell (DNA).

genetics (jĕnĕt′ĭks) [Gk. *genesis*, descent] That part of biology dealing with heredity and variation.

genus (jē′nŭs) [L. *genus*, race] A group of closely related species.

gills (gĭl) [M.E. *gille*, gill] Filamentous or plate-like outgrowths serving as respiratory organs in many aquatic animals.

gill slits (gĭl) [M.E. *gille*, gill] A series of openings in the walls of the pharynx—persistent in lower chordates, transitory in higher groups.

gingival (jĭnjĭ′val) [L. *gingivae*, gums] The gums which attach to the maxillae and mandibles.

ginglymoid (gĭng′glĭmoid) [Gk. *gingglymos*, hinge-joint; *eidos*, form] Constructed like a hinge joint.

germ cell (jĕrm) [L. *germen*, bud] A reproductive cell.

glabella (glăbĕl′a) [L. *glober*, bald] The space of forehead between superciliary ridges.

glands of Littré (A. Littré, French surgeon) Branching tubules or glands in the walls of the cavernous urethra.

glia (glī′ă) [Gk. *glia*, glue] A cell of the neuroglia.

glomerulus (glō-mĕr′-ū-lŭs) [L. *globulus*, small globe] A network of blood capillary within Bowman's capsule.

glossopharyngeal (glŏs′ŏfărĭn′jĕăl) [Gk. *glossa*, tongue; *pharynx*, gullet] The ninth cranial nerve.

glottis (glŏt′ĭs) [Gk. *glotta*, tongue] The slit between the vocal folds which marks the opening of the larynx.

goiter (goi′tĕr) [Fr. *goitre*] An enlargement of the thyroid gland.

Golgi material [Golgi, Italian histologist] Network of fibrils or a series of membranous structures commonly found close to cell center.

gonad (gŏn'ăd) [Gk. *gone*, birth] A sexual gland, male or female.

gubernaculum (gū'bërnăk'ŭlŭm) [L. *gubernaculum*, rudder] A ligament which connects the testis and epididymis to the inside of the scrotal swelling.

gustatory (gŭs'tātörĭ) [L. *gustare*, to taste] Pertaining to sense of taste.

gut (gŭt) [A.S. *gut*, channel] Intestine or part thereof, according to structure of animal.

gyrus (jī'rŭs) [L. *gyrus*, circle] A cerebral convolution; a ridge between two grooves.

hair (hăr) [A.S. *haer*] A thread-like or filamentous outgrowth of epidermis of animals.

hamulus (hăm'ūlŭs) [L. *hamulus*, little hook] A bony process shaped like a hook.

haploid (hăp'loid) [Gk. *haploos*, simple; *eidos*, form] Having the number of chromosomes characteristic of mature germ cells; half the number of the somatic cells.

haustrum (haws'trum) [L. *haustor*, drawer] The recess made by one of the sacculations of the intestinal wall.

Haversian canal [Havers, English anatomist, 1650–1702] A canal in bone tissue through which blood vessels pass.

Harversian systems [Havers, English anatomist] These consist of a central Haversian canal around which are concentric circles of bony matrix, the concentric lamellae.

hemopoiesis (hē'möpoiē'sĭs) [Gk. *haima*, blood; *poiesis*, making] The process of blood cell formation.

hemopoietic (hē'möpoiĕt'ĭk) [Gk. *haima*, blood; *poietiikos*, productive] Blood-forming tissue.

Henle's loop [Friedrich Gustav Jakob Henle, German anatomist, 1809–1885] Loop in a kidney tubule within apical portion of pyramid.

hepatic ducts (hēpăt'ĭk dŭkt) [L. *hepar*, liver; *ducere*, to lead] Paired ducts of the major liver lobes.

hepatitis (hĕp'a-tī'tĭs) [Gk. *hepat*, liver; *itis*, inflammation] Inflammation of the liver.

hernia (her'ne-ah) [L.] A rupture or separation of some part of the abdominal wall; the protrusion of a viscus from its normal position.

hilus (hī'lŭs) [L. *hilum*, trifle] Small notch, opening, or depression, usually where blood vessels enter.

histology (hĭstŏl'ŏjĭ) [Gk. *histos*, tissue; *logos*, discourse] The science of plant and animal tissues.

holocrine (hol'o-krin) [Gk. *holos*, whole; *krino*, to separate] A type of gland in which the secreting cells and their secretions constitute the glandular product. Example: sebaceous glands.

homo (hō'mö) [Gk. *homos*, alike, same] Meaning the same or alike; the genus of man, when capitalized.

horizontal ((hŏr'ĭzŏn'tăl) [Gk. *horizon*, bounding] A plane at right angles to both the sagittal and the frontal planes dividing the body into superior and inferior portions.

hormones (hôrmōn'z) [Gk. *hormaein*, to excite] Secretions of ductless glands.

hyaline cartilage (hī'ălĭn) [Gk. *hylos*, glass] A cartilage, with a homogeneous matrix, having a translucent appearance.

hydrocele (hī'drösēl) [Gk. *hydro*, water; *koilos*, hollow] A condition in which fluid accumulates in the tunica vaginales.

hydrocephalus (hī-dro-sĕfahlŭs) [Gk. *hydro*, water; *kephalus*, head] Accumulation in excess of cerebrospinal fluid as a result of closure of the median and lateral apertures of the fourth ventricle.

hymen (hī'mĕn) [Gk. *hymen*, membrane] Thin fold of mucous membrane at the orifice of the vagina.

hypaxial (hĭpak'sĭăl) [Gk. *hypo*, under; L. *axis*, axis] Below the axis; ventral muscle mass.

hyperfunction (hi'per) [Gk. *hyper*, over, above] Overactivity.

hypodermis (hī pödër'mĭs) [Gk. *hypo*, under; L. *dermis*, skin] Subcutaneous tissue separating integument from underlying muscle.

hypofunction (hi-po) [Gk. *hypo*, below, under] Underactivity.

hyponychium (hī'pönik'ĭum) [Gk. *hypo*, under; *onyx*, nail] A thickened layer of stratum corneum at the distal end of digit under the free edge of nail.

hypophysis (hīpŏf'ĭsĭs) [Gk. *hypo*, under; *physis*, growth] An endocrine gland which rests in the hypophyseal fossa of the sella turcica.

hypospadias (hi-po-spa'de-as) [Gk. *hypo*, under; *spadionos*, no generative powers] A condition in which the external urethral orifice is located on the underside of the penis or in the perineum.

hypothalamus (hī'pöthăl'ămŭs) [Gk. *hypo*, under; *thalamos*, chamber] Region below thalamus; structures forming greater part of floor of third ventricle.

ileum (ĭl'ĕŭm) [L. *ileum*, groin] The terminal two-thirds of the small intestine; about 12 feet.

implantation (ĭm'plăntā shŭn) [L. *in*, into; *plantare*, to plant] Transplanting an organ or part to an abnormal position or the embedding of the embryo into the lining of the uterus.

incisor (ĭnsī'zör) [L. *incisus*, cut into] One of the front, chisel-shaped teeth, primarily for cutting.

inclusions (in'klōō zhŭn) [L. *inclusus*, confined, shut up] The non-living particles of the cytoplasm.

induction (ĭndŭk′shŭn) [L. *inductio*, causing to occur] Act or process of causing to occur.

infundibulum (ĭnfŭndĭb′ŭlŭm) [L. *infundibulum*, funnel] Any funnel-shaped organ or structure; as the free end of the uterine tube.

ingestion (ĭnjĕs′chŏn) [L. *ingestus*, taken in] The taking in of food at the mouth.

inhibition (ĭn′hĭbĭsh′ŏn) [L. *inhibere*, to prohibit] Prohibition or checking of an action already commenced.

inspiration (ĭnspĭrā′shŭn) [L. *inspirare*, to inhale] The act of drawing air into the lungs.

insulin (ĭn′sūlĭn) [L. *insula*, island] Product of island cells of pancreas which regulates carbohydrate metabolism and blood sugar levels.

integument (ĭntĕg′ūmĕnt) [L. *integumentum*, covering] The protective covering over entire body, the skin and its derivatives.

intercalated discs (in-ter′kah-lāt-ed) [L. *intercalatus*, interposed, inserted] Short lines or stripes extending across the fibers of heart muscle.

intermediate cell mass (ĭn′tĕrmē′dĭat) [L. *inter*, between; *medius*, middle] A constricted mass of mesoderm between the somites and the lateral mesoderm.

internode (ĭn′tĕrnōd) [L. *inter*, between; *nodus*, knot] Segments between nodes, as in nerve fibers.

interoceptor (ĭn′tĕrösĕp′tŏr) [L. *internus*, inside; *caper*, to take] A receptor which receives stimuli from the internal environment.

interphase (ĭn′tĕrfāz) [L. *inter*, between; Gk. *phase*, appearance, show] Resting stage between two mitotic divisions.

interstitial cells (ĭn′tĕrstĭsh′ĭăl) [L. *inter*, between; *sistere*, to set] Hormone-producing cells found in mature testis.

intervertebral discs (ĭn′tĕrvĕr′tĕbrăl dĭsk) [L. *inter*, between; *vertebra*, vertebra; Gk. *diskos*, disc] Pads of fibrocartilage between the bodies of the vertebrae.

intramembranous (ĭn′tra′mĕm′brănŭs) [L. *intra*, within; *membrane*, membrane] Bones formed directly in a fibrous membrane.

inversion (ĭnvĕr′shŭn) [L. *invertere*, to turn upside down] Rotation of the foot which turns the sole inward.

iris (ī′ris) [L. *iris*, rainbow] A thin, circular, muscular diaphragm of the eye; contains the eye color.

irradiation (ir-ra′de-a′shun) [L. *in*, into; *radiare*, to emit rays] The phenomenon in which nervous activity is spread to other levels and across the spinal cord.

irritability (ĭr′ĭtăbĭl′ĭtĭ) [L. *irritare*, to provoke] The capacity to respond to stimulation.

Islands of Langerhans (lahng′er hanz) [Paul Langerhans, German pathologist and anatomist, 1839–1915] Endocrine glands of the pancreas which secrete the hormone insulin.

isometric (i-sō-mĕt′rĭk) [Gk. *iso*, equal; *metron*, measure] Of equal measure; a muscle contraction in which the tension increases, but the length of the muscle remains the same

isotonic (ĭsōtŏn′ĭk) [Gk. *isos*, equal; *tonos*, strain] Of equal tension: a muscle contraction in which the fibers shorten but the tension remains the same; a solution having the same osmotic pressure as blood plasma.

isthmus (ĭsth′mŭs) [Gk. *isthmos*, neck] A narrow structure connecting two large parts.

jejunum (jējoon′ŭm) [L. *jejunus*, empty] The central portion of the small intestine.

kyphosis (kī′fō′sĭs) [Gk. *kyphos*, humpback] Abnormal posterior convexity of the vertebral column—hunchback.

labial frenulum (lā′bĭăl fren′ulum) [L. *labium*, lip; *frenum*, bridle] Medial folds of mucous membrane between the gums and the inner surface of the lips.

labial glands (lā′bĭăl) [L. *labium*, lip] Glands of the submucosa of the lips which secrete mucus into the vestibule.

labia majora (lā′bĭa ma′jōr-ā) [L. *labium*, lip; *magnus*, great] Two large folds of skin which constitute the outer lips of the vulva; homologous to scrotum of male.

labia minora (lā′bĭă mi′nōr-ā) [L. *labium*, lip; *minor*, less, smaller] Two small folds lying between the labia majora.

labyrinth (lăb′ĭrĭnth) [L. *labyrinthus*, labyrinth] The complex internal ear; bony and membranous labyrinths comprise vestibule, cochlea, and semicircular canals.

lacrimal gland (lăk′rĭmăl) [L. *lacrima*, tear] One of the glands of the eye found on upper lateral side of the orbit.

lacteals (lăk′tealz) [L. *lac*, milk] Lymphatic vessels of small intestine.

lacuna (lăkū′nă) [L. *lacuna*, cavity] A space between cells.

lamina (lăm′ĭnah) [L. *lamina*, plate] A thin, flat plate or layer.

lanugo (lănū′gö) [L. *lanugo*, wool] The first hair to appear in the fetus.

laryngeal prominence (larin′jeal) [Gk. *larynx*, upper part of windpipe] The result of the angle formed by two laminae of the thyroid cartilage—commonly Adam's apple.

lateral (lăt′ĕrăl) [L. *latus*, side] Those structures farther to the sides, away from the midline.

leukocytes (lū′kōsīt) [Gk. *leukos*, white; *kytos*, hollow] Colorless blood corpuscles.

ligament (lĭg′amĕnt) [L. *ligamentum*, bandage] A fibrous band of connective tissue holding two or more bones in articulation.

lingual frenulum (ling′gwal frĕn′ūlŭm) [L. *lingua*, tongue; *frenulum*, bridle] A fold of mucous membrane which connects the tongue to the floor of the mouth.

lipase (lip′ās) [Gk. *lipes*, fat] An enzyme produced by the chief cells of the stomach which acts on fats; also an enzyme of the pancreas.

lordosis (lôr dō′sĭs) [Gk. *lordos*, bent so as to be convex in front] Abnormal posterior concavity of vertebral column—hollow or swing back.

lumen (lū′mĕn) [L. *lumen*, light] The cavity of a tubular structure or hollow organ.

lunula (lū′nūlă) [L. *lunula*, small moon] The whitish area at the proximal end of the nail.

lymph (lĭmf) [L. *lympha*, water] A tissue fluid confined to vessels and nodes of the lymphatic system.

lymphocyte (lĭm′fōsīt) [L. *lympha*, water; Gk. *kytos*, hollow] A small mononuclear colorless corpuscle of blood and lymph.

lymphoid (tissue) (lĭm′foĭd) [L. *lympha*, water; Gk. *eidos*, form] A reticular connective tissue infiltrated with lymphocytes.

lymph node (lĭmf nōd) [L. *lympha*, water; *nodus*, knob] Smsll oval collection of lymphatic tissue interposed in the course of lymphatic vessels.

macroglia (măk′rōglĭ′ă) [Gk. *makros*, large; *glia*, glue] Supporting cells of neuroglia which are ectodermal in origin.

macrophage (măk′rōfāj) [Gk. *makros*, large; *phagein*, to eat] Connective tissue cell which ingests and stores microscopic particles; common in loose connective tissue.

macula (măk′ūlă) [L. *macula*, spot] One of the sensitive areas in the walls of the saccule and utricle.

macula lutea (măk′ūlă lū′teă) [L. *macula*, spot; *luteus*, orange-yellow] An oval, yellowish thickened area found in the center of the posterior part of retina; contains the fovea centralis, the area of keenest vision.

mammary (măm′ărĭ) [L. *mamma*, breast] Specialized integumentary glands, characteristic of class Mammalia.

marrow (măr′ō) [A.S. *mearg*, pith] Special tissue, related to blood and connective tissue, found in medullary cavities of bones.

mastication (măs′tĭkā′shŭn) [L. *masticare*, to chew] Process of chewing food with teeth until reduced to small pieces or pulp.

matrix (māt′rĭks) [L. *mater*, mother] Ground substance of connective tissue.

maturation (măt′ūrā′shŭn) [L. *maturus*, ripe] Completion of germ-cell development, by which the chromosome number is reduced by one-half (dipoid to haploid).

meatus (mēā′tŭs) [L. *meatus*, passage] A short canal.

medial (mē′dĭal) [L. *medius*, middle] Structures of the body nearer the midline.

mediastinum (mē′dĭăstĭ′nŭm) [L. *mediastinus*, middle] Cleft between right and left pleura in and near median sagittal thoracic plane.

mediastinum testis (mē′dĭăstĭ′nŭm) [L. *mediastinus*, middle] Incomplete vertical septum of testis.

medulla (mĕdŭl′ă) [L. *medulla*, marrow, pith] Central part of an organ or tissue as the medulla of the adrenal gland.

medulla oblongata (mĕdŭl′ă ob-long-ga′tah) [L. *medulla*, marrow, pith] Posterior portion of brain, continuous with spinal cord, which houses fourth ventricle.

medullary cavity (mĕdŭl′ărĭ) [L. *medulla*, pith] The cavity within long bones.

medullary sheath (mĕdŭl′ărĭ shēth) [L. *medullaris*, pith; A.S. *sceth*, shell or pod] The thick covering of myelin which surrounds myelinated nerve fibers.

megakaryocytes (mĕg′ăkăr′ĭōsīt) [Gk. *megas*, large; *karyon*, nut; *kytos*, hollow] Large ameboid cells of bone marrow.

meiosis (mīō′sĭs) [Gk. *meion*, smaller] A special type of nuclear division occurring during the formation of germ cells which results in the reduction of the chromosome number (diploid) by one-half (haploid).

melanin (mĕl′ănĭn) [Gk. *melas*, black] Black or dark-brown pigment.

melanocytes (mĕl′anosīt) [Gk. *melas*, black; *kytos*, hollow] Pigmented cells located in the basal cells of the spinosum, which produce melanin.

meninges (mĕnĭn′jēz) [Gk. *meningx*, membrane] Singular, meninx. Membranes which enclose the spinal cord and continue through foramen magnum to cover the brain.

meniscus (mēnĭs′kus) [Gk. *meniskos*, little moon] Interarticular fibrocartilage found in joints exposed to violent concussion; as the knee joint.

menopause (mē′nöpôz) [Gk. *men*, month; *pausi*, ending] The period when childbearing ceases.

menstruation (mĕn′strooā′shŭn) [L. *mensis*, month; *struere*, to flow] The monthly discharge of blood and epithelial cells from the uterus.

menstruum (mĕn′stroo-um) [L. *menstruus*, menstruous] The superficial endometrium, blood and glandular secretions discharged during menstruation.

merocrine (mer′o-krīn) [Gk. *meros*, part; *krino*, to separate] A type of gland in which there is no cell destruction in the production of the secretion.

mesencephalon (mĕs′ĕnsĕf′ălŏn) [Gk. *mesos*, middle; *en*, in; *kephale*, head] The midbrain.

mesenchyme (mĕsĕng′kĭm) [Gk. *mesos*, middle; *engchein*, to pour in] A primitive, diffuse, embryonic tissue derived largely from the mesoderm.

mesentery (mĕs′ĕntĕri) [L. *mesenterium*, mesentery] A peritoneal fold serving to hold viscera in position; specifically, it supports the hollow organs of the digestive tube.

mesoderm (mĕs′ŏdĕrm) [Gk. *mesos*, middle; *derma*, skin] Embryonic layer between ectoderm and endoderm.

mesonephric duct (mĕs′ŏnĕf′rĭk) [Gk. *mesos*, middle; *nephros*, kidney] The duct of the early vertebrate kidney which in man contributes to the ductus deferens.

mesosalpinx (mĕs′ŏsăl′pĭngks) [Gk. *mesos*, middle; *salpingx*, trumpet] The portion of the broad ligament enclosing and supporting the uterine tube.

mesothelium (mĕs′ŏthē-lĭŭm) [Gk. *mesos*, middle; *thele*, nipple] An epithelial tissue lining coelomic cavities, covering the surfaces of mesenteries and omenta, and forming the outermost layer of many of the viscera.

mesovarium (mĕs′ŏvā′rĭŭm) [Gk. *mesos*, middle; L. *ovarium*, ovary] The mesentery which supports the ovary.

metabolism (mĕtăb′ŏlĭzm) [Gk. *metabola*, change] All chemical changes, constructive and destructive, by which protoplasm uses and transforms materials.

metacarpus (mĕt′ăkăr′pŭs) [Gk. *meta*, after; *karpos*, wrist] The collective name for the five bones which support the palm of the hand.

metamere (mĕt′ămēr) [Gk. *meta*, after; *meros*, part] A body segment.

metaphase (mĕt′ăfāz) [Gk. *meta*, after; *phainein*, to appear] The stage in mitosis in which split chromosomes are arranged on equatorial plane.

metatarsus (mĕt′ă târ′sŭs) [Gk. *meta*, after; L. *tarsus*, ankle] Collective name for the five bones supporting the region of the foot between the ankle and the digits.

metencephalon (mĕt′ĕnsĕf′ălŏn) [Gk. *meta*, after; *en*, in; *kephale*, head] The forward portion of hindbrain.

microcephalus (mī′krŏsĕfăl′ŭs) [Gk. *mikros*, small; *kephale*, head] A condition resulting in an abnormally small head; more specifically, a small cranial cavity—1350 cc or less.

microglia (mīkrŏg′lĭă) [Gk. *mikros*, small; *glia*, glue] Small phagocytic cells of mesodermal origin in gray and white matter of central nervous system.

microvilli (mīkrö vĭl ī) [Gk. *mikros*, small; L. *villus*, shaggy hair] Very tiny protoplasmic projections of epithelial cells; visible individually only with the use of the electron microscope. Make up striated and brush borders.

micturition (mĭk′tūrĭsh′ŭn) [L. *mengere*, to void water] The emptying of the bladder through the urethra.

mitochondria (mi′tokon′dria) [Gk. *mitos*, thread; *chondros*, grain] Spherical or rod-shaped organelles of the cytoplasm, which are the centers of catabolic enzyme activity.

mitosis (mĭtō′sĭs) [Gk. *mitos*, thread] Indirect nuclear division.

modiolus (mödĭ′ŏlŭs) [L. *modiolus*, small measure] The conical-shaped central axis of the cochlea of the ear.

monocytes (mŏn′ösīts) [Gk. *monos*, alone; *kytos*, hollow] Among the largest of the leukocytes having phagocytic properties.

mons pubis (monz pu′bis) [L. *mons*, mountain; *pubis*, mature] A fatty eminence in front of the symphysis pubis.

mucin (mū′sĭn) [L. *mucus*] Protein material produced by mucous cells.

mucous membrane (mū′kŭs) [L. *mucus*] Lines all hollow organs and cavities which open upon the skin surface of the body.

multipennate (mul-tipĕn-āt) [L. *multus*, many; *penna*, contour feather] Muscle fiber arrangement in which the fibers converge to many tendons.

muscle (mŭs'ĕl) [L. *musculus*, muscle] Generally characterized by high degree of contractility.

myelencephalon (mī'ĕlĕnsĕf'ălŏn) [Gk. *myelos*, marrow; *en*, in; *kephale*, head] The lower part of hindbrain.

myelin (mī'ĕlĭn) [Gk. *myelos*, marrow] A white fatty material forming medullary sheath of nerve fibers.

myoblast (mī'ōblăst) [Gk. *mys*, muscle; *blastos*, bud] A cell which develops into muscle fiber.

myocardium (mī'ōkar'dĭum) [Gk. *mys*, muscle; *kardia*, heart] The thick muscular layer of the heart wall.

myoepithelial (mī'ōĕpĭthe'lĭăl) [Gk. *mys*, muscle; *epi*, upon; *thele*, nipple] Contractile cells of epithelial origin; present in some glands such as mammary, salivary and those of eyelids.

myofibril (mī'ōfībrĭl) [Gk. *mys*, muscle; L. *fibrilla*, small fiber] Contractile fibril within muscle fiber.

myogenesis (mī'ōjĕn'ĕsĭs) [Gk. *mys*, muscle; *logos*, discourse] Origin and development of muscles.

myology (mīŏl'ōjĭ)]Gk. *mys*, muscle; *logos*, discourse] The study of the muscular system.

myoneural junction (mī'ōnū'răl) [Gk. *mys*, muscle; *neuron*, nerve] That point at which terminal nerve branches make connection with muscle fibers.

myxedema (mĭk'sĕde'mă) [Gk. *myxo*, slime, mucus; *edema*, swelling] A hypothyroid in adults characterized by small thyroid, slow pulse, dry and wrinkled skin, dull mentality, sluggishness, low basal metabolism, baggy swellings of face and hands.

nares (nā'rĕz) [L. *nares*, nostrils] The openings into the nasal cavities.

nephritis (ne-fri'tus) [Gk. *nephros*, kidney; *itis*, inflammation] Inflammation of the kidney tubules.

nephron (nĕf'rŏn) [Gk. *nephros*, kidney] Structural and functional unit of the kidney, including the renal corpuscle, convoluted tubules, and Henle's loop.

nerve (nerv) [L. *nervus*, sinew] Bundles of nerve fibers coursing together outside central nervous system.

nervous (tissue) (nĕr'vŭs) [L. *nervus*, sinew] Composed of cells specialized in the properties of irritability and conductivity.

neural tube (nū'răl) [Gk. *neuron*, nerve] The structure made by the union of the neural folds.

neurilemma (nū'rĭ'lĕm'a) [Gk. *neuron*, nerve; *lemm*, husk, sheath] Single layer of flattened cells found on fibers of peripheral and autonomic system (also sheath of Schwann).

neurofibrils (nū'rŏfī'brĭlz) [Gk. *neuron*, nerve; L. *fibrilla*, fine fiber] Delicate structures which extend through the nerve cell and into the processes—also involved in conduction.

neuroglia (nūrŏg'lĭă) [Gk. *neuron*, nerve; *glia*, glue] Supporting and protecting tissue of the central nervous system consisting, in part, of macroglial and microglial cells and their processes.

neuromere (nū'rōmēr) [Gk. *neuron*, nerve; *meros*, part] Spinal segment.

neuromuscular spindles (nū'rōmŭs'kūlăr) [Gk. *neuron*, nerve; L. *musculus*, muscle] Receptors present in most skeletal muscles.

neuron (nū'rŏn) [Gk. *neuron*, nerve] A complete nerve cell with outgrowths constituting the basic structural unit of nervous system.

neurohypophysis (nū'rŏhipŏf'isis) [Gk. *neuron*, nerve; *hypo*, under; *phyein*, to grow] The posterior portion of the hypophysis.

neurotendinous spindles (nū'rōtĕn'dinus) [Gk. *neuron*, nerve; L. *tendere*, to stretch] Proprioceptors associated with tendons near their junctions with muscle fibers.

neutrophils (nū'trŏfĭl) [L. *neuter*, neither; Gk. *philein*, to love] Abundant leukocytes having a nucleus of three to five lobes and fine granules which stain light orchid with Wright' s stain.

Nissl bodies [Franz Nissl, neurologist in Heidelberg, 1860–1919] Angular protein particles found in cytoplasm of nerve cell; related to cell metabolism.

nodes of Ranvier [Louis Antoine Ranvier, French pathologist, 1835–1922] Constrictions at intervals of the medullary sheath of nerve fiber.

norepinephrine (nor"ep-e-nef'rin) A hormone of adrenal medulla; differs from epinephrine in absence of an N-methyl group.

notochord (nō'tōkôrd) [Gk. *noton*, back; *chorde*, cord] A semistiff axial rod in the middorsal line between the chordate nerve cord and the dorsal aorta.

nucleolus (nukle'olus) [L. *nucleolus*, little kernel] A rounded mass occurring in nucleus.

nucleus (nu'kleus) [L. *nucleus*, kernel] The controlling center of the cell. A group of nerve cell bodies in the central nervous system.

nucleus pulposus (nū'klĕŭs) [L. *nucleus*, kernel] The soft core of an intervertebral disc; a remnant of notochord.

omentum (ōmĕn'tŭm) [L. *omentum*, fold] A fold of mesentery either free or acting as a connection between organs.

ontogeny (ŏntŏj'ĕnĭ) [Gk. *on*, being; *genisis*, descent] The entire life history of the individual.

40

optic chiasma (ŏp′tĭk kĭăz′mă) [Gk. *opsis*, sight; *chiasma*, cross] The point of decussation of optic nerves; anterior to the infundibulum.

ora serrata (o′rah ser-a′tah) [L. *oralis*, mouth; *serratus*, saw-like] The jagged anterior margin of the retina near the ciliary body where its nervous portions cease.

organ of Corti [Alfonso Corti, Italian histologist, 1822–1888] The spiral organ on the inner portion of the basilar membrane; contains the vital acoustic cells and their supporting cells; the true receptor for hearing.

organelles (ôrgănĕlz) [Gk. *organon*, instrument] The living particles of the cytoplasm.

organology (ôr′gănŏl′ŏjĭ) [Gk. *organon*, instrument; *logos*, discourse] The study of organs as they are developed from tissues.

orgasm (ôr′găzm) [Gk. *organ*, to swell] The climax of sexual excitement.

orifice (ŏr′ĭfĭs) [L. *os*, mouth; *facere*, to make] An opening or aperture of a tube of duct.

ossicle (ŏs′ĭkl) [L. *os*, bone] Any small bone; specifically, the small bones of the middle ear, malleus, incus, and stapes.

osteoblasts (ŏs′tëöblăst) [Gk. *osteon*, bone; *blastos*, bud] Bone-forming cells.

osteoclasts (ŏs′tëöklăst) [Gk. *osteon*, bone; *klan*, to break] Bone-destroying cells.

osteoid tissue (ŏs′tëiod) [Gk. *osteon*, bone; *eidos*, form] Young bone previous to calcification; pre-osteal tissue.

osteology (ŏs′teŏl′ōjĭ) [Gk. *osteon*, bone; *logos*, discourse] The study of bones; structure, nature, and development.

osteomalacia (os′te-o-mah-la′she-ah) [Gk. *osteon*, bone; *malakos*, soft] Condition in which bones become soft and pliable due to vitamin D deficiency.

otoconium (ō′tōkō′nĭŭm) [Gk. *oto*, ear; *konis*, sand] Crystal of calcium carbonate attached to hairs of maculae.

ovary (ō′vărĭ) [L. *ovarium*, ovary] The female reproductive and endocrine organ producing ova and hormones.

ovulation (ōvūlā′shŭn) [Gk. *ovum*, egg; *latum*, borne away] The emission of the egg from the ovary.

ovum (ō′vŭm) [L. *ovum*, egg] A female germ cell; mature egg cell.

Pacinian corpuscles (pa sin′e-an kor′pus-l) [Filippo Pacini, Italian anatomist; 1812–1883; L. *corpus′- culum*, little body] The largest of the nerve end organs of the subcutaneous layer of the skin—nerve fibril covered by a series of concentric lamellae.

palate (păl′ĕt) [L. *palatum*, palate] The roof of the mouth.

palmar (păl′măr) [L. *palma*, palm of hand] Pertaining to the palm of the hand; palmar surface.

pampiniform plexus (pămpĭn′ĭfôrm) [L. *pampinus*, tendril; *forma*, shape] A large plexus formed by the somatic veins of the testes.

pancreas (păn′krĕăs) [Gk. *pan*, all; *kreas*, flesh] A compound tubulo-alveolar gland, with exocrine and endocrine functions.

panniculus adiposus (pănĭk′ūlŭs ăd′ĭpōsŭs) [L. *pannus*, cloth; *adeps*, fat] The name given superficial fascia where fat is abundant.

papilla(ae) (păpĭl′ă) [L.] A small nipple-shaped elevation.

papillary layer (pap′ilari) [L. *papilla*, nipple] Outer layer of the dermis, characterized by numerous projections into epidermis.

papillary plexus (păp′ĭlarĭ plĕk′sŭs) [L. *papilla*, nipple; *plexus*, interwoven] Network of arteries at the level of the papillary layer.

paraganglia (păr′ăgăng′glĭă) [Gk. *para*, beside; *ganglion*, swell] Autonomic ganglia and plexuses.

paranasal sinus (parah nā′zăl) [Gk. *para*, beside; L. *nasus*, nose] Spaces in the maxillary, frontal, sphenoid and ethmoid bones; open into nasal passageways.

parasagittal (pără′săjĭt′ăl) [L. *para*, beside; *sagitta*, arrow] Any plane parallel to the median plane.

parasympathetic (păr′ăsĭmpăthĕt′ĭk) [Gk. *para*, beside; *sym*, with; *pathos*, feeling] One of the two divisions of the visceral efferent system of the autonomic nervous system; craniosacral portion.

parathyroid (pără′thī′roid) [Gk. *para*, beside; *thyreos*, shields; *eidos*, form] One of four small endocrine glands embedded in posterior side of thyroid.

parotid gland (par-ot′id) [L. *para*, beside; A.S. *eare*, ear] Paired salivary glands lying below and in front of ear and opening into mouth.

pars distalis (parz dis tal′is) The anterior lobe or bulk of adenohypophysis of pituitary.

pars intermedia (parz in-ter-me′de-ă) The portion of adenohypophysis between the pars distalis and neurohypophysis.

pars tuberalis (parz tu-ber-al′is) Small mass of tissue along the infundibulum from the adenohypophysis of pituitary.

parturition (pârtūrĭsh'ŭn) [L. *parturire*, to bring forth] The act or process of birth.

pedicle (pĕd'ĭkĕl) [L. *pediculus*, small foot] A backward-projecting vertebral process.

pelvis (pĕl'vĭs) [L. *pelvis*, basin] The bony cavity formed by pelvic girdle along with coccyx and sacrum; also the cavity in the kidney at the superior end of the ureter.

penis (pē'nĭs) [L. *penis*] The male copulatory organ.

pennate (pĕn'āt) [L. *penna*, contour feather] Muscle fiber arrangement in which the fibers are attached to tendon in a feather-like manner.

pepsin (pĕp'sĭn) [Gk. *pepsis*, digesting] An enzyme secreted by chief cells of stomach which changes the proteins into proteoses and peptones.

perforating fibers (pĕr'fŏrāt'ĭng) [L. *perforare*, to bare through] Fibers of the inner layer of periosteum which are continuous with those in the matrix of the bone.

perforations (pĕr'fŏr'āshŭn) [L. *perforare*, to bare through] Pores or openings.

perichondrium (pĕr'ĭkôn'drĭŭm) [Gk. *peri*, round; *chondros*, cartilage] A fibrous membrane that covers cartilage.

perikaryon (pĕr'ĭkăr'ĭŏn) [Gk. *peri*, round; *karyon*, nucleus] A nerve cell body, as distinct from its axon and dendrites.

perilymph (pĕr'ĭlĭmf) [Gk. *peri*, round; L. *lympha*, water] A fluid separating membranous from osseous labyrinth of ear.

perimysium (pĕr'ĭmĭz'ĭŭm) [Gk. *peri*, around; *mys*, muscle] Layer of loose connective tissue covering fasciculi of muscle fibers.

perineum (pĕr'ĭnē'ŭm) [Gk. *perinaion*, part between anus and scrotum] The region of the outlet of the pelvis.

perineurium (pĕr'ĭnū'rĭŭm) [Gk. *peri*, round; *neuron*, nerve] Areolar connective tissue forming outer wrapping of fasciculi of nerve fibers.

periosteum (pĕr'ŏs'tĕŭm) [Gk. *peri*, around; *osteom*, bone] A fibrous membrane around bone.

perirenal (per-e-re'nal) The area around the kidney.

peristalsis (pĕr'ĭstăl'sĭs) [Gk. *peri*, round; *stellein*, to place] Progressing waves of contraction along a muscular tube by action of circular muscles; moves material through tube.

Peyer's patches [Johann Konrad Peyer, Swiss anatomist, 1653–1712] Oval patches of aggregated lymph follicles on walls of ileum.

phagocyte (făg'ŏsīt) [Gk. *phagein*, to eat; *kytos*, hollow] A colorless blood corpuscle, or other cell which ingests foreign particles.

phagocytic macrophage cells (făg'ŏsĭt-ĭc măk'rŏfăj) [Gk. *phagein*, to eat; *kytos*, hollow; *makros*, large; *phagein*, to eat] Large cells which are able to ingest bacteria and particulate material.

phalanx, phalanges, plu. (făl'ăngks; fălăn'jēz) [Gk. *phalangx*, line of battle] The bones of the digits.

phallus (făl'ŭs) [Gk. *phallos*, penis] The embryonic structure which becomes penis or clitoris.

pharyngeal pouches (fărĭn'jĕăl) [Gk. *pharyngx*, gullet] Evaginations of the lateral pharyngeal walls.

pharynx (far'ingks) [Gk. *pharyngx*, gullet] The anterior part of alimentary canal following the buccal cavity.

photopic (fo-top'ik) Vision in the light; an eye which is light-adapted; vision with cones.

phylogeny (fīlŏj'ĕnĭ) [Gk. *phyllon*, race; *genesis*, origin, descent] History of development of species or race.

physiology (fĭzĭŏl'ŏjĭ) [Gk. *physis*, nature; *logos*, discourse] That part of biology dealing with functions of organisms.

pia mater (pī'ă mā'tēr) [L. *pia mater*, kind mother] The innermost meninx which is a delicate membrane closely investing brain and spinal cord.

pineal body (pin'ē-al) [L. *pinea*, pine cone] A median outgrowth from roof of diencephalon.

pinna (pĭn'ă) [L. *pinne*, feather] Auricle or outer ear.

pitocin (pi-to'sin) A hormone of the posterior lobe of the hypophysis.

pitressin (pi-tres'in) A hormone of the posterior lobe of the hypophysis.

pituicyte (pĭtū'ĭsĭt) [L. *pituita*, phlegm; Gk. *kytos*, hollow] A modified glial cell of the hypophysis.

pituitary gland (pĭtū'ĭtarĭ) [L. *pituita*, phlegm] The hypophysis.

placenta (plăsĕn'tă) [L. *placenta*, flat cake] A double structure derived in part from maternal tissue and in part from the embryo.

plantar (plăn'tăr) [L. *planta*, sole of foot] Refers to the sole of the foot.

plasma (plăz'mă) [Gk. *plasma*, form] The fluid or intercellular part of blood.

pleura (ploor'ă) [Gk. *pleura*, side] A serous membrane lining thoracic cavity and investing lung.

plexus (plĕk'sŭs) [L. *plexus*, interwoven] A network of interlacing blood vessels or nerves.

plica circularis (plī'ka) [L. *plicare*, to fold] Valve-like fold of the mucosa and submucosa which projects into the intestinal lumen.

polar bodies (pō′lăr) [Gk. *polos*, pivot] Cells divided off from ovum during maturation.

pons (pŏnz) [L. *pons*, bridge] A structure connecting two parts, as the pons of the hindbrain.

postnatal (pōst′nā′tăl) [L. *post*, after; *natus* (past. part.), to be born] The period after birth.

premolar (prēmō′lăr) [L. *prae*, before; *mola*, mill] Premolars or bicuspids are found between the canines and molars.

prenatal (prē′nā′tăl) [L. *prae*, before; *natus* (past. part.), to be born] Preceding birth.

prepuce (pre′pūs) Part of integument of penis which leaves surface at the neck and is folded upon itself.

prime mover A muscle directly responsible for change in position of a part.

primitive groove A mass of rapidly proliferating cells on the dorsal side of the embryo in which lies in the primitive streak.

primitive node Thickened mass of proliferating cells at the cephalic end of the primitive streak.

primitive streak (prĭm′ĭtĭv) [L. *primitivus*, early] A mass of rapidly proliferating cells on the dorsal side of the embryo.

process (pros′es) [L. *processus*, process] A broad designation for any bony prominence or prolongation.

progesterone (prōjĕs′tĕrōn) [L. *pro*, for; *gestare*, to bear] Hormone of the corpus luteum.

pronation (prōnā′shŭn) [L. *pronare*, to bend forward] Medial rotation of the forearm which brings the palm of the hand downward.

prophase (prō′fāz) [Gk. *pro*, before; *phasis*, appearance] The first stage in mitosis.

proprioception (prō′prĭōsĕp′shŭn) [L. *proprius*, one's own; *capere*, to take] Sensations which convey position and movements of joints and muscles, hence of the parts of the body.

proprioceptor (prō′prĭōsĕp′tŏr) [L. *proprius*, one's own; *capere*, to take] A receptor which receives stimuli from the muscle tissue and tendons and enables us to orient the body and its parts.

prosencephalon (prŏs′ĕnsĕf′ălŏn) [Gk. *pro*, before; *engkephalos*, brain] The most anterior enlargement of the brain, the forebrain.

prostate (prŏs′tāt) [L. *pro*, before; *stare*, to stand] A muscular and glandular organ found ventral to the rectum and inferior to the urinary bladder.

prostatic utricle (prŏstăt′ĭk ū′trĭkl) [L. *pro*, before; *stare*, to stand; *utriculus*, small bag] Small recess in urethral crest, homologous to uterus of female.

protoplasm (prō′tōplăzm) [Gk. *protos*, first; *plasma*, form] The living substance in all plant and animal bodies.

protozoa (prō′tö zō′ăh) [Gk. *protos*, first; *zoon*, animal] Phylum of unicellular animal organisms.

proximal (prŏk′sĭmăl) [L. *proximus*, next] Nearest body, center or base of attachment.

ptyalin (tī′ălĭn) [Gk. *ptyalon*, saliva] The enzyme of saliva which initiates carbohydrate digestion.

pudendum (pūděn′dŭm) [L. *pudere*, to be ashamed] External female genitalia.

Purkinje fibers (pur-kin′jez) [Johannes Evangelista Purkinje, Bohemian physiologist, 1787–1869] Specialized muscular fibers in the subendocardial tissue forming an important part of the intrinsic conduction mechanism of the heart.

pyknotic (pĭknŏt′ĭk) [Gk. *pyknosis*, condensation] Small irregular nucleus of degenerated cells.

pyloric glands (pīlŏr′ik) [Gk. *pyloros*, gate-keeper] Short, branched, tubular gastric glands found in the pyloric portion of the stomach.

pyramid (pĭr′ămĭd) [L. *pyramis*, pyramid] A conical structure, protuberance or eminence; as pyramids of kidneys.

radiate (rā′dĭăt) [L. *radius*, ray] Muscle fiber arrangement in which the fibers converge from a broad area to a common tendinous point.

ramus (rā′mŭs) [L. *ramus*, branch] Any branch-like structure.

raphe (rā′fē) [Gk. *rhaphe*, seam] A ridge or seam-like structure.

Rathke's pouch [Martin H. Rathke, German anatomist, 1793–1860] A diverticulum of ectoderm from the roof of the stomodaeum.

receptors (rĕsĕp′tŏr) [L. *recipere*, to receive] Sense organs which receive stimuli from the environment.

rectal columns (rĕk′tăl) [L. *rectus*, straight] Folds of mucosa and muscle tissue of the upper portion of anal canal.

rectum (rĕk′tŭm) [L. *rectus*, straight] The continuation of the digestive tract from the pelvic colon to anal orifice.

reflex (rē′flĕks) [L. *reflectere*, to turn back] An involuntary response to stimulus.

renal columns (rē′năl) [L. *ren*, kidney] Projections of cortical arches between pyramids of kidneys.

renal corpuscle (rē′năl) [L. *ren*, kidney] The glomerulus and Bowman's capsule of a nephron.

renal fascia (rē′năl făsh′ĭă) [L. *ren*, kidney; *fascia*, band] A part of the subserous fascia supporting the kidney.

rennin (ren′in) An enzyme secreted by the chief cells of the stomach which curdles milk.

respiration (rĕs'pĭrā'shŭn) [L. *re*, again; *spirare*, to breathe] Interchange of oxygen and carbon dioxide between an organism and its surrounding medium.

rete testis (rē'tē) [L. *rete*, net] Network of tubes formed by the tubuli recti.

reticular formation (rĕtĭk'ūlâr) [L. *reticulum*, small net] Minute nerve network extending through central part of brain stem.

retina (rĕt'ĭnă) [L. *rete*, net] The nervous coat which forms the inner layer of the eyeball; contains rod and cone cells.

retroperitoneal (rĕt'röpĕr'ĭtönē'ăl) [L. *retro*, backwards; Gk. *peri*, round; *teinein*, to stretch] Behind peritoneum.

rhombencephalon (rômb'ĕnsĕf'ălŏn) [Gk. *rhombos*, wheel; *engkephalos*, brain] Hindbrain.

rickets (rĭk'ĕts) A condition of bones caused by a vitamin D deficiency in which normal ossification does not take place; results in deformation of bones.

rod One of the dim light receptors of the retina.

root (root) [A.S. *wyrt*, root] That part of the tooth which embeds in the bony alveolus.

rotation (rötā'shŭn) [L. *rota*, wheel] Movement of a bone around an axis, either its own or that of another.

ruga (roog'ă) [L. *ruga*, wrinkle] Prominent fold of the mucosa and submucosa of the stomach lining.

saccule (săk'ūl) [L. *sacculus*, small bag] The lower and smaller of the two chambers of the vestibular portions of the membranous labyrinth.

sagittal (săjĭt'ăl) [L. *sagitta*, arrow] The median plane or any plane parallel to it which divides the body into right and left parts.

salivary glands (săl'ĭvărĭ) [L. *saliva*, spittle] The three glands of the mouth region involved in the production and secretion of saliva.

sapiens (sā'pĭĕns [L. *sapiens*, wise] The species name of man.

sarcostyles (sâr'köstīl) [Gk. *sarx*, flesh; *stylos*, pillar] A fibril of muscular tissue.

sarcolemma (sâr'kölĕm'a) [Gk. *sarx*, flesh; *lemma*, skin] An elastic, membranous sheath enclosing muscle cells.

sarcoplasm (sâr'köplăzm) [Gk. *sarx*, flesh; *plasma*, mold] The interstitial substance between fibrils of muscle fibers.

scala tympani (skā'lă) [L. *scala*. ladder] The lower portion of the divided canal of the cochlea.

scala vestibuli (skā'lă) [L. *scala*, ladder] The upper portion of the divided canal of the cochlea.

sclera (sklē'ră) [Gk. *skleros*, hard] The outer fibrous tunic of the eyeball.

sclerotomes (sklēr'ötōm) [Gk. *skleros*, hard; *tome*, cutting] Mesenchymatous tissue from somites destined to form vertebrae.

scoliosis (skō lī'osĭs) [Gk. *skoliosis*, crookedness] An abnormal lateral curvature of the vertebral column.

scotopic (sko-top'ik) Pertains to vision in the dark; a dark-adapted eye; vision with the rods of the retina.

scrotum (skrō'tŭm) [L. *scrotum*] A medial pouch of loose skin which contains testes in mammals.

sebaceous (sēbā'shus) [L. *sebum*, tallow] Epithelial gland which secretes sebum.

secretin (sēkrē'tĭn) [L. *secernere*, to separate] A hormone which initiates the secretions of pancreatic juices.

secretion (sēkrē'shŭn) [L. *secretio*, separation] Substance or fluid which is separated and elaborated by cells or glands.

segmentation (sĕg'mentā'shŭn) [L. *segmentum*, piece] The division or splitting into segments or parts.

semicircular canals (sĕm'ĭsĕr'kŭlăr) [L. *semi*, half; *circulus*, circle] Three bony canals in mammals, lying posterior to the vestibule which serve in maintaining equilibrium.

seminal vesicle (sĕm'ĭnăl ves'ikl) [L. *semen*, seed; *vesicula*, bladder] A convoluted and saccular outgrowth of the ductus deferens, behind bladder; produces fluid for sperms.

seminiferous tubules (sĕmĭnĭf'ĕrŭs) [L. *semen*, seed; *ferre*, to carry] The structure in which the spermatozoa and seminal fluids are produced.

septum (sĕp'tŭm) [L. *septum*, partition] A partition of connective tissue separating two cavities or masses.

serous membrane (sē'rŭs) [L. *serum*, serum] Lines the celomic cavities; contributes to mesenteries and omenta, and covers the outer surfaces of related organs.

sesamoid (sĕs'ămoid) [Gk. *sesamon*, sesame; *eidos*, form] A bone developed within a tendon and near a joint.

sino-atrial node (si-no-a'tre-al nōd) Collection of specialized myocardial cells in right atrial wall which initiates heart beat (derived in part from sinus venosus).

sinusoid (sī'nŭsoid) [L. *sinus*, curve; Gk. *eidos*, form] A minute blood space in organ tissue, as the liver.

sinus venosus (sī′nŭs veno′sus) [L. *sinus*, curve; *vena*, vein] A receiving chamber for veins entering the heart in the lower vertebrates and in the embryos of higher vertebrates.

somatopleure (sō′mătöploor) [Gk. *soma*, body; *pleura*, side] The outer wall of the mesoderm, together with the ectoderm against which it lies.

somites (sō′mĭt) [Gk. *soma*, body] Blocks formed by clefts in the thickened embryonic mesoderm on either side of the neural tube and notochord.

species (spē′shēz) [L. *species*, particular kind] A systematic unit including geographic races and varieties, included in a genus.

spermatogonium (spěr′mătögō′nĭum) [Gk. *sperma*, seed; *gonos*, offspring] Sex cells derived from cords of epithelial cells of the testes.

spermatozoon (spěr′mătözō′ŏn) [Gk. *sperma*, seed; *zoon*, animal] A male reproductive cell.

sphincter (sfĭng′ktěr) [Gk. *sphinggein*, to bind tight] A muscle which contracts and closes an orifice.

spinal ganglion (spī′năl găng′glĭŏn) [L. *spina*, spine; Gk. *ganglion*, little tumor] An aggregate of nerve cell bodies on the dorsal root of the spinal nerve.

spine (spĭn) [L. *spina*, spine] A more or less sharp projection.

splanchnocranium (splăngk′nökrā′nĭum) [Gk. *splangchnon*, entrail; *kranion*, skull] Jaws and visceral arches of the skull.

splanchnopleure (splăngk′nöploor) [Gk. *splangchnon*, entrail; *pleura*, side] The inner wall of the mesoderm with the adjacent endoderm.

stimulus (stĭm′ūlŭs) [L. *stimulare*, to incite] An environmental change or an act which produces reaction in a receptor or in an irritable tissue.

stomodeum (stöm′ödē′ŭm) [Gk. *stoma*, mouth; *hodaios*, on the way] Anterior invaginated portion of embryonic gut.

stratum corneum (strā′tŭm kor′ne-um) [L. *stratum*, layer; *cornu*, horn or horny] The outermost layer of the epidermis of the skin.

stratum granulosum (strā′tŭm gran-u-lo′sum) [L. *stratum*, layer; *granulum*, small grain] A layer of the epidermis below the stratum lucidum; granular in appearance.

stratum lucidum (strā′tŭm lu′sid um) [L. *stratum*, layer; *lucidus*, clear] The clear layer between the stratum corneum and stratum granulosum of the skin.

stratum spinosum (strā′tŭm spi′nos um) [L. *stratum*, layer; *spinose*, spinous] The inner, growing layer of the epidermis which rests on the papillary surface of the dermis.

subarachnoid space (sŭbărăk′noid) [L. *sub*, under; Gk. *arachne*, spider's web; *eidos*, form] A wide space surrounding the spinal cord and its pia mater; under the arachnoid.

subdural space (sŭbdū′răl) [L. *sub*, under; *durus*, hard] A space containing a small amount of fluid below the dura mater.

sublingual gland (sŭblĭng′gwăl) [L. *sub*, under; *lingua*, tongue] A salivary gland found in a fold of mucous membrane in the floor of the mouth.

submaxillary gland (sub′maksil′ari) [L. *sub*, under; *maxilla*, jaw] A salivary gland which lies below the body of the mandible and mylohyoid muscle; opens into mouth; now called the submandibular gland.

subserous fascia (sŭbsē′rŭs) [L. *sub*, under; *serum*, whey] Fascia present beneath a serous membrane.

sudoriferous (sū′dorĭf′erus) [L. *sudor*, sweat; *ferre*, to carry] Simple coiled tubules commonly called sweat glands.

sulcus (sul′kus) [L. *sulcus*, furrow] A groove.

summation (sum-a′shun) The process by which changes produced by stimuli are accumulated until threshold is reached.

supination (sūpĭnā′shŭn) [L. *supinus*, bent backward] Lateral rotation of the forearm which brings the palm of the hand upward.

sustentacular cells (sŭstěntăk′ūlăr) [L. *sustentaculum*, prop, support] Supporting cells.

suture (sū′tūr) [L. *sutura*, seam] Line of junction of two bones immovably connected; as in the skull.

sympathetic (sĭmpăthět′ĭk) [Gk. *syn*, with; *pathos*, feeling] One of the divisions of the visceral efferent systems of the autonomic nervous system; thoracolumbar portion.

symphysis (sĭm′fĭsĭs) [Gk. *symphysis*, a growing together] Permanent cartilaginous joints.

synapse (sĭnăps) [Gk. *synapsis*, union] The area of functional continuity between neurons.

synapsis (sĭnăp′sĭs) [Gk. *synapsis*, union] The union of homologous chromosomes.

synchondrosis (sĭn′kŏndrō′sis) [Gk. *syn*, with; *chondros*, cartilage] Temporary cartilaginous joint.

syncytium ((sĭnsĭt′ĭŭm) [Gk. *syn*, with; *kytos*, hollow] Condition in which no membrane separates the cells.

syndesmosis (sĭn'dĕsmō'sĭs) [Gk. *syndesmos,* ligament] Articulations with fibrous tissue between the bones.

synergist (syn'ergist) [Gk. *synergos,* cooperate] Muscles which combine with prime movers and fixation muscles in movement.

synostosis (sin-os-to'sis) [Gk. *syn,* with, together; *osteon,* bone] The union of adjacent bones by means of osseous matter.

synovial joints (sĭnō'vĭal) [Gk. *syn,* with; L. *ovum,* egg] Joints characterized by one or more synovial cavities.

synovial membranes (sĭnō'vĭal) [Gk. *syn,* with; L. *ovum,* egg] Lining of articular capsules which secretes a synovia (synovial fluid).

systole (sĭs'tōlē) [Gk. *systole,* drawing together] Contraction phase of the heart beat.

tactile corpuscles (tăk'tĭl kôr'pŭsël) [L. *tanger,* to touch; L. *corpusculus,* small body] Specialized nerve endings of the papillae; receptors of touch.

tactile corpuscles of Meissner (Meissner's corpuscles) [Georg Meissner, German histologist, 1829–1905] Receptors for light touch occurring in the papillae of the corium.

taeniae coli (tē'nē-e) [L. *taenia,* ribbon] Three bands formed by the longitudinal muscle fibers of the large intestine.

tarsal (târ'săl) [Gk. *tarsos,* sole of foot] Pertains to tarsus bones, or to certain glands of the tarsal region of the eyelids.

telencephalon (tĕl'ĕnsĕf'ălŏn) [Gk. *tele,* far; *engkephalos,* brain] Anterior terminal segment of brain.

telereceptor (tele'rĕcĕp'tôr) [Gk. *tele,* far; *recipere,* to receive] Distance receptors.

telophase (tĕl'ōfāz) [Gk. *telos,* end; *phasis,* aspect] Final phase of mitosis in which the nuclei become reestablished.

tendon (tĕn'dŏn) [L. *tendere,* to stretch] A white fibrous cord connecting a muscle with another structure, usually bone.

testis (tĕs'tĭs) [L. *testis,* testicle] Male reproductive and endocrine organ producing spermatozoa and male sex hormones.

testosterone (tĕs'tōstē'rŏn) [L. *testis,* testicle; Gk. *stear,* suet] A hormone produced by the interstitial cells of the mature testis.

tetany (tĕt'ăh-ne) [Gk. *tetanos,* stretched] A condition marked by muscular spasms, sharp flexion of wrist and ankle joints, cramps, and convulsions; due to abnormal calcium metabolism.

tetrad (tĕt'răd) [Gk. *tetros,* four] A quadruple group of chromatids in meiosis.

thalamus (thăl'ămŭs) [Gk. *thalamos,* receptacle] One of two large nuclear masses which form lateral walls of diencephalon forming an important sensory center of brain.

thyroid (thī'roid) [Gk. *thyra,* door; *eidos,* form] An endocrine gland which lies in the neck region; produces thyroxin.

thyroid cartilage (thī'roid kâr'tĭlĕj) [Gk. *thyra,* door; *eidos,* form; L. *cartilago,* cartilage] The largest single cartilage of the larynx.

thyrotrophic (thī'rōtrŏf'ĭk) [Gk. *thyra,* door; *trophe,* nourishment] Hormone secreted by hypophysis which governs release and production of thyroxin.

thyroxin (thī'rŏksĭn) [Gk. *thyra,* door; *oxys,* sharp] A hormone of the thyroid gland.

tidal air (tīd-al) The air exchange in normal breathing.

tonsil (tŏn'sĭl) [L. *tonsilla,* tonsil] Aggregates of lymphatic follicles in the pharynx.

tonsilar crypts (tŏn'si-lar kripts) [L. *tonsilla,* tonsil; *crypta,* hidden] Pit-like depressions of the epithelium covering the free surface of the palatine tonsils.

tonus (tōn'ŭs) [Gk. *tonos,* tension] A constant state of partial contraction or tension.

torus (tō'rŭs) [L. *torus,* swelling] A firm prominence or marginal ridge.

trabeculae (trăbĕk'ūlē) [L. *trabecula,* little beam] Septa of connective tissue or muscle extending from a capsule or wall into the enclosed substance or cavity of an organ as in lymph nodes, trabeculae carneae of heart, etc.

trachea (trăkē'ă) [L. *trachia,* windpipe] A fibroelastic tube found at the level of the sixth cervical vertebra to the fifth thoracic vertebra; carries air to and from lungs.

tracheal rings (trăkē'ăl) [L. *trachia,* windpipe] Cartilaginous rings in the mesenchyme of the trachea which prevent collapsing of the tube.

tract (trăkt) [L. *trahere,* to draw] An organized system of nerve fibers within the central nervous system.

transverse (trănsvĕrs') [L. *transversus,* across] (same as horizontal) A plane at right angles to both the sagittal and frontal planes, dividing the body into superior and inferior portions.

trigone (trī'gōn) [Gk. *trigonon,* triangle] A small triangular area in the urinary bladder between the orifices of the ureters and urethra.

trochanter (trōkăn'tēr) [Gk. *trochanter*, runner] A very large, usually blunt, process.

trochlea (trŏk'lēă) [Gk. *trochilia*, pulley] A pulley-like structure through which a tendon passes.

trochlear (trŏk'lēăr) [Gk. *trochilia*, pulley] The fourth cranial nerve; also a pulley.

trophoblast (trŏf'ŏblăst) [Gk. *trophe*, nourishment; *blastos*, bud] The outer layer of cells of the morula.

tuber cinerium (tu'ber sin-e're-um) [L. *tuber*, knob; *cinereus*, ashen-hued] A rounded eminence of gray matter forming part of the inferior surface of the hypothalamus between the mammillary bodies and the optic chiasma; the infundibulum arises from its under-surface.

tubercle (tū'bĕrkĕl) [L. *tuberculum*, small hump] Usually a small, rounded eminence.

tuberosity (tū'bĕrŏs'ĭtĭ) [L. *tuber*, hump] Usually, a large, rounded eminence.

tubuli recti (tu'buli) [L. *tubulus*, small tube; *rectus*, straight] The less convoluted, nearly straight ducts of the seminiferous tubules.

tunica adventitia (tū'nĭkă ad-ven-tish'-e-ah) [L. *tunica*, coating; *ad*, to; *venire*, to come] The outer tunic of various tubular structures, as arteries, veins, esophagus, uterine tubes, and ductus deferens.

tunica externa (tu'nik-ah ex'terna) Outer layer of wall of artery or vein.

tunica intima (tu'nik-ah in'tima) The innermost coat of wall of artery or vein.

tunica media (tu'nik-ah me'dia) Intermediate coat of the wall of an artery or vein.

tunica vaginalis (tū'nĭkă văj'inăl-is) [L. *tunica*, coating; *vagina*, sheath] A double-walled sac covering most of the testis, enclosing a part of the celom between its walls.

tympanum (tĭm'pănŭm) [Gk. *tympanon*, drum] The tympanic cavity or middle ear.

umbilical cord (ŭm'bĭlĭ'kăl) [L. *umbilicus*, navel] The cord formed from the yolk sac and the body stalk, which connects the embryo with the placenta and carries the umbilical arteries and veins.

unipennate (ū'nĭ pen'āt) [L. *unus*, one; *penna*, contour feather] Muscle fiber arrangement in which the fibers are attached to one side of a tendon.

ureter (ūrē'tēr) [Gk. *oureter*, ureter] Duct conveying urine from kidney to bladder or cloaca.

urethra (ūrē'thră) [Gk. *ourethra*, urine] Duct from the urinary bladder to body surface which carries urine.

urethral crest An elevated area on the posterior wall of the prostatic urethra.

urogenital diaphragm (uro-jen'-i-tal) The muscular layer in the floor of the pelvis; forms the external sphincter of the urethra.

urogenital triangle The region of the pelvic floor below the symphysis pubis.

uterine tube (ū'tĕrĭn) [L. *uterus*, womb] The upper portion of oviduct (Fallopian tubes).

uterus (ū'tĕrŭs) [L. *uterus*, womb] Single, hollow, muscular organ which lies between urinary bladder and the sigmoid colon.

utricle (ū'trikl) [L. *utriculus*, small bag] The larger of the two chambers of the vestibular portions of the membranous labyrinth.

uvula (ū'vūlă) [L. *uva*, grape] A small conical projection of the soft palate.

vacuoles (vak'ūōl) [L. *vacuus*, empty] Spaces in cell protoplasm containing air sap, or partially digested food.

vagina (văjī'nă) [L. *vagina*, sheath] Canal leading from uterus to vestibule.

vasa vasorum (va'sah vaso'rum) [L. *vas*, vessel; *vasorum*, genit. of vas] Nutrient vessels for larger arteries and veins.

veins (vānz) [L. *vena*, vein] Vessels which convey blood to or toward the heart.

ventricle (vĕn'trikl) [L. *ventriculus*, belly] A cavity or chamber, as in heart or brain; the dispensing chamber of the heart.

ventricular folds (vĕntrĭk'ūlăr) [L. *ventriculus*, belly] Lower free border of the vestibular membranes attaching to inside angle of thyroid cartilage and to arytenoids; "false" vocal cords.

venule (vēn'ŭl) [L. *venula*, vein] Small vessel conducting venous blood from capillaries to vein.

vermis (vĕr'mĭs) [L. *vermis*, worm] Narrow median portion of the cerebellum separating the two cerebellar hemispheres.

vestibular membrane One of a pair of mucous membranes between the lateral border of the lower epiglottis and arytenoid cartilages.

vestibule (vĕs'tĭbūl) [L. *vestibulum*, passage] A cavity leading into another cavity or passage; the vestibule of the internal ear or mouth.

villus (vĭl'ŭs) [L. *villus*, shaggy hair] Minute finger-like projections of the mucosa into the lumen of the small intestine.

vitreous body (vĭt'rĕŭs) [L. *vitreus*, glassy] A transparent, semigelatinous substance filling large cavity of the eye behind the lens.

vocal folds (vō'kăl) [L. *vox*, voice] Mucous membrane folds involved in sound production located on the inferior margin of the vestibule of the larynx; "true" vocal cords.

volar (vō'lăr) [L. *vola*, palm of hand] Anterior surfaces of the hands or forearms.

Volkmann's canals (A. W. Volkmann, German physiologist, 1830–1889] Minute passages which penetrate the compact bone.

vulva (vŭl'vă) [L. *vulva*, vulva] The external female genitalia.

white matter   The substance found in spinal cord and brain composed of bundles of medullated nerve fibers.

yolk sac (yōk săk) [A.S. *geoloca*, yellow part; L. *saccus*, sack] A membranous sac attached to embryo and containing yolk.

zona pellucida (zō'nă pĕ-lū'sĭda) [Gk. *zone*, girdle; L. *pellucidus*, clear, transparent] Thick, transparent membrane surrounding ovum.

zygote (zī'gōt) [Gk. *zygote*, yolked] Cell formed by union of two gametes; fertilized egg.

Index

When a number of pages are listed for any index entry, the page number or numbers on which the item is defined or treated in some detail are placed in **boldface**. Pages on which structures are illustrated are shown in *italics*. Many structures which are represented only in the illustrations are listed in the index in the belief that much can be learned from a good picture.

;